THE THEORY OF FUSION SYSTEMS

Fusion systems are a recent development in finite group theory and sit at the intersection of algebra and topology. This book is the first to deal comprehensively with this new and expanding field, taking the reader from the basics of the theory right to the state of the art.

Three motivational chapters, indicating the interaction of fusion and fusion systems in group theory, representation theory, and topology are followed by six chapters that explore the theory of fusion systems themselves. Starting with the basic definitions, the topics covered include: weakly normal and normal subsystems; morphisms and quotients; saturation theorems; results about control of fusion; and the local theory of fusion systems. At the end, there is also a discussion of exotic fusion systems.

Designed for use as a text and reference work, this book is suitable for graduate students and experts alike.

David A. Craven is a Junior Research Fellow in the Mathematical Institute at the University of Oxford.

Already published
 85 J. Carlson, S. Müller-Stach & C. Peters *Period mappings and period domains*
 86 J. J. Duistermaat & J. A. C. Kolk *Multidimensional real analysis, I*
 87 J. J. Duistermaat & J. A. C. Kolk *Multidimensional real analysis, II*
 89 M. C. Golumbic & A. N. Trenk *Tolerance graphs*
 90 L. H. Harper *Global methods for combinatorial isoperimetric problems*
 91 I. Moerdijk & J. Mrčun *Introduction to foliations and Lie groupoids*
 92 J. Kollár, K. E. Smith & A. Corti *Rational and nearly rational varieties*
 93 D. Applebaum *Lévy processes and stochastic calculus (1st Edition)*
 94 B. Conrad *Modular forms and the Ramanujan conjecture*
 95 M. Schechter *An introduction to nonlinear analysis*
 96 R. Carter *Lie algebras of finite and affine type*
 97 H. L. Montgomery & R. C. Vaughan *Multiplicative number theory, I*
 98 I. Chavel *Riemannian geometry (2nd Edition)*
 99 D. Goldfeld *Automorphic forms and L-functions for the group GL(n,R)*
100 M. B. Marcus & J. Rosen *Markov processes, Gaussian processes, and local times*
101 P. Gille & T. Szamuely *Central simple algebras and Galois cohomology*
102 J. Bertoin *Random fragmentation and coagulation processes*
103 E. Frenkel *Langlands correspondence for loop groups*
104 A. Ambrosetti & A. Malchiodi *Nonlinear analysis and semilinear elliptic problems*
105 T. Tao & V. H. Vu *Additive combinatorics*
106 E. B. Davies *Linear operators and their spectra*
107 K. Kodaira *Complex analysis*
108 T. Ceccherini-Silberstein, F. Scarabotti & F. Tolli *Harmonic analysis on finite groups*
109 H. Geiges *An introduction to contact topology*
110 J. Faraut *Analysis on Lie groups: An introduction*
111 E. Park *Complex topological K-theory*
112 D. W. Stroock *Partial differential equations for probabilists*
113 A. Kirillov, Jr *An introduction to Lie groups and Lie algebras*
114 F. Gesztesy *et al.* *Soliton equations and their algebro-geometric solutions, II*
115 E. de Faria & W. de Melo *Mathematical tools for one-dimensional dynamics*
116 D. Applebaum *Lévy processes and stochastic calculus (2nd Edition)*
117 T. Szamuely *Galois groups and fundamental groups*
118 G. W. Anderson, A. Guionnet & O. Zeitouni *An introduction to random matrices*
119 C. Perez-Garcia & W. H. Schikhof *Locally convex spaces over non-Archimedean valued fields*
120 P. K. Friz & N. B. Victoir *Multidimensional stochastic processes as rough paths*
121 T. Ceccherini-Silberstein, F. Scarabotti & F. Tolli *Representation theory of the symmetric groups*
122 S. Kalikow & R. McCutcheon *An outline of ergodic theory*
123 G. F. Lawler & V. Limic *Random walk: A modern introduction*
124 K. Lux & H. Pahlings *Representations of groups*
125 K. S. Kedlaya *p-adic differential equations*
126 R. Beals & R. Wong *Special functions*
127 E. de Faria & W. de Melo *Mathematical aspects of quantum field theory*
128 A. Terras *Zeta functions of graphs*
129 D. Goldfeld & J. Hundley *Automorphic representations and L-functions for the general linear group, I*
130 D. Goldfeld & J. Hundley *Automorphic representations and L-functions for the general linear group, II*
131 D. A. Craven *The theory of fusion systems*
132 J. Väänänen *Models and games*
133 G. Malle & D. Testerman *Linear algebraic groups and finite groups of Lie type*

The Theory of Fusion Systems

An Algebraic Approach

DAVID A. CRAVEN
University of Oxford

CAMBRIDGE
UNIVERSITY PRESS

CAMBRIDGE UNIVERSITY PRESS

Cambridge, New York, Melbourne, Madrid, Cape Town,
Singapore, São Paulo, Delhi, Tokyo, Mexico City

Cambridge University Press
The Edinburgh Building, Cambridge CB2 8RU, UK

Published in the United States of America by Cambridge University Press, New York

www.cambridge.org
Information on this title: www.cambridge.org/9781107005969

First published 2011

Printed in the United Kingdom at the University Press, Cambridge

A catalogue record for this publication is available from the British Library

Library of Congress Cataloguing in Publication data
Craven, David A.
The theory of fusion systems : an algebraic approach / David A. Craven.
p. cm. – (Cambridge studies in advanced mathematics ; 131)
Includes bibliographical references and index.
ISBN 978-1-107-00596-9 (hardback)
1. Finite groups. 2. Representations of algebras. 3. Algebraic topology.
I. Title. II. Series.
QA177.C73 2011
512′.23–dc22   2011006354

ISBN 978-1-107-00596-9 Hardback

Contents

Preface   *page* ix

PART I   MOTIVATION   1

1   Fusion in finite groups   3
  1.1   Control of fusion   4
  1.2   Normal p-complements   8
  1.3   Alperin's fusion theorem   12
  1.4   The focal subgroup theorem   16
  1.5   Fusion systems   20
  Exercises   24

2   Fusion in representation theory   26
  2.1   Blocks of finite groups   27
  2.2   The Brauer morphism and relative traces   34
  2.3   Brauer pairs   38
  2.4   Defect groups and the first main theorem   41
  2.5   Fusion systems of blocks   47
  Exercises   54

3   Fusion in topology   55
  3.1   Simplicial sets   56
  3.2   Classifying spaces   66
  3.3   Simplicial and cosimplicial objects   74
  3.4   Bousfield–Kan completions   78
  3.5   The centric linking systems of groups   83
  3.6   Constrained fusion systems   86
  Exercises   89

PART II THE THEORY   91

4   Fusion systems   93
  4.1   Saturated fusion systems   94
  4.2   Normalizing and centralizing   101
  4.3   The equivalent definitions   105
  4.4   Local subsystems   108
  4.5   Centric and radical subgroups   117
  4.6   Alperin's fusion theorem   121
  4.7   Weak and strong closure   127
  Exercises   131

5   Weakly normal subsystems, quotients, and morphisms   134
  5.1   Morphisms of fusion systems   135
  5.2   The isomorphism theorems   141
  5.3   Normal subgroups   148
  5.4   Weakly normal subsystems   150
  5.5   Correspondences for quotients   160
  5.6   Simple fusion systems   171
  5.7   Soluble fusion systems   181
  Exercises   186

6   Proving saturation   188
  6.1   The surjectivity property   189
  6.2   Reduction to centric subgroups   193
  6.3   Invariant maps   201
  6.4   Weakly normal maps   205
  Exercises   212

7   Control in fusion systems   215
  7.1   Resistance   216
  7.2   Glauberman functors   221
  7.3   The ZJ-theorems   227
  7.4   Normal p-complement theorems   232
  7.5   The hyperfocal and residual subsystems   236
  7.6   Bisets   252
  7.7   The transfer   260
  Exercises   267

8   Local theory of fusion systems   270
  8.1   Normal subsystems   271
  8.2   Weakly normal and normal subsystems   275

	8.3	Intersections of subsystems	280
	8.4	Constraint and normal subsystems	287
	8.5	Central products	295
	8.6	The generalized Fitting subsystem	299
	8.7	L-balance	308
	Exercises		314
9	**Exotic fusion systems**		**317**
	9.1	Extraspecial p-groups	318
	9.2	The Solomon fusion system	326
	9.3	Blocks of finite groups	330
	9.4	Block exotic fusion systems	335
	9.5	Abstract centric linking systems	343
	9.6	Higher limits and centric linking systems	349
	Exercises		356
	References		358
	Index of notation		364
	Index		366

Preface

It is difficult to pinpoint the origins of the theory of fusion systems: it could be argued that they stretch back to Burnside and Frobenius, with arguments about the fusion of p-elements of finite groups. Another viewpoint is that it really started with the theorems on fusion in finite groups, such as Alperin's fusion theorem, or Grün's theorems.

We will take as the starting point the important paper of Solomon [Sol74], which proves that, for a Sylow 2-subgroup P of $\mathrm{Spin}_7(3)$, there is a particular pattern of the fusion of involutions in P that, while not internally inconsistent, is not consistent with living inside a finite group. This is the first instance where the fusion of p-elements looks fine on its own, but is incompatible with coming from a finite group.

Unpublished work of Puig during the 1990s and even before (some of which is collected in [Pui06]), together with work of Alperin–Broué [AB79], is the basis for constructing a fusion system for a p-block of a finite group. It was with Puig's work where the axiomatic foundations of fusion systems started, and where some of the fundamental notions begin. It cannot be overestimated how much the current theory of fusion systems owes to Puig, both in originating the definition and related notions, and in furthering the theory.

Various results that could be considered part of local finite group theory (the study of p-subgroups, normalizers, conjugacy, and so on) were extended to p-blocks of finite groups during the 1990s and early part of the twenty-first century, but at the time were not viewed as taking place in the more general setting of fusion systems. With this theory now becoming more popular, more and more results are being cast in this language, and extended to this area.

The internal theory of fusion systems, starting from these foundations, has developed rapidly, and in many respects has mimicked the theory of finite groups, with normal subsystems, quotients, the generalized Fitting subsystem, composition series, soluble fusion systems, and so on. However, there is also a topological aspect to this theory.

Along with the representation theory, topology has played an important role in the development of the theory: Benson [Ben98a] constructed a topological space that should be the 2-completed classifying space of a finite group whose fusion pattern matched that which Solomon considered. Since such a group does not exist, this space can be thought of as the shadow cast by an invisible group. Benson predicted that this topological space is but one facet of a general theory, a prediction that was confirmed with the development of p-local finite groups.

Although we will not cover the topic of p-local finite groups here (we only meet the definition in Chapter 9), they can be thought of as some data describing a p-completed classifying space of a fusion system. In the case where the fusion system arises from a finite group, the corresponding p-local finite group describes the normal p-completed classifying space.

In this direction, we have Oliver's proof [Oli04] [Oli06] of the Martino–Priddy conjecture [MP96], which states that two finite groups have homotopy equivalent p-completed classifying spaces if and only if the fusion systems are isomorphic. The topological considerations have fuelled development in the algebraic aspects of fusion systems and vice versa, and the two viewpoints are somewhat intertwined. Having said that, we will not deal with the topological theory here beyond that which is given in Part I, and concentrate on the more algebraic aspects.

As this is a young subject, still in development, the foundations of the theory have not yet been solidified; indeed, there is some debate as to the correct *definition* of a fusion system! (It should be noted that the definitions are all equivalent, and so the choice is only apparent.) The definition of a 'normal' subsystem is also under discussion, and which definition is used often indicates the intended applications of the theory. Here we have made a choice based upon the evidence available now; this might change as time goes on. Since group theorists, representation theorists, and topologists all converge on this area, there are several different conventions and styles, as well as approaches.

The first three chapters are preliminary in nature, and deal with group theory, representation theory, and topology. The first chapter is essentially a run-through of the theory of fusion in finite groups, giving for example the p-complement theorems of Frobenius, Burnside, and Glauberman–Thompson. The second chapter introduces the representation theory aspects, and in particular develops the block theory needed to construct the fusion system associated to a p-block of a finite group. The third chapter develops the topological methods used in the theory, but since the main thrust of this work is the algebraic theory of fusion systems, we necessarily skip over many of the details in this chapter.

The remaining six chapters deal with the theory of fusion systems. The fourth chapter starts by defining fusion systems and in particular saturated fusion systems, then constructing local subsystems, proving Alperin's fusion theorem, and introducing strongly closed subgroups. In Chapter 5, we start looking at the normal and quotient structure of fusion systems, introducing morphisms, quotients, weakly normal and characteristic subsystems, the centre, and so on. The sixth chapter deals with methods used to prove saturation, and introduces weakly normal maps as well.

Chapter 7 deals with topics around control of fusion, with analogues of the Glauberman–Thompson normal p-complement theorem, Glauberman's ZJ-theorem, and the two normal subgroups $O^p(G)$ and $O^{p'}(G)$, the former of which is the object of the theory of transfer. After proving the existence of a certain kind of biset associated to any saturated fusion system in Section 7.6, we use the biset to develop the transfer for a fusion system.

Chapter 8 focuses on work of Aschbacher which attempts to translate some aspects of local finite group theory into the domain of fusion systems. We prove here that, for constrained fusion systems, there is a one-to-one correspondence between the normal subsystems of the fusion system and the normal subgroups of the associated model. Other highlights include a description of the generalized Fitting subsystem of a fusion system, and the proof of L-balance for fusion systems, which is considerably easier than the proof of the corresponding theorem for finite groups.

The final chapter consists of questions about exotic fusion systems (i.e., fusion systems that do not come from groups), with a few details on some of the known exotic fusion systems, theorems on which exotic fusion systems do not come from *blocks* of finite groups, and Oliver's conjecture relating modular representation theory of p-groups to the existence and uniqueness of centric linking systems. A solution to this conjecture would remove the requirement of the classification of the finite simple groups for the proof of the Martino–Priddy conjecture.

The choice of definitions and conventions has been influenced by the background of the author: as I am a group theorist and group representation theorist, the conventions here will be the standard group theory conventions, rather than topology conventions. In particular, homomorphisms will be composed from left to right. The only chapter where this will be relaxed is Chapter 3, the topological chapter; the reason for this is that to keep left-to-right notation would go against every other topology book in existence, and require writing functors on the right, something that I, even as a group theorist, cannot bring myself to do.

It remains for me to thank various people, most notably Adam Glesser for reading much of this work and for being a sounding board for various ideas, mathematical, pedagogical and notational. Thank you to George Raptis for reading Chapter 3, and explaining some of the topological ideas to an algebraist, making the exposition in that chapter considerably clearer. Proof reading and valuable comments were given by (in alphabetical order): Tobias Barthel, Michael Collins, Radha Kessar, Bob Oliver, Oscar Randal-Williams, George Raptis, Raphaël Rouquier, Jason Semeraro and Matt Towers. Any errors that remain in this work are, of course, my own.

David A. Craven, Oxford

Part I

Motivation

1

Fusion in finite groups

The fusion of elements of prime power order in a finite group is the source of many deep theorems in finite group theory. In this chapter we will briefly survey this area, and use this theory to introduce the notion of a fusion system of a finite group.

As this is the first chapter, we introduce some basic notation and terminology that we will use throughout the text. If G is a finite group and g is an element of G, we denote by c_g the conjugation map $c_g : G \to G$ given by $c_g : x \mapsto x^g = g^{-1}xg$. If H is a subgroup of G, then by c_H we mean the natural map $\mathrm{N}_G(H) \to \mathrm{Aut}(H)$ sending $g \in \mathrm{N}_G(H)$ to c_g. If K is another subgroup of G, the *automizer* of H (or rather, $\mathrm{N}_H(K)$) in K, denoted $\mathrm{Aut}_H(K)$, is the image of H under c_K, and is naturally isomorphic to $\mathrm{N}_H(K)/\mathrm{C}_H(K)$, so we will often identify $\mathrm{Aut}_H(K)$ and $\mathrm{N}_H(K)/\mathrm{C}_H(K)$. Write $\mathrm{Inn}(K) = \mathrm{Aut}_K(K)$, and by $\mathrm{Out}(K)$ and $\mathrm{Out}_H(K)$ we mean $\mathrm{Aut}(K)/\mathrm{Inn}(K)$ and $\mathrm{Aut}_H(K)\,\mathrm{Inn}(K)/\mathrm{Inn}(K)$ respectively.

If $\phi : H \to L$ is some isomorphism, then there is an induced map $\mathrm{Aut}(H) \to \mathrm{Aut}(L)$ such that, for $\psi \in \mathrm{Aut}(H)$,

$$\psi \mapsto \phi^{-1}\psi\phi.$$

If ψ is an element of $\mathrm{Aut}(H)$, we denote its image under this map as ψ^ϕ, and we will normally simply use ϕ to describe this induced map; no confusion should arise because the domains are different. If confusion could arise however, or we want to emphasize that it is this map we are considering, we will denote it by c_ϕ, since it is an analogue of the conjugation map c_g.

If x and y are elements of G, then x and y are *H-conjugate* if there is some h in H such that $x^h = y$, and we extend this definition to subgroups in the obvious way.

If π is a set of primes, then π' denotes all primes not in π. For any finite group G, $O_\pi(G)$ denotes the largest normal π-subgroup, and $O^\pi(G)$ denotes the smallest normal subgroup whose quotient is a π-group. As usual, if $\pi = p$ we simply write $O_p(G)$ and $O^p(G)$ respectively. (A standard fact that we will use often is that $O^\pi(G)$ contains all elements of π'-order in G.) The *exponent* (denoted $\exp(G)$) is the lowest common multiple of all orders of all elements in the group.

A *p-local subgroup* is a subgroup of G of the form $N_G(Q)$, for Q a non-trivial p-subgroup of G. Finally, the set of Sylow p-subgroups of a finite group G is denoted by $\mathrm{Syl}_p(G)$. We remind the reader that homomorphisms act on the right.

1.1 Control of fusion

We begin with a famous theorem.

Theorem 1.1 (Burnside) *Let G be a finite group, and let P be a Sylow p-subgroup of G. Suppose that P is abelian. Let x and y be two elements in a Sylow p-subgroup P of G. If x and y are G-conjugate then they are $N_G(P)$-conjugate.*

Proof Let x and y be elements of P, and suppose that there is some $g \in G$ such that $x^g = y$, via $c_g : x \mapsto y$. We have that

$$P^g \leq C_G(x)^g = C_G(x^g) = C_G(y),$$

and so both P and P^g are Sylow p-subgroups of $C_G(y)$. Thus there exists $h \in C_G(y)$ such that $P^{gh} = P$. Therefore $gh \in N_G(P)$ and we have

$$x^{gh} = (x^g)^h = x^g,$$

and so $c_{gh} = c_g$ on x, as required. $\square$

This theorem is a statement about the fusion of P-conjugacy classes in G.

Definition 1.2 Let G be a finite group, let H and K be subgroups of G with $H \leq K$, and let x and y be elements of H.

(i) If x and y are not conjugate in H, then x and y are *fused* in K if they are conjugate by an element of K. Similarly, two subgroups or two conjugacy classes of H are fused in K if they satisfy the obvious condition.

(ii) The subgroup K *controls weak fusion in H with respect to G if*, whenever x and y are fused in G, they are fused in K. (This is equivalent to the fusion of conjugacy classes.)

(iii) The subgroup K *controls G-fusion in H* if, whenever two subgroups A and B are conjugate via a conjugation map $c_g : A \to B$ for some $g \in G$, then there is some $k \in K$ such that c_g and c_k agree on A. (This is stronger than simply requiring any two subgroups conjugate in G to be conjugate in K.)

In the literature, control of weak fusion is often called 'control of fusion', and control of G-fusion is often called 'control of strong fusion'. However, when we get to fusion systems, control of weak fusion will be much less important than control of G-fusion.

It is easy to see that if K controls G-fusion in H, then K controls weak fusion in H with respect to G. The next example proves that the converse is not true.

Example 1.3 Let K be the group $\mathrm{GL}_3(2)$, which acts naturally on a 3-dimensional vector space V over $\mathbb{F}_2$. The order of $\mathrm{GL}_3(2)$ is 168, and so there are elements of orders 3 and 7 in K. In fact, we can find a subgroup of K of order 21. More specifically, let

$$x = \begin{pmatrix} 0 & 1 & 0 \\ 1 & 0 & 1 \\ 1 & 0 & 0 \end{pmatrix} \quad \text{and} \quad y = \begin{pmatrix} 1 & 1 & 0 \\ 1 & 0 & 0 \\ 0 & 0 & 1 \end{pmatrix}.$$

The element x has order 7, y has order 3, and y normalizes $\langle x \rangle$, so that $\langle x, y \rangle$ has order 21.

The vector space V is simply an elementary abelian group of order 8, and it is not difficult to see that if P is an elementary abelian group of order 8 then $\mathrm{Aut}(P) = \mathrm{GL}_3(2)$. Hence the subgroup $\langle x, y \rangle$ of K becomes a subgroup of $\mathrm{Aut}(P)$ of order 21. Let G denote the semidirect product of P by this subgroup of $\mathrm{Aut}(P)$, a (soluble) group of order 168; we will identify x and y with their counterparts in $\mathrm{Aut}(P)$, and in G. The element x of order 7 acts non-trivially on P, and so must permute the seven *involutions* – i.e., elements of order 2 – transitively. In particular, the subgroup $H = \langle P, x \rangle$ also has the property that all involutions of P are H-conjugate, and so H controls weak fusion in P with respect to G.

However, it is not difficult to prove that H does not control G-fusion in P, since there are subgroups of P of order 4 that are G-conjugate but not H-conjugate (and we leave it as an exercise to find such subgroups).

Thus a subgroup H controlling weak fusion in P with respect to G does not necessarily control G-fusion in P.

Given these definitions, Theorem 1.1 has the following restatement.

Theorem 1.4 *Let G be a finite group, and let P be a Sylow p-subgroup of G. If P is abelian, then the normalizer $N_G(P)$ controls weak fusion in P with respect to G.*

In the proof of Theorem 1.1, if we replace x and y by subsets A and B of P, then we get that if $c_g : A \to B$ is a map in G then there is some $h \in N_G(P)$ such that $c_g = c_h$ on A. In other words, we get the following theorem.

Theorem 1.5 *Let G be a finite group, and let P be a Sylow p-subgroup of G. If P is abelian, then the normalizer $N_G(P)$ controls G-fusion in P.*

What we are saying is that any fusion inside a Sylow p-subgroup P of a finite group must take place inside its normalizer, at least if P is abelian. In general, this is not true.

Example 1.6 Let G be the group $GL_3(2)$, the simple group of order 168. This group has a dihedral Sylow 2-subgroup P, generated by the two matrices

$$x = \begin{pmatrix} 1 & 1 & 0 \\ 0 & 1 & 0 \\ 0 & 0 & 1 \end{pmatrix} \quad \text{and} \quad y = \begin{pmatrix} 1 & 0 & 0 \\ 0 & 1 & 1 \\ 0 & 0 & 1 \end{pmatrix}.$$

Note that x and y are both involutions. In $GL_3(2)$, all of the twenty-one involutions are conjugate, but this is not true in $N_G(P)$, since we claim that $N_G(P) = P$. To see this, let

$$z = \begin{pmatrix} 1 & 0 & 1 \\ 0 & 1 & 0 \\ 0 & 0 & 1 \end{pmatrix},$$

the central involution in P. Notice that, since z has twenty-one conjugates in G, $C_G(z)$ has order 8, so that $C_G(z) = P$. Hence $C_G(P)$ is a 2-group. Since $\text{Aut}(P)$ is a 2-group (as P is dihedral), and $\text{Aut}_G(P) = N_G(P)/C_G(P)$, we see that $N_G(P)$ is also a 2-group; thus $N_G(P) = P$. In fact, there is no 2-local subgroup H for which x and y are H-conjugate. However, all is not lost; let $Q_1 = \langle x, z \rangle$, and let $N_1 = N_G(Q_1)$. The subgroup N_1 is isomorphic with the symmetric group on four letters, and so inside here Q_1 has the property that all of its non-identity elements are conjugate in the overgroup N_1; therefore x and z are conjugate in N_1.

Similarly, write $Q_2 = \langle y, z \rangle$ and $N_2 = N_G(Q_2)$. The same statements apply, and so y and z are conjugate inside N_2. Thus x and y are conjugate, via z, inside 2-local subgroups.

This idea of fusion of p-elements not being controlled by a single subgroup, but two elements being conjugate 'in stages' by a collection of subgroups is important, and is the basis of Alperin's fusion theorem, which we shall see in Section 1.3.

The notions of fusion and control of fusion (particularly the stronger 'control of G-fusion'), are interesting for us, and we will explore the fusion and control of fusion in Sylow p-subgroups of finite groups, and more abstractly with the notion of fusion systems. For a group, we give the definition of a fusion system now.

Definition 1.7 Let G be a finite group and let P be a Sylow p-subgroup of G. The *fusion system* of G on P is the category $\unlhd (G)$, whose objects are all subgroups of P and whose morphisms are given by

$$\mathrm{Hom}_{\mathcal{F}_P(G)}(A, B) = \mathrm{Hom}_G(A, B),$$

the set of all (not necessarily surjective) maps $A \to B$ induced by conjugation by elements of G. The composition of morphisms is composition of maps.

This definition is meant to capture the notion of fusion of p-elements and p-subgroups in the group G. We give an example of a fusion system now.

Example 1.8 Let G be the group $\mathrm{GL}_3(2)$, considered in Example 1.6, and let P be the Sylow 2-subgroup given there, with the elements x, y and z as given. The subgroup P is isomorphic with D_8, so $\mathcal{F}_P(P)$ is simply all of the conjugation actions given by elements of P. For example, we have the (not surjective) map $\phi : \langle x \rangle \mapsto \langle x, z \rangle$ sending x to xz; this is realized by conjugation by y.

Consider the fusion system $\mathcal{F}_P(G)$, which contains $\mathcal{F}_P(P)$. We will simply describe the bijective maps in $\mathcal{F}_P(G)$, since all injective maps in $\mathrm{Hom}_G(A, B)$ are bijections followed by inclusions. There are bijections $\langle g \rangle \to \langle h \rangle$, where g and h are involutions. The two elements of order 4 are conjugate in P, so there is a map $\langle xy \rangle \to \langle xy \rangle$ sending xy to

$(xy)^3$. Finally, there are maps involving the Klein four-subgroups. Let $Q_1 = \langle x, z \rangle$ and $Q_2 = \langle y, z \rangle$, as before.

We first consider the maps in $\mathrm{Hom}_{\mathcal{F}_P(G)}(Q_1, Q_1) = \mathrm{Aut}_{\mathcal{F}_P(G)}(Q_1)$. Since $\mathrm{N}_G(Q_1)$ is the symmetric group S_4, and $\mathrm{C}_G(Q_1) = Q_1$, we must have that $\mathrm{Aut}_{\mathcal{F}_G(P)}(Q_1) = \mathrm{Aut}_G(Q_1)$ is isomorphic with S_3, and so is the full automorphism group. (Similarly, $\mathrm{Aut}_{\mathcal{F}_P(G)}(Q_2) = \mathrm{Aut}(Q_2)$.) If ϕ is any map $Q_1 \to Q_2$ in $\mathcal{F}_P(G)$, then by composing with a suitably chosen automorphism of Q_2, we get all possible isomorphisms $Q_1 \to Q_2$. This would include the map ϕ where $\phi : x \mapsto y$ and $\phi : z \mapsto z$; then x and y would be conjugate in $\mathrm{C}_G(z) = \mathrm{N}_G(P) = P$, and this is not true. Therefore there are no maps between Q_1 and Q_2. (One may prove this more easily using the fact that Q_1 and Q_2 stabilize different maximal flags, but the above argument is more in keeping with the rest of the book.)

This shows that, although all of the non-identity elements in Q_1 are conjugate to all non-identity elements in Q_2 in $\mathcal{F}_P(G)$, the subgroups Q_1 and Q_2 are not isomorphic in $\mathcal{F}_P(G)$. This is why we take all *subgroups* of P in the fusion system, rather than merely all elements.

The fusion system is meant to capture the concept of control of fusion, and indeed it does.

Proposition 1.9   *Let G be a finite group and let P be a Sylow p-subgroup of G. Let H be a subgroup of G containing P. The subgroup H controls G-fusion in P if and only if $\mathcal{F}_P(G) = \mathcal{F}_P(H)$.*

Proof   This is essentially a restatement of the definition of control of G-fusion, and which maps $\phi : A \to B$ lie in the fusion system. The details are left to the reader.   □

We have the following corollary of this proposition, our first result about fusion systems proper.

Corollary 1.10   *Let G be a finite group and let P be a Sylow p-subgroup of G. If P is abelian, then*

$$\mathcal{F}_P(G) = \mathcal{F}_P(\mathrm{N}_G(P)).$$

1.2 Normal p-complements

One of the first applications of fusion of finite groups was to the question of whether a group has a normal p-complement.

Definition 1.11 A finite group G has a *normal p-complement*, or is said to be *p-nilpotent*, if $O_{p'}(G) = O^p(G)$, i.e., $G = H \rtimes P$, where $H = O_{p'}(G)$ and P is a Sylow p-subgroup of G.

The first results on the question of whether a finite group has a normal p-complement are from Burnside and Frobenius. Burnside's theorem is generally proved as an application of transfer, which we shall meet briefly in Chapter 7 (but see, for example, [Asc00, Section 37], [Gor80, Section 7.3], or [Ros78, Chapter 10], and also Section 7.7).

Frobenius's normal p-complement theorem is a set of three conditions, each equivalent to the existence of a normal p-complement. Modern proofs of this theorem use, along with the transfer, some machinery from the theory of fusion in finite groups, like Grün's first theorem or Alperin's fusion theorem. We will state this normal p-complement theorem but not prove it until Section 1.4.

Theorem 1.12 (Frobenius's normal p-complement theorem) *Let G be a finite group, and let P be a Sylow p-subgroup of G. The following are equivalent:*

(i) *G possesses a normal p-complement;*

(ii) *$\mathcal{F}_P(G) = \mathcal{F}_P(P)$;*

(iii) *every p-local subgroup of G possesses a normal p-complement;*

(iv) *for every p-subgroup Q of G, $\mathrm{Aut}_G(Q)$ is a p-group.*

This is not, of course, exactly what Frobenius proved, but instead of $\mathcal{F}_P(G) = \mathcal{F}_P(P)$ there was a statement about conjugacy in the Sylow p-subgroup, which is easily equivalent.

From this result, we will deduce Burnside's normal p-complement theorem, which is a sufficient, but not necessary, condition to having a p-complement.

Theorem 1.13 (Burnside's normal p-complement theorem) *Let G be a finite group, and let P be a Sylow p-subgroup of G. If $P \leq Z(N_G(P))$ then G possesses a normal p-complement.*

Proof Since $P \leq Z(N_G(P))$, we must have that P is abelian. Therefore, $\mathcal{F}_P(G) = \mathcal{F}_P(N_G(P))$ by Corollary 1.10. Furthermore, since P is central in $N_G(P)$, we see that $\mathcal{F}_P(N_G(P)) = \mathcal{F}_P(P)$, and so by Frobenius's normal p-complement theorem, G possesses a normal p-complement, as claimed. □

We can quickly derive a result of Cayley from Frobenius's normal p-complement theorem as well, proving that no simple group has a cyclic Sylow 2-subgroup.

Corollary 1.14 (Cayley)   *Let G be a finite group of even order, and let P be a Sylow 2-subgroup of G. If P is cyclic, then G has a normal 2-complement.*

Proof   Notice that, if Q is any cyclic 2-group of order 2^m, then $|\operatorname{Aut}(Q)|$ is itself a 2-group. (It is the size of the set

$$\{x \mid 0 < x < 2^m,\ x \text{ is prime to } 2^m\},$$

which has size 2^{m-1}.) Thus $\operatorname{Aut}_G(Q)$ is a 2-group for all subgroups Q of G, since Q is cyclic. Hence by Frobenius's normal p-complement theorem, G possesses a normal 2-complement, as claimed. (Alternatively, since $\mathrm{C}_G(P) = \mathrm{N}_G(P)$, one may use Burnside's normal p-complement theorem.) □

Example 1.15   We return to our familiar example, where $G = \mathrm{GL}_3(2)$ and P is the Sylow 2-subgroup considered above. Since $\mathcal{F}_P(G)$ is not $\mathcal{F}_P(P)$, we should have that $\operatorname{Aut}_{\mathcal{F}_P(G)}(Q)$ is not a 2-group, for some $Q \leq P$. As we saw, the automizers in G of Q_1 and Q_2, the Klein four subgroups of P, have order 6, confirming Frobenius's theorem in this case.

While Frobenius's normal p-complement theorem was a breakthrough, Thompson's normal p-complement theorem was a significant refinement. The original theorem of Thompson [Tho64] proved that, for odd primes, G possesses a normal p-complement if two particular subgroups possess normal p-complements. (Note that Thompson proved an earlier normal p-complement theorem in [Tho60].) Glauberman [Gla68a] refined this further, proving that, for odd primes, G possesses a normal p-complement if *one* particular p-local subgroup possesses a normal p-complement! Both Thompson's and Glauberman's results use the Thompson subgroup, which we will define now.

Definition 1.16   Let P be a finite p-group, and let $\mathscr{A}$ denote the set of all elementary abelian subgroups of P of maximal order. The *Thompson subgroup*, $J(P)$, is defined to be the subgroup generated by all elements of $\mathscr{A}$.

There are several similar definitions of the Thompson subgroup in the literature, but this one will be fine for our purposes. We are now in a position to state the theorems; for a proof of the second theorem, see also [Gor80, Theorem 8.3.1].

Theorem 1.17 (Thompson [Tho64]) *Let G be a finite group, and let P be a Sylow p-subgroup of G, where p is an odd prime. The group G has a normal p-complement if and only if* $N_G(J(P))$ *and* $C_G(Z(P))$ *have normal p-complements.*

Theorem 1.18 (Glauberman–Thompson [Gla68a]) *Let G be a finite group, and let P be a Sylow p-subgroup of G, where p is an odd prime. The group G has a normal p-complement if and only if* $N_G(Z(J(P)))$ *has a normal p-complement.*

Note that both of these theorems were originally proved using different versions of the Thompson subgroup (and, indeed, different from each other as well) and some modifications need to be made in order for the original proofs to be valid for our version of $J(P)$.

It may seem very surprising that a single p-local subgroup controls whether the whole group possesses a normal p-complement, but this is indeed the case. This theorem tells us that, with $N = N_G(Z(J(P)))$, if $\mathcal{F}_P(N) = \mathcal{F}_P(P)$, then $\mathcal{F}_P(N) = \mathcal{F}_P(G)$. Thus one way of looking at this theorem is that it gives a sufficient condition for N to control G-fusion in P.

In fact, this happens much more often. Glauberman's ZJ-theorem is a sufficient condition for this subgroup N given above to control G-fusion in P. It holds for odd primes, and for every group that does not involve a particular group $Qd(p)$, as a subquotient. Let p be a prime, and let $Q = C_p \times C_p$: this can be thought of as a 2-dimensional vector space, and so $SL_2(p)$ acts on this group in a natural way. Define $Qd(p)$ to be the semidirect product of Q and $SL_2(p)$.

Example 1.19 In the case where $p = 2$, the group $Qd(p)$ has a normal elementary abelian subgroup of order 4, and is the semidirect product of this group and $SL_2(2) = S_3$. Hence, $Qd(2) = S_4$, the symmetric group on four letters.

Proposition 1.20 *Let G be the group $Qd(p)$, and let P be a Sylow p-subgroup of G. Then* $\mathcal{F}_P(G) \neq \mathcal{F}_P(N)$, *where* $N = N_G(Z(J(P)))$.

Proof The Sylow p-subgroup of $SL_2(p)$ is cyclic, of order p, and so P is non-abelian of order p^3. Since P is a split extension of $C_p \times C_p$ by C_p, it has exponent p by Exercise 1.3. As every subgroup of index p is elementary abelian (and of maximal order), the Thompson subgroup of

P is all of P, and so $\mathrm{Z}(J(P)) = \mathrm{Z}(P)$. Write Q for the normal subgroup $C_p \times C_p$ in the semidirect product, and N for $\mathrm{N}_G(\mathrm{Z}(P))$. [This is equal to $\mathrm{N}_G(P)$, but we do not need this.]

As $\mathrm{SL}_2(p)$ acts on a 2-dimensional vector space in such a way that any non-zero vector can be mapped to any other, we see that all non-identity elements of Q are G-conjugate. This cannot be true in N however since $\mathrm{Z}(P)$, which has order p and lies inside Q, is normal in this subgroup. Hence $\mathcal{F}_P(G) \neq \mathcal{F}_P(N)$, as claimed. $\qquad\square$

Thus if $G = Qd(p)$, then the subgroup N considered above does not control G-fusion in P. The astonishing thing is that $Qd(p)$ is the only obstruction to the statement.

Theorem 1.21 (Glauberman's ZJ-theorem [Gla68a]) *Let p be an odd prime, and let G be a finite group with no subquotient isomorphic with $Qd(p)$ (i.e., G has no subgroup H such that $Qd(p)$ is a quotient of H). Let P be a Sylow p-subgroup of G, and write $N = \mathrm{N}_G(\mathrm{Z}(J(P)))$. Then $\mathcal{F}_P(N) = \mathcal{F}_P(G)$.*

Many of the results given above have analogues for fusion systems. Some are almost direct translations but Glauberman's ZJ-theorem in particular requires a bit of thought to be converted adequately to fusion systems. The reason for this is that the condition of the theorem – that $Qd(p)$ is not involved in the group – needs to be separated from the language of groups.

1.3 Alperin's fusion theorem

Alperin's fusion theorem is one of the fundamental results on fusion in finite groups, and in some sense gives justification to the goal of local finite group theory. One of the main ideas in finite group theory, during the 1960s in particular, is that the structure of p-local subgroups, especially for the prime 2, should determine the global structure of a finite simple group, or more generally an arbitrary finite group, in some sense. We saw an example of p-local control, at least for odd primes, in the Glauberman–Thompson theorem on normal p-complements, which said that whether a finite group G possesses a normal p-complement or not (a *global* property) depends only on what happens in one particular p-local subgroup (a *local* property).

Alperin's fusion theorem is the ultimate justification of this approach, at least in terms of fusion of p-elements, because it tells you that if x and

y are two elements of a Sylow p-subgroup P, then you can tell whether x and y are conjugate in G by only looking at p-local subgroups. Example 1.6 shows that fusion in a Sylow p-subgroup need not be controlled by any single p-local subgroup, but we proved there that, once one took the right collection of p-local subgroups, we could determine conjugacy, by repeatedly conjugating an element inside different p-local subgroups until we reached our target. Alperin's fusion theorem states that this behaviour occurs in *every* finite group. Moreover, the p-local subgroups we need are a very restricted subset.

Definition 1.22 Let G be a finite group, and let P and Q be Sylow p-subgroups of G. We say that $R = P \cap Q$ is a *tame intersection* if both $N_P(R)$ and $N_Q(R)$ are Sylow p-subgroups of $N_G(R)$.

Examples of tame intersections are when the intersection is of index p in one (and hence both) of the Sylow subgroups, and more generally if the intersection is normal in both Sylow subgroups.

Theorem 1.23 (Alperin's fusion theorem [Alp67]) *Let G be a finite group, and let P be a Sylow p-subgroup of G. Let A and B be two subsets of P such that $A = B^g$. There exist Sylow p-subgroups $Q_1, \ldots, Q_n$ of G, elements $x_1, \ldots, x_n$ of G, and an element $y \in N_G(P)$ such that*

(i) $g = x_1 x_2 \ldots x_n y$;
(ii) $P \cap Q_i$ *is a tame intersection for all i*;
(iii) x_i *is a p-element of $N_G(P \cap Q_i)$ for all i*;
(iv) $A^{x_1 x_2 \cdots x_i}$ *is a subset of $P \cap Q_{i+1}$ for all $0 \leq i \leq n - 1$*.

For proofs of this theorem, see the original paper of Alperin [Alp67], or treatments in [Asc00, (38.1)] and [Gor80, Theorem 7.2.6]. We will not prove this theorem here, and instead give a much shorter proof of a weaker result, Theorem 1.29, closer in spirit to the theory of fusion systems. For most purposes the weaker theorem is sufficient.

In [Alp67], Alperin goes on to show that if one relaxes statement (i) in the theorem to simply '$a^{x_1 \cdots x_n y} = a^g$ for all $a \in A$', then one may impose the extra condition that, writing $R = P \cap Q$, we have that $C_P(R) \leq R$.

Definition 1.24 Let G be a finite group and let P be a Sylow p-subgroup of G.

(i) A *family* is a collection of pairs (Q, X), where Q is a subgroup of P and X is a subset of $N_G(Q)$.

(ii) A family F is a *weak conjugation family* if, whenever A and B are
subsets of P with $A^g = B$ for some $g \in G$, there exist elements
$(Q_1, X_1), (Q_2, X_2), \ldots, (Q_n, X_n)$ of F and elements
$x_1, \ldots, x_n, y$ of G such that

   (a) $a^{x_1 \cdots x_n y} = a^g$ for all $a \in A$;

   (b) x_i is an element of X_i for all i and $y \in N_G(P)$;

   (c) $A^{x_1 x_2 \cdots x_i} \subseteq Q_{i+1}$ for all $0 \leq i \leq n-1$.

(iii) A weak conjugation family F is a *conjugation family* if, in addition,
we have $x_1 \ldots x_n y = g$ for some choice of the (Q_i, X_i), x_i, and y.

Alperin's fusion theorem states that if F_t is the family (R, X), where
$R = P \cap Q$ is a tame intersection of P and $Q \in \mathrm{Syl}_p(G)$ and X is the set
of p-elements of $N_G(R)$, then F_t is a conjugation family. Let F_c denote
the subset of F_t consisting only of pairs (R, X) such that $C_P(R) \leq R$.
Alperin proves the following theorem in [Alp67].

Theorem 1.25 (Alperin [Alp67])   *The family F_c given above is a weak
conjugation family.*

Goldschmidt [Gol70] examined Alperin's proof more closely, and noticed that a refinement of the theorem was possible, further reducing the
subgroups needed. To state this restriction, we first need the definition
of a strongly p-embedded subgroup. Let G be a finite group with $p \mid |G|$.
A subgroup M of G is *strongly p-embedded* in G if M contains a Sylow
p-subgroup of G, and $M \cap M^g$ is a p'-group for all $g \in G \setminus M$.

Theorem 1.26 (Goldschmidt [Gol70])   *Let G be a finite group and
let P be a Sylow p-subgroup of G. Let F denote the family of all pairs
$(R, N_G(R))$, where R is a subgroup of P for which the following four
conditions hold:*

  (i) *R is a tame intersection $P \cap Q$, where $Q \in \mathrm{Syl}_p(G)$;*

  (ii) *$C_P(R) \leq R$;*

  (iii) *R is a Sylow p-subgroup of $O_{p',p}(N_G(R))$, the preimage in $N_G(R)$
of $O_p(N_G(R)/O_{p'}(N_G(R)))$;*

  (iv) *$R = P$ or $\mathrm{Out}_G(R) = N_G(R)/R\,C_G(R)$ has a strongly p-embedded
subgroup.*

Then F is a weak conjugation family.

We will not prove either of Theorems 1.25 or 1.26 here. What we will
prove is that a certain related family is a conjugation family.

Definition 1.27 Let G be a finite group, and let P be a Sylow p-subgroup of G. A subgroup Q of P is *extremal* in P with respect to G if $N_P(Q)$ is a Sylow p-subgroup of $N_G(Q)$.

Traditionally, an extremal subgroup is defined to be one for which $N_P(Q)$ is maximal amongst its G-conjugates inside P; we will show in Proposition 1.38 that these statements are equivalent. Notice that a tame intersection $P \cap Q$ is extremal in both P and Q with respect to G. We have the following easy lemma.

Lemma 1.28 *Let G be a finite group and let P be a Sylow p-subgroup of G. Let Q be a subgroup of P. There exists an extremal subgroup R of G that is G-conjugate to Q.*

Proof Let S be a Sylow p-subgroup of $N_G(Q)$ containing $N_P(Q)$. Since P is a Sylow p-subgroup of G, there is some $g \in G$ such that $S^g \leq P$; we claim that $R = Q^g$ is extremal. To see this, notice that $N_G(Q)^g = N_G(R)$, so the Sylow p-subgroups of $N_G(Q)$ and $N_G(R)$ have the same order. Since S is a Sylow p-subgroup of $N_G(Q)$, S^g is a Sylow p-subgroup of $N_G(R)$. However, $S^g \leq P$, so clearly $S^g = N_P(R)$, as required. $\qquad\square$

Theorem 1.29 (Alperin's fusion theorem) *Let G be a finite group, and let P be a Sylow p-subgroup of G. The set of pairs (Q, X), where Q is an extremal subgroup of P with respect to G, and X is the set of p-elements of $N_G(Q)$, is a conjugation family.*

Proof Let A and B be subsets of P such that $A^g = B$ for some $g \in G$. Write $A \xrightarrow{g} B$ if there exist, for $1 \leq i \leq n$, extremal subgroups Q_i of P with respect to G, p-elements $x_i \in N_G(Q_i)$, and $y \in N_G(P)$, such that

(i) $g = x_1 x_2 \ldots x_n y$;
(ii) $A^{x_1 x_2 \ldots x_i} \subseteq Q_{i+1}$ for all $0 \leq i \leq n-1$.

We will show that if A and B are two subsets of P and $g \in G$ is such that $A^g = B$, then $A \xrightarrow{g} B$. Note that if $A \xrightarrow{g} B$ and $C \subseteq A$, then $C \xrightarrow{g} C^g$, and if $A \xrightarrow{g} B$ and $B \xrightarrow{h} C$, then $A \xrightarrow{gh} C$ and $B \xrightarrow{g^{-1}} A$.

We may assume that A and B are subgroups of P, since $A \xrightarrow{g} B$ if and only if $\langle A \rangle \xrightarrow{g} \langle B \rangle$. We proceed by induction on $m = |P : A|$. If $m = 1$ then $A = P$, whence $g \in N_G(P)$, and $P \xrightarrow{g} P$, so we may suppose that $m > 1$.

Let C be an extremal subgroup of P with respect to G such that $A^k = C$ for some $k \in G$, which exists by Lemma 1.28. If we can show that $A \xrightarrow{k} C$ then similarly $B \xrightarrow{g^{-1}k} C$, whence $A \xrightarrow{g} B$, as needed.

Since A is a proper subgroup of P, $A < \mathrm{N}_P(A)$. As $\mathrm{N}_G(A)^k = \mathrm{N}_G(C)$, there exists $h \in \mathrm{N}_G(C)$ such that $\mathrm{N}_P(A)^{kh}$, a p-subgroup of $\mathrm{N}_G(C)$, is contained in $\mathrm{N}_P(C)$, a Sylow p-subgroup of $\mathrm{N}_G(C)$. Thus by induction $\mathrm{N}_P(A) \xrightarrow{kh} \mathrm{N}_P(A)^{kh}$, and so $A \xrightarrow{kh} C$.

In fact, since all Sylow p-subgroups of any finite group X are conjugate in $\mathrm{O}^{p'}(X)$, the subgroup generated by p-elements of X, we may choose h to be a product of p-elements of $\mathrm{N}_G(C)$, and so $C \xrightarrow{h^{-1}} C$. This proves that $A \xrightarrow{k} C$, since $A \xrightarrow{kh} C \xrightarrow{h^{-1}} C$, completing the proof. $\square$

In fact, that this set is a weak conjugation family follows from Alperin's fusion theorem for fusion systems. Indeed, the more difficult result of Goldschmidt, Theorem 1.26, replacing 'tame intersection' with 'extremal', follows at once from Theorem 4.51, which we will prove using the theory of fusion systems.

1.4 The focal subgroup theorem

One important question in group theory is whether a finite group G has a quotient group of order a power of p. Of course, this is equivalent to asking whether there is a quotient group of order p, which is an *abelian* p-factor group. The reason that this might be helpful is that we know how to detect abelian quotients, via the derived subgroup. Since G/G' is an abelian group, there is a maximal p'-subgroup of it, whose quotient is a p-group, and so there is a normal subgroup $H = G' \mathrm{O}^p(G) \trianglelefteq G$ such that G/H is an abelian p-group, and H is minimal subject to this constraint. Clearly, $G = PH$, and so $G/H \cong P/(P \cap H)$. Since $|H : G'|$ is prime to p, we must have that $P \cap H = P \cap G'$, and hence

$$G/(G' \mathrm{O}^p(G)) \cong P/(P \cap G');$$

one may detect the abelian p-quotient by the Sylow p-subgroup P and the intersection $P \cap G'$, called the *focal subgroup* of P in G. The focal subgroup theorem allows us to calculate this intersection. (This theorem was independently proved by Brauer in [Bra53].)

Theorem 1.30 (Focal subgroup theorem, D. Higman [Hig53]) *If G is a finite group with Sylow p-subgroup P, then*

$$P \cap G' = \langle [x, g] : x, x^g \in P, \ g \in G \rangle$$
$$= \langle x^{-1}y : x, y \in P, \ y = x^g \text{ for some } g \in G \rangle$$
$$= \langle x^{-1}(x\phi) : x \in P, \ \phi \in \mathrm{Hom}_{\mathcal{F}_P(G)}(\langle x \rangle, P) \rangle.$$

This last interpretation is purely in terms of the fusion system; in other words, the question of what the quotient $G/G' \, O^p(G)$ is – the maximal abelian p-quotient – is determined by the fusion system $\mathcal{F}_P(G)$. This theorem is a simple application of the transfer homomorphism, and a proof may be found in, for example, [Gor80, Theorem 7.3.4].

The next result on the generation of the focal subgroup is sometimes useful.

Proposition 1.31 *Let P be a Sylow p-subgroup of the finite group G. The focal subgroup $P \cap G'$ is generated by the subgroups $[R, N_G(R)]$, as R ranges over all non-trivial extremal subgroups R of P with respect to G.*

Proof Let S be the subgroup generated by the subgroups of the form $[R, N_G(R)]$, where R is a non-trivial extremal subgroup of P as in the statement. Certainly S is contained within G', and since $[R, N_G(R)] \leq R \leq P$, we have that $S \leq P \cap G'$.

By the focal subgroup theorem, if we can show that $a^{-1}b$ lies in S for all a and b in P that are G-conjugate, we are done. We will use Alperin's fusion theorem to prove this result: by this theorem, there are extremal subgroups R_i of P with respect to G, and elements $x_1, \ldots, x_r$, with $x_i \in N_G(R_i)$ for all i, such that

$$(x_1 \ldots x_i)^{-1}a(x_1 \ldots x_i) \in R_{i+1}$$

for all $i < r$, and $x_1 \ldots x_r = g$. (Here we have used the fact that P is extremal, and so may be one of the R_i.)

Let $a_0 = a$ and $a_i = a_{i-1}^{x_i}$, so that $a_r = b$; then both a_{i-1} and a_i lie in R_i (since a_{i-1} lies in R_i by assumption and x_i normalizes R_i), and $a_{i-1}^{x_i} = a_i$, with $x_i \in N_G(R_i)$. Therefore

$$a_{i-1}^{-1}a_i = a_{i-1}^{-1}x_i^{-1}a_{i-1}x_i \in [R, N_G(R_i)],$$

and since

$$a^{-1}b = a_0^{-1}a_r = (a_0^{-1}a_1) \ldots (a_{r-1}^{-1}a_r),$$

we have that $a^{-1}b \in S$, as needed. □

Note that if one uses Theorem 1.23 rather than Theorem 1.29, one may replace extremal subgroups with tame intersections.

Using the focal subgroup theorem, we may prove Frobenius's normal p-complement theorem. There are many ways to do this; see, for example, [Gor80, Theorem 7.4.5], [Asc00, (39.4)], [Rob96, 10.3.2], or [Ros78, 10.47] for four different, still fusion-theoretic, ways.

Let G be a finite group and let P be a Sylow p-subgroup of G. Recall that Frobenius's theorem states that the following are equivalent:

(i) G possesses a normal p-complement;
(ii) $\mathcal{F}_P(G) = \mathcal{F}_P(P)$;
(iii) every p-local subgroup of G possesses a normal p-complement;
(iv) for every p-subgroup Q of G, $\mathrm{Aut}_G(Q)$ is a p-group.

Some of these implications are obvious, but one in particular requires a little work. If a group possesses a normal p-complement, i.e.,

$$G = O_{p'}(G) \rtimes P,$$

where $P \in \mathrm{Syl}_p(G)$, then in this case, it is easy to see that $O_{p'}(G)$ consists of all p'-elements of G. We are now able to tackle the main part of the proof, that (iv) implies (i), which thanks to the focal subgroup theorem is considerably easier.

Lemma 1.32   *Let G be a finite group. If $\mathrm{Aut}_G(Q)$ is a p-group for every non-trivial p-subgroup Q of G, then G possesses a normal p-complement.*

Proof   We will show firstly that G has a non-trivial p-factor group. Let P be a Sylow p-subgroup of G; we will prove that $P \cap G' = P'$, and so G has a non-trivial p-factor group. By Exercise 1.6(i), $\mathcal{F}_P(G) = \mathcal{F}_P(P)$, and so the G-conjugates of elements of P are simply the P-conjugates of elements of P. The focal subgroup theorem says that

$$\begin{aligned}
P \cap G' &= \langle x^{-1}y \; : \; y = x^g \text{ for some } g \in G \rangle \\
&= \langle x^{-1}y \; : \; y = x^g \text{ for some } g \in P \rangle \\
&= \langle [x,g] \; : \; x,g \in G \rangle \\
&= P'.
\end{aligned}$$

Since $P' < P$, we see that G has a non-trivial p-factor group.

From here, it is easy to prove the theorem. Let H be the normal subgroup $G' O^p(G)$, which we now know is a proper subgroup of G. By induction, if the hypothesis passes to H, then the theorem is proved, since H has a characteristic subgroup N for which p does not divide N, and both $|H : N|$ and $|G : H|$ are powers of p. Note that, if Q is a p-subgroup of $P \cap H$, then

$$N_H(Q) = H \cap N_G(Q), \qquad C_H(Q) = H \cap C_G(Q),$$

and so $N_H(Q)/C_H(Q)$ is a p-group, since $N_G(Q)/C_G(Q)$ is. Hence the lemma is proved. $\square$

The rest of the implications will be left as Exercise 1.6.

Just as the maximal abelian p-factor group can be written, via the second isomorphism theorem, as a quotient of P by some naturally defined subgroup, the same is true for the maximal p-factor group. Indeed, since $G/\operatorname{O}^p(G)$ is a p-group, we have that $G = P\operatorname{O}^p(G)$, and so

$$G/\operatorname{O}^p(G) \cong P/(P \cap \operatorname{O}^p(G)).$$

Just as the maximal abelian p-factor group can be detected in the fusion system, the same is true for the maximal p-factor group. This the content of Puig's hyperfocal subgroup theorem. Define the *hyperfocal subgroup* of P in G to be the subgroup $P \cap \operatorname{O}^p(G)$.

Theorem 1.33 (Hyperfocal subgroup theorem, Puig [Pui00]) *Let G be a finite group, and let P be a Sylow p-subgroup of G. Then*

$$\begin{aligned}
P \cap \operatorname{O}^p(G) &= \langle [x,g] \,:\, x \in Q \leq P,\, g \in \operatorname{N}_G(Q),\, g \text{ has } p'\text{-order}\rangle \\
&= \langle x^{-1}(x\phi) \,:\, x \in Q \leq P,\, \phi \in \operatorname{O}^p(\operatorname{N}_G(Q))\rangle \\
&= \langle x^{-1}(x\phi) \,:\, x \in Q \leq P,\, \phi \in \operatorname{O}^p(\operatorname{Aut}_G(Q))\rangle \\
&= \langle x^{-1}(x\phi) \,:\, x \in Q \leq P, \phi \in \operatorname{O}^p(\operatorname{Aut}_{\mathcal{F}_P(G)}(Q))\rangle.
\end{aligned}$$

Proof Firstly notice that the subgroups on the right-hand side are all the same, and so it suffices to check the theorem for any one of them; write S for this subgroup of P. Firstly, $S \leq P \cap \operatorname{O}^p(G)$, since if Q is a subgroup of P, if $x \in Q$ and $y \in \operatorname{O}^p(\operatorname{N}_G(Q))$, then $[x,y] \in \operatorname{O}^p(G)$, and it also lies in Q. Therefore it remains to show that $P \cap \operatorname{O}^p(G)$ is contained in S.

Let $H = \operatorname{O}^p(G)$ and $Q = P \cap H$; since H has no non-trivial p-quotients, $Q \cap H' = Q$, and by Proposition 1.31 this subgroup is generated by the subgroups $[R, \operatorname{N}_H(R)]$, where R is a subgroup of Q extremal in Q with respect to H.

Also, $\operatorname{N}_H(R)$ is generated by $\operatorname{O}^p(\operatorname{N}_H(R))$ and a Sylow p-subgroup of $\operatorname{N}_H(R)$, such as $\operatorname{N}_Q(R)$ (as R is extremal). Therefore, $[R, \operatorname{N}_H(R)]$ is generated by $[R, \operatorname{O}^p(\operatorname{N}_H(R))]$ and $[R, \operatorname{N}_Q(R)] \leq Q'$. Hence Q is generated by subgroups of the form $[R, \operatorname{O}^p(\operatorname{N}_H(R))]$, each of which is contained within S, and Q', so $Q = \langle S, Q'\rangle$; since S is a normal subgroup of Q, this means that Q/S is perfect, a contradiction unless Q/S is trivial, as needed. $\square$

If it can be shown directly that in a fusion system $\mathcal{F}_P(G)$, the hyperfocal subgroup is trivial if and only if the fusion system is $\mathcal{F}_P(P)$, then

Theorem 1.33 would imply Frobenius's normal p-complement theorem. (This is exactly the statement of Frobenius's normal p-complement theorem, since G has a normal p-complement if and only if $P \cap O^p(G) = 1$.) In fact, we will prove a wide-ranging generalization of this statement, Theorem 7.51.

1.5 Fusion systems

Having defined a fusion system of a finite group, we now turn to defining a fusion system in general. Like fusion systems of finite groups, this takes place over a finite p-group, and like fusion systems of finite groups, it involves injective homomorphisms between subgroups of groups. Since we have no underlying group from which to draw our morphism sets, we need to make some compatibility conditions on the morphisms.

We will return to the definition of fusion systems in Chapter 4, and repeat what we do here, with more detail. This section can therefore be thought of as a preview of what is to come.

Definition 1.34   Let P be a finite p-group. A *fusion system* $\mathcal{F}$ on P is a category whose objects are all subgroups of P, and whose morphisms $\operatorname{Hom}_{\mathcal{F}}(Q, R)$ are subsets of all injective homomorphisms $Q \to R$, where Q and R are subgroups of P, with composition of morphisms given by the usual composition of homomorphisms. The sets $\operatorname{Hom}_{\mathcal{F}}(Q, R)$ should satisfy the following three axioms:

  (i) for each $g \in P$ with $Q^g \leq R$, the associated conjugation map $c_g : Q \to R$ is in $\operatorname{Hom}_{\mathcal{F}}(Q, R)$;
 (ii) for each $\phi \in \operatorname{Hom}_{\mathcal{F}}(Q, R)$, the associated isomorphism $Q \to Q\phi$ lies in $\operatorname{Hom}_{\mathcal{F}}(Q, Q\phi)$;
(iii) if $\phi \in \operatorname{Hom}_{\mathcal{F}}(Q, R)$ is an isomorphism, then its inverse $\phi^{-1} : R \to Q$ lies in $\operatorname{Hom}_{\mathcal{F}}(R, Q)$.

We write $\operatorname{Aut}_{\mathcal{F}}(Q)$ for the set (in fact group) $\operatorname{Hom}_{\mathcal{F}}(Q, Q)$.

If there is an isomorphism in $\mathcal{F}$ between two subgroups Q and R of P then we say that Q and R are *$\mathcal{F}$-isomorphic*. We will unravel the definition of a fusion system slightly now: the first condition requires that all morphisms in $\mathcal{F}_P(P)$ lie in $\mathcal{F}$; the second condition says that if one map ϕ with domain Q is in the fusion system then so is the induced isomorphism $\bar{\phi} : Q \to Q\phi$; and the final axiom requires that a map ϕ in $\mathcal{F}$ being an isomorphism in $\mathcal{F}$ (in the sense of category theory) is identical to ϕ being an isomorphism in the group-theoretic sense, this last condition ensuring that being $\mathcal{F}$-isomorphic is an equivalence relation. If x and y are elements of P, and there is a map ϕ in $\mathcal{F}$ such that ϕ maps x to y, then x and y are *$\mathcal{F}$-conjugate*.

The next proposition is clear, and its proof is left as an exercise.

Proposition 1.35 *Let G be a finite group and let P be a Sylow p-subgroup of G. Then $\mathcal{F}_P(G)$ is a fusion system on P.*

The concept of a fusion system is a little loose for good theorems to be proved about it, and we prefer to deal with *saturated* fusion systems. To define a saturated fusion system, we need to define the concept of fully centralized and fully normalized subgroups.

Definition 1.36 Let P be a finite p-group, and let Q be a subgroup of P. Let $\mathcal{F}$ be a fusion system on P.

(i) The subgroup Q is *fully centralized* if, whenever $\phi : Q \to R$ is an isomorphism in $\mathcal{F}$, we have that

$$|\,C_P(Q)| \geq |\,C_P(R)|.$$

(ii) The subgroup Q is *fully normalized* if, whenever $\phi : Q \to R$ is an isomorphism in $\mathcal{F}$, we have that

$$|\,N_P(Q)| \geq |\,N_P(R)|.$$

We now come to the definition of a saturated fusion system. One of the conditions concerns extensions of maps in the fusion system, so we will discuss these first. Let $\phi : Q \to P$ be an injective homomorphism: as we have said at the start of this chapter, this induces a map from $\mathrm{Aut}(Q)$ to $\mathrm{Aut}(Q\phi)$. We would like the image of $\mathrm{Aut}_P(Q)$ under this map to be a subgroup of $\mathrm{Aut}_P(Q\phi)$, but in general this won't be true, and so we consider the preimage of $\mathrm{Aut}_P(Q\phi)$ under this map c_ϕ. This is some subgroup X of $\mathrm{Aut}_P(Q)$, and this has a corresponding subgroup Y in $N_P(Q)$ containing $C_P(Q)$, since

$$\mathrm{Aut}_P(Q) \cong N_P(Q)/\,C_P(Q)$$

through the standard isomorphism taking $g \in N_P(Q)$ to c_g. This subgroup Y is the largest subgroup of $N_P(Q)$ to which ϕ may be extended, and will be denoted by N_ϕ. Another characterization of N_ϕ is

$$N_\phi = \{\, g \in N_P(Q) \,:\, \text{there exists } h \in N_P(R) \text{ with } (Qc_g)\phi = (Q\phi)c_h \,\},$$

in other words, all $g \in \mathrm{N}_P(Q)$ such that there is some $h \in \mathrm{N}_P(Q\phi)$, with $(x^g)\phi = (x\phi)^h$ for all $x \in Q$.

Definition 1.37 Let P be a finite p-group. A fusion system $\mathcal{F}$ on P is *saturated* if

(i) $\mathrm{Aut}_P(P)$ is a Sylow p-subgroup of $\mathrm{Aut}_{\mathcal{F}}(P)$, and

(ii) every morphism $\phi : Q \to P$ in $\mathcal{F}$ such that $Q\phi$ is fully normalized extends to a morphism $\bar{\phi} : N_\phi \to P$.

Saturated fusion systems have a lot more structure, and are a lot closer to the fusion systems arising from finite groups. We will prove that every fusion system arising from a finite group is saturated, but before we do that, we will need a characterization of fully normalized subgroups for fusion systems of finite groups.

Proposition 1.38 *Let G be a finite group, and let P be a Sylow p-subgroup of G. A subgroup Q of P is fully $\mathcal{F}_P(G)$-normalized if and only if $\mathrm{N}_P(Q)$ is a Sylow p-subgroup of $\mathrm{N}_G(Q)$.*

Proof We begin by noting the following. Let R be a Sylow p-subgroup of $\mathrm{N}_G(Q)$ containing $\mathrm{N}_P(Q)$; there is an element $g \in G$ such that $R^g \leq P$, and so

$$R^g \leq P \cap \mathrm{N}_G(Q)^g = P \cap \mathrm{N}_G(Q^g) = \mathrm{N}_P(Q^g).$$

Thus

$$|\mathrm{N}_P(Q)| \leq |R| \leq |\mathrm{N}_P(Q^g)|.$$

If $|\mathrm{N}_P(Q)| \geq |\mathrm{N}_P(Q^g)|$ for all $g \in G$ – i.e., Q is fully $\mathcal{F}_P(G)$-normalized – then $R = \mathrm{N}_P(Q)$, and so is a Sylow p-subgroup of $\mathrm{N}_G(Q)$. Conversely, if $\mathrm{N}_P(Q)$ is a Sylow p-subgroup of $\mathrm{N}_G(Q)$, then $|\mathrm{N}_P(Q)| = |R|$, and this is the order of a Sylow p-subgroup of $\mathrm{N}_G(Q^g) = \mathrm{N}_G(Q)^g$, so that $|\mathrm{N}_P(Q)| \geq |\mathrm{N}_P(Q^g)|$. Hence we get the result. $\square$

Note that a similar result holds for centralizers, and the proof is very similar. Proposition 1.38 simply states that Q is fully $\mathcal{F}_P(G)$-normalized if and only if Q is extremal in P with respect to G. We will therefore no longer consider extremal subgroups, and simply refer to fully normalized subgroups.

Theorem 1.39 *Let G be a finite group and let P be a Sylow p-subgroup of G. The fusion system $\mathcal{F}_P(G)$ is saturated.*

Proof Since $\mathrm{Aut}_{\mathcal{F}}(P) = \mathrm{N}_G(P)/\mathrm{C}_G(P)$, and the image of P in this quotient group is a Sylow p-subgroup of $\mathrm{Aut}_{\mathcal{F}}(P)$, the first axiom is satisfied. Thus, let Q be a subgroup of P, let $\phi : Q \to P$ be a morphism in $\mathcal{F}$, and suppose that $Q\phi$ is fully normalized. Since ϕ is induced by conjugation, there is some $g \in G$ such that $\phi = c_g$ on Q. In this case, the set N_ϕ is given by

$$N_\phi = \{\, x \in \mathrm{N}_P(Q) \,:\, \text{there is } y \in \mathrm{N}_P(Q^g) \text{ with } Qc_{xg} = Qc_{gy} \,\}.$$

Thus $x \in N_\phi$ if and only if $xgy^{-1}g^{-1}$ centralizes Q. In this case $g^{-1}xgy^{-1}$ centralizes Q^g, and so $x^g = hy$, for some $h \in \mathrm{C}_G(Q^g)$. Thus

$$(N_\phi)^g \leq \mathrm{N}_P(Q^g)\,\mathrm{C}_G(Q^g).$$

Since N_ϕ is a p-subgroup of G, and, by Proposition 1.38, $\mathrm{N}_P(Q^g)$ is a Sylow p-subgroup of $\mathrm{N}_G(Q^g)$, there is some element a of $\mathrm{C}_G(Q^g)$ such that $(N_\phi)^{ga}$ is contained within $\mathrm{N}_P(Q^g)$.

Define $\theta : N_\phi \to P$ by $x\theta = x^{ga}$, for all $x \in N_\phi$. Since $a \in \mathrm{C}_G(Q^g)$, θ extends ϕ, and so this is the map required by the definition. $\quad\square$

We end the chapter with a discussion of the Solomon fusion system, which is one of the foundations of the subject. In [Sol74], Solomon considered the Sylow 2-subgroup of the finite group $\mathrm{Spin}_7(3)$; the group $\mathrm{Spin}_7(q)$ is a central extension of order 2 of the 7-dimensional orthogonal group $\Omega_7(q)$. Let G be the group $\mathrm{Spin}_7(3)$, and let $H = G/\langle z \rangle$, where z is the central involution of G; then H has a Sylow 2-subgroup isomorphic with that of A_{12}. Solomon proved the following.

Theorem 1.40 (Solomon [Sol74]) *There does not exist a finite group K with the following properties:*

(i) *a Sylow 2-subgroup P of K is isomorphic with that of $\mathrm{Spin}_7(3)$;*
(ii) *$\mathcal{F}_P(\mathrm{Spin}_7(3)) \subseteq \mathcal{F}_P(K)$;*
(iii) *all involutions in K are conjugate.*

Theorems such as these are often proved using analysis on the 2-local structure of the group. Solomon attempted this, but found that no contradiction could be reached this way; he was forced to find another way. The reason for this is the following.

Theorem 1.41 (Levi–Oliver [LO02]) *Let P be isomorphic with the Sylow 2-subgroup of $\mathrm{Spin}_7(3)$. There exists a saturated fusion system $\mathcal{F}$ on P such that $\mathcal{F}_P(\mathrm{Spin}_7(3)) \subseteq \mathcal{F}$ and all involutions are $\mathcal{F}$-conjugate.*

This fusion system is an example of an *exotic* fusion system, i.e., one that is not the fusion system of a finite group. The interesting thing about this is that the fusion system is also simple, a term that will not be defined until Chapter 5.

Let q be an odd prime power such that $q \equiv \pm 3 \bmod 8$, and let P be a Sylow 2-subgroup of $\mathrm{Spin}_7(q)$ (which is the same as that of $\mathrm{Spin}_7(3)$). Solomon actually showed that there does not exist a finite group having P as a Sylow 2-subgroup, with a single conjugacy class of involutions, and such that another technical condition on centralizers holds, which we will examine later.

Levi and Oliver [LO02] proved that, for all odd q, there exists a saturated fusion system on the Sylow 2-subgroup of $\mathrm{Spin}_7(q)$ that has the above properties. Furthermore, these are also examples of the simple fusion systems that we mentioned before. They are the only known examples of simple, exotic fusion systems that do not arise from finite groups for the prime 2, although for odd primes many simple exotic fusion systems are known.

Exercises

1.1 Let G be a finite group, and let H be a subgroup of G. We say that H is a *trivial intersection* subgroup or *TI subgroup* if, whenever $g \in G$, either $H^g = H$ or $H^g \cap H = 1$. Prove that if the Sylow p-subgroups of G are TI, then $\mathrm{N}_G(P)$ controls G-fusion in P.

1.2 Using Burnside's normal p-complement theorem, prove that if G is a simple group and p is the smallest prime dividing $|G|$, then either p^3 divides $|G|$ or $p = 2$ and $12 \mid |G|$.

1.3 Let p be a prime. Prove that there are exactly five groups of order p^3: three abelian, two non-abelian. If p is odd, prove that, of the two non-abelian p-groups of order p^3, one has exponent p^2 and one has exponent p. Deduce that if P is a non-abelian group of order p^3 and exponent p then all maximal subgroups of P are elementary abelian. Prove that if P is non-abelian and is a split extension of $C_p \times C_p$ by C_p, then P is of exponent p.

1.4 (Grün's first theorem) Let G be a finite group and let P be a Sylow p-subgroup of G. Prove that

$$P \cap G' = \langle\, P \cap \mathrm{N}_G(P)',\ P \cap Q' \ :\ Q \in \mathrm{Syl}_p(G)\,\rangle.$$

[Hint: mimic the proof of Proposition 1.31.]

1.5 Determine the Thompson subgroup for dihedral, semidihedral, generalized quaternion and abelian p-groups.

1.6 Let G be a finite group, and let P be a Sylow p-subgroup of G. Lemma 1.32 proves one implication of Frobenius's normal p-complement.

   (i) Prove that, if for all non-trivial p-subgroups Q of G we have that $\mathrm{Aut}_G(Q)$ is a p-group, then $\mathcal{F}_P(G) = \mathcal{F}_P(P)$.

   (ii) Prove the rest of the implications involved in Frobenius's normal p-complement theorem.

1.7 Let G be a finite group and let P be a Sylow p-subgroup of G. Prove that the fusion systems $\mathcal{F}_P(G)$ and $\mathcal{F}_Q(G/\mathrm{O}_{p'}(G))$ are isomorphic, where $Q = P\,\mathrm{O}_{p'}(G)/\mathrm{O}_{p'}(G)$.

1.8 Let $\mathcal{F}$ be a saturated fusion system on the finite abelian p-group P. Show that any morphism ϕ may be extended to an automorphism of P that lies in $\mathcal{F}$. (This property will later be called resistance. This exercise shows that abelian p-groups are resistant.)

2

Fusion in representation theory

Modular representation theory has two major facets: module theory and block theory. Every p-block of a finite group has a fusion system attached to it, and it turns out that this fusion system is saturated. Even to get to the point where we are able to define the fusion system of a block takes us on a whistle-stop tour of modular representation theory.

We begin in the first section by defining the concept of a p-block of a finite group: a p-block of a group is an indecomposable two-sided ideal summand of the group algebra kG, where k is an algebraically closed field of characteristic p. After examining their properties and developing some theory, we move on to the Brauer morphism and relative traces. The Brauer morphism Br_P is of fundamental importance, and allows us to pass from groups to centralizers (and normalizers) of p-subgroups. Relative traces are a useful tool for proving results about blocks, and we will use them extensively here.

Armed with these tools, we define and study Brauer pairs in Section 2.3. These will form the elements of the fusion system of a block. A Brauer pair is a pair (P, e), where P is a p-subgroup of G and e is a block idempotent of $k\, \mathrm{C}_G(P)$. We deal with conjugation and inclusion of Brauer pairs, which are needed to define a fusion structure on them.

In Section 2.4 we introduce defect groups and Brauer's first main theorem, which sets up a bijection between blocks of kG with defect group D and blocks of $\mathrm{N}_G(D)$ with defect group D. This allows us to prove that there is a unique fusion system associated to a block, by proving that all maximal Brauer pairs associated to a given block are conjugate.

Finally, we introduce the fusion system of a block, and prove that it is saturated. We end by considering blocks with trivial defect group, the principal block, and nilpotent blocks.

Much of the material here is standard: treatments for Brauer pairs start from the original article of Alperin and Broué [AB79], and a subsequent article by Broué and Puig [BP80], to the book of Thévenaz [Thé95] and the article of Kessar [Kes07].

2.1 Blocks of finite groups

If K is an algebraically closed field of characteristic 0, say the complex numbers $\mathbb{C}$, then we may form the group algebra KG for any finite group G. Artin–Wedderburn theory tells us that KG is isomorphic (as a K-algebra) to a direct sum of matrix algebras. This direct sum can be found by decomposing KG into a direct sum of indecomposable two-sided ideals; a two-sided ideal B is said to be *indecomposable* if, whenever $B = B_1 \oplus B_2$ may be written as the direct sum of two other two-sided ideals, then either B_1 or B_2 is 0.

More generally, we will let R be a commutative ring and consider R-algebras. We recall a few facts about these objects, which may be found in, for example, [Dor72, Section 39]. We focus on the case where R is Noetherian; in this case, any finitely generated R-free (i.e., free as an R-module) R-algebra A is Noetherian. We will always assume that R-algebras are free as R-modules. Any finitely generated module for a Noetherian R-algebra A is the direct sum of a finite number of indecomposable A-modules.

Lemma 2.1 *Let R be a Noetherian ring and let A be an R-algebra. There is a decomposition*

$$A = \bigoplus_{i=1}^{r} B_i$$

of A into indecomposable two-sided ideals B_i.

This follows because a right ideal of A is a submodule of A, thought of as a right A-module, and similarly for left ideals, so a two-sided ideal summand is an indecomposable (A, A)-bimodule summand, and there is a decomposition of A into finitely many of these for the same reasons as there are decompositions into finitely many indecomposable summands for any finitely generated A-module. Any indecomposable two-sided ideal summand of A is a *block* of A.

Related to two-sided ideals are idempotents. Recall that an *idempotent* of a ring is an element $e \neq 0$ such that $e^2 = e$, and e is *central* if e lies in the centre of the ring. Two idempotents e and e' are called *orthogonal* if $ee' = e'e = 0$, and finally an idempotent e is *primitive* if e cannot be written as the sum of two orthogonal idempotents. If e is a central idempotent, then e is primitive if it is primitive in $Z(A)$, rather than being primitive in A itself, so that if e can be written as $e = e_1 + e_2$, then the e_i are not central.

The fundamental result that links central idempotents and two-sided ideal summands is the next proposition. If e is a central idempotent, then $Ae = eA$ is a two-sided ideal. This sets up a correspondence between blocks and idempotents, as follows.

Proposition 2.2   *Let R be a Noetherian ring, and let A be a finitely generated R-algebra.*

(i) *There is a one-to-one correspondence between decompositions of A into two-sided ideals and decompositions of 1 as the sum of pairwise orthogonal, central idempotents, with a decomposition $1 = e_1 + \cdots + e_r$ in correspondence with $A = Ae_1 \oplus \cdots \oplus Ae_r$, and a decomposition $A = B_1 \oplus \cdots \oplus B_r$ in correspondence with the decomposition of 1 as an element of A in this ideal decomposition.*

(ii) *If B is a block, with $A = B \oplus B'$ and $1 = e + e'$ as the corresponding decomposition, then e is a primitive central idempotent, and if e is a primitive central idempotent then Ae is a block of A.*

Proof   Let $1 = e_1 + \cdots + e_r$ be a decomposition of 1 into a sum of central idempotents, and let $B_i = Ae_i$, a two-sided ideal of A. If xe_i is an element of Ae_i, we must show that it does not lie in $B_i' = \sum_{j \neq i} Ae_j$. However, since e_i is an idempotent, $xe_i = xe_i^2$, and so $xe_i \in B_ie_i = 0$, since the e_i are pairwise orthogonal. Hence the sum of the Ae_i is direct. Certainly $1 \in \sum Ae_i$, and so A is the direct sum of the Ae_i, completing the proof of one direction.

Conversely, let $A = B_1 \oplus \cdots \oplus B_r$ be a decomposition of A into two-sided ideals, and let $1 = e_1 + \cdots + e_r$ be the induced decomposition of 1. We will show that each e_i is a central idempotent, and that the e_i are pairwise orthogonal. This latter statement is easy, since e_ie_j lies in $B_i \cap B_j$, which is zero if $i \neq j$. Decomposing 1 as the sum of the e_i, we have

$$\sum_{i=1}^{r} e_i = 1 = 1^2 = \left(\sum_{i=1}^{r} e_i \right)^2 = \sum_{i=1}^{r} e_i^2,$$

with the last equality true because the cross-products e_ie_j are all zero.

By the uniqueness of the decomposition of 1, each e_i is an idempotent. Finally, if $x \in A$, then $1 \cdot x = x \cdot 1 = x$, and we get

$$\sum_{i=1}^{r} e_i x = 1 \cdot x = x \cdot 1 = \sum_{i=1}^{r} x e_i,$$

and again, by uniqueness of decompositions, $x e_i = e_i x$ for all $x \in A$ and all i, proving that the e_i are central. This completes the proof of (i).

If B is not a block, then there is a refinement of $A = B \oplus B'$ to $A = B_1 \oplus B_2 \oplus B'$, and so by (i) the idempotent e corresponding to B cannot be primitive, as $e = e_1 + e_2$, where e_i corresponds to B_i. Conversely, if e is not primitive, then $e = e_1 + e_2$ and $1 = e_1 + e_2 + (1 - e)$ is a refinement of the decomposition $1 = e + (1 - e)$, proving that Ae is not a block, finishing (ii). $\qquad\square$

We know that there is a decomposition of A into blocks, and therefore there is a decomposition of 1 into primitive central idempotents. We will use the correspondence to prove the following result.

Proposition 2.3 *Let A be a finitely generated R-algebra, and let $A = Ae_1 \oplus \cdots \oplus Ae_r$ be a decomposition of A into blocks. If $B = Ae$ is a block of A, then $B = Ae_i$ for some i. In particular, the blocks of A are precisely the Ae_i, and the primitive central idempotents are precisely the e_i.*

Proof We will prove that e is one of the e_i, and note that the rest follows. To see this, write

$$e = e \cdot 1 = ee_1 + ee_2 + \cdots + ee_r.$$

Since e and each e_i are central idempotents, so is each ee_i. The ee_i are orthogonal, as if $i \neq j$ then $(ee_i)(ee_j) = e^2 e_i e_j = 0$. As e is primitive, we must have that $ee_i = e$ for some i. Since $e_i = ee_i + (1 - e)e_i$ is an orthogonal decomposition of e_i into central idempotents, we must have that $(1 - e)e_i = 0$, so that $e_i = ee_i = e$, as required. $\qquad\square$

Although we call the ideals Ae_i the blocks of A, and the e_i the *block idempotents*, it should be noted that many authors, particularly modern ones, switch this, and call the e_i 'blocks' and the Ae_i 'block ideals'. Of course, since there is a correspondence between them it does not matter which one is called what, although we will generally use the block idempotents in what follows. Recall that the *Jacobson radical* of a ring R, denoted by $J(R)$, is the intersection of all maximal right ideals; there are numerous other characterizations of $J(R)$.

Proposition 2.4   *Let G be a finite group and let k be an algebraically closed field of characteristic p. Let B be a block of kG, and let b be its block idempotent. The centre of B is a local ring, and $Z(B)/J(Z(B)) \cong k$, with $\{J(Z(B)) + b\}$ a k-basis for this quotient.*

For a proof, see for example [PD77, Lemma 4.1]. A brief sketch is as follows: $Z(kG)b$ is a $Z(kG)$-module, and is a summand of a free $Z(kG)$-module. The module $Z(kG)b$ is indecomposable since b is a primitive idempotent, and so the top of $Z(kG)$ is simple. Hence $Z(kG)b/J(Z(kG)b)$ is a simple k-algebra, and so is isomorphic with k, as k is algebraically closed, and the only simple commutative finite-dimensional k-algebra is k.

Before we leave the subject of idempotents and ideals, we will need Rosenberg's lemma, which is an important tool in proving results about where primitive idempotents lie in sums of ideals (see, for example [Ben98b, Lemma 1.7.10]).

Lemma 2.5 (Rosenberg's lemma)   *Let A be a finite-dimensional k-algebra, and let e be a primitive idempotent of A. If e lies in a sum of two-sided ideals, then e lies in (at least) one of them.*

Now that we have a reasonable idea of what blocks are – ideal summands in the group algebra – we look at modules. In characteristic 0, the theory is well known. One of the main results in this case is Maschke's theorem.

Theorem 2.6 (Maschke's theorem)   *Let G be a finite group, and let K be a field of characteristic 0, or of characteristic p for some $p \nmid |G|$. If M is a finite-dimensional KG-module, then M is semisimple, i.e., it is the direct sum of simple KG-modules.*

This, combined with the fact that the number of simple KG-modules (up to isomorphism) is equal to the number of conjugacy classes in G, implies that understanding the representation theory is in some sense a finite problem for a given group. Indeed, the dimensions of the simple KG-modules are the same as the degrees of the matrix algebras that constitute the blocks of the algebra KG. (In fact, if KG itself is considered as a KG-module by right multiplication, we see that KG is isomorphic with the direct sum of n copies of each simple module of dimension n, which results in the well-known statement that the group order is equal to the sum of the squares of the dimensions of the irreducible representations.)

This contrasts significantly with the case of a field k of characteristic $p \mid |G|$; in this case, there are still only finitely many simple kG-modules, but it is not the same number.

Theorem 2.7 (Brauer) *Let G be a finite group, and let k be an algebraically closed field of characteristic p. The number of simple kG-modules is equal to the number of conjugacy classes of elements whose order is prime to p.*

The real problem is the fact that Maschke's theorem no longer holds, and there are indecomposable modules that are not simple. In general – if the Sylow p-subgroups are not cyclic – there are infinitely many isomorphism classes of indecomposable modules. While we cannot write the group algebra as a direct sum of matrix algebras using Artin–Wedderburn theory, we may write the group algebra as the sum of the blocks.

We can also relate the indecomposable, and in particular simple, modules for a group to the blocks. Let M be an indecomposable kG-module and let $1 = \sum e_i$ be the decomposition of 1 into the block idempotents. Since M is a right A-module, and the e_i are orthogonal, we have

$$M = M \cdot 1 = M \cdot e_1 \oplus M \cdot e_2 \oplus \cdots \oplus M \cdot e_r;$$

since M is indecomposable, there is a unique i such that $M \cdot e_i \neq 0$ (in fact $M \cdot e_i = M$), and for all other j, $M \cdot e_j = 0$. We say that M *belongs* to Ae_i. We extend the terminology, and say that a sum of modules belonging to a particular block also belongs to the block.

If N is a submodule of M then N belongs to Ae_i also, since $N \cdot e_i = N$, and similarly all quotients of M belong to Ae_i. In particular, all composition factors of M – the simple modules inside M – belong to Ae_i. This partitions the indecomposable, and simple, modules among the blocks. To see that every block has some indecomposable modules, and hence some simple modules, we notice that Ae_i, viewed as a kG-module, belongs to the block Ae_i, and hence all composition factors of Ae_i belong to Ae_i.

In order to pass between characteristic 0 and characteristic p representations, we need a third object, a ring with the field k (of characteristic p) as a quotient and the field K (of characteristic 0) as field of fractions. This ring is normally denoted $\mathcal{O}$ in the literature; the three objects K, $\mathcal{O}$ and k that we are considering will have the following properties, designed to make the theory work properly:

(i) $\mathcal{O}$ is a local principal ideal domain, K is its field of fractions, and k is the quotient $\mathcal{O}/J(\mathcal{O})$ by the unique maximal ideal $J(\mathcal{O})$ of $\mathcal{O}$;

(ii) $\mathcal{O}$ is complete with respect to the $J(\mathcal{O})$-adic topology;

(iii) K contains a primitive $|G|$th root of unity;

(iv) k is algebraically closed.

In the literature, a triple $(K, \mathcal{O}, k)$ with K, $\mathcal{O}$ and k as above is a *p-modular system*. We will make no suggestion as to why such triples $(K, \mathcal{O}, k)$ exist, but instead refer the reader to [Ser79, Theorem II.3.1, Theorem II.5.3]. We may assume that the intersection of the powers $J(\mathcal{O})^n$ of the radical $J(\mathcal{O})$ is trivial.

We need to pass between idempotents of $\mathcal{O}G$ and idempotents of kG. It is clear that the quotient map $\pi : \mathcal{O} \to k$ extends to a ring homomorphism $\mathcal{O}G \to kG$. Indeed, if A is any $\mathcal{O}$-algebra, we denote by $\bar{A}$ the image of the extension of the map π to $\mathcal{O}G$, i.e., $\bar{A} = A/J(\mathcal{O})A$, called the *reduction modulo p*. One direction of this bijection between idempotents of $\mathcal{O}G$ and KG is fairly easy.

Proposition 2.8 *Let A be an $\mathcal{O}$-algebra, and let $\bar{A}$ denote the reduction modulo p of A. If $e_1, \ldots, e_n$ are orthogonal idempotents of A such that $\sum e_i = 1$ then $\bar{e}_1, \ldots, \bar{e}_n$ are orthogonal idempotents of $\bar{A}$ such that $\sum \bar{e}_i = 1$.*

Proof Firstly the images of the e_i clearly satisfy $\bar{e}_i^2 = \bar{e}_i$, $\bar{e}_i \bar{e}_j = 0$ if $i \neq j$, and the sum of the $\bar{e}_i$ is 1. It remains to show that $\bar{e}_i$ is non-zero, i.e., to show that no e_i lies in R, the kernel of the extension of π to A. However,

$$\bar{e}_i = \bar{e}_i^n \in R^n,$$

and as A is a free $\mathcal{O}$-module,

$$\bigcap_{n=1}^{\infty} R^n = \bigcap_{n=1}^{\infty} J(\mathcal{O})^n \cdot A = 0,$$

since the intersection of the $J(\mathcal{O})^n$ is zero. This proves the proposition. $\qquad \square$

Passing back from kG to $\mathcal{O}G$ is more difficult, and would take us too long to prove. The converse to the previous proposition is true, however; for a proof, see for example [PD77, Theorem 3.4A] or [Ben98b, Theorem 1.9.4].

Proposition 2.9 *Let A be an $\mathcal{O}$-algebra, and let $\bar{A}$ denote the reduction modulo p of A. If $\bar{e}_1, \ldots, \bar{e}_n$ are orthogonal idempotents of $\bar{A}$ such that $\sum \bar{e}_i = 1$ then there exist orthogonal idempotents $e_1, \ldots, e_n$ in A such that $\sum e_i = 1$ and the image of e_i is $\bar{e}_i$ under π.*

Corollary 2.10 *There is a bijection between the blocks of $\mathcal{O}G$ and kG, given by reduction modulo p.*

Proof We first note that it is easy to see that reduction modulo p maps the centre of $\mathcal{O}G$ surjectively onto $Z(kG)$. The corollary is then an obvious consequence of Propositions 2.8 and 2.9, applied to the collections of all blocks of $\mathcal{O}G$ and kG, and $A = Z(\mathcal{O}G)$. $\qquad\square$

By passing from $\mathcal{O}G$ to KG, the blocks of $\mathcal{O}G$ break up as direct sums of matrix algebras, and this allows us to associate every irreducible representation in characteristic 0 to a p-block of G.

We will need to lift idempotents in the following slightly different situation to that given above (see [Ben98b, Corollary 1.7.4]).

Proposition 2.11 *Let I be a nilpotent ideal in a finite-dimensional k-algebra A. If $e_1, \ldots, e_n$ are primitive orthogonal idempotents of A such that $\sum e_i = 1$, then $\bar{e}_1, \ldots, \bar{e}_n$ are primitive orthogonal idempotents of A/I such that $\sum \bar{e}_i = 1$. Conversely, if $\bar{e}_1, \ldots, \bar{e}_n$ are primitive orthogonal idempotents of A/I such that $\sum \bar{e}_i = 1$, then there exist orthogonal idempotents $e_1, \ldots, e_n$ in A such that $\sum e_i = 1$ and the image of e_i is $\bar{e}_i$ under $A \mapsto A/I$.*

There is, of course, one obvious kG-module for any finite group G, namely the trivial module. The p-block to which the trivial module belongs is the *principal block*. This is, in some sense, the most complicated block in a finite group, in terms of the 'defect', a concept that we will introduce later. The opposite of this is a block of defect zero, which shall actually a matrix algebra just like the characteristic 0 case, as we shall see in Theorem 2.39. In general, not all blocks are like this, but quite a few times these exist: the most well-known case is a group of Lie type (e.g., $\mathrm{SL}_n(q)$) where the characteristic of k is p and $q = p^n$ for some n. In this case, the Steinberg module lies in a block that is a matrix algebra. This defect in some sense measures how far a block is from being a matrix algebra. In the next few sections we will develop the idea of Brauer pairs, and get to the idea of a defect group for a block of a finite group, before defining the fusion system of a block.

2.2 The Brauer morphism and relative traces

We begin this section by defining the Brauer morphism, and giving one of its most important properties. Put simply, the Brauer morphism is a projection map in the sense of linear algebra. Let P be a p-subgroup of G; the *Brauer morphism* $\mathrm{Br}_P : kG \to k\, \mathrm{C}_G(P)$ is the surjective k-linear map

$$\sum_{g \in G} \alpha_g g \mapsto \sum_{g \in \mathrm{C}_G(P)} \alpha_g g.$$

If H is a subgroup of G then conjugation by elements of H may be extended linearly to a conjugation action on kG; if X is a set with an action of G on it, then an element $x \in X$ is H-*stable* if $x^h = x$ for all $h \in H$. The H-stable elements of X are denoted by X^H; if $X = kG$ we include some brackets and write $(kG)^P$.

Proposition 2.12 *Let G be a finite group and let P be a p-subgroup of G. The map Br_P is multiplicative when restricted to the P-stable elements $(kG)^P$ of kG.*

Proof A k-basis for $(kG)^P$ is the set of all P-class sums of elements of G. Let $\mathcal{X}$ denote the set of all P-conjugacy classes of G, and if $X \in \mathcal{X}$, let $\hat{X}$ denote its class sum. Let g be an element of $\mathrm{C}_G(P)$, and consider the multiplicity of g in the product set $X \cdot Y$, where X and Y are in $\mathcal{X}$. We claim that either this multiplicity is 0, or $|X| = |Y|$ and this multiplicity is $|X|$.

Suppose that this claim is true. Since P-conjugacy classes have size either 1 or a multiple of p, we see that $\mathrm{Br}_P(\hat{X})$ is $\hat{X}$ if X is a singleton set (lying inside $\mathrm{C}_G(P)$) and 0 otherwise. Thus $\mathrm{Br}_P(\hat{X})\, \mathrm{Br}_P(\hat{Y})$ is 0 unless both X and Y are singleton sets, in which case it is $\hat{X}\hat{Y}$. Conversely, $\mathrm{Br}_P(\hat{X}\hat{Y})$ is $\hat{X}\hat{Y}$ if both X and Y are singleton sets, and is 0 if at least one of them is not, by the claim. Hence Br_P is multiplicative on the basis elements of $(kG)^P$, and so by linearity is multiplicative.

It remains to prove the claim. Let $g \in \mathrm{C}_G(P)$, and suppose that g appears in the product set $X \cdot Y$; write n for the number of distinct ways of making g from one element of X and one of Y. If $|X| = |Y| = 1$ then certainly $n = 1$, and so our claim is true in this case. If $g = xy$, then

$$g = g^h = x^h y^h$$

for every $h \in P$, and since h runs over all elements of P, all elements of

X and all elements of Y appear. Hence the map from X to Y sending x^h to y^h is a bijection, so $|X| = |Y| = n$. $\qquad\square$

The Brauer morphism therefore is a surjective algebra homomorphism

$$\mathrm{Br}_P : (kG)^P \to k\,\mathrm{C}_G(P),$$

for any p-subgroup P. (It remains surjective because if $x \in \mathrm{C}_G(P)$ then the P-conjugacy class containing x is simply $\{x\}$; hence $k\,\mathrm{C}_G(P) \leq (kG)^P$.) This leads to an important corollary, which will be crucial in what follows.

Corollary 2.13 *Let G be a finite group and let P be a p-subgroup of G.*

(i) *If e is a central idempotent of kG then $\mathrm{Br}_p(e)$ is either 0 or a central idempotent of $k\,\mathrm{C}_G(P)$.*

(ii) *If b is a block idempotent of kG such that $\mathrm{Br}_P(b) \neq 0$ then $\mathrm{Br}_P(b) = b_1 + \cdots + b_r$ is a sum of block idempotents of $k\,\mathrm{C}_G(P)$. If e is a block idempotent of $k\,\mathrm{C}_G(P)$, then $\mathrm{Br}_P(b)e \neq 0$ if and only if $e = b_i$ for some i, and in this case $\mathrm{Br}_P(b)b_i = b_i$.*

Proof Since Br_P is multiplicative on $(kG)^P \supseteq Z(kG)$, $\mathrm{Br}_P(b)$ is an idempotent, and since Br_P is surjective, $\mathrm{Br}_P(b) \in Z(k\,\mathrm{C}_G(P))$, proving (i).

Any central idempotent is a sum of block idempotents, so we may write $\mathrm{Br}_P(b) = b_1 + \cdots + b_r$ for block idempotents b_i. Clearly, if e is a block idempotent of $k\,\mathrm{C}_G(P)$ then either $b_i e = e = b_i$ or $b_i e = 0$, and this proves (ii). $\qquad\square$

Next, we define the concept of the relative trace.

Definition 2.14 Let G be a finite group acting on an abelian group X under addition, and let H be a subgroup of G. The *relative trace* is the map $\mathrm{Tr}_H^G : X^H \to X^G$, given by

$$\mathrm{Tr}_H^G(x) = \sum_{g \in T} x^g,$$

where T is a right transversal to H in G.

If Y is a subset of X^H, then we denote by $\mathrm{Tr}_H^G(Y)$ the image of Y under the relative trace map. If M is a group with a G-action, then we consider the group algebra kM as an abelian group with a G-action by extending linearly, and the image of $(kM)^H$ under the trace map Tr_H^G will be denoted by $(kM)_H^G$.

The following lemma is elementary, and its proof is left as Exercise 2.1.

Lemma 2.15   *Let H and L be subgroups of the finite group G, and suppose that G acts on the abelian group X. Let T be a set of representatives of the (H, L)-double cosets of G.*

(i) *If $H \leq L$ and $x \in X^H$, then*

$$\mathrm{Tr}_H^G(x) = \mathrm{Tr}_L^G(\mathrm{Tr}_H^L(x)).$$

(ii) *If $H \leq L$, $x \in X^H$ and $y \in X^L$, then*

$$\mathrm{Tr}_H^L(xy) = \mathrm{Tr}_H^L(x)y \;\text{ and }\; \mathrm{Tr}_H^L(yx) = y\,\mathrm{Tr}_H^L(x).$$

(iii) *If $x \in X^H$ and $y \in X^L$, then*

$$\mathrm{Tr}_H^G(x) = \sum_{t \in T} \mathrm{Tr}_{H^t \cap L}^L(x^t) \;\text{ and }\; \mathrm{Tr}_H^G(x)\,\mathrm{Tr}_L^G(y) = \sum_{t \in T} \mathrm{Tr}_{H^t \cap L}^G(x^t y).$$

The next proposition gives a basis for the image $(kX)_P^G$ of $(kX)^P$ under the trace map; the proof is left as an exercise, namely Exercise 2.2.

Proposition 2.16   *Let G be a finite group acting on a group X, and let P be a p-subgroup of G. The subset $(kX)_P^G$ is a two-sided ideal of $(kX)^G$ and has as k-basis the sums of all elements lying in the G-orbit of a given element $x \in X$, for which P contains a Sylow p-subgroup of $C_G(x)$ (up to G-conjugacy).*

This proposition gives us a corollary on the structure of $(kG)_P^G$, for Sylow p-subgroups of G, as well as on what happens under conjugation.

Corollary 2.17   *Let G be a finite group acting on a group X, and let P be a p-subgroup of G.*

(i) *If $g \in G$ then $((kX)^P)^g = (kX)^{(P^g)}$, and hence $(kX)_P^G = (kX)_{P^g}^G$.*
(ii) *If P is a Sylow p-subgroup of G, then $\mathrm{Z}(kG) = (kG)_P^G$.*

Proof   If x lies in $(kX)^P$ then x^g lies in $(kX)^{(P^g)}$, since for $a \in P$, we have $(x^g)^{(g^{-1}ag)} = x^{ag} = x^g$ as $x^a = x$. Hence the first equation holds, and the second equation follows from the first. (Notice that we do not require that P is a p-subgroup of G for this to hold.)

To see (ii), notice that, if P is a Sylow p-subgroup of G, then P contains a Sylow p-subgroup of $C_G(x)$ up to conjugacy, for any $x \in G$. Hence all G-conjugacy class sums lie in $(kG)_P^G$, completing the proof. $\square$

This allows us to prove the following result.

Theorem 2.18 *Let G be a finite group G, and let e be a block idempotent of G. There exists a p-subgroup D such that, for any subgroup $H \leq G$, we have that $e \in (kG)_H^G$ if and only if H contains a conjugate of D. The subgroup D is unique, up to G-conjugacy.*

Proof Let P be a Sylow p-subgroup of G; by Corollary 2.17, $Z(kG) = (kG)_P^G$, and so there exists a p-subgroup Q of G such that $e \in (kG)_Q^G$. Consider the collection of all p-subgroups of G such that $e \in (kG)_Q^G$, and let D be an element of the collection of minimal order.

Let H be a subgroup of G such that a conjugate E of D is contained in H. By Corollary 2.17, $e \in (kG)_E^G$. By the transitivity of relative traces (Lemma 2.15(i)),

$$e \in (kG)_E^G = \mathrm{Tr}_H^G \left((kG)_D^H \right) \leq \mathrm{Tr}_H^G \left((kG)^H \right) = (kG)_H^G.$$

Hence $e \in (kG)_H^G$, proving one direction. Conversely, suppose that $e \in (kG)_H^G$. Let a be an element of $(kG)^D$ such that $e = \mathrm{Tr}_D^G(a)$ and let b be an element of $(kG)^H$ such that $e = \mathrm{Tr}_H^G(b)$. Since e is an idempotent, we have

$$e = e \cdot e = \left(\mathrm{Tr}_D^G(a) \right) \left(\mathrm{Tr}_H^G(b) \right) = \sum_{t \in T} \mathrm{Tr}_{D^t \cap H}^G (a^t b) \in \sum_{t \in T} (kG)_{D^t \cap H}^G,$$

where T is a set of (H, D)-double coset representatives in G. By Rosenberg's lemma, since e lies in a sum of two-sided ideals, it lies in one of them, say $(kG)_R^G$, where $R = D^t \cap H$. By minimality of D we must have $D^t \leq H$, completing the proof.

The uniqueness of D up to conjugacy follows at once: if D and E are two p-subgroups of G with the required property, then a conjugate of D is contained in E and vice versa, so D and E are conjugate. Since $(kG)_D^G = (kG)_E^G$ by Corollary 2.17, D has the required property if and only if E does. $\square$

The final result of this section relates relative traces and the Brauer morphism.

Proposition 2.19 *If G is a finite group and H is a subgroup of G, and P is a p-subgroup of H, then*

$$\ker(\mathrm{Br}_P) \cap (kG)^H = \sum_{Q \in \mathcal{Q}} (kG)_Q^H,$$

where $\mathcal{Q}$ is the set of all subgroups of H not containing a subgroup H-conjugate to P.

Proof   Let Q be a subgroup in $\mathcal{Q}$. Let X denote the subset of $x \in G$ such that Q contains a Sylow p-subgroup of $C_H(x)$. Clearly $(kG)_Q^H \subseteq (kG)^H$, and Proposition 2.16 states that $(kG)_Q^H$ is generated by the H-orbit sums of the elements in X. Let x be an element of X; if $P \leq C_H(x^h)$ for some $h \in H$ then some H-conjugate of P lies in Q, since Q contains a Sylow p-subgroup of $C_H(x)$, contradicting $x \in X$. Hence $P \not\leq C_H(x^h)$ for any $h \in H$, so that $x^h \not\leq C_H(P)$ for any $h \in H$. In particular, therefore, $x \in \ker(\mathrm{Br}_P)$, proving that the term on the left-hand side contains the term on the right-hand side.

To prove the converse, suppose that $a \in \ker(\mathrm{Br}_P) \cap (kG)^H$; we proceed by induction on the number of elements of G in the support of a. Let x be in the support of a, with coefficient α in a, i.e., so that $a - \alpha x$ does not have x in its support; then $\mathrm{Br}_P(x) = 0$, so that $x \notin C_G(P)$. This means that if Q is a Sylow p-subgroup of $C_H(x)$, then Q does not contain P. Since $a \in (kG)^H$, x^h lies in the support of a for all $h \in H$, and so $\mathrm{Br}_P(x^h) = 0$ also; hence no H-conjugate of Q contains P, and so $Q \in \mathcal{Q}$. Also, as $a \in (kG)^H$, for all $h \in H$, $a - \alpha x^h$ does not have x^h in the support, and so if $\hat{x}$ denotes the H-orbit sum of x then $\alpha \hat{x} \in (kG)_Q^H$ and $a - \alpha \hat{x}$ has fewer elements in its support, so that by induction a lies in $\sum_{Q \in \mathcal{Q}} (kG)_Q^H$, as needed. $\square$

2.3 Brauer pairs

A Brauer pair is a very powerful concept, and the basis of fusion systems of blocks. As well as being useful, it is quite easy to define. The idea is that a Brauer pair is a p-subgroup of G, together with a p-block idempotent of its centralizer.

Definition 2.20   Let G be a finite group and let p be a prime dividing $|G|$. A *Brauer pair* is an ordered pair (Q, e), where Q is a p-subgroup of G and e is a block idempotent of $k \, C_G(Q)$. Denote by $\mathcal{B}(Q)$ the set of block idempotents of $k \, C_G(Q)$.

Since G acts by conjugation on the set of all p-subgroups and (transporting from Q to Q^g) maps bijectively the set of all primitive central idempotents of $k \, C_G(Q)$ to the primitive central idempotents of $k \, C_G(Q^g)$, we see that G acts by conjugation on the set of all Brauer pairs. Denote by $\mathrm{N}_G(Q, e)$ the set of elements that stabilize the Brauer pair (Q, e) under the conjugation action.

Lemma 2.21 *Let G be a finite group, and let $Q \trianglelefteq R$ be p-subgroups of G. If e is a block idempotent of $k\, C_G(R)$, then there is a unique R-stable block idempotent f of $k\, C_G(Q)$ such that*

$$\mathrm{Br}_R(f)e = e.$$

If f' is any other R-stable block idempotent of $k\, C_G(Q)$, then $\mathrm{Br}_R(f')e = 0$.

Proof Suppose that f is a block idempotent of $k\, C_G(Q)$ such that $\mathrm{Br}_R(f)e = e$, and let $f' \neq f$ be any other R-stable block idempotent in $\mathcal{B}(Q)$. We have that

$$\mathrm{Br}_R(f')e = \mathrm{Br}_R(f')\,(\mathrm{Br}_R(f)e) = \mathrm{Br}_R(f'f)e = 0,$$

since Br_R is multiplicative on $(kG)^R$; thus $\mathrm{Br}_R(f)\,\mathrm{Br}_R(f') = \mathrm{Br}_R(ff')$.

It remains to show that there is such an idempotent f. Notice that if $g \in G$ and $r \in R$ then $\mathrm{Br}_R(g^r) = \mathrm{Br}_R(g)$ since $g \in C_G(R)$ if and only if $g^r \in C_G(R)$, and so $\mathrm{Br}_R(f^r) = \mathrm{Br}_R(f)$ for any block idempotent $f \in \mathcal{B}(Q)$. The group R acts by conjugation on the $\mathcal{B}(Q)$, and any block that is not R-stable is in an orbit of length a multiple of p. Since k has characteristic p and $\mathrm{Br}_R(f) = \mathrm{Br}_R(f^r)$, we have (taking the decomposition of $1 \in k\, C_G(Q)$ into block idempotents)

$$1 = \mathrm{Br}_R(1) = \sum_{b \in \mathcal{B}(Q)} \mathrm{Br}_R(b) = \sum_{b \in \mathcal{B}(Q)^R} \mathrm{Br}_R(b).$$

Using this decomposition, we get

$$e = 1 \cdot e = \sum_{b \in \mathcal{B}(Q)^R} \mathrm{Br}_R(b)e.$$

This means that $\mathrm{Br}_R(b)e \neq 0$ for some e; write f for one of these idempotents. By Corollary 2.13, $\mathrm{Br}_R(f)e = e$, completing the proof. $\qquad\square$

This allows us to define a partial order relation on the set of all Brauer pairs, often called the *inclusion* of Brauer pairs.

Definition 2.22 Let (Q, f) and (R, e) be Brauer pairs.

(i) Define $(Q, f) \trianglelefteq (R, e)$ if $Q \trianglelefteq R$, the block idempotent f is R-stable, and $\mathrm{Br}_R(f)e = e$.

(ii) Define $\leq$ to be the transitive extension of $\trianglelefteq$.

The relation $\leq$ on Brauer pairs is clearly reflexive, and anti-symmetric, and by definition transitive. By Lemma 2.21, given $Q \leq R$ and a block idempotent $e \in \mathcal{B}(R)$, there is some Brauer pair (Q, f) such that $(Q, f) \leq$

(R, e). We would like this to be unique, and we will prove this now, with a helpful lemma to begin with, which deals with the situation where $Q \lhd R$. The reason why uniqueness might not be guaranteed is that there are many subnormal chains between a subgroup Q and R, and going up different ones might produce a different answer.

Lemma 2.23   *Let G be a finite group, and let R be a p-subgroup of G. Let P and Q be normal subgroups of R with $P \leq Q$. Suppose that e is a block idempotent of $k\,C_G(R)$, and write f_1 and f_2 for the (unique) elements of $\mathcal{B}(P)^R$ and $\mathcal{B}(Q)^R$ such that $\operatorname{Br}(f_i)e = e$. If f is the (unique) element of $\mathcal{B}(P)^Q$ such that $\operatorname{Br}_Q(f)f_2 = f_2$, then $f = f_1$.*

Consequently, if $P \lhd R$ and $e \in \mathcal{B}(R)$, then there is a unique Brauer pair (P, f) such that $(P, f) \leq (R, e)$.

Proof   We will show that f is R-stable, and that $\operatorname{Br}_R(f)e = e$. Let x be an element of R; then $f^x \in \mathcal{B}(P)$. We know that f is Q-stable, and so f^x is Q-stable also (as $Q \lhd R$). Thus $\operatorname{Br}_Q(f^x) = \operatorname{Br}_Q(f)^x$. However, f_2 is R-stable, and so

$$\operatorname{Br}_Q(f^x)f_2 = \operatorname{Br}_Q(f)^x f_2^x = f_2^x = f_2,$$

and so $f^x = f$, proving R-stability (via Lemma 2.21).

To see that $\operatorname{Br}_R(f)e = e$, note that

$$\operatorname{Br}_R(f)e = \operatorname{Br}_R(f)\operatorname{Br}_R(f_2)e = \operatorname{Br}_R(\operatorname{Br}_Q(f)f_2)e = \operatorname{Br}_R(f_2)e = e.$$

Therefore, by Lemma 2.21, we see that $f = f_1$. This proves that if $(P, f_1) \lhd (Q, f_2) \lhd (R, e)$, and $(P, f) \lhd (R, e)$, then $(P, f) = (P, f_1)$, i.e., that if $(P, f_1) \lhd (Q, f_2) \lhd (R, e)$ then $(P, f_1) \lhd (R, e)$ whenever $P \lhd R$.

We now suppose that there is some other Brauer pair $(P, f') \leq (R, e)$, so there is some chain

$$(P, f') \lhd (Q_1, f_1') \lhd \cdots \lhd (Q_r, f_r') \lhd (R, e);$$

by repeated use of the previous part (on $(P, f') \lhd (Q_1, f_1') \lhd (Q_2, f_2')$ and so on), we see that $(P, f') \lhd (R, e)$, proving the second part of the lemma.   □

This lemma proves the first step of the following result.

Theorem 2.24   *Let G be a finite group, and let $Q \leq R$ be p-subgroups of G. If (R, e) is a Brauer pair, then there is a unique Brauer pair (Q, f) such that $(Q, f) \leq (R, e)$.*

Proof We proceed by induction on the index $|R : Q|$. The case where $|R : Q| = p$ is covered by Lemma 2.23, so our induction starts. Also by Lemma 2.23, each subnormal series provides exactly one chain of inclusions of Brauer pairs. The standard chain of taking iterated normalizers yields one (Q, f) such that $(Q, f) \leq (R, e)$. Let

$$Q = R_0 \lhd R_1 \lhd \cdots \lhd R_s \lhd R$$

be a subnormal series. We have $R_1 \leq \mathrm{N}_R(Q)$, and so by induction there is a unique Brauer pair $(R_1, f') \leq (R, e)$ and a unique Brauer pair $(\mathrm{N}_R(Q), f'') \leq (R, e)$. By construction $(Q, f) \lhd (\mathrm{N}_R(Q), f'')$. Since $R_1 \leq \mathrm{N}_R(Q)$, by induction $(R_1, f') \leq (\mathrm{N}_R(Q), f'')$ because both are less than (R, e). Finally, again by induction, since $(Q, f) \lhd (\mathrm{N}_R(Q), f'')$ and $(R_1, f') \leq (\mathrm{N}_R(Q), f'')$, we have $(Q, f) \lhd (R_1, f')$, completing the proof. $\square$

Therefore if (P, e) is a Brauer pair then there is, for each $Q \leq P$, a unique Brauer pair (Q, e_Q) contained in (P, e). The collection of all Brauer pairs contained in (P, e) forms a poset that is identical to the poset of subgroups of P. While it is true that for $R \leq Q$ there is a unique (R, e_R) contained in a given (Q, f), the same is not true in the other direction, and (R, e_R) might well be contained in more than one pair (Q, f) for a given subgroup Q.

We will see later on that the largest subgroups of P containing a given Brauer pair are important, and will be the defect groups of the blocks of kG.

2.4 Defect groups and the first main theorem

Having built up the machinery of Brauer pairs, we are in a position to start developing the structure of the fusion system of a block. This starts with the notion of a b-Brauer pair.

Definition 2.25 Let G be a finite group and let b be a block idempotent of kG. A b-*Brauer pair* is a Brauer pair (R, e) such that $\mathrm{Br}_R(b)e = e$, i.e., such that, when $\mathrm{Br}_R(b)$ is expressed as a sum of block idempotents in $\mathcal{B}(R)$, e is one of the terms.

A *maximal* b-Brauer pair is a b-Brauer pair (D, e) such that $|D|$ is maximal. The subgroup D is a *defect group* of the block b.

It is easy to see that a b-Brauer pair is a pair (R, e) such that $(1, b) \trianglelefteq$ (R, e), or equivalently $(1, b) \leq (R, e)$, via Theorem 2.24. We have the following trivial lemma on defect groups and b-Brauer pairs.

Lemma 2.26 *Let G be a finite group, let b be a block idempotent of kG, and let Q be a p-subgroup of G.*

(i) *There is a b-Brauer pair (Q, e) if and only if $\mathrm{Br}_Q(b) \neq 0$.*
(ii) *The subgroup Q is a defect group if and only if $\mathrm{Br}_Q(b) \neq 0$ and, for all R properly containing a conjugate of Q, we have $\mathrm{Br}_R(b) = 0$.*

We will define a conjugation structure on all Brauer pairs, and in fact on all b-Brauer pairs, so that the collection of b-Brauer pairs included in a fixed maximal b-Brauer pair becomes a fusion system.

Notice that if b is a block idempotent of kG then b is fixed under conjugation (as b is central) and so G acts on the set of all b-Brauer pairs by conjugation. In particular, if D is a defect group of b then so is D^g, and so the set of defect groups of a block is a union of conjugacy classes of p-subgroups of G.

Theorem 2.27 *Let G be a finite group, and let b be a block idempotent of kG. A minimal p-subgroup P such that $b \in (kG)_P^G$ is a defect group of b. Furthermore, G acts transitively by conjugation on the set of all defect groups of b.*

Proof Firstly, the minimal p-subgroups P of G such that $b \in (kG)_P^G$ are described in Theorem 2.18, where in particular it is shown that they form a single G-conjugacy class of p-subgroups, and hence G acts transitively on them, proving the second assertion subject to the first.

Let D denote a minimal p-subgroup of G such that $b \in (kG)_D^G$. We will prove that, for a p-subgroup Q of G, we have that $\mathrm{Br}_Q(b) \neq 0$ if and only if Q is contained in D. Since the defect groups of b are maximal p-subgroups such that $\mathrm{Br}_Q(b) \neq 0$, this will prove the result.

Suppose that $\mathrm{Br}_Q(b) = 0$; by Proposition 2.19, since $b \in Z(kG)$, we have that b lies in the sum $\sum_{R \in \mathcal{R}} (kG)_R^G$, where $\mathcal{R}$ is the collection of all p-subgroups not containing a conjugate of Q. By Rosenberg's lemma, b lies in $(kG)_R^G$, where R is some p-subgroup not containing a conjugate of Q; hence Q is not contained in D, proving one direction.

Conversely, suppose that $\mathrm{Br}_Q(b) \neq 0$. In this case, again by Proposition 2.19, b does not lie in any $(kG)_R^G$, where R does not contain a conjugate of Q. Therefore, if $b \in (kG)_S^G$ for some p-subgroup S, then S contains a conjugate of Q. In particular, since $b \in (kG)_D^G$, we see that D contains a conjugate of Q, as claimed. $\square$

Note that the defect groups of a block therefore form a conjugacy class of p-subgroups of G. We will actually show that G acts transitively on the set of maximal b-Brauer pairs. To do this, we first need to consider the case when P is a normal p-subgroup of G.

Proposition 2.28   *Let G be a finite group and let P be a normal p-subgroup of G. Let b be a block idempotent of kG.*

(i) *The defect groups of b contain P, and $b \in (k\, C_G(P))^G$.*

Write $b = b_1 + b_2 + \cdots + b_r$, where $b_i \in \mathcal{B}(P)$, using Corollary 2.13.

(ii) *For each i we have that $b = \mathrm{Tr}^G_{H_i}(b_i)$, where $H_i = \mathrm{N}_G(D_i, b_i)$ (for D_i a defect group of b_i) is the stabilizer of b_i.*

(iii) *If b has defect group P, then the Brauer pairs (P, b_i) are the maximal b-Brauer pairs, and G acts transitively on the set of maximal b-Brauer pairs.*

Proof   Firstly, since P is a normal p-subgroup of G, it acts trivially on every simple kG-module ([Fei82, Corollary III.2.13], or using Theorem 2.7 and Clifford's theorem [PD77, Theorem 2.2A]), and therefore P acts trivially on $kG/J(kG)$. If Q is a proper subgroup of P then we claim that $(kG)^P_Q$ annihilates every simple kG-module. This is true simply because if $a \in (kG)^Q$ and $g \in P$ then both a and a^g act the same on any simple module, and so $(kG)^P_Q$ annihilates any simple module, since k has characteristic p. Thus $(kG)^P_Q$ is contained within $J(kG)$ and so, by Proposition 2.19, the kernel of Br_P on $(kG)^P$ is contained within $J(kG)$. Since the image of Br_P on $(kG)^P$ is $k\, C_G(P)$, we get that

$$(kG)^P = k\, C_G(P) + J(kG)^P,$$

and so

$$Z(kG) = (k\, C_G(P))^G + J(Z(kG)).$$

Therefore b lies in $(k\, C_G(P))^G$ as b is an idempotent.

We are now in a position to show that a defect group of b contains P. A basis for $(k\, C_G(P))^G$ is the set of G-class sums of elements of $C_G(P)$, and if X is a G-conjugacy class of $C_G(P)$, with class sum $\hat{X}$, and $x \in X$, then a Sylow p-subgroup of $C_G(x)$ contains P. If b has defect group D then $b \in (kG)^G_D$, and by Proposition 2.16 the ideal $(kG)^G_D$ has as k-basis all G-orbit sums $\ddot{Y}$, where D contains a Sylow p-subgroup of $C_G(y)$ (for a representative y from $\hat{Y}$). Therefore for any x in the support of b, $C_G(x)$ contains P and is contained in D, so $P \leq D$, completing the proof of (i).

We now prove (ii). Let T be a transversal to $H = H_1$ in G; if $t \in T$ then $b^t = b$, so that $(\sum b_i)^t = \sum b_i$. Furthermore, b_1^t is a block idempotent of $k\,C_G(P)$ since conjugation by t induces an automorphism of $k\,C_G(P)$, and so $b_1^t = b_i$ for some i. Therefore we may order $T = \{t_1, \ldots, t_r\}$ so that $b_1^{t_i} = b_i$, and therefore $b = \mathrm{Tr}_H^G(b_1)$, as claimed.

Finally, suppose that b has defect group P. Since $b \in k\,C_G(P)$, we have $\mathrm{Br}_P(b) = b$, and since $b = b_1 + \cdots + b_r$ is a decomposition of b into block idempotents of $k\,C_G(P)$, we have that (P, b_i) are the maximal b-Brauer pairs (as b has defect group P). Since G acts transitively on the b_i by (ii), G acts transitively on the maximal b-Brauer pairs, proving (iii). □

We now need to move from the case where P is normal in G to the general case. We begin with a proposition.

Proposition 2.29 *Let G be a finite group, and let P be a p-subgroup of G. If $x \in (kG)^P$, then*

$$\mathrm{Br}_P\left(\mathrm{Tr}_P^G(x)\right) = \mathrm{Br}_P\left(\mathrm{Tr}_P^{N_G(P)}(x)\right) = \mathrm{Tr}_P^{N_G(P)}\left(\mathrm{Br}_P(x)\right).$$

Hence, Br_P induces a surjective algebra morphism $\mathrm{Br}_P : (kG)_P^G \to (k\,C_G(P))_P^{N_G(P)}$.

Proof If $x \in (kG)^P$ then by Lemma 2.15 we have

$$\mathrm{Tr}_P^G(x) = \sum_{t \in T} \mathrm{Tr}_{P^t \cap N_G(P)}^{N_G(P)}(x^t),$$

where T is a set of $(P, N_G(P))$-double coset representatives. By Proposition 2.19, $\ker(\mathrm{Br}_P) \cap (kG)^{N_G(P)}$ is spanned by all ideals of the form $(kG)_Q^{N_G(P)}$, where Q does not contain P. If $P^t \neq P$ for $t \in T$, then $|P^t \cap N_G(P)| < |P|$, and so the term corresponding to t in the above sum lies in the kernel of Br_P. Hence

$$\mathrm{Br}_P\left(\mathrm{Tr}_P^G(x)\right) = \mathrm{Br}_P\left(\mathrm{Tr}_P^{N_G(P)}(x)\right),$$

proving the first equality. The proof of the second equality is Exercise 2.3.

To see the final statement, since $\mathrm{Br}_P\left(\mathrm{Tr}_P^G(x)\right) = \mathrm{Tr}_P^{N_G(P)}\left(\mathrm{Br}_P(x)\right)$, the image of $(kG)_P^G$ under Br_P is exactly $(k\,C_G(P))_P^{N_G(P)}$. Since Br_P, restricted to $(kG)^P$, is an algebra morphism, this completes the proof. □

Brauer's first main theorem allows us to pass from blocks of kG with defect group P to blocks of $k\,\mathrm{N}_G(P)$ with defect group P, in the sense that there is a bijection between these two sets of blocks, induced by the Brauer morphism.

Theorem 2.30 (Brauer's first main theorem) *Let G be a finite group and let P be a p-subgroup of G. The map Br_P induces a bijection between the blocks of kG with defect group P and the blocks of $k\,\mathrm{N}_G(P)$ with defect group P.*

Proof Let E' denote the set of block idempotents of kG whose defect groups are contained in (a conjugate of) P, and let E be the subset of E' consisting of those idempotents whose defect groups are (conjugate to) P. By Exercise 2.4, $(kG)_P^G$ is a direct sum of kGe for $e \in E'$ and a nilpotent ideal N. Write A for the direct sum of kGe for $e \in E$ and N. In other words, we have

$$(kG)_P^G = N \oplus \bigoplus_{e \in E'} kGe \quad \text{and} \quad A = N \oplus \bigoplus_{e \in E} kGe \,\Big|\, (kG)_P^G.$$

Write A' for the sum of kGe for $e \in E' \setminus E$, so that $(kG)_P^G = A \oplus A'$. Since all block idempotents $e \in E' \setminus E$ satisfy $\mathrm{Br}_P(e) = 0$ by Lemma 2.26, the kernel A' of the projection $(kG)_P^G \to A$ lies in the kernel of the surjective map $\mathrm{Br}_P : (kG)_P^G \to (k\,\mathrm{C}_G(P))_P^{\mathrm{N}_G(P)}$, given in Proposition 2.29. Therefore Br_P induces a surjective algebra homomorphism $\mathrm{Br}_P' : A \to (k\,\mathrm{C}_G(P))_P^{\mathrm{N}_G(P)}$.

We claim that the kernel I of Br_P' is a nilpotent ideal. If $x \in I$, we have that

$$x = y + \sum_{e \in E} x_e e,$$

where $y \in N$. The element x is nilpotent if and only if each x_e is nilpotent, and since $xe = x_e e$ (as e is a primitive idempotent and all other x_f and y lie in different blocks of $\mathrm{Z}(kG)$), we have that $x_e e \in I$. Hence if $I \cap kGe$ is nilpotent for each $e \in E$, I is a nilpotent ideal. However, since kGe is local, and $e \notin I$ as $\mathrm{Br}_P(e) \neq 0$, we must have that $I \subseteq J(kGe)$, as needed.

Since I is nilpotent, we may use the lifting of idempotents, Proposition 2.11, to get that the primitive idempotents of A are in bijection via Br_P' with the primitive idempotents of $(k\,\mathrm{C}_G(P))_P^{\mathrm{N}_G(P)}$. Let e be a block idempotent in E, and let $f = \mathrm{Br}_P(e)$, a primitive idempotent in $(k\,\mathrm{C}_G(P))_P^{\mathrm{N}_G(P)}$. We claim that f is a block idempotent of

$k \, N_G(P)$. It is clearly a central idempotent; suppose that $f = f_1 + f_2$ is a sum of central idempotents of $k \, N_G(P)$ with $f_1 f_2 = 0$. By Proposition 2.28, the f_i lie in $(k \, C_G(P))^{N_G(P)}$, and since $(k \, C_G(P))_P^{N_G(P)}$ is an ideal of $(k \, C_G(P))^{N_G(P)}$, each f_i lies in $(k \, C_G(P))_P^{N_G(P)}$. This contradicts the fact that f is primitive in $(k \, C_G(P))_P^{N_G(P)}$, so f is a block idempotent. It remains to show that f has defect group P; however, since P is certainly contained in the defect group of f by Proposition 2.28, and $f \in (k \, C_G(P))_P^{N_G(P)}$ implies that P contains a defect group of f, we must have that P is a defect group for f. The same argument shows that all block idempotents of $k \, N_G(P)$ with defect group P lie in $(k \, C_G(P))_P^{N_G(P)}$.

Hence Br_P' induces a bijection between primitive idempotents of A and those of $(k \, C_G(P))_P^{N_G(P)}$; the primitive idempotents of A are precisely the block idempotents of kG with defect group P, and the primitive idempotents of $(k \, C_G(P))_P^{N_G(P)}$ are precisely the block idempotents of $k \, N_G(P)$ with defect group P. This completes the proof. $\qquad \square$

This gives us the following corollary.

Corollary 2.31 *Let G be a finite group. If b is a block idempotent of kG, then G acts transitively on the set of maximal b-Brauer pairs.*

Proof By Theorem 2.27, G acts transitively on the defect groups of b, so we may fix such a defect group D. The map Br_P induces a bijection between block idempotents of kG with defect group D and those of $k \, N_G(D)$ with defect group D, and $N_G(D)$ acts transitively on the maximal $\mathrm{Br}_P(b)$-Brauer pairs, which are just the maximal b-Brauer pairs. This completes the result. $\qquad \square$

As well as this piece of information, we need to know something about the group $N_G(D, e_D)$, where (D, e_D) is a maximal b-Brauer pair.

Proposition 2.32 *Let G be a finite group, and let b be a block idempotent. If (D, e) is a maximal b-Brauer pair, then $N_G(D, e)/D \, C_G(D)$ is a p'-group.*

Proof Without loss of generality we may assume that $G = N_G(D)$. As in Proposition 2.28, write $b = b_1 + \cdots + b_r$, where the b_i are the blocks of $k \, C_G(D)$ and the (D, b_i) are the maximal b-Brauer pairs. Let H_i denote the stabilizer of b_i, so $H_i = N_G(D, b_i)$. Since the b_i are all G-conjugate, so are the H_i, and so if p divides $|H_i/D \, C_G(D)|$ for some i then it does so for all i.

Suppose that $p \mid |H_1/D \, C_G(D)|$; then

$$\mathrm{Tr}^{H_i}_{D \, C_G(D)}(b_i) = 0$$

for all i, and hence $\mathrm{Tr}^G_{D \, C_G(D)}(b_i) = 0$. As $\mathrm{Br}_D(b) = b = b_1 + \cdots + b_r$, the b_i project onto a k-basis for $\mathrm{Z}(k \, C_G(D)b)/J(\mathrm{Z}(k \, C_G(D)b))$ by Proposition 2.4, and we get that

$$(\mathrm{Z}(k \, C_G(D)b))^G_{D \, C_G(D)} \subseteq \left(J(\mathrm{Z}(k \, C_G(D)b))\right)^G_{D \, C_G(D)}.$$

Since the Jacobson radical of a k-algebra is invariant under algebra automorphisms, we actually have

$$(\mathrm{Z}(k \, C_G(D)b))^G_{D \, C_G(D)} \subseteq \left(J(\mathrm{Z}(k \, C_G(D)b))\right).$$

Since $(k \, C_G(D))^{D \, C_G(D)}_D \subseteq \mathrm{Z}(k \, C_G(D))$, the transitivity of relative trace maps yields

$$(k \, C_G(D)b)^G_D \subseteq \left(J(\mathrm{Z}(k \, C_G(D)b))\right).$$

However, $b \in (k \, C_G(D)b)^G_D$ as D is a defect group of b, and so $b \in J(\mathrm{Z}(k \, C_G(D)b))$, contradicting the fact that b is an idempotent. Hence $p \nmid |H/D \, C_G(D)|$, as claimed. $\qquad \square$

2.5 Fusion systems of blocks

Having spent a long time discussing Brauer pairs and defect groups, we are at last at a point where we can introduce the fusion system of a block. If b is a block idempotent, then the fusion system of b is a fusion system on a defect group D of b, consisting of the b-Brauer pairs (P, e_P) with $P \le D$.

Definition 2.33 Let G be a finite group and let k be a field of characteristic p. Let b be a block idempotent of kG, and let (D, e_D) denote a maximal b-Brauer pair. Denote by $\mathcal{F} = \mathcal{F}_{(D,e_D)}(G, b)$ the category whose objects are the subgroups of D, and whose morphisms sets are described below. Let Q and R be subgroups of D, and e_Q and e_R be the unique block idempotents such that $(Q, e_Q) \le (D, e_D)$ and $(R, e_R) \le (D, e_D)$. The set $\mathrm{Hom}_{\mathcal{F}}(Q, R)$ is the set of conjugation maps $c_x : Q \to R$ for all elements x in G such that $(Q, e_Q)^x \le (R, e_R)$. The category $\mathcal{F}$ is the *fusion system* of the block kGb and of the block idempotent b.

Although we have used the notation $\mathcal{F}_{(D,e_D)}(G, b)$ to denote this category, suggesting that it is a fusion system, we have not actually proved it.

Theorem 2.34 *Let G be a finite group, let b be a block idempotent of kG, and let (D, e_D) denote a maximal b-Brauer pair. The category $\mathcal{F}_{(D,e_D)}(G, b)$ is a fusion system on D. Furthermore, if (E, e) is another maximal b-Brauer pair, then $\mathcal{F}_{(D,e_D)}(G, b)$ and $\mathcal{F}_{(E,e)}(G, b)$ are isomorphic.*

Proof Let $\mathcal{F} = \mathcal{F}_{(D,e_D)}(G, b)$ and let $g \in D$. If $(Q, e_Q) \leq (D, e_D)$, then $(Q^g, e_Q^g) \leq (D^g, e_D^g) = (D, e)$. By the uniqueness of inclusion of Brauer pairs, $(Q^g, e_Q^g) = (Q^g, e_{Q^g})$, and so $\mathrm{Hom}_{\mathcal{F}}(Q, P)$ contains the map c_g. Therefore $\mathcal{F}$ satisfies the first axiom of a fusion system.

By definition if ϕ is a morphism in $\mathcal{F}$ so is the associated isomorphism, and if an isomorphism is in $\mathcal{F}$ then so is its inverse, and so $\mathcal{F}$ is a fusion system on D, as required. To see the second part, simply note that, by Corollary 2.31, G acts transitively by conjugation on the set of maximal b-Brauer pairs. □

In fact, the fusion system of a block is saturated; we will prove this soon, but we first need to know something about Brauer pairs that are fully $\mathcal{F}$-normalized.

Lemma 2.35 *Let G be a finite group and let b be a block idempotent with defect group D. Write $\mathcal{F} = \mathcal{F}_{(D,e_D)}(G, b)$ for the fusion system of b. If (R, e_R) is a fully $\mathcal{F}$-normalized b-Brauer pair, then $(\mathrm{N}_D(R), e_{\mathrm{N}_D(R)})$ is a maximal Brauer pair for the group $\mathrm{N}_D(R) \mathrm{C}_G(R)$.*

Proof Write $N = \mathrm{N}_D(R)$ and $H = \mathrm{N}_D(R) \mathrm{C}_G(R)$, and suppose that (S, e) is a Brauer pair such that $(N, e_N) \leq (S, e)$ in H. Clearly this is also true in G, and since (N, e_N) is a b-Brauer pair, this means that (S, e) is a b-Brauer pair. Since (D, e_D) is a maximal b-Brauer pair, and G acts transitively on the set of maximal b-Brauer pairs, there is $g \in G$ such that $(S^g, e^g) \leq (D, e_D)$. Chaining our inequalities together, we get

$$(R^g, e_R^g) \leq (S^g, e^g) \leq (D, e_D).$$

Since $S \leq \mathrm{N}_D(R) \mathrm{C}_G(R)$, $R^g \trianglelefteq S^g$, but by hypothesis R is fully $\mathcal{F}$-normalized, so that $|\mathrm{N}_D(R)| \geq |\mathrm{N}_D(R^g)|$. Hence $|S| \leq |\mathrm{N}_D(R)|$, and since $\mathrm{N}_D(R) \leq S$, we get equality, as claimed. □

Theorem 2.36 *Let G be a finite group, let b be a block idempotent of kG, and let (D, e_D) denote a maximal b-Brauer pair. The fusion system $\mathcal{F} = \mathcal{F}_{(D,e_D)}(G, b)$ is saturated.*

Proof Fix a maximal b-Brauer pair (D, e_D), and for $Q \leq D$, let (Q, e_Q) denote the unique Brauer pair such that $(Q, e_Q) \leq (D, e_D)$. Note that $\mathrm{Aut}_{\mathcal{F}}(Q) = \mathrm{N}_G(Q, e_Q)/\mathrm{C}_G(Q)$. Proposition 2.32 shows that $\mathrm{Aut}_D(D)$ is a Sylow p-subgroup of $\mathrm{Aut}_{\mathcal{F}}(D)$, so that the first axiom of saturation is satisfied.

Now let (Q, e_Q) be a b-Brauer pair, and let $\phi : Q \to R$ be an $\mathcal{F}$-isomorphism such that R is fully $\mathcal{F}$-normalized; write (R, e_R) for the b-Brauer pair included in (D, e_D). Let $g \in G$ be an element such that $c_g|_Q = \phi$. We clearly have

$$N_\phi = \mathrm{N}_D(Q) \cap \mathrm{N}_{D^{g^{-1}}}(Q)\, \mathrm{C}_G(Q) \quad \text{and} \quad N_\phi^g = \mathrm{N}_{D^g}(R) \cap \mathrm{N}_D(R)\, \mathrm{C}_G(R).$$

Writing $H = \mathrm{N}_D(R)\, \mathrm{C}_G(R)$ and $N = N_\phi$, the fact that $(Q, e_Q) \leq (N, e_N)$ in G implies that $(R, e_R) \leq (N^g, e_{N^g})$ in G, and so therefore in H. However, we also have that $(R, e_R) \leq (\mathrm{N}_D(R), e_{\mathrm{N}_D(R)})$ in H, and so $(\mathrm{N}_D(R), e_{\mathrm{N}_D(R)})$ and (N^g, e_{N^g}) are both b'-Brauer pairs for some block b' of H. By Lemma 2.35, $(\mathrm{N}_D(R), e_{\mathrm{N}_D(R)})$ is a maximal b'-Brauer pair for H and hence, since all maximal b-Brauer pairs for H are H-conjugate, there is some $h \in H$ such that

$$(N^g, e_{N^g})^h \leq (\mathrm{N}_D(R), e_{\mathrm{N}_D(R)})$$

in H, and hence also in G. Since $H = \mathrm{N}_D(R)\, \mathrm{C}_G(R)$, $h = h'h''$ for $h' \in \mathrm{N}_D(R)$ and $h'' \in \mathrm{C}_G(R)$; the element h' stabilizes $e_{\mathrm{N}_D(R)}$, and therefore we may assume that $h \in \mathrm{C}_G(R)$. Hence

$$\psi : N \to \mathrm{N}_P(R), \qquad \psi : x \mapsto x^{gh}$$

is a morphism in $\mathcal{F}$ and $\psi|_Q = \phi$, proving the second axiom of saturation. Thus $\mathcal{F}$ is saturated, as claimed. $\qquad\square$

Now that we have proved that the fusion system of a block is saturated, a natural question to ask is whether it is the fusion system of a finite group. Of course, if the defect group is not a Sylow p-subgroup, this can only be a different finite group, but the question remains.

Conjecture 2.37 *Let G be a finite group and let b be a block idempotent of kG. There exists a finite group H such that the fusion system of b is isomorphic to the fusion system of H.*

This conjecture is widely believed, but as yet unsolved. A discussion of this conjecture, and the progress towards it, is in Chapter 9.

The defect group, and more generally the fusion system of a block, control in some sense the structure of the block, although many of the methods by which they do this are both mysterious and conjectural at the moment. One example of such control is the following.

Conjecture 2.38 (Brauer's $k(B)$ conjecture)   *Let G be a finite group and let B be a block of kG, with defect group D. If $k(B)$ denotes the number of simple KG-modules belonging to B, then $k(B) \leq |D|$.*

This is just one of the many conjectures surrounding the modular representation theory of finite groups, and the ways in which the p-subgroups and normalizers of p-subgroups control the characters and representations of the group.

In the case where $D = 1$, Brauer's $k(B)$ conjecture suggests that B should possess a single ordinary irreducible character. In other words, if B is a block of kG, then we know (by lifting of idempotents) that B lifts to a block of $\mathcal{O}G$, and that this block becomes a single matrix algebra when $\mathcal{O}$ is extended to K. In other words, it suggests that blocks with trivial defect group are matrix algebras.

Theorem 2.39   *Suppose that G is a finite group and let b be a block idempotent with trivial defect group. Write $B = kGb$.*

(i) *There is a single kG-module S belonging to B, and it is projective and simple.*

(ii) *The ideal B is isomorphic with a matrix algebra $M_{\dim S}(k)$.*

Furthermore, if b is any block idempotent of kG such that kGb is a matrix algebra, then b has trivial defect group.

Proof Suppose that b is a block idempotent with trivial defect group. By Theorem 2.27, $b = \mathrm{Tr}_1^G(x)$ for some $x \in kG$. Let M and N be kG-modules belonging to B, and suppose that $\phi : M \to N$ is a surjective homomorphism. We will construct a splitting for ϕ, proving that every kG-module belonging to B is projective. Once we have done that, the fact that every module is projective implies that B is semi-simple. Since it is also indecomposable, it must be a simple algebra, and so a matrix algebra over k (since we assume that k is algebraically closed). That a matrix algebra has a single simple module of dimension that of the matrix algebra is a standard fact about matrix algebras.

Let $\theta : N \to M$ be a splitting of ϕ as a map of vector spaces (i.e., $\theta\phi = \mathrm{id}_N$). For each $y \in N$, define

$$y\bar{\theta} = \sum_{g \in G}(yg^{-1}x)\theta g;$$

then $\bar{\theta}$ is a kG-module splitting, as required. To see this, notice firstly that it is easily seen to be a kG-module homomorphism, so we need that $\bar{\theta}\phi$ acts trivially on N. To see this, we calculate:

$$y(\bar{\theta}\phi) = \left(\sum_{g \in G}(yg^{-1}x)\theta g\right)\phi$$

$$= \left(\sum_{g \in G}(yg^{-1}x)\theta\phi\right)g$$

$$= \sum_{g \in G}y \cdot g^{-1}xg$$

$$= yb = y.$$

The converse is omitted.   □

Apart from blocks with trivial defect group, the $k(B)$ conjecture is known for blocks with only a few types of defect group, such as blocks with cyclic, dihedral, semidihedral, and generalized quaternion defect groups. Approaching the problem from another direction – placing restrictions on the underlying finite group rather than the defect group of the block – it was recently proved that the $k(B)$ conjecture holds whenever G is a finite p-soluble group and B is a block of kG [GMRS04]. In general the conjecture has resisted attacks.

The opposite case to blocks with trivial defect group is blocks whose defect groups are Sylow p-subgroups of G. While there are not always blocks with trivial defect group in a finite group, there is always a block with defect groups the Sylow p-subgroups of G, namely the principal block. Recall that the *augmentation ideal* of kG is the ideal consisting of all elements of kG which, when written as $\sum_g \alpha_g g$, have the property that $\sum_g \alpha_g = 0$. In general, the *augmentation* of an element $x \in kG$ is the sum of the coefficients – i.e., if $x = \sum_g \alpha_g g$ then the augmentation of x is $\sum_g \alpha_g$ – so that the augmentation ideal of kG consists of all elements of kG with augmentation 0. It is the kernel of the unique homomorphism $kG \to k$.

Lemma 2.40 *Let G be a finite group and let B_0 denote the principal block of kG (i.e., the block to which the trivial module belongs). The defect groups of B_0 are the Sylow p-subgroups of G.*

Proof Let D denote a defect group of B_0, and let $b_0 = \sum_g \beta_g g$ be the block idempotent of B_0. Since g acts trivially on k for all $g \in G$ (where k denotes the trivial module), and so does b_0 (since the 1 of kG acts trivially and 1 is the sum of the blocks of kG), the augmentation of b_0 must be 1. As D is a defect group of b_0, we may write $b_0 = \mathrm{Tr}_D^G(a)$ for some $a = \sum_g \alpha_g g \in (kG)^D$. The augmentation of a is equal to the augmentation of a^g for any $g \in G$, so

$$1 = \sum_g \beta_g = \sum_{t \in T} \left(\sum_g \alpha_g \right)^t = |G : D| \sum_g \alpha_g.$$

In particular, $|G : D|$ cannot be divisible by p, completing the proof. $\square$

One corollary of the proof of this lemma is the following.

Corollary 2.41 *Let G be a finite group. The principal block idempotent b_0 is the only block idempotent that does not lie in the augmentation ideal of kG. In particular, $\mathrm{Br}_P(b_0)$ is non-zero for any p-subgroup P.*

Proof The only point that needs to be shown is that $\mathrm{Br}_P(b_0)$ is non-zero for any p-subgroup P; this follows because the map $\mathrm{Br}_P : (kG)^P \to k\,\mathrm{C}_G(P)$ is surjective, and so the image of the augmentation ideal of kG, restricted to $(kG)^P$, maps onto the augmentation ideal of $k\,\mathrm{C}_G(P)$. In particular, since b_0 lies outside the augmentation ideal of kG, $\mathrm{Br}_P(b_0)$ must also lie outside the augmentation ideal of $k\,\mathrm{C}_G(P)$, as needed. $\square$

Let b_0 denote the principal block idempotent of kG, and let (Q, e_Q) denote the Brauer pair where Q is a non-trivial p-subgroup of G and e_Q is the principal block of $k\,\mathrm{C}_G(Q)$. We claim that (Q, e_Q) is a b_0-Brauer pair. Since b_0 lies outside of the augmentation ideal of kG, $\mathrm{Br}_Q(b_0)$ – which is a sum of block idempotents of $k\,\mathrm{C}_G(Q)$ – lies outside of the augmentation ideal of $k\,\mathrm{C}_G(Q)$. As all other block idempotents of $k\,\mathrm{C}_G(Q)$ lie in the augmentation ideal, e_Q must appear as a term in the expression of $\mathrm{Br}_P(b_0)$ as a sum of block idempotents of $k\,\mathrm{C}_G(Q)$. Therefore $\mathrm{Br}_Q(b_0)e_Q = e_Q$, proving that (Q, e_Q) is a b_0-Brauer pair.

Furthermore, we claim that if $Q \leq R$ are p-subgroups of P, and e_Q and e_R are the principal block idempotents of $k\,\mathrm{C}_G(Q)$ and $k\,\mathrm{C}_G(R)$ respectively, then $(Q, e_Q) \leq (R, e_R)$; since both are b_0-Brauer pairs, by the uniqueness of inclusion of Brauer pairs this must hold.

Theorem 2.42 *Let G be a finite group with Sylow p-subgroup P, and let b and e denote the principal block idempotents of kG and $k\mathrm{C}_G(P)$ respectively. The fusion systems $\mathcal{F} = \mathcal{F}_{(P,e)}(G,b)$ and $\mathcal{F}_P(G)$ are isomorphic.*

Proof Let e_P denote the principal block idempotent of $k\mathrm{C}_G(P)$; by the above discussion, the elements of $\mathcal{F}$ are the Brauer pairs (Q,e_Q) where $Q \le P$ and e_Q is the principal block idempotent of $k\mathrm{C}_G(Q)$. Let Q and R be subgroups of P, and let $g \in G$ be an element such that $Q^g = R$. We must show that $e_Q^g = e_R$. Since conjugation by g induces an isomorphism $k\mathrm{C}_G(Q) \to k\mathrm{C}_G(R)$, the image of e_Q is a block idempotent of $k\mathrm{C}_G(R)$ not lying in the augmentation ideal, and hence is e_R. Therefore $c_g : Q \to R$ lies in $\mathcal{F}$, as needed.   □

We will end with a (very) brief discussion of one particular type of block, namely a nilpotent block.

Definition 2.43 Let G be a finite group, and let b be a block idempotent of G, with defect group D. If the fusion system of the block b is $\mathcal{F}_D(D)$ then b is *nilpotent*.

Nilpotent blocks are of considerable importance in the representation theory of finite groups, and their theory has been extensively developed. A fundamental result in this theory is the following.

Theorem 2.44 (Puig [Pui88, (1.4.1)]) *Let G be a finite group and let B be a nilpotent block of G with defect group P. There exists an integer n such that $B \cong M_n(k) \otimes_k kP$, where $M_n(k)$ denotes the matrix algebra of $n \times n$ matrices over k.*

If B is a nilpotent block then $B/J(B)$ is a simple k-algebra [Pui88, (1.9.1)], and so in particular nilpotent blocks have a unique simple kG-module. The structure of the unique simple module in a nilpotent block is a difficult question: the source M (i.e., the unique – up to conjugation by $\mathrm{N}_G(P)$ – indecomposable kP-module M such that $M \uparrow^G$ has as a summand the simple kG-module belonging to B) of the simple module in a nilpotent block B is known to be endo-permutation (i.e., $M \otimes M^*$ is a permutation module) but conjecturally it has finite order in the Dade group of such modules (see [Dad78] for these definitions). At the moment this conjecture is still open.

Exercises

2.1 Prove Lemma 2.15.

2.2 Prove Proposition 2.16.

2.3 Let G be a finite group and let P be a p-subgroup of G. Show that, for $x \in (kG)^P$,

$$\mathrm{Br}_P\left(\mathrm{Tr}_P^{N_G(P)}(x)\right) = \mathrm{Tr}_P^{N_G(P)}\left(\mathrm{Br}_P(x)\right).$$

2.4 Let A be a finite-dimensional commutative k-algebra. Let $E = \{e_1, \ldots, e_n\}$ be the block idempotents of A. Show that every ideal I of A has the form

$$I = \bigoplus_{e \in E'} Ae \oplus N,$$

where $E' \subseteq E$ and N is a nilpotent ideal. Consequently, the set of primitive idempotents of I is E'. (Hint: every block of A is a local ring.)

Now suppose that G is a finite group, and let $A = kG$. If $I = (kG)_P^G$ for some p-subgroup P of G, show that E' in this case is the set of all block idempotents whose defect groups are contained in a conjugate of P.

2.5 Let G and H be finite groups. Let $\{b_i \mid i \in I\}$ be the block idempotents of kG and let $\{e_j \mid j \in J\}$ be the block idempotents of kH. Prove that $\{b_i \otimes e_j \mid i \in I, j \in J\}$ is the set of block idempotents of $k(G \times H) = kG \otimes kH$.

2.6 Let P be a finite p-group. Show that kP has a single block, with block idempotent 1.

3

Fusion in topology

We have seen the idea of fusion systems in both group theory and representation theory, but there is another approach, through topology. Here, a major idea is to understand the fusion system via decompositions of the classifying space of a finite group (and vice versa).

One of the highlights of the topological approach to this theory is a topological criterion for when two finite groups have isomorphic fusion systems. The Martino–Priddy conjecture [MP96], proved by Bob Oliver [Oli04] [Oli06], establishes this precisely.

Theorem 3.1 (Martino–Priddy conjecture)   *If G and H are finite groups, with Sylow p-subgroups P and Q respectively, then $\mathcal{F}_P(G) \cong \mathcal{F}_Q(H)$ if and only if $BG_p^\wedge$ and $BH_p^\wedge$ are homotopy equivalent.*

In Chapter 1, we defined the fusion system $\mathcal{F}_P(G)$, and in this chapter we will define the classifying space BG and its p-completion $BG_p^\wedge$. In doing so, we will discuss simplicial sets, nerves of categories, the geometric realization of a simplicial set, and finally the R-completion functor, where R is a commutative ring.

The topics involved in this chapter are significantly different from the rest of the work. As such, if one is only interested in the internal structure of fusion systems then this chapter is largely superfluous, but at least a basic understanding of the topological aspects of this theory might be useful for balance.

The main result from this approach that we will need in the later chapters is Theorem 3.70, which proves that a certain class of fusion system, *constrained* fusion systems, always comes from finite groups, and furthermore such a group is in a certain sense unique.

We start with simplicial sets, which are a combinatorial way of thinking about topological spaces. After considering a few constructions, including getting a CW-complex from each simplicial set and vice versa, we move on to classifying spaces of groups, which are simplicial sets whose fundamental group is isomorphic with that particular group. For finite groups, the Bousfield–Kan p-completion is a useful construction that isolates in some sense the p-part of the classifying space. We will construct it in Section 3.4, and give some of its properties. The last two sections describe centric linking systems for group fusion systems, and consider constrained fusion systems.

The algebraic topology that we need here is fairly elementary: we don't need CW-complexes except for motivation and comparison, and so the reader need not be familiar with those. Simplicial complexes are not strictly necessary either, although the reader will probably find the constructions fairly difficult to comprehend without some understanding of these. Homology and homotopy are used as well, although they aren't central to the text.

We remind the reader that, in this chapter, all functions will be composed right-to-left, rather than left-to-right. The reason for this is that there are almost no category theorists who write functors on the right, and doing so here would only lead to confusion.

3.1 Simplicial sets

Simplicial sets are a combinatorial way of doing homotopy theory. Classical algebraic topology uses topological spaces and CW-complexes, and homotopy equivalences. A more convenient notion of equivalence is one that can be tested using an algebraic invariant.

Definition 3.2 Let X and Y both be topological spaces. A *weak homotopy equivalence* $f : X \to Y$ is a continuous map such that f induces an isomorphism on all homotopy groups with respect to all choices of basepoint.

Any topological space X has a 'CW-approximation', a CW-complex Y and a weak homotopy equivalence $Y \to X$. Furthermore, a weak homotopy equivalence between CW-complexes is a homotopy equivalence. An important point here is that two topological spaces merely having the same homotopy groups does not guarantee that they are weakly homotopy equivalent: we require a continuous map inducing an isomorphism of homotopy groups. (For example, if S^n denotes the n-sphere and $\mathbb{R}P^n$ denotes n-dimensional real projective space, then $S^3 \times \mathbb{R}P^2$ and

$S^2 \times \mathbb{R}P^3$ have the same homotopy groups but are not weakly homotopy equivalent.)

The combinatorial approach is to replace topological spaces by *simplicial sets* and CW-complexes by *fibrant* simplicial sets (also known as Kan complexes); we shall meet the former in this section and the latter in the next. In this section we will define simplicial sets and consider some of the basic constructions, such as products and mapping spaces, the nerve of a category, and the geometric realization of a simplicial set. We delay a description of how simplicial sets and topological spaces are essentially 'the same' until the next section, because we need to define homotopy groups for simplicial sets first.

Let $[n]$ denote the set of all integers between 0 and n inclusive, together with the usual ordering $\leq$. A morphism $[n] \to [m]$ is a function between the two sets that respects the ordering $\leq$ on the sets (so that a morphism need not necessarily be injective). The collection of all such sets $[n]$ for $n \geq 0$, together with the order-preserving morphisms, is a category that we will denote by $\mathbf{\Delta}$. Since we will be interested in functors from this category to various categories, for a combinatorial description of such functors we need a presentation of the category $\mathbf{\Delta}$, which we will get now.

Firstly, we need to define some maps between $[n]$ and $[n+1]$, and between $[n]$ and $[n-1]$. Define $d^i : [n] \to [n+1]$ (for $0 \leq i \leq n+1$) by the injection that does not have i in the image, and define $s^i : [n] \to [n-1]$ (for $0 \leq i \leq n-1$) to be the surjection that sends both i and $i+1$ to i.

Proposition 3.3 *The functions d^i and s^i satisfy the relations*

$$d^j d^i = d^i d^{j-1} \ (i < j), \qquad s^{j-1} s^i = s^i s^j \ (i < j),$$

$$s^j d^i = \begin{cases} d^i s^{j-1} & i < j \\ \mathrm{id} & i = j, j+1 \\ d^{i-1} s^j & i > j+1. \end{cases}$$

Furthermore, these relations determine the category $\mathbf{\Delta}$ uniquely.

The proof of this proposition is Exercise 3.1. Thus we have a combinatorial description of the category $\mathbf{\Delta}$, which may be used when checking that a map is a functor. We will give the technical definition of a simplicial set now, and then give a more intuitive (for algebraists at least) definition afterward.

Definition 3.4   A *simplicial set* is a functor $X : \mathbf{\Delta}^{\mathrm{op}} \to \mathsf{Set}$, i.e., a contravariant functor from $\mathbf{\Delta}$ to Set, where Set is the category of sets.

Since a simplicial set is a functor, we can simply consider the images of the elements of $\mathbf{\Delta}$. Let X be a simplicial set: the image of $[n]$ under X (just a set) is denoted by X_n, and there are functions $d_i : X_n \to X_{n-1}$ (called *face maps*) and $s_i : X_n \to X_{n+1}$ (called *degeneracy maps*). The elements of X_n are called *n-simplices*; if $x \in X_n$ is an n-simplex and x lies in the image of one of the s_i, then x is *degenerate*, and if not then x is *non-degenerate*.

It might be helpful here to think of a simplicial complex: here the non-degenerate elements of X_n are the usual n-simplices, and the maps d_i are the maps sending an n-simplex to its ith face. The existence of degenerate simplices in a simplicial set is why simplicial sets are in some sense 'better' than simplicial complexes: by adding in phantom n-simplices – $(n-1)$-simplices masquerading as n-simplices – many constructions become easier.

By our combinatorial description of the category $\mathbf{\Delta}$, we know that the face and degeneracy maps satisfy the relations given in Proposition 3.3, except read left-to-right rather than right-to-left (because the functor is contravariant). For future reference, we write them explicitly here in the form of a lemma.

Lemma 3.5   *In the category* $\mathbf{\Delta}^{\mathrm{op}}$, *the maps* d_i *and* s_i *(corresponding to* d^i *and* s^i *respectively) satisfy the relations*

$$d_i d_j = d_{j-1} d_i \ (i < j), \qquad s_i s_{j-1} = s_j s_i \ (i < j),$$

$$d_i s_j = \begin{cases} s_{j-1} d_i & i < j \\ \mathrm{id} & i = j, j+1 \\ s_j d_{i-1} & i > j+1. \end{cases}$$

Example 3.6   Let Y be a simplicial complex, and let Y_0 denote the vertices of Y, which will be labelled with non-negative numbers so that they inherit an order. The n-simplices of Y can be identified with subsets of the non-negative numbers $\mathbb{N}$ of size $n+1$ (these will form the non-degenerate simplices of a simplicial set X). Instead of considering subsets of $\mathbb{N}$, we think of Y_n as increasing sequences of $n+1$ non-negative integers; given an n-simplex y, there are $n+1$ faces, which are obtained by deleting one element of the corresponding sequence in Y_n. This yields $n+1$ face maps d_i, from n-simplices to $(n-1)$-simplices.

To construct the simplicial set X corresponding to Y, we take as the set of n-simplices all *weakly* increasing sequences x of $n + 1$ elements of Y_0 such that there is some $m \leq n$ and $y \in Y_m$ with all terms of the sequence x lying in y, with the face maps d_i as above. The degeneracy map s_i acting on an n-simplex x simply repeats the ith element of the sequence. (Since these sequences are only weakly increasing they do not represent a genuine $(n + 1)$-simplex of the simplicial complex Y.)

As an example, consider the simplicial complex Y of the line (i.e., a standard 1-simplex), so that $Y_0 = \{0, 1\}$, and $Y_1 = \{01\}$. (We will write sequences as strings since our integers are all below 10.) The corresponding simplicial set X has sets $X_0 = \{0, 1\}$, $X_1 = \{00, 01, 11\}$ (the first and last being degenerate), $X_2 = \{000, 001, 011, 111\}$ (all of which are degenerate) and so on. Not every simplicial set can be obtained in such a way, and so the collection of simplicial sets is strictly larger than the collection of simplicial complexes.

If X and Y are simplicial sets, we need to be able to talk about maps between them.

Definition 3.7 Let X and Y be simplicial sets. A *simplicial map* $f : X \to Y$ is a natural transformation from $X : \Delta^{\mathrm{op}} \to \mathsf{Set}$ to $Y : \Delta^{\mathrm{op}} \to \mathsf{Set}$. More explicitly, it is a sequence of maps $f_n : X_n \to Y_n$ such that f commutes with the face and degeneracy maps, in the sense that $d_i f_n = f_{n-1} d_i$ and $s_i f_n = f_{n+1} s_i$, where the d_i and s_i are the face and degeneracy maps for both X and Y.

The category of simplicial sets, denoted sSet, has as objects all simplicial sets and as morphisms all simplicial maps.

Simplicial sets might be easy to define, but in order to do anything interesting with them we will need some constructions. The first one is to construct a simplicial set given any small category, the nerve of a category.

Definition 3.8 Let $\mathscr{C}$ be a small category. Let X_0 be the set of all elements of $\mathscr{C}$, and define X_n to be the set of all sequences of n composable morphisms in $\mathscr{C}$. Thus the elements of X_n can be thought of as chains

$$x = \left(\quad c_0 \xrightarrow{\ \phi_1\ } c_1 \xrightarrow{\ \phi_2\ } c_2 \xrightarrow{\quad\quad} \cdots \xrightarrow{\quad\quad} c_{n-1} \xrightarrow{\ \phi_n\ } c_n \quad \right),$$

where the c_i are objects of $\mathscr{C}$ and the ϕ_i are arrows in $\mathscr{C}$. If the element x of X_n is as above, the maps d_i and s_i act as follows: $s_i(x)$ is the chain above with an identity arrow inserted into the ith position; and $d_i(x)$ is the chain with the ith and $(i + 1)$th arrow composed, unless $i = 0$ or $i = n$, in which case the first or last arrow respectively is removed. The object consisting of the X_n, d_i and s_i is the *nerve* of $\mathscr{C}$ and denoted by $\mathcal{N}\mathscr{C}$.

The nerve of a category can be thought of as a simplicial complex built out of the arrows of $\mathscr{C}$ (with 0-simplices the objects of $\mathscr{C}$), with the n-simplices consisting of n composable, *non-identity* arrows. By thinking of simplicial sets and allowing degenerate n-simplices, we no longer require the condition that the arrows are not the identity.

Proposition 3.9 *Let $\mathscr{C}$ be a small category. The collection of X_n, with face and degeneracy maps d_i and s_i as in Definition 3.8, is a simplicial set.*

Proof We merely need to show that the face and degeneracy maps d_i and s_i satisfy the relations in Lemma 3.5. This is a simple exercise and left to the reader.   □

The other thing we need to know is how taking nerves affects functors between categories.

Proposition 3.10 *Taking nerves is a functor from the category* Cat *of small categories to* sSet, *namely,*

$$(\mathcal{N}\mathscr{C})_n = \mathrm{Hom}_{\mathsf{Cat}}([n], \mathscr{C}),$$

where we view $[n]$ as a category in the obvious way (see Example 3.11).

Proof See Exercise 3.2.   □

Example 3.11 A poset may be turned into a category by keeping the same objects as before, and replacing the relation $x \leq y$ with a unique arrow $x \to y$. (The composition of morphisms in this category is forced.) Doing this to the poset $[n]$ gives a category with objects the integers $0 \leq i \leq n$, and a unique arrow $i \to j$ if and only if $i \leq j$. We may take the nerve $\mathcal{N}[n]$ of this category, and we get a simplicial set whose i-simplices are just all weakly increasing sequences of $i + 1$ integers between 0 and n.

Now consider the n-dimensional simplex Δ^n as a simplicial complex (i.e., a triangle for $n = 2$, a tetrahedron for $n = 3$, and so on). We will take the simplicial set X constructed in Example 3.6, with the integers 0 to n labelling the $n+1$ vertices of Δ^n. The i-simplices of X are simply all weakly increasing sequences of integers between 0 and n, and so this simplicial set coincides with $\mathcal{N}[n]$. This simplicial set will also be denoted by Δ^n, and is the simplicial set analogue of the standard n-simplex in classical homotopy theory.

[When we construct the geometric realization of a simplicial set, we will be able to construct a CW-complex from a simplicial set, and in the case of $\mathcal{N}[n]$ we will recover the simplicial complex Δ^n from the simplicial set Δ^n, so that no confusion arises from using the same notation.]

Two more constructions of new simplicial sets from others will now be given in this section: product simplicial sets and function spaces.

If A and B are CW-complexes, then $A \times B$ is usually, but not always, a CW-complex using the normal product topology. (Whitehead proved that if either A or B is locally compact, then $A \times B$ is a CW-complex, for example.) In full generality, we must define a new topology on $A \times B$, making a set open if and only if the intersection with every compact subset of $A \times B$, under the *product* topology, is open. Using this topology, $A \times B$ becomes a CW-complex. (This is usually called the *Kelley product*.) For most CW-complexes of interest, the usual product and the Kelley product coincide.

The definition for simplicial sets is rather easier: if X and Y are simplicial sets, then $(X \times Y)_n$ is simply $X_n \times Y_n$.

Definition 3.12 Let X and Y be simplicial sets. The *product*, denoted by $X \times Y$, is given by

$$(X \times Y)_n = X_n \times Y_n,$$

and if (x, y) is an n-simplex in $(X \times Y)_n$ then $d_i(x, y) = (d_i(x), d_i(y))$ and $s_i(x, y) = (s_i(x), s_i(y))$.

Since the face and degeneracy maps for X and Y satisfy the relations of Lemma 3.5, the d_i and s_i for $X \times Y$ also satisfies the required relations. Notice that the product of two degenerate n-simplices need not be a degenerate n-simplex.

Proposition 3.13 *Let $\mathscr{C}$ and $\mathscr{D}$ be small categories. The two simplicial sets $\mathcal{N}\mathscr{C} \times \mathcal{N}\mathscr{D}$ and $\mathcal{N}(\mathscr{C} \times \mathscr{D})$ are isomorphic.*

Proof One way of seeing this is that $\mathrm{Hom}_{\mathsf{sSet}}([n], -)$ preserves products; we will give a more concrete proof now. Let $N = \mathcal{N}(\mathscr{C} \times \mathscr{D})$; an n-simplex x in N is a sequence of n composable morphisms from $\mathscr{C} \times \mathscr{D}$, and this can be thought of as two sequences of n composable morphisms, one from $\mathscr{C}$ and one from $\mathscr{D}$. Since each of these is an n-simplex in $\mathcal{N}\mathscr{C}$ and $\mathcal{N}\mathscr{D}$

respectively, we see that x is an element of $(\mathcal{NC} \times \mathcal{ND})_n = \mathcal{NC}_n \times \mathcal{ND}_n$. Since the n-simplices are the same in both of N and $\mathcal{NC} \times \mathcal{ND}$, it suffices to check that the face and degeneracy maps are the same. We will show that the degeneracy maps are the same, and note that the case of the face maps is very similar.

Let $x = (x', x'')$ be an n-simplex of N, where $x' \in \mathcal{NC}_n$ and $x'' \in \mathcal{ND}_n$. The map $s_i(x)$ inserts an identity morphism in the ith place, and this is the same as adding an identity map in the ith place of both x' and x'', whence $s_i(x) = (s_i(x'), s_i(x''))$. This is the same as the degeneracy map on $\mathcal{NC} \times \mathcal{ND}$, and so the two degeneracy maps agree, as claimed.

$\square$

We now move on to function spaces. If X and Y are topological spaces, then the set $\mathrm{Map}(X, Y)$ of continuous maps forms a topological space with respect to the compact-open topology. (This means: for all compact subsets $U \subseteq X$ and open subsets $V \subseteq Y$, let $C(U, V)$ be the set of all maps f in $\mathrm{Map}(X, Y)$ such that $f(U) \subseteq V$. The sets $C(U, V)$ are the generating sets for the compact-open topology.) However, homotopy theory needs CW-complexes, and so we want $\mathrm{Map}(X, Y)$ to be a CW-complex if X and Y are both CW-complexes. We need to alter the topology, like with products, to get a nice result, and even then we don't *quite* get that $\mathrm{Map}(X, Y)$ is a CW-complex. We say that a subset of $\mathrm{Map}(X, Y)$ is open if it has open intersection with any subset that is open in the compact-open topology. With this definition, we get the following theorem.

Theorem 3.14 (Milnor, 1958)   *If X and Y are CW-complexes then* $\mathrm{Map}(X, Y)$ *is homotopy equivalent to a CW-complex.*

We are interested in simplicial sets rather than CW-complexes, and so we need a definition of $\mathrm{Map}(X, Y)$ that should be a simplicial set for simplicial sets X and Y. Before we give the definition, we need to recall that the m-simplices of Δ^n may be thought of as all weakly increasing sequences of $m + 1$ integers between 0 and n. With this, there are maps $d^i : \Delta^n \to \Delta^{n+1}$ and maps $s^i : \Delta^n \to \Delta^{n-1}$. The map d^i maps the vertex j to j if $j < i$ and $j + 1$ if $j \geq i$ (i.e., it misses out the vertex i) and s^i maps j to j if $j \leq i$ and $j - 1$ if $j \geq i + 1$ (i.e., it maps both vertices i and $i + 1$ to i). These induce maps on all simplices of Δ^n; notice that they satisfy the relations of the category $\boldsymbol{\Delta}$.

Definition 3.15 Let X and Y be simplicial sets. The *function space* $\mathrm{Map}(X, Y)$ of simplicial sets has, as n-simplices, all simplicial maps between $X \times \Delta^n$ and Y. The face and degeneracy maps on $f \in \mathrm{Map}(X, Y)_n$ are given by premultiplication by the maps (id, d^i) and (id, s^i).

More precisely, if f is a map in $\mathrm{Map}(X, Y)_n$, then $f : X \times \Delta^n \to Y$, and we have

$$f \circ (\mathrm{id}, d^i) : X \times \Delta^{n-1} \to Y, \qquad f \circ (\mathrm{id}, s^i) : X \times \Delta^{n+1} \to Y.$$

Since we are multiplying on the right, this reverses the relationships between the d^i and s^i, so that they satisfy the simplicial identities needed for face and degeneracy maps.

We will need function spaces in a few sections' time, but we include one nice result on them, relating products and function spaces via an adjunction.

Proposition 3.16 *Let X, Y, and Z be simplicial sets. There is a natural isomorphism*

$$\mathrm{Map}(X \times Y, Z) \to \mathrm{Map}(Y, \mathrm{Map}(X, Z)).$$

As the final topic in this section, we need to describe how simplicial sets should be thought of as spaces in their own right. We mentioned that if X is a simplicial set then the X_n should be thought of as n-simplices, but we will now make this explicit, by associating to every simplicial set X a CW-complex $|X|$, where the non-degenerate elements of X_n become the n-simplices in $|X|$.

We begin by producing a geometric realization $|\Delta^n|$ of the simplicial n-simplex Δ^n: this is given by

$$\{(a_0, \dots, a_n) \in \mathbb{R}^{n+1} \mid a_i \geq 0, \ \sum_{i=0}^{n} a_i = 1\}.$$

(This is an n-dimensional triangle in $(n + 1)$-dimensional space. It is much more convenient to define it this way than attempt to define an n-dimensional triangle in n-dimensional space directly. The reader should convince herself that this actually *does* give the standard n-simplex in algebraic topology.)

There are two (homotopy equivalent) ways to construct the geometric realization, either by considering all simplices and both the face and degeneracy maps, or by considering the non-degenerate simplices and just the face maps. We can associate to each *non-degenerate n-simplex* the geometric realization above, but we need to know how to 'sew' them together in order to reflect the face maps d_i present in X. Thus we need some way of stitching together simplicial n-simplices: an order-preserving map $f : [m] \to [n]$ gives rise to a map $f_\sharp$ from $|\Delta^m|$ to $|\Delta^n|$ by specifying the jth co-ordinate of the image to be the sum of a_i, as i runs over all elements whose image is j itself.

Using all simplices, we can also construct the geometric realization $|X|$: to each n-simplex x we associate the cell $|\Delta^n|$, and we glue cells together via the order-preserving maps f in Δ and the associated linear maps $f_\sharp$ between cells. This is the general idea, and we give a formal definition now.

Definition 3.17   Let X be a simplicial set. Define

$$|X| = \coprod_n (X_n \times |\Delta^n|)/\sim,$$

where each X_n is a discrete set, and $\sim$ is the equivalence relation generated by

$$(f^*x, u) \sim (x, f_\sharp u),$$

where f is a relation in Δ and f^* is the induced relation on X. The CW-complex $|X|$ is the *geometric realization* of X.

This association is functorial, as we shall now see.

Proposition 3.18   *Let X and Y be simplicial sets. If $f : X \to Y$ is a simplicial map, then there is a corresponding continuous map $|f| : |X| \to |Y|$ such that $|\ |$ becomes a functor.*

Proof   Let $|f|$ be defined on n-simplices in the obvious way, by mapping $|x|$ to $|f(x)|$. Since f is a simplicial map, it commutes with the face and degeneracy maps, and hence is well defined. A function between two CW-complexes is continuous if and only if it is on each cell (see for example [Swi75, Proposition 5.5]), and we have defined it to be so. (We omit the proof that $|\ |$ respects composition of morphisms.) Clearly $|\ |$ sends identity maps to identity maps, and hence $|\ |$ forms a functor. $\square$

One of the reasons for the degenerate n-simplices in simplicial sets is to make products work well. Taking products and geometric realizations commute, as the following proposition states.

Proposition 3.19   *Let X and Y be simplicial sets. The two CW-complexes $|X| \times |Y|$ (with the Kelley product) and $|X \times Y|$ are homeomorphic.*

If $\mathscr{C}$ is a small category, then instead of writing $|\mathcal{N}\mathscr{C}|$ for the geometric realization of the nerve of $\mathscr{C}$, we will simply write $|\mathscr{C}|$.

Proposition 3.20 *Let $\mathscr{C}$ and $\mathscr{D}$ be small categories. If $F : \mathscr{C} \to \mathscr{D}$ is a functor, then F induces a continuous map, $|F|$, between $|\mathscr{C}|$ and $|\mathscr{D}|$.*

Proof Taking nerves and geometric realizations are both functors, so the composition is a functor. □

Along with categories come functors and natural transformations, so having dealt with functors, we turn to natural transformations.

Proposition 3.21 *Let $\mathscr{C}$ and $\mathscr{D}$ be small categories, and let F and F' be two functors $\mathscr{C} \to \mathscr{D}$. If there is a natural transformation relating F and F' then the continuous maps $|F|$ and $|F'|$ are homotopic.*

Proof Firstly, we note that a natural transformation $\phi : F \to F'$ induces a functor $H : \mathscr{C} \times [1] \to \mathscr{D}$; the category $\mathscr{C}' = \mathscr{C} \times [1]$ is the set of all ordered pairs (x, i), where $x \in \mathscr{C}$ and $i \in \{0, 1\}$, together with morphisms (f, g), where f is a morphism in $\mathscr{C}$ and g is a morphism in $[1]$.

The nerve $\mathcal{N}(\mathscr{C} \times [1])$ is simply $\mathcal{N}\mathscr{C} \times \Delta^1$ by Proposition 3.13, and the geometric realization of Δ^1 is (homeomorphic to) the unit interval $[0, 1]$. Thus the geometric realization of the nerve of $\mathscr{C} \times [1]$ is (the Kelley product) $|\mathscr{C}| \times [0, 1]$.

Now everything is clear: since the functor H induces a continuous map $|H| : |\mathscr{C}| \times [0, 1] \to |\mathscr{D}|$ by Proposition 3.20, we see that H induces a homotopy between the continuous maps $|\mathscr{C}| \to |\mathscr{D}|$ evaluated at 0 and $|\mathscr{C}| \to |\mathscr{D}|$ evaluated at 1, i.e., between $|F|$ and $|F'|$, as claimed. □

If $F : \mathscr{C} \to \mathscr{D}$ is an equivalence of categories, then $|F| : |\mathscr{C}| \to |\mathscr{D}|$ is a homotopy equivalence. More generally, if F has a right or left adjoint, then $|F|$ has a right or left homotopy inverse.

Corollary 3.22 *If $\mathscr{C}$ has an initial or terminal object, then $|\mathscr{C}|$ is contractible.*

Proof It is easy to see that if $\mathscr{C}$ has a terminal object x, there is a natural transformation $F \to F'$, where F is the identity functor on $\mathscr{C}$ and F' is the functor sending all elements to x. By Proposition 3.21, this means that there is a homotopy from the identity map on $|\mathscr{C}|$ to the constant map on $|\mathscr{C}|$, proving that $|\mathscr{C}|$ is contractible. To prove the result for initial objects, we simply note that $|\mathscr{C}^{\mathrm{op}}| = |\mathscr{C}|$. □

Having given a functor from simplicial sets to topological spaces, we now need one the other way round. This is the singular simplices functor, and is defined now.

Definition 3.23 Let X be a topological space. A *singular n-simplex* of X is a continuous map $|\Delta^n| \to X$. The *singular simplices functor*

$$\mathrm{Sing}(X) : \Delta^{\mathrm{op}} \to \mathsf{Set},$$

is given by sending $[n]$ to the set of singular n-simplices of X, and a map $f : [m] \to [n]$ is sent to composition of the map $|\Delta^n| \to X$ with the map $f_\sharp : |\Delta^m| \to |\Delta^n|$.

Notice that since the composition is on the right this does define a functor from Δ^{op}, and so $\mathrm{Sing}(X)$ forms a simplicial set. To see that $\mathrm{Sing}(-)$ is a functor from Top to sSet, we notice that a continuous map of topological spaces gives rise to a map of singular simplices, and so $\mathrm{Sing}(-)$ is actually a functor (since it obviously sends identity morphisms to identity morphisms and respects compositions).

The geometric realization and singular simplices functors are related by the following theorem.

Theorem 3.24 *The singular simplices functor is right adjoint to the geometric realization functor, so that, if X is a simplicial set and Y is a topological space, then*

$$\mathrm{Hom}_{\mathsf{Top}}(|X|, Y) \cong \mathrm{Hom}_{\mathsf{sSet}}(X, \mathrm{Sing}(Y)).$$

In a sense that can be made precise, although we will not do so until the next section, as far as homotopy theory is concerned it doesn't really matter whether one uses topological spaces or simplicial sets. Note that this adjunction yields a map $X \to \mathrm{Sing}(|X|)$ for a simplicial set X (setting $Y = |X|$), and we will use this map in the next section.

3.2 Classifying spaces

The classifying space of a group G is a topological space X that can be associated with G.

Definition 3.25 Let G be a group. A *classifying space* for G is a path-connected CW-complex X such that $\pi_1(X) = G$ and the universal cover $\tilde{X}$ of X is contractible.

Uniqueness of the classifying space cannot be possible because of the fact that the conditions are invariant under homotopies. Apart from this obstruction, however, it is unique (see [Ben98c, Section 2.2]).

Proposition 3.26 *Let G be a group. If X and Y are classifying spaces for G, then X and Y are homotopy equivalent.*

Because of this proposition, we will tend to talk of *the* classifying space, even though it is only defined up to homotopy. By BG we denote a simplicial set such that $|BG|$ is the classifying space for G, although we will conflate the two, and refer to BG as the classifying space as well as $|BG|$. The (contractible) universal cover of BG will be denoted by EG.

We have yet to prove that such a space exists, of course. A general construction is given now.

Definition 3.27 Let G be a finite group. Let $\mathscr{B}(G)$ denote the category with one object $*$, and whose morphism set is given by

$$\text{Hom}_{\mathscr{B}(G)}(*, *) = G,$$

with the associated multiplication from G. Let $\mathscr{E}(G)$ denote the category with object set G, and such that for any two objects g and h there is a unique morphism $g \to h$. (This is action by the element hg^{-1}.)

The category $\mathscr{E}(G)$ has a free action of G on it by right multiplication, and it is not difficult to see that the orbit groupoid, $\mathscr{E}(G)/G$, is simply $\mathscr{B}(G)$. We claim that the geometric realization of the nerve of $\mathscr{B}(G)$, namely $|\mathscr{B}(G)|$, is $|BG|$, and the geometric realization of the nerve of $\mathscr{E}(G)$ is $|EG|$.

(A proof of this is not difficult and we sketch it now, assuming a result we will prove later: the element $1 \in \mathscr{E}(G)$ is an initial object (in fact, every object in $\mathscr{E}(G)$ is), and so $|\mathscr{E}(G)|$ is contractible by Corollary 3.22. In Proposition 3.37 we prove that $\pi_1(|\mathscr{B}(G)|) = G$, and so it remains to show that the universal cover of $|\mathscr{B}(G)|$ is $|\mathscr{E}(G)|$. Since G acts freely on $\mathscr{E}(G)$, it acts freely on $|\mathscr{E}(G)|$; hence $|\mathscr{E}(G)| \to |\mathscr{E}(G)|/G$ is a covering map. We may naturally identify the image $|\mathscr{E}(G)|/G$ with $|\mathscr{B}(G)|$, completing the proof.)

Let us describe the simplicial set obtained from $\mathscr{B}(G)$ properly, rather than simply saying that it is the nerve. Let $BG_0 = \{*\}$, $BG_1 = G$, $BG_2 = G \times G$, and in general BG_n is the n-fold Cartesian product of G. The degeneracy map $s_i : BG_n \to BG_{n+1}$ involves inserting a 1 in the ith place, and the face map $d_i : BG_n \to BG_{n-1}$ is defined by

$$d_i(g_1, g_2, \ldots, g_n) = \begin{cases} (g_2, \ldots, g_n) & i = 0 \\ (g_1, \ldots, g_{i-1}, g_{i+1}g_i, g_{i+2}, \ldots, g_n) & 1 \leq i \leq n-1 \\ (g_1, \ldots, g_{n-1}) & i = n. \end{cases}$$

(Here, as always in the chapter, the multiplication is right-to-left.) It is a simple exercise to check that this is the nerve of $\mathscr{B}(G)$.

Example 3.28   Here we will give the classifying spaces of some groups. We will build these as topological spaces, so in truth we are really giving examples of $|BG|$.

   (i) Consider the group $\mathbb{Z}$; this is the fundamental group of the circle, S^1. If the universal cover of S^1 is contractible, then S^1 is the classifying space of $\mathbb{Z}$. Since the universal cover of S^1 is the real line $\mathbb{R}$, and this is contractible, $B\mathbb{Z}$ is indeed homotopy equivalent to S^1.

  (ii) Consider the cyclic group C_2 and the n-sphere S^n. The map sending x to $-x$ (i.e., the antipodal map) is an action of C_2 on S^n, and the orbit space is real projective space $\mathbb{R}P^n$. Since S^n is simply connected, this means that the fundamental group of $\mathbb{R}P^n$ is C_2. However, S^n is not contractible, so these are not classifying spaces of C_2. However, $\mathbb{R}P^\infty$, the union of the $\mathbb{R}P^n$, still has a C_2-action and the union of spheres *is* contractible, and hence BC_2 is homotopy equivalent to $S^\infty/C_2 = \mathbb{R}P^\infty$.

 (iii) We have that $B(G \times H)$ and $BG \times BH$ are homotopy equivalent, and hence if E_{2^n} denotes the elementary abelian group of order 2^n, then BE_{2^n} is the Kelley product of n copies of $\mathbb{R}P^\infty$.

Classifying spaces of groups are very nice topological spaces; for example, their higher homotopy groups are trivial (see Proposition 3.37). In order to compute homotopy groups of simplicial sets combinatorially (which are the homotopy groups of the corresponding geometric realization), one has to have a special type of simplicial set, called fibrant; fibrant simplicial sets are also known as *Kan complexes*. We give two definitions of this concept. The first is a technical description of the condition, requiring that there be 'enough' simplices in the simplicial set.

Definition 3.29   Let X be a simplicial set, with face maps $d_i : X_n \to X_{n-1}$ and degeneracy maps $s_i : X_n \to X_{n+1}$. We say that X is *fibrant* if, for each n and $k \leq n+1$, and n-simplices $x_0, x_1, \ldots, x_{k-1}, x_{k+1}, \ldots, x_{n+1}$

such that $d_{j-1}x_i = d_i x_j$ for all $i < j$ with $i < j$ and $i, j \neq k$, there is an $(n+1)$-simplex y such that $d_i y = x_i$ for all $i \neq k$.

The idea here is that if the n-simplices $x_0, x_1, \ldots, x_{k-1}, x_{k+1}, \ldots, x_{n+1}$ look as though they are (all but one of) the faces of some $(n+1)$-simplex y, then there is such a simplex; the condition $d_{j-1}x_i = d_i x_j$ comes from the face relation $d_i d_j = d_{j-1} d_i$. The alternative definition is more intuitive and diagrammatical.

Definition 3.30 The (n, k)-*horn* Λ_k^n (for $0 \leq k \leq n$) is the simplicial complex obtained from Δ^n by removing the single n-simplex and the kth face (i.e., $(n-1)$-simplex) of Δ^n. We also consider Λ_k^n to be a simplicial set as in Example 3.6.

The fibrant condition for X can now be stated as follows: for all $n \geq 0$ and for $0 \leq k \leq n$, given the inclusion $\Lambda_k^n \to \Delta^n$ and a map $\Lambda_k^n \to X$, there is a map $\Delta^n \to X$ making the diagram

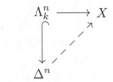

commute. The map $\Delta^n \to X$ is referred to as a *filling* of the horn.

We should note that the $(n+1)$-simplex y in the definition of fibrant is not uniquely determined, i.e., the induced map $\Delta^n \to X$ is not uniquely determined. In a broad sense, the idea is that being a fibrant simplicial set guarantees that there are 'lots' of simplices, enough at least to do homotopy theory, as we shall see below.

Lemma 3.31 (Moore) *Let X be a simplicial set. If each X_n is a group and the face maps d_i are group homomorphisms, then X is fibrant.*

In the next section we will define simplicial objects other than simplicial sets; this lemma proves that simplicial groups are fibrant. In general, there are simplicial sets that are not fibrant; for example, the simplicial n-simplex for $n \geq 1$. However, for any simplicial set X there is a fibrant simplicial set Y and a map $X \to Y$ that induces a weak homotopy equivalence on the geometric realizations of X and Y. We will define the homotopy groups of *fibrant* simplicial sets later in this section. For simplicial sets that are not fibrant, one method of defining homotopy groups is to define them to be the homotopy groups of the geometric realization. There are methods to define homotopy groups without using the geometric realization though; either via Kan's Ex^∞ functor (see [GJ99, Section III.4]) or via the small-object argument (see [Hov99, Section 2.1.2]). We also have the analogue of a weak homotopy equivalence in the category of simplicial sets.

Definition 3.32   Let X and Y be simplicial sets. A *weak equivalence* $f : X \to Y$ is a simplicial map such that f induces an isomorphism on all homotopy groups with respect to all choices of basepoint.

We arrive at the main proposition that tells us that classifying spaces are nice objects to study. For a proof, see [GJ99, Lemma I.3.5].

Proposition 3.33   *For any group G, the simplicial set BG is fibrant.*

As promised, we will now introduce homotopy theory for fibrant simplicial sets. We start with the definition of a homotopy between simplices.

Definition 3.34   Let X be a simplicial set, and let $*$ be a basepoint in X_0. Also write $*$ for the element $s_0(*)$ for each X_n inductively, and set $Z_n = \{x \in X_n \mid d_i x = * \text{ for all } i = 0, 1, \ldots, n\}$. We say that x and x' in Z_n are *homotopic* (written $x \sim x'$) if there is an $(n+1)$-simplex y (a *homotopy* from x to x') such that

$$
d_i y = \begin{cases} * & i < n \\ x & i = n \\ x' & i = n+1. \end{cases}
$$

It is now that the requirement that the simplicial set be fibrant is important.

Lemma 3.35   *Let X be a simplicial set. If X is fibrant then $\sim$ is an equivalence relation.*

Proof   Let x be an element of Z_n. Since $d_{n+1} s_n$ is the identity, we see that $s_n x$ is a homotopy from x to itself. To prove that $\sim$ is symmetric and transitive we will use the fact that X is fibrant. Let y be a homotopy from x to x', so that $d_{n+1} y = x'$. Since X is fibrant, there is some $(n+2)$-simplex z such that $d_i z = *$, y, and $s_n x$ for $i < n$, $i = n$, and $i = n+1$ respectively. (These $n+2$ choices can be checked to satisfy the hypothesis of the fibrant condition.) The $(n+1)$-simplex $d_{n+2} z$ has the property that (for $i \leq n + 1$)

$$d_i(d_{n+2}z) = d_{n+1}(d_i z) = \begin{cases} d_{n+1}* & i < n \\ d_{n+1}y & i = n \\ d_{n+1}(s_n x) & i = n+1 \end{cases} = \begin{cases} * & i < n \\ x' & i = n \\ x & i = n+1, \end{cases}$$

proving that $d_{n+2}z$ is a homotopy from x' to x.

Finally, suppose that $x \sim x'$ and $x' \sim x''$, and let y and y' be homotopies $x' \sim x$ (which exists since $\sim$ is symmetric) and $x' \sim x''$; let z be an $(n+2)$-simplex such that $d_i z = *$, y, and y' for $i < n$, $i = n$, and $i = n + 1$ respectively. (Again, this exists by the fibrant condition.) The $(n+1)$-simplex $d_{n+2}z$ has the property that (for $i \leq n + 1$)

$$d_i(d_{n+2}z) = d_{n+1}(d_i z) = \begin{cases} d_{n+1}* & i < n \\ d_{n+1}y & i = n \\ d_{n+1}y' & i = n+1 \end{cases} = \begin{cases} * & i < n \\ x & i = n \\ x'' & i = n+1, \end{cases}$$

proving that $d_{n+2}z$ is a homotopy from x to x''. This completes the proof. $\qquad\square$

Thus for fibrant simplicial sets we can define a quotient.

Definition 3.36 Let X be a fibrant simplicial set. The *homotopy groups* of X are defined by $\pi_n(X, *) = Z_n/\sim$ for $n \geq 1$, and $\pi_0(X) = Z_0/\sim$ (this is simply a set).

Of course, we need to define the group structure on $\pi_n(X, *)$ for $n \geq 1$, which we will do now, and then prove that it is well defined. Let x and x' be elements of Z_n, and define $x \cdot x'$ to be the n-simplex $d_n z$, where z is any $(n+1)$-simplex satisfying $d_i z = *$, x, and x' for $i < n - 1$, $i = n - 1$ and $i = n + 1$ respectively (so that $k = n$ in the fibrant condition). Essentially, this is the 'composition' of x and x', as in the following diagram for $n = 1$.

$$\begin{array}{ccc} 0 & \xrightarrow{\ x\ } & 2 \\ {\scriptstyle x'}\Big\uparrow & \nearrow & \\ & \diagup\ {\scriptstyle x \cdot x'} & \\ 1 & & \end{array}$$

Firstly, we need to prove that if z and z' are two different $(n+1)$-simplices satisfying the conditions, then $d_n z \sim d_n z'$. By the fibrant condition, there is some $w \in X_{n+2}$ such that

$$d_i w = \begin{cases} * & i < n-1 \\ s_n(d_{n-1}z) & i = n-1 \\ z & i = n+1 \\ z' & i = n+2. \end{cases}$$

We notice that $d_n w$ is a homotopy from $d_n z$ to $d_n z'$, proving what we need.

Next, we need to prove that $x \cdot x'$ is well defined up to homotopy. Hence let y' be an n-simplex such that $y' \sim x'$. We will show that $x \cdot y' \sim x \cdot x'$. Let z' be an $(n+1)$-simplex satisfying $d_i z' = *$, x, and y' for $i < n-1$, $i = n-1$, and $i = n+1$ respectively. We have that $d_n z' = x \cdot y'$, so we must show that $d_n z$ and $d_n z'$ are homotopic.

Let u be a homotopy $y' \sim x'$, so that $d_n u = y'$ and $d_{n+1} u = x'$. The fibrant condition on X means that there is some $w \in X_{n+2}$ such that

$$d_i w = \begin{cases} * & i < n-1 \\ s_n x & i = n-1 \\ z' & i = n \\ u & i = n+2. \end{cases}$$

We claim that $v = d_{n+1} w$ satisfies the conditions that $d_i v = *$, x, and x' for $i < n-1$, $i = n-1$, and $i = n+1$ respectively, so that $d_n v \sim d_n z$ by the first part. To see this, we calculate:

$$d_i(d_{n+1}w) = d_n * = *;$$
$$d_{n-1}(d_{n+1}w) = d_n(d_{n-1}w) = x;$$
$$d_n(d_{n+1}w) = d_n(d_n w) = d_n z';$$
$$d_{n+1}(d_{n+1}w) = d_{n+1}(d_{n+2}w) = d_{n+1}u = x'.$$

Thus $d_n v \sim d_n z$; but we also see here that $d_n v = d_n z'$, so we have that $d_n z \sim d_n z'$, as claimed. Finally, by the symmetric nature of $\sim$, we get that the definition of $x \cdot x'$ is independent of the choice of the representative of the class containing x, and so $x \cdot x'$ is well defined.

If X is *path connected* (see Exercise 3.5) then $\pi_i(X, *)$ does not depend on the basepoint $*$, and so we simply write $\pi_i(X)$.

Proposition 3.37 *Let G be a group. We have that $\pi_1(BG) = G$ and $\pi_n(BG) = 1$ for all $n > 1$.*

Proof Choose $* = 1$. Firstly, if $i > 1$ then $Z_i = \{*\}$, since if $(g_1, \ldots, g_i)$ lies in Z_i then $d_j(g_1, \ldots, g_i) = (1, \ldots, 1)$, and this means that each g_i is trivial. This leaves the case $i = 1$, when of course $Z_1 = BG_1 = G$. We need to check that $\sim$ is trivial and also that the multiplication is the same.

Suppose that $g_1 \sim g_2$, so that there is a 2-simplex $x = (h_1, h_2)$ such that $d_0 x = 1$, $d_1 x = g_1$ and $d_2 x = g_2$; then $x = (g_1, 1)$ and $g_1 = g_2$, so that $\sim$ is trivial. Finally, we note that $y = (g_2, g_1)$ is a 2-simplex such that $d_0 y = g_1$ and $d_2 y = g_2$, and so $d_1 y = g_2 g_1$ is the product of g_1 by g_2, completing the proof. $\square$

If X is fibrant, then we have the following important theorem.

Theorem 3.38 *Let X be a fibrant simplicial set, and let $*$ be a base-point in X_0. Let $*'$ denote the image of $*$ in $|X|$. For all $n \geq 1$, we have*

$$\pi_n(X, *) \cong \pi_n(|X|, *').$$

For a proof of this theorem, see [May92, Theorem 16.1]. Because of this theorem, we have defined homotopy groups entirely combinatorially for the geometric realizations of fibrant simplicial sets. (See also Exercise 3.6.) If X is not fibrant then either we can use Kan's Ex^∞ functor or the small-object argument, as we mentioned before, but for brevity we use the fact that $\mathrm{Sing}(Y)$ is fibrant for all Y, and define $\pi_n(X, *)$ to be $\pi_n(\mathrm{Sing}(|X|), *')$, where $*'$ is the image of the point $*$ through both functors. This makes the map $X \mapsto \mathrm{Sing}(|X|)$ defined at the very end of Section 3.1 into a weak equivalence, and so every simplicial set has a *fibrant replacement*, i.e., a fibrant simplicial set weakly equivalent to it.

In Section 3.1 we said that topological spaces and simplicial sets are 'the same', for some suitable sense of the phrase. We will now briefly describe in what way this is meant. The following theorem of Whitehead describes why CW-complexes are important in homotopy theory; this result was mentioned in Section 3.1.

Theorem 3.39 (Whitehead) (i) *Every weak homotopy equivalence between CW-complexes is a homotopy equivalence.*

(ii) *Every topological space X is weakly homotopy equivalent to a CW-complex. In particular, it is weakly homotopy equivalent to the geometric realization of its singular simplices, $|\mathrm{Sing}(X)|$.*

Paralleling Whitehead's theorem, we have the following result. (There is a combinatorial definition of homotopy equivalence for simplicial sets, but it is a little complicated for this survey, so we will simply note that $f : X \to Y$ is a homotopy equivalence if and only if $|f| : |X| \to |Y|$ is a homotopy equivalence.)

Theorem 3.40   (i) *Every weak equivalence between fibrant simplicial sets is a homotopy equivalence.*

(ii) *Every simplicial set is weakly equivalent to a fibrant simplicial set. In particular, it is weakly equivalent to the singular simplices of its geometric realization.*

The sense in which the homotopy theories of topological spaces and simplicial sets are the same uses the theory of *model categories*. We do not have the space to discuss them here, but to every model category there are associated a collection of weak equivalences, a collection of distinguished objects called *fibrant cofibrant objects*, to which every object is weakly equivalent, and a *homotopy category* in which weak equivalences are inverted, in some category-theoretic way. Topological spaces and simplicial sets are both model categories, with the weak equivalences given above and the fibrant cofibrant objects being CW-complexes and fibrant simplicial sets respectively. The two model categories are *Quillen equivalent*, which means that doing homotopy theory in one is essentially the same as doing homotopy theory in the other. One consequence of this is that the associated homotopy categories of the two model categories are equivalent as categories.

For more details on this area, see (for example) [May92] for a constructive approach to homotopy theory for simplicial sets, [GJ99] for simplicial homotopy theory from a 'modern', more abstract setting and some information on model categories, and [Hov99] for further details on model categories themselves.

From now on, since simplicial sets and topological spaces are 'the same', when we say 'space' we mean a simplicial set.

3.3 Simplicial and cosimplicial objects

A simplicial set is a functor from Δ^{op} to Set. By replacing the category Set with an arbitrary category $\mathscr{C}$, we get a simplicial object in $\mathscr{C}$. We denote a simplicial object by $X.$, because it is contravariant. (This suggests that we shall meet the covariant version later, which we shall.)

Definition 3.41 Let $\mathscr{C}$ be a category. A *simplicial object* in $\mathscr{C}$ is a functor $\Delta^{\mathrm{op}} \to \mathscr{C}$. The category of all simplicial objects in $\mathscr{C}$ is denoted by $s\mathscr{C}$.

A simplicial object X. in $\mathscr{C}$ can be thought of as an object such that the X_n are objects in $\mathscr{C}$, and the maps d_i and s_i are morphisms in $\mathscr{C}$ satisfying the relations. A cosimplicial object is the same as a simplicial object, except this time the functor is covariant rather than contravariant.

Definition 3.42 Let $\mathscr{C}$ be a category. A *cosimplicial object* in $\mathscr{C}$ is a functor $\Delta \to \mathscr{C}$. The category of all cosimplicial objects in $\mathscr{C}$ is denoted by $c\mathscr{C}$.

The concrete way of thinking about cosimplicial objects $X^{\cdot}$ in $\mathscr{C}$ is that they consist of objects X^n in $\mathscr{C}$, with *coface* and *codegeneracy* maps d^i and s^i; these satisfy the relations of Proposition 3.3 (i.e., the same relations as Δ).

Common categories to take cosimplicial and simplicial objects in are sets, groups, R-modules for some ring R, and spaces. In this case, we simply call them 'simplicial groups', 'cosimplicial spaces', and so on. In the previous section, Lemma 3.31 stated that if, in a simplicial set X, the X_n have a group structure with the d_i group homomorphisms then X is fibrant. In other words, we have the following result.

Proposition 3.43 *All simplicial groups are fibrant. In particular, if R is a ring, then all simplicial R-modules are fibrant.*

Let X be a space and R be a commutative ring. Eventually we want to construct a functor from spaces to themselves that behaves nicely with respect to R-homology, so we need to understand R-homology. In the next section we will define simplicial R-modules $R \otimes X$ and RX, whose homotopy groups are the R-homology groups and reduced R-homology groups of X with coefficients in R respectively. This proposition confirms that the homotopy groups of $R \otimes X$ and RX are 'easy' to compute.

We now turn our attention to cosimplicial objects. We will almost exclusively be taking cosimplicial objects in the category sSet of spaces (i.e., simplicial sets), and so we will deal with *cosimplicial spaces*. (The category of cosimplicial spaces will be denoted csSet.)

There are two different ways of thinking about cosimplicial spaces: the first is as a functor from Δ to sSet, as we have described; however, since sSet is itself a category of functors, we can think of a cosimplicial space as a functor from $\mathbf{\Delta}^{\mathrm{op}} \times \mathbf{\Delta}$ to Set.

Suppose that $X^{\cdot}$ is a cosimplicial space, and let m and n be natural numbers: write X^n_m for the value of $X^{\cdot}$ on $([m], [n])$, where $X^{\cdot}$ is thought of as a functor $\mathbf{\Delta}^{\mathrm{op}} \times \mathbf{\Delta} \to$ Set. Thus $X^{\cdot}$ is a sequence of simplicial sets X^n (whose mth set is X^n_m) with a map $X^n \to X^\ell$ whenever there is an order-preserving map $[n] \to [\ell]$, which is the same as giving coface and codegeneracy maps satisfying the right relations.

One can envisage a cosimplicial space as points in the top-right quadrant of the plane:

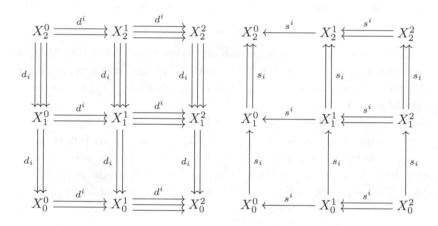

(Here we have drawn the face and coface maps on the left diagram, and the degeneracy and codegeneracy maps on the right diagram.) This diagram is commutative in the following sense: since the d^i and s^i are simplicial maps between the simplicial sets X^n and $X^{n\pm1}$, we see that $d^i d_j = d_j d^i$, and so on.

Example 3.44 The *cosimplicial simplex* $\Delta^{\cdot}$ is a cosimplicial space: the nth space of $\Delta^{\cdot}$ is the simplicial n-simplex Δ^n (so our notation is consistent). The coface and codegeneracy maps d^i and s^i are induced by the maps $d^i : [n] \to [n+1]$ and $s^i : [n] \to [n-1]$.

Recall the definition of the function space $\mathrm{Map}(X, Y)$ for two spaces X and Y: its n-simplices are all simplicial maps between $X \times \Delta^n$ and Y. If $X^{\cdot}$ and $Y^{\cdot}$ are cosimplicial spaces, we want to consider cosimplicial space maps between $X^{\cdot} \times \Delta^n$ and $Y^{\cdot}$. However, $X^{\cdot} \times \Delta^n$ isn't yet a cosimplicial space, so we cannot do this.

Definition 3.45 Suppose that $X^{\cdot}$ is a cosimplicial space and Y is a space. We will turn $X^{\cdot} \times Y$ into a cosimplicial space by defining

$$(X^{\cdot} \times Y)^n_m = X^n_m \times Y_m.$$

The face and degeneracy maps are just the same as in the standard product space, and the coface and codegeneracy maps simply act on the first co-ordinate.

Now that we have turned $X^{\cdot} \times \Delta^n$ into a cosimplicial space, we can consider the set

$$\mathrm{Hom}_{\mathsf{csSet}}(X^{\cdot} \times \Delta^n, Y^{\cdot})$$

of morphisms between cosimplicial spaces.

Definition 3.46 Let $X^{\cdot}$ and $Y^{\cdot}$ be cosimplicial spaces. The *function space* $\mathrm{Map}(X^{\cdot}, Y^{\cdot})$ is a simplicial set (not a cosimplicial space), with as n-simplices the sets

$$\mathrm{Map}(X^{\cdot}, Y^{\cdot})_n = \mathrm{Hom}_{\mathsf{csSet}}(X^{\cdot} \times \Delta^n, Y^{\cdot}).$$

We endow $\mathrm{Map}(X^{\cdot}, Y^{\cdot})$ with the structure of a space in exactly the same way as we did for the function space between simplicial sets.

It turns out that this is the 'right' definition, at least because it shares the adjointness properties of the function space of spaces.

Lemma 3.47 *Let $X^{\cdot}$ and $Z^{\cdot}$ be cosimplicial spaces and Y be a space. There is a natural isomorphism of sets*

$$\mathrm{Hom}_{\mathsf{sSet}}(Y, \mathrm{Map}(X^{\cdot}, Z^{\cdot})) \to \mathrm{Hom}_{\mathsf{csSet}}(X^{\cdot} \times Y, Z^{\cdot}).$$

The last definition that we need in this section is that of the total space. It will use all of our previous definitions!

Definition 3.48 Let $X^{\cdot}$ be a cosimplicial space. The *total space* of $X^{\cdot}$, $\mathrm{Tot}\, X^{\cdot}$, is the space

$$\mathrm{Tot}\, X^{\cdot} = \mathrm{Map}(\Delta^{\cdot}, X^{\cdot}).$$

The R-completion of a space X will be the total space of some cosimplicial space $R^{\cdot} X$ built out of X; this cosimplicial space is constructed in a way that will be reminiscent of a free resolution.

We should note that the total space construction is functorial, in the sense that if $f : X^{\cdot} \to Y^{\cdot}$ is a map of cosimplicial spaces, then there is an induced map

$$\mathrm{Map}(\Delta^{\cdot}, X^{\cdot}) \to \mathrm{Map}(\Delta^{\cdot}, Y^{\cdot}),$$

obtained by concatenation with f. We will show in the next section that the construction $X \to R^{\cdot}X$ is also functorial, and hence taking R-completions is functorial.

3.4 Bousfield–Kan completions

In this section, R is always a commutative ring. The Bousfield–Kan R-completion is a difficult construction to understand; it will effectively isolate the 'R-part' of a space, in the sense that completing with respect to R induces an isomorphism in R-homology. In our motivating example of the classifying space, the completion with respect to $\mathbb{F}_p$ (called p-completion) contains all of the information about fusion at the prime p and nothing more, in the sense that the Martino–Priddy conjecture (Theorem 3.1) proves that two classifying spaces have the same p-completions if and only if their associated groups have the same fusion system at the prime p.

In order to construct the R-completion, we first need two constructions, mentioned in the previous section, that produce simplicial R-modules, given a simplicial set X and a ring R. The homotopy groups of these two constructions will be the homology and reduced homology groups of X with coefficients in R, so they are of independent interest.

Let X be a space. Define $R \otimes X$ to be the simplicial set whose n-simplices are all formal finite linear combinations $\sum r_i x_i$, where $x_i \in X_n$ and the coefficients r_i lie in R. The face and degeneracy maps d_i and s_i are the formal linear combinations of the face and degeneracy maps for the individual n-simplices in X_n, so that if $x = \sum_i r_i x_i$ is an n-simplex, we have

$$d_i(x) = \sum_j r_j d_i(x_j), \quad s_i(x) = \sum_j r_j s_i(x_j).$$

The ring R acts on the left by multiplication, and this clearly commutes with the face and degeneracy maps, so that the d_i and s_i become morphisms in the category of R-modules; thus $R \otimes X$ becomes a simplicial R-module.

Inside this, we will take $(RX)_n$ to be the subset of $(R \otimes X)_n$ defined by

$$(RX)_n = \left\{ \sum_i r_i x_i \in (R \otimes X)_n \mid \sum_i r_i = 1 \right\}.$$

The left R-action is not quite the same: choose a basepoint $* \in X_0$, and as before write $*$ for $s_0(*)$, so that $*$ is an n-simplex for all n (degenerate if $n \geq 1$). Each $(RX)_n$ is a free R-module on the basis $X_n \setminus \{*\}$ with left multiplication by elements of R, because the coefficient of $*$ in an arbitrary n-simplex of RX is determined by the others, and this is the only constraint on the coefficients. Again, the face and degeneracy maps are morphisms in the category of R-modules, and so RX becomes a simplicial R-module. (It is a simplicial set because it inherits the face and degeneracy maps from $R \otimes X$, the image of an element of RX under either map clearly also lying in RX.)

Because $R \otimes X$ and RX are simplicial R-modules, by Proposition 3.43 they are fibrant, and so their homotopy groups are easy to describe combinatorially. Our statements about homotopy and homology earlier in this section are given now.

Definition 3.49   Let X be a space. The *homology groups* of X with coefficients in R are defined to be

$$H_n(X; R) = \pi_n(R \otimes X), \quad \tilde{H}_n(X; R) = \pi_n(RX).$$

The result suggested by the notation is that these homology groups are the same as the homology groups of $|X|$.

Theorem 3.50   *Let X be a space. The homology groups of $|X|$ satisfy*

$$H_n(|X|; R) \cong H_n(X; R), \quad \tilde{H}_n(|X|; R) \cong \tilde{H}_n(X; R).$$

We will use the construction of RX to build up a cosimplicial space $R^\cdot X$ corresponding to X and a ring R. This cosimplicial space will be the object that we will take the total space of to get the R-completion.

For any space X we have an obvious simplicial map $\phi_X : X \to RX$ given by $\phi_X : x \mapsto 1 \cdot x$. While there is no obvious map $RX \to X$, we *can* make a map $\psi_X : RRX \to RX$ by multiplication in R:

$$\psi_X : \left(\sum_i r_i \left(\sum_j r'_{ij} x_{ij} \right) \right) \mapsto \sum_{i,j} (r_i r'_{ij}) x_{ij}.$$

Writing $R(-)$ for the construction sending X to RX, we actually see that $R(-)$ is a functor from sSet to sSet, acting on simplicial maps $\phi : X \to Y$ in the obvious way to get a simplicial map $R(\phi) : RX \to RY$. We write $R^2 X$ for RRX, and in general $R^n X$ for $R(R^{n-1} X)$.

We want to turn the collection of $R^n X$ for $n \geq 1$ into a cosimplicial space, so we need maps $R^n X \to R^{n+1} X$ and $R^n X \to R^{n-1} X$. We have a map $\phi_{R^{n-i+1}X} : R^{n-i+1}X \to R^{n-i}X$ for any $0 \le i \le n$, and so applying the functor $R(-)$ to this map i times yields $n + 1$ different maps $d^i : R^n X \to R^{n+1}X$.

In the opposite direction, we have a function $\psi_{R^{n-i}X} : R^{n-i}X \to R^{n-i+1}X$ for any $0 \le i \le n - 2$, and so applying the functor $R(-)$ to this map i times yields $n - 1$ different maps $s^i : R^n X \to R^{n-1}X$. The maps d^i and s^i describe the coface and codegeneracy maps for a cosimplicial space, whose n-cosimplices will be the set $R^{n+1}X$.

Definition 3.51 Let X be a space. The cosimplicial space $R^{\cdot}X$ has as n-cosimplices $R^{n+1}X$ (so that the 0-cosimplices are RX), with coface maps $d^i : (R^{\cdot}X)^n \to (R^{\cdot}X)^{n+1}$ and codegeneracy maps $s^i : (R^{\cdot}X)^n \to (R^{\cdot}X)^{n-1}$ given above.

With this definition, we may now give the R-completion.

Definition 3.52 Let X be a space. The R-*completion* of X is the space

$$R_\infty X = \mathrm{Tot}(R^{\cdot}X).$$

If $R = \mathbb{F}_p$, we write $X_p^\wedge$ for $R_\infty X$.

If $f : X \to Y$ is a simplicial map, then f induces a map of cosimplicial spaces $R^{\cdot}X \to R^{\cdot}Y$ in the obvious way; since taking total spaces is functorial, this means that R_∞ is a functor. What is more difficult to see is that there is a simplicial map $X \to R_\infty X$. The first step of this is to augment the cosimplicial space $R^{\cdot}X$.

Definition 3.53 Let $X^{\cdot}$ be a cosimplicial space. An *augmentation* of $X^{\cdot}$ is a pair $(X^{\cdot}, X^{-1})$, together with a simplicial map $d^0 : X^{-1} \to X^0$, such that $d^0 d^0 = d^1 d^0$ as maps $X^{-1} \to X^1$.

The reason for introducing it here is that $R^{\cdot}X$ is augmented, with augmentation $(R^{\cdot}X, X)$; the map d^0 is simply the map $\phi : X \to RX$ that we saw earlier. If $(X^{\cdot}, X^{-1})$ is an augmented cosimplicial space, then there is an induced (simplicial) map from X^{-1} to $\mathrm{Tot}\,X^{\cdot}$, and hence there is a map $X \to R_\infty X$.

Having now constructed the R-completion of a simplicial set X, the reader might well see why it could be considered difficult to get an intuition for the R-completion of a space, and so people often treat completion as a 'black box', i.e., as a functor $R_\infty : X \to R_\infty X$ that has a list of properties. Some of these properties we will discuss now. By and large they will be far too difficult to prove, and we will simply give a reference to the original proof in [BK72].

In order to describe their properties, we will split all spaces into two collections, R-good and R-bad.

Definition 3.54 A space X is *R-complete* if the map $X \to R_\infty X$ is a weak equivalence. A space X is *R-good* if the map $X \to R_\infty X$ is an R-homology equivalence, and X is *R-bad* if it is not R-good.

It turns out that the R-completion of a space is always fibrant [BK72, I.4.2], and so a weak equivalence between R-completions is a homotopy equivalence; hence completing an R-completion of a space gives you back a homotopy-equivalent space. One of the fundamental properties about R-completions is the following.

Proposition 3.55 (Bousfield–Kan [BK72, Lemma I.5.5]) *Let X and Y be spaces. If $f : X \to Y$ is a continuous map of spaces, then f is an R-homology equivalence if and only if $R_\infty f : R_\infty X \to R_\infty Y$ is a homotopy equivalence.*

If X is R-good, then the map $f : X \to R_\infty X$ is an R-homology equivalence, and so the map $R_\infty f : R_\infty X \to R_\infty(R_\infty X)$ is a homotopy equivalence; this means that R-completing an R-good space gives you an R-complete space. In fact, the following is true.

Proposition 3.56 (Bousfield–Kan [BK72, Proposition I.5.2]) *Let X be a space. The following are equivalent:*

  (i) *X is R-good;*
 (ii) *$R_\infty X$ is R-complete;*
(iii) *$R_\infty X$ is R-good.*

Therefore if X is R-bad then completing it will always result in an R-bad space; so all spaces are either R-good or R-bad, and R-completing does not mix the two collections.

Also, if X is R-good, and $f : X \to Y$ is an R-homology equivalence, then $R_\infty X$ is R-complete, and $R_\infty f$ is a homotopy equivalence. Therefore the map $Y \to R_\infty Y$ must be an R-homology equivalence, and so Y is R good as well.

Thus being R-good is invariant under R-homology equivalences. We also need some conditions on whether various spaces are R-good, for interesting rings R (like $R = \mathbb{Z}$ and $R = \mathbb{F}_p$).

Theorem 3.57 (Bousfield–Kan [BK72, Prop. VII.4.3.iii, Prop. VII.5.1])
Let X be a space.

(i) *If X is connected and $\pi_n(X)$ is finite for all $n \geq 1$, then X is R-good for any subring R of $\mathbb{Q}$.*
(ii) *If X is connected and $\pi_1(X)$ is finite, then X is $\mathbb{F}_p$-good.*

This means that classifying spaces of finite groups, whose homotopy groups are finite by Proposition 3.37, are $\mathbb{F}_p$-good for any prime p and $\mathbb{Z}$-good. (Recall that we write $X_p^\wedge$ for $(\mathbb{F}_p)_\infty X$.) The idea is to find an analogue for all finite groups of the fact that nilpotent groups G are the direct products of their Sylow p_i-subgroups P_i, and so

$$BG \simeq BP_1 \times BP_2 \times \cdots \times BP_n.$$

For non-nilpotent groups this cannot be done, but we have the related decomposition given below.

Proposition 3.58 *Let X be a connected space such that $\pi_n(X)$ is finite for all $n \geq 1$. The natural map*

$$\mathbb{Z}_\infty X \to \prod_p X_p^\wedge$$

is a homotopy equivalence.

Proof   This is a special case of [BK72, Proposition VII.4.1].   □

If $X = BG$ then we get

$$\mathbb{Z}_\infty BG \simeq \prod_p BG_p^\wedge;$$

the classifying space of a finite group is $\mathbb{Z}$-complete if and only if the group is nilpotent [BS08, Remark 4.3.3], and it is $\mathbb{F}_p$-complete if and only if the group is a p-group. Thus the decomposition above is a generalization of the decomposition of BG where G is nilpotent.

We end this section discussing homology and cohomology. We know that the map $BG \to \mathbb{Z}_\infty BG$ induces an isomorphism in $\mathbb{Z}$-homology, and that the map $BG \to BG_p^\wedge$ induces an isomorphism in $\mathbb{F}_p$-homology.

Theorem 3.59 (See [BS08, Theorem 2.4.8])   *Let R be a PID, and X and Y be spaces. If $f : X \to Y$ is an R-homology equivalence then it is an R-cohomology equivalence. If $X = BG$ for G finite, and $R = \mathbb{Z}$ or $\mathbb{F}_p$, then the converse also holds.*

This theorem has the following obvious corollary.

Corollary 3.60 *Let G be a finite group, and let p be a prime. The map $BG \to \mathbb{Z}_\infty BG$ induces isomorphisms in $\mathbb{Z}$-homology and $\mathbb{Z}$-cohomology, and the map $BG \to BG_p^\wedge$ induces isomorphisms in $\mathbb{F}_p$-homology and $\mathbb{F}_p$-cohomology.*

Generally, one may treat the p-completion functor as being a 'black box', which satisfies the various properties above. One more corollary that we will need later is the following criterion.

Corollary 3.61 *Let X and Y be $\mathbb{F}_p$-good spaces. There is a homotopy equivalence $X_p^\wedge \simeq Y_p^\wedge$ if and only if there exists a space Z and $\mathbb{F}_p$-homology equivalences*

$$X \leftarrow Z \to Y.$$

3.5 The centric linking systems of groups

The classifying space BG, and its p-completion $BG_p^\wedge$, are fundamental objects of study in the topological side of the theory. The Martino–Priddy conjecture states that $BG_p^\wedge$ is completely determined by $\mathcal{F}_P(G)$, although this proof is difficult and (at present) relies on the classification of the finite simple groups. One can therefore think of the p-completed classifying space $BG_p^\wedge$ as being the classifying space of the fusion system.

If $\mathcal{F}$ is a fusion system that does not come from a finite group, however, there is no classifying space $BG_p^\wedge$ to consider; the missing link between fusion systems and classifying spaces is centric linking systems. These are categories, with a similar construction to the fusion system of a finite group, whose p-completed nerve is homotopy equivalent to $BG_p^\wedge$ when $\mathcal{F} = \mathcal{F}_P(G)$. One may think of them as lying half-way between fusion systems and p-completed classifying spaces.

In this section we describe centric linking systems associated to a group fusion system. For arbitrary saturated fusion systems there is an abstract definition of a centric linking system that we will meet in Chapter 9, but we will not describe it here. (The centric linking systems for groups that we describe are centric linking systems in the sense of the abstract definition in Chapter 9.)

Definition 3.62 Let G be a finite group with Sylow p-subgroup P. A p-subgroup $Q \leq P$ is *p-centric* if

$$C_G(Q) = Z(Q) \times O_{p'}(C_G(Q)).$$

We will use the notation of, for example, [BLO03b], and denote the subgroup $O_{p'}(C_G(Q))$ by $C'_G(Q)$. If Q and R are subgroups of P, define the *transporter* between Q and R, denoted by $T_G(Q, R)$, to be

$$T_G(Q, R) = \{x \in G \mid x^{-1}Qx \leq R\}.$$

(In many places this is denoted by $N_G(Q, R)$, but we will use the notation $N_{\mathcal{F}}(Q, R)$ in the sequel, so we change this to avoid confusion.) It is easy to see that two elements g and h of $T_G(Q, R)$ define the same element of $\mathcal{F}_P(G)$ if and only if $gh^{-1} \in C_G(Q)$. Also, if $g \in T_G(Q, R)$, then for all $x \in C_G(Q)$, we see that $xg \in T_G(Q, R)$. It makes sense then to 'collapse $T_G(Q, R)$ on the left' by $C_G(Q)$. We write

$$\mathrm{Hom}_{\mathcal{F}_P(G)}(Q, R) = T_G(Q, R) / C_G(Q),$$

even though there is no formal quotient group to speak of. (Indeed, this is very much closer to the topological notion of quotienting, which is where one formally identifies various points.)

In the case where Q is a centric subgroup, we not only have $C_G(Q)$ to quotient by, but also this other natural subgroup, $C'_G(Q)$. We could also, of course, not identify morphisms at all.

Definition 3.63 Let G be a finite group and let P be a Sylow p-subgroup of G. The *centric linking system*, $\mathcal{L}^c_P(G)$, is a category whose objects are all p-centric subgroups of P, and whose morphism sets are given by

$$\mathrm{Hom}_{\mathcal{L}^c_P(G)}(Q, R) = T_G(Q, R) / C'_G(Q).$$

The *transporter system*, $\mathcal{T}_P(G)$, is the category whose objects are all subgroups of P, and whose morphism sets are given by

$$\mathrm{Hom}_{\mathcal{T}_P(G)}(Q, R) = T_G(Q, R).$$

In the case of $\mathcal{F}_P(G)$, we are identifying elements that factor through any element of $C_G(Q)$, and in the case of $\mathcal{L}^c_P(G)$, we are identifying elements that factor through a p'-element of $C_G(Q)$. Thus the same homomorphism $Q \to R$ will be labelled by different elements of G.

The main results linking the centric linking system of a finite group and its p-completed classifying space are given in [BLO03a]. In particular, their Theorem A is the main result in this direction.

Theorem 3.64 (Broto–Levi–Oliver [BLO03a, Theorem A]) *Let G and H be finite groups, with Sylow p-subgroups P and Q respectively. The* *p-completed classifying spaces $BG_p^\wedge$ and $BH_p^\wedge$ are homotopy equivalent if and only if $\mathcal{L}_P^c(G)$ and $\mathcal{L}_Q^c(H)$ are equivalent (as categories).*

The proof uses different techniques for each direction: to prove that a homotopy equivalence between p-completed classifying spaces implies that the centric linking systems are equivalent, Broto, Levi, and Oliver construct a centric linking system $\mathcal{L}_p^c(X)$ for a space X, a homotopy invariant of X, and then prove that the categories $\mathcal{L}_P^c(G)$ and $\mathcal{L}_p^c(BG_p^\wedge)$ are equivalent. Therefore, if $BG_p^\wedge$ and $BH_p^\wedge$ are homotopy equivalent, the categories $\mathcal{L}_P^c(G)$ and $\mathcal{L}_Q^c(H)$ are equivalent.

The converse involves proving the following result, alluded to earlier.

Proposition 3.65 (Broto–Levi–Oliver [BLO03a, Proposition 1.1]) *Let G be a finite group and let P be a Sylow p-subgroup of G. Then*

$$BG_p^\wedge \simeq |\mathcal{L}_P^c(G)|_p^\wedge.$$

The proof of this proposition employs Corollary 3.61. Let α denote the map $\mathcal{T}_P(G) \to \mathscr{B}(G)$ given by sending each object to o_G and each morphism to the corresponding morphism in $\mathscr{B}(G)$, so that a morphism and its image are labelled by the same element. The induced continuous map $|\alpha| : |\mathcal{T}_P(G)| \to BG$ is an $\mathbb{F}_p$-homology equivalence, and consequently

$$|\mathcal{T}_P(G)|_p^\wedge \simeq BG_p^\wedge.$$

(This is proved in Lemma 1.2 of [BLO03a].) Having proved that there is an $\mathbb{F}_p$-homology equivalence between the p-completed nerve of the transporter category and BG, we now need a map from the full subcategory $\mathcal{T}_P^c(G)$ of $\mathcal{T}_P(G)$ on the p-centric subgroups, to $\mathcal{L}_P^c(G)$, that induces an isomorphism of $\mathbb{F}_p$-homology. The obvious surjection performs this function [BLO03a, Lemma 1.3], and hence, by Corollary 3.61,

$$|\mathcal{L}_P^c(G)|_p^\wedge \simeq BG_p^\wedge.$$

Therefore if $\mathcal{L}_P^c(G)$ and $\mathcal{L}_Q^c(H)$ are equivalent, $BG_p^\wedge$ and $BH_p^\wedge$ are equivalent, proving the other half of Theorem 3.64.

Now we know that the p-completed classifying space of a finite group is determined by the centric linking system and vice versa, but the Martino–Priddy conjecture states that $BG_p^\wedge$ – and hence now the centric linking system – should be determined actually by the fusion system.

This necessitates a theorem that says something like the following: two groups G and H possess the same fusion system (at the prime p) if and only if they possess the same centric linking system. One way to attack this problem is to define a centric linking system associated to an arbitrary (saturated) fusion system $\mathcal{F}$, and then prove that centric linking systems exist and are unique.

For centric linking systems associated with abstract fusion systems (rather than group fusion systems), an obstruction theory regarding their existence and uniqueness has been developed [BLO03b, Section 3]. In general, the uniqueness, and even existence, of centric linking systems is open at present, although for odd primes some work has been done in this direction; see Section 9.6 for more details.

Of course, for group fusion systems the existence of centric linking systems is obvious, since we have defined them, but the uniqueness is not guaranteed from the outset. By reducing the question to simple groups, Oliver used the classification of the finite simple groups to prove that the relevant obstruction groups vanish, and proved the following result.

Theorem 3.66 (Oliver [Oli04] [Oli06]) *If $\mathcal{F}$ is a saturated fusion system that is the fusion system of a finite group, then there is a unique (up to equivalence) centric linking system associated with $\mathcal{F}$.*

Theorems 3.64 and 3.66 together prove the Martino–Priddy conjecture.

3.6 Constrained fusion systems

The local theory of finite groups in some sense can be split into two parts: the theory of p-constrained groups and the theory of groups that aren't p-constrained.

Definition 3.67 Let p be a prime, let G be a finite group, and write $H = G/\mathrm{O}_{p'}(G)$. We say that G is *p-constrained* if

$$\mathrm{C}_H(\mathrm{O}_p(H)) \leq \mathrm{O}_p(H).$$

Since constrained groups play such an important role in the local theory of finite groups, it is essential to find an analogue for fusion systems. By Exercise 1.7, the groups G and $G/\mathrm{O}_{p'}(G)$ have the same fusion system, and so we don't need to worry about this technicality at least. We need to find the analogue of the subgroup $\mathrm{O}_p(G)$, and the analogue of $\mathrm{C}_G(Q)$ for some subgroup Q, or at least the condition that $\mathrm{C}_G(Q) \leq Q$. This section, like Section 1.5, acts as a preview of some notions to come, and so it is not strictly necessary to follow the definitions here, as they will be repeated later.

If a subgroup H of a finite group G is normal and g is any element of G, then any conjugation map $c_g : Q \to R$ extends to a map $c_g : HQ \to HR$ and, since H is normal, the restriction of c_g to H acts as an automorphism. This is exactly the condition that we want.

Definition 3.68 Let $\mathcal{F}$ be a fusion system on a finite p-group P. A subgroup Q of P is *normal* in $\mathcal{F}$ if, whenever $\phi : R \to S$ is a morphism in $\mathcal{F}$, there is a map $\psi : QR \to QS$ in $\mathcal{F}$, with $\psi|_Q$ an automorphism of Q and $\psi|_R = \phi$.

It turns out that there is a largest normal subgroup of a fusion system, which we will denote by $O_p(\mathcal{F})$; this is our analogue of $O_p(G)$. We need the analogue of $C_G(O_p(G)) \leq O_p(G)$ as well; the actual condition that we want is that $O_p(\mathcal{F})$ is $\mathcal{F}$-centric, but this isn't defined until Section 4.5. In the special case of this subgroup, we can make do with the simple statement that $C_P(O_p(\mathcal{F})) \leq O_p(\mathcal{F})$. If $O_p(\mathcal{F})$ satisfies this condition, we say that $\mathcal{F}$ contains a *normal, centric* subgroup. We make the next definition only for *saturated* fusion systems.

Definition 3.69 Let $\mathcal{F}$ be a saturated fusion system on a finite p-group P. We say that $\mathcal{F}$ is *constrained* if $C_P(O_p(\mathcal{F})) \leq O_p(\mathcal{F})$.

The fundamental theorem on constrained fusion systems is that they are always fusion systems of finite groups, and in fact we have much more than that.

Theorem 3.70 ([BCGLO05, Proposition C]) *Let $\mathcal{F}$ be a saturated fusion system on a finite p-group P. If $\mathcal{F}$ is constrained then there is a unique finite group G, which has P as a Sylow p-subgroup, such that the following three conditions hold:*

(i) $\mathcal{F} = \mathcal{F}_P(G)$;
(ii) $O_{p'}(G) = 1$;
(iii) $C_G(O_p(G)) \leq O_p(G)$.

This group G is called the model *for $\mathcal{F}$.*

This theorem says that a constrained fusion system is modelled by an essentially unique, p-constrained finite group. In the conditions above, the first one just says that G has fusion system $\mathcal{F}$, the third says that G is p-constrained, and the second says that G is p'-reduced.

Definition 3.71 Let G be a finite group, and let π be a set of primes. We say that G is *π-reduced* if $O_\pi(G) = 1$.

We will often state Theorem 3.70 as 'a constrained fusion system is modelled by a unique p-constrained, p'-reduced finite group'. This unique group will generally be denoted by $L^{\mathcal{F}}$. It is difficult to overstate the importance of this theorem in the theory of fusion systems, and it is in some sense disappointing (at least for algebraists) that there is no algebraic proof of this theorem. However, since the only method we can currently envisage involves constructing an extension of one group by another (namely $Z(O_p(\mathcal{F}))$ by $\mathrm{Aut}_{\mathcal{F}}(O_p(\mathcal{F}))$) it is not surprising that cohomology comes into the picture somewhere.

Example 3.72 Let $\mathcal{F}$ be a saturated fusion system on an abelian p-group P. By Exercise 1.8, every morphism in $\mathcal{F}$ extends to an automorphism of P that lies in $\mathcal{F}$, and so P is a normal subgroup of $\mathcal{F}$. Therefore $O_p(\mathcal{F}) = P$. Obviously $C_P(P) \leq P$, and so $\mathcal{F}$ is constrained. Thus every saturated fusion system on an abelian p-group is the fusion system of a unique p-constrained group.

Let $G = A_5$, the alternating group on five letters. This has an abelian Sylow 2-subgroup P, and it is not difficult to see that the only (non-trivial) morphisms of the fusion system $\mathcal{F}_P(G)$ are those of order 3, permuting the three non-trivial elements of P transitively. However, G is not 2-constrained, and so G is *not* the unique 2-constrained model for $\mathcal{F}_P(G)$; this is the alternating group A_4.

In general, if $A = \mathrm{Aut}_{\mathcal{F}}(P)$, then the model of the saturated fusion system $\mathcal{F}$ on P is $P \rtimes A$ (see Exercise 3.7).

We will not prove Theorem 3.70 here, as the proof is involved. However, we will make some remarks about the situation. Let G be a p-constrained group with $O_{p'}(G) = 1$, write $Q = O_p(G)$. Since $Z(Q) = C_G(Q) \leq Q$, we see that G is an extension of $Z(Q)$ by $\mathrm{Aut}_G(Q)$. If $\mathcal{F}$ denotes the fusion system $\mathcal{F}_P(G)$, then $\mathrm{Aut}_{\mathcal{F}}(Q) = \mathrm{Aut}_G(Q)$. We have that G is an extension

$$1 \to Z(Q) \to G \to \mathrm{Aut}_G(Q) \to 1,$$

and this determines some 2-cocycle γ in $H^2(\mathrm{Aut}_G(Q), Z(Q))$.

Since the group $\mathrm{Aut}_{\mathcal{F}}(Q) = \mathrm{Aut}_G(Q)$ is determined by the fusion system $\mathcal{F}$, what we are really saying is that if γ and γ' are 2-cocycles in $H^2(\mathrm{Aut}_{\mathcal{F}}(Q), Z(Q))$ such that the associated group extensions G and H both have fusion system $\mathcal{F}$, then $G \cong H$.

In fact, even more is true. The map in cohomology

$$H^2(\mathrm{Aut}_{\mathcal{F}}(Q), \mathrm{Z}(Q)) \to H^2(\mathrm{Aut}_P(Q), \mathrm{Z}(Q))$$

is injective, since $\mathrm{Aut}_P(Q)$ is a Sylow p-subgroup of $\mathrm{Aut}_{\mathcal{F}}(Q)$ [Ben98b, Corollary 3.6.18]. Therefore the group G is determined up to isomorphism by the group $\mathrm{Aut}_{\mathcal{F}}(Q)$ and the p-group extension

$$1 \to \mathrm{Z}(Q) \to \mathrm{N}_P(Q) \to \mathrm{Aut}_P(Q) \to 1.$$

This particular statement will be important in Section 7.3.

Another useful corollary to Theorem 3.70 is the following.

Corollary 3.73 *Let $\mathcal{F}$ be a saturated fusion system on a finite p-group P, and suppose that $\mathcal{F}$ is constrained. If G is the unique p-constrained, p'-reduced group with $\mathcal{F} = \mathcal{F}_P(G)$, and H is another finite group with $|H| < |G|$, then H cannot have fusion system $\mathcal{F}$. In particular, if $H < G$ then H does not have fusion system $\mathcal{F}$.*

Exercises

3.1 Prove Proposition 3.3.

3.2 Prove Proposition 3.10.

3.3 Let X be a fibrant simplicial set. Using similar methods to the proof that the multiplication on $\pi_n(X)$ is well defined for a simplicial set X, prove that $\pi_n(X)$ is an abelian group if $n > 1$.

3.4 A horn Λ_i^n is *inner* if $1 \leq i \leq n-1$, and if $i = 0, n$ then it is *outer*. Let $\mathcal{C}$ be a small category. Prove that fillings of inner horns for $N\mathcal{C}$ exist and are unique.

3.5 Let X be a fibrant simplicial set, with basepoints $*$ and $*'$. Prove that $\pi_0(X)$ is in bijection with the path-connected components of $|X|$ and that, if $\pi_0(X)$ has exactly one element, then $\pi_i(X, *) = \pi_i(X, *')$ for all $i > 0$. (If $\pi_0(X)$ has exactly one element we say that X is *path connected*.)

3.6 Let X be a fibrant simplicial set, and let $*$ denote a basepoint. Identify $*$ with its image in $|X|$. Convince yourself that $\pi_1(X, *) \cong \pi_1(|X|, *)$.

3.7 Let $\mathcal{F}$ be a saturated fusion system on the abelian p-group P, and write $A = \mathrm{Aut}_{\mathcal{F}}(P)$. Prove that the unique p-constrained, p'-reduced group G such that $\mathcal{F} = \mathcal{F}_P(G)$ is $P \rtimes A$.

Part II

The theory

4

Fusion systems

In Chapter 1, we were briefly introduced to the definition of a saturated fusion system, which perhaps did not seem particularly natural or easy to follow at the time. Here we will give a simplified definition of saturated fusion systems, not involving fully normalized subgroups, but focusing on how maps extend. It is hoped that this new definition, due to Roberts and Shpectorov [RoS09], is more intuitive than the previous definitions, due to Puig [Pui06], Broto, Levi, and Oliver [BLO03b], and Stancu [Sta06]. (Note that they are all equivalent.)

Because the previous 'standard' definitions appear throughout the literature, and the concepts involved in these definitions are important in the subject, we will introduce the notions of fully normalized and fully centralized subgroups in Section 4.2. Here we will start proving our first theorems about fusion systems, and saturated fusion systems in particular: saturated fusion systems offer the structurally rich definition necessary for a satisfying theory to be developed.

Section 4.3 sets out to prove the equivalence of the different definitions that abound in the literature. One aspect of this array of possible definitions is that some are easier to use in different situations, and so it can be helpful to have these different definitions to hand: proving that a fusion system *is* saturated might benefit from a simpler criterion, whereas proving that a fusion system *isn't* saturated could use one of the more *a priori* stronger conditions.

Section 4.4 introduces the centralizer and normalizer subsystems, both crucial to the local theory of fusion systems; this mirrors the importance of normalizers in the local theory of finite groups. Section 4.5 introduces the concepts of centric, radical, and essential subgroups, before these are used in Section 4.6 to prove the version for fusion systems of Alperin's fusion theorem.

We end with the concept of weak and strong closure for fusion systems, which is a straight generalization of that for finite groups. Strong closure in particular is of great importance in the theory of fusion systems because of the idea of quotients, which have as kernels strongly closed subgroups.

4.1 Saturated fusion systems

One of the ideas of a fusion system is to provide an abstract framework for arguments concerning fusion in finite groups. By removing the ambient group, and being left with a theory of fusion systems, it might make it clearer why things are true. Another goal is to unite somewhat the theory of finite groups and of blocks of finite groups, via the fusion systems of them.

Definition 4.1   Let P be a finite p-group. A *fusion system* $\mathcal{F}$ on P is a category whose objects are all subgroups of P, and whose morphisms $\mathrm{Hom}_{\mathcal{F}}(Q, R)$ are sets of injective homomorphisms $Q \to R$, with composition of morphisms given by the usual composition of homomorphisms. The sets $\mathrm{Hom}_{\mathcal{F}}(Q, R)$ should satisfy the following three axioms:

(i) for each $g \in P$ with $Q^g \leq R$, the associated conjugation map $c_g : Q \to R$ is in $\mathrm{Hom}_{\mathcal{F}}(Q, R)$;

(ii) for each $\phi \in \mathrm{Hom}_{\mathcal{F}}(Q, R)$, the associated isomorphism $Q \to Q\phi$ lies in $\mathrm{Hom}_{\mathcal{F}}(Q, Q\phi)$;

(iii) if $\phi \in \mathrm{Hom}_{\mathcal{F}}(Q, R)$ is an isomorphism, then its inverse $\phi^{-1} : R \to Q$ lies in $\mathrm{Hom}_{\mathcal{F}}(R, Q)$.

If $Q \to R$ is an isomorphism in $\mathcal{F}$, then we say that Q and R are *$\mathcal{F}$-conjugate* or *$\mathcal{F}$-isomorphic*.

The idea of a fusion system is to provide a model for conjugation of p-subgroups of a fixed Sylow p-subgroup of a finite group. In this case, the first axiom is obvious, since it simply states that the conjugation already present in the Sylow p-subgroup P should exist in any fusion system $\mathcal{F}$. The second and third axioms are there essentially to make the notion of $\mathcal{F}$-conjugacy into an equivalence relation, which is of course true for G-conjugacy in a finite group G.

Notice that these axioms aren't particularly restrictive: what we mean is that given any collection of injective homomorphisms between subgroups of P, we can get a fusion system containing it by adding inverses, restrictions, and conjugation maps by elements of P.

This considerable flexibility in the structure of a fusion system is not good in terms of making things look like groups: for example, suppose that $\mathcal{F}$ is a fusion system on $P = D_8$ coming from a finite group G. Since $\mathrm{Aut}_G(P) = \mathrm{N}_G(P)/\mathrm{C}_G(P)$ and $|\mathrm{Aut}(P)| = 8$, we must have that $|\mathrm{Aut}_G(P)| = 4$. However, there are plenty of fusion systems $\mathcal{F}$ on P for which $\mathrm{Aut}_{\mathcal{F}}(P) = \mathrm{Aut}(P)$, for example the following fusion system.

Definition 4.2 Let P be a finite p-group. The *universal* fusion system on P, denoted by $\mathcal{U}(P)$, is the fusion system for which $\mathrm{Hom}_{\mathcal{U}(P)}(Q, R)$ consists of *all* injective homomorphisms $Q \to R$.

As with all algebraic objects, we must have a notion of a 'subobject', and a notion of 'generation'. We will define these now.

Definition 4.3 Let $\mathcal{F}$ be a fusion system on a finite p-group P.

(i) A *subsystem* $\mathcal{E}$ of $\mathcal{F}$ is a subcategory that is itself a fusion system, in the sense that it satisfies the three axioms to be a fusion system. If the objects of $\mathcal{E}$ are all subgroups of a subgroup Q of P then $\mathcal{E}$ is *defined* on Q or simply to be on Q.

(ii) If $\mathcal{E}_1$ and $\mathcal{E}_2$ are subsystems of $\mathcal{F}$, defined on subgroups Q_1 and Q_2 of P respectively, then the *intersection* is the subsystem of $\mathcal{F}$ with objects all subgroups of $Q = Q_1 \cap Q_2$ and as morphisms all morphisms that lie in both $\mathcal{E}_1$ and $\mathcal{E}_2$. It will be denoted as usual by $\mathcal{E}_1 \cap \mathcal{E}_2$.

(iii) Let $\mathcal{C}$ be a collection of morphisms from $\mathcal{F}$ between subgroups of a subgroup Q of P. The fusion system *generated by* $\mathcal{C}$ is the intersection $\mathcal{E}$ of all subsystems of $\mathcal{F}$ that contain all morphisms in the collection $\mathcal{C}$. In other words, $\mathcal{E}$ is the subcategory of $\mathcal{F}$ consisting of all morphisms in $\mathcal{C}$ and $\mathcal{F}_Q(Q)$, and all morphisms obtained from these by composition, restriction of the codomain, and taking inverses. The subsystem generated by $\mathcal{C}$ is denoted by $\langle \mathcal{C} \rangle$.

If we want fusion systems to look like groups, then we need to get rid of fusion systems such as $\mathcal{U}(D_8)$, and this is the reasoning behind the definition of saturated fusion systems. We will return to these later, but for now let's consider some of the very first properties of fusion systems.

Proposition 4.4 *Let $\mathcal{F}$ be a fusion system on a finite p-group P, and let $\phi : Q \to R$ be a morphism in $\mathcal{F}$. Let S be any subgroup of Q,* *and let T be any subgroup of P containing $S\phi$. There is a morphism $\psi : S \to T$ such that ϕ and ψ agree on S. Thus, given a morphism, one may restrict the domain arbitrarily, and extend or constrict the codomain to any overgroup of the image of the restriction.*

Proof If S is a subgroup of Q, then there is an inclusion map $\iota : S \to Q$, coming from the fusion system $\mathcal{F}_P(P)$. The composition $\iota \circ \phi$ also lies in $\mathcal{F}$, since compositions of maps in $\mathcal{F}$ also lie in $\mathcal{F}$, and this composition is the restriction of ϕ to S. We have already mentioned that one may constrict or extend the codomain as claimed in the proposition.   □

When we defined the notion of $\mathcal{F}$-conjugacy, we were lax about which direction the map went; we simply said that Q and R were $\mathcal{F}$-conjugate. There is a reason behind this.

Proposition 4.5 *Let $\mathcal{F}$ be a fusion system on a finite p-group P. The relation of $\mathcal{F}$-conjugacy is an equivalence relation on the set of all subgroups of P.*

Proof Certainly $\mathcal{F}$-conjugacy is reflexive since all identity maps, and in particular the identity automorphism, are in $\mathcal{F}$. It is symmetric since, by axiom (iii) of a fusion system, if $\phi : Q \to R$ is an isomorphism in $\mathcal{F}$, so is $\phi^{-1} : R \to Q$. Finally, since $\mathcal{F}$ is a category it is closed under composition of maps, and therefore if $\phi : Q \to R$ and $\psi : R \to S$ are $\mathcal{F}$-isomorphisms so is their composition. Therefore $\mathcal{F}$-conjugacy is an equivalence relation, as claimed.   □

Now suppose that Q and R are two $\mathcal{F}$-conjugate subgroups of P, and suppose that there are two different isomorphisms $\phi, \psi : Q \to R$. Since the inverses of ϕ and ψ are in $\mathcal{F}$, the compositions $\phi \circ \psi^{-1}$ and $\phi^{-1} \circ \psi$ are in $\mathcal{F}$: the first is an automorphism of Q, and the second is an automorphism of R. This means that any two $\mathcal{F}$-isomorphisms between two subgroups differ by an automorphism, either of the domain or of the image. There appears therefore to be a relationship between the automorphisms of $\mathcal{F}$-conjugate subgroups. We also note that the automorphisms of a subgroup carry a natural group structure; the relationship between the automorphism groups of $\mathcal{F}$-conjugate subgroups is total, in the following sense.

Proposition 4.6 *Let $\mathcal{F}$ be a fusion system on a finite p-group P. If Q and R are $\mathcal{F}$-conjugate subgroups under an isomorphism ϕ, then $\mathrm{Aut}_{\mathcal{F}}(Q)$ and $\mathrm{Aut}_{\mathcal{F}}(R)$ are isomorphic under the map $\psi \mapsto \phi^{-1}\psi\phi$.*

Proof Let $\phi : Q \to R$ be an isomorphism in $\mathcal{F}$. For $\psi \in \text{Aut}_{\mathcal{F}}(Q)$, denote by ψ^ϕ the automorphism

$$\phi^{-1}\psi\phi \in \text{Aut}_{\mathcal{F}}(R).$$

We claim that the map $f : \psi \mapsto \psi^\phi$ is an isomorphism. Indeed, it has an inverse $f' : \text{Aut}_{\mathcal{F}}(R) \to \text{Aut}_{\mathcal{F}}(Q)$ given by

$$f' : \theta \mapsto \phi\theta\phi^{-1}.$$

Firstly, notice that the domains and codomains of f and f' are correct, since the compositions of morphisms in $\mathcal{F}$ are also in $\mathcal{F}$. Next, we see that both $f \circ f'$ and $f' \circ f$ are the identity on their respective domains, and so both f and f' are bijections. It is clear that f is a homomorphism since

$$(\psi\theta)^\phi = \phi^{-1}\psi\theta\phi = \phi^{-1}\psi\phi\phi^{-1}\theta\phi = \psi^\phi\theta^\phi,$$

and so f is an isomorphism, as claimed. □

The idea is that the p-automorphisms of a subgroup should come from the Sylow p-subgroup itself, and the p'-automorphisms should come from the rest of the group. However, this cannot be true as stated, since Proposition 4.6 will give us some problems here; it is not hard to imagine two $\mathcal{F}$-conjugate subgroups – Q and R – whose automizers – $\text{N}_P(Q)/\text{C}_P(Q)$ and $\text{N}_P(R)/\text{C}_P(R)$ – are different. (For example, take the universal fusion system on the 2-group $P = D_8 \times C_4$, and consider the C_4 direct factor Q and the cyclic subgroup R of index 2 inside the D_8 factor; $\text{Aut}_P(R)$ has order 2 and $\text{Aut}_P(Q)$ is trivial, whereas in $\mathcal{U}(P)$ the two are conjugate, so $|\text{Aut}_{\mathcal{F}}(R)| = |\text{Aut}_{\mathcal{F}}(Q)|$.) By Proposition 4.6, if Q has a smaller automizer in P than R, it will inherit some p-automorphisms from R.

It seems as though we should demand, therefore, that the subgroup Q whose automizer is maximal amongst the subgroups $\mathcal{F}$-conjugate to it should have the property that this automizer, $\text{Aut}_P(Q)$, is a Sylow p-subgroup of $\text{Aut}_{\mathcal{F}}(Q)$. This motivates the following definition.

Definition 4.7 Let $\mathcal{F}$ be a fusion system on a finite p-group P. A subgroup Q of P is *fully $\mathcal{F}$-automized* if $\text{Aut}_P(Q) \cong \text{N}_P(Q)/\text{C}_P(Q)$ is a Sylow p-subgroup of $\text{Aut}_{\mathcal{F}}(Q)$. (If the fusion system involved is obvious, we will merely talk about fully automized subgroups, rather than fully $\mathcal{F}$-automized subgroups.)

It is not clear that an $\mathcal{F}$-conjugacy class of subgroups should have an element that is fully automized; indeed, for arbitrary fusion systems it is not true as, for example, the ambient p-group itself may not be fully automized, in the universal system $\mathcal{U}(D_8)$, for example. The definition of a saturated fusion system includes the statement that every $\mathcal{F}$-conjugacy class of subgroups contains a fully automized member.

In Proposition 4.4 we showed that any morphism in a fusion system may be restricted to any subgroup of the domain. The converse, that a morphism may be extended to any overgroup of the domain, is not only false, but well beyond what is reasonable. For example, if one takes the symmetric group $G = S_4$ with Sylow 2-subgroup P, then conjugation by $(2, 3, 4)$ acts by cycling the three elements $(1, 2)(3, 4)$, $(1, 3)(2, 4)$, and $(1, 4)(2, 3)$, so has order 3; this conjugation action does not lift to an automorphism of P, firstly because $(2, 3, 4)$ does not fix a Sylow 2-subgroup, and secondly because $\mathrm{Aut}(P)$ has order 8. Thus this conjugation action cannot lift to P, so we cannot expect to be able to lift maps completely, even in fusion systems of groups.

To understand extensions of morphisms, we need to get a better handle on the automorphism groups of the subgroups of P. Our conventions on notation for automorphisms are those of Chapter 1, so see the start of that chapter. If X is a subset of $\mathrm{Aut}(P)$ then we denote by X^ϕ the image of X under this induced map ϕ.

Notice firstly that if an isomorphism ϕ in $\mathcal{F}$ with domain Q has an extension $\bar{\phi}$ with domain $R > Q$, then $\mathrm{N}_R(Q) > Q$ and so ϕ extends to a subgroup of $\mathrm{N}_P(Q)$. Therefore ϕ has an extension to a proper overgroup in P if and only if it extends to a proper overgroup in $\mathrm{N}_P(Q)$. However, extensions inside $\mathrm{N}_P(Q)$ are much easier to understand, as we shall see.

Proposition 4.8 *Let $\mathcal{F}$ be a fusion system on a finite p-group P, and let Q and R be $\mathcal{F}$-conjugate subgroups, via an isomorphism ϕ. Suppose that ϕ extends to a morphism $\bar{\phi} : S \to P$ in $\mathcal{F}$, with $S \leq \mathrm{N}_P(Q)$. The image of $\bar{\phi}$ is contained within $\mathrm{N}_P(R)$, and we have that*

$$Sc_Q \leq \mathrm{Aut}_P(Q) \cap \mathrm{Aut}_P(R)^{\phi^{-1}}.$$

Proof Let x be an element of S. For all $g \in Q$ we have $g^x \in Q$, and thus $(g\bar{\phi})^{x\bar{\phi}} = (g^x)\bar{\phi} \in R$; therefore $x\bar{\phi} \in \mathrm{N}_P(R)$. This proves that the image of $\bar{\phi}$ is in $\mathrm{N}_P(R)$.

In particular, $(x\bar{\phi})c_R \in \mathrm{Aut}_P(R)$, and therefore

$$(Sc_Q)^\phi = (S\bar{\phi})c_R \leq \mathrm{Aut}_P(R).$$

Since $Sc_Q \leq \mathrm{Aut}_P(Q)$ obviously, we have that

$$Sc_Q \leq \mathrm{Aut}_P(Q) \cap \mathrm{Aut}_P(R)^{\phi^{-1}},$$

as claimed. $\qquad\qquad\qquad\qquad\qquad\qquad\qquad\qquad\qquad\qquad\qquad$ □

What this proposition says is that some isomorphism $\phi : Q \to R$ need not extend to a given overgroup, even to a nice overgroup like $\mathrm{N}_P(Q)$, and that the extension of ϕ within $\mathrm{N}_P(Q)$ is controlled strictly. Denote by N_ϕ the preimage under the map c_Q of the subgroup $\mathrm{Aut}_P(Q) \cap \mathrm{Aut}_P(R)^{\phi^{-1}}$. Inside $\mathrm{N}_P(Q)$, the map ϕ may extend only as far as N_ϕ.

Definition 4.9   Let $\mathcal{F}$ be a fusion system on a finite p-group P. A subgroup Q is said to be *receptive* if every isomorphism ϕ whose image is Q is extensible to N_ϕ.

This subgroup N_ϕ has an alternative definition: if $\phi : Q \to R$ is an isomorphism, then $g \in N_\phi$ if and only if $c_g = gc_Q$, mapped under ϕ, may be written as $c_h = hc_R \in \mathrm{Aut}_P(R)$ for some $h \in \mathrm{N}_P(R)$. In other words,

$$N_\phi = \{\, g \in \mathrm{N}_P(Q) : \text{there exists } h \in \mathrm{N}_P(R) \text{ with } (Qc_g)\phi = (Q\phi)c_h \,\}.$$

Notice that it is immediate that $Q\,\mathrm{C}_P(Q)$ is always a subgroup of N_ϕ, so that

$$Q\,\mathrm{C}_P(Q) \leq N_\phi \leq \mathrm{N}_P(Q).$$

We use this definition to find some receptive subgroups in group fusion systems $\mathcal{F}_P(G)$.

Proposition 4.10   *Let G be a finite group with a Sylow p-subgroup P, and let Q be a subgroup of P. If $\mathrm{N}_P(Q)$ is a Sylow p-subgroup of $\mathrm{N}_G(Q)$ then Q is receptive.*

Proof   Suppose that $\mathrm{N}_P(Q)$ is a Sylow p-subgroup of $\mathrm{N}_G(Q)$, and let $\phi : R \to Q$ be a morphism in $\mathcal{F}_P(G)$. Since ϕ is induced by conjugation, there is some $g \in G$ such that $\phi = c_g$ on R. In this case, the set N_ϕ is given by

$$N_\phi = \{\, x \in \mathrm{N}_P(R) : \text{there exists } y \in \mathrm{N}_P(Q) \text{ with } Rc_{xg} = Rc_{gy} \,\}.$$

Thus $x \in N_\phi$ if and only if $xgy^{-1}g^{-1}$ centralizes R for some y, and then $g^{-1}xgy^{-1}$ centralizes $R^g = Q$; we see that $x^g = hy$ for some $h \in \mathrm{C}_G(Q)$. Thus

$$(N_\phi)^g \leq \mathrm{N}_P(Q)\,\mathrm{C}_G(Q).$$

Since N_ϕ is a p-subgroup, and $N_P(Q)$ is a Sylow p-subgroup of $N_G(Q)$, there is some element a of $C_G(Q)$ such that $(N_\phi)^{ga}$ is contained within $N_P(Q)$.

Define $\theta : N_\phi \to P$ by $x\theta = x^{ga}$, for all $x \in N_\phi$. Since $a \in C_G(Q)$, θ extends ϕ, and so this is the map required by the definition. $\qquad\square$

Thus, in a finite group, by Propositions 1.38 and 4.10 every $\mathcal{F}_P(G)$-conjugacy class of subgroups of a Sylow p-subgroup P of G contains a receptive subgroup. This is almost the second axiom in the definition of a saturated fusion system, but we need an extra condition to make things work. Earlier, we said that every $\mathcal{F}$-conjugacy class must contain a fully automized member in a saturated fusion system, and so we will combine this statement with the statement that every $\mathcal{F}$-conjugacy class of subgroups contains a receptive member to produce the following definition.

Definition 4.11 Let $\mathcal{F}$ be a fusion system on a finite p-group P. We say that $\mathcal{F}$ is *saturated* if every $\mathcal{F}$-conjugacy class of subgroups of P contains a subgroup that is both receptive and fully automized.

In Exercise 4.1, the reader is invited to show that this is *not* equivalent to merely requiring that every conjugacy class contains a fully automized member and a receptive member, since there are fusion systems satisfying this condition that are not saturated.

If G is a finite group, then $\mathcal{F}_P(G)$ would satisfy this requirement if we could show that Q is fully automized whenever $N_P(Q)$ is a Sylow p-subgroup of $N_G(Q)$.

Theorem 4.12 *Let G be a finite group, and let P be a Sylow p-subgroup of G. The fusion system $\mathcal{F}_P(G)$ is saturated.*

Proof We need to show that every $\mathcal{F}_P(G)$-conjugacy class of subgroups contains a fully automized, receptive member. Let R be a subgroup of P, let $\bar{P}$ be a Sylow p-subgroup of G containing a Sylow p-subgroup of $N_G(R)$, and let $g \in G$ be such that $\bar{P}^g = P$. Write $Q = R^g$, and notice that $N_P(Q)$ is a Sylow p-subgroup of $N_G(Q)$, since $N_{\bar{P}}(R)$ is a Sylow p-subgroup of $N_G(R)$; by Proposition 4.10 we know that Q is receptive. It remains to show that Q is fully automized. Note that $\mathrm{Aut}_G(Q) \cong N_G(Q)/C_G(Q)$ has a Sylow p-subgroup

$$N_P(Q)\, C_G(Q)/C_G(Q),$$

which is isomorphic to

$$N_P(Q)/(C_G(Q) \cap N_P(Q)) = N_P(Q)/C_P(Q),$$

whence $\operatorname{Aut}_P(Q) \cong N_P(Q)/C_P(Q)$ is a Sylow p-subgroup of $\operatorname{Aut}_G(Q) = \operatorname{Aut}_{\mathcal{F}_P(G)}(Q)$, as needed. $\qquad\qquad\square$

4.2 Normalizing and centralizing

The idea of a fully automized subgroup is to find a subgroup in an $\mathcal{F}$-conjugacy class that is maximal with respect to some property. In Proposition 4.10 we saw that being receptive for groups was implied by having the largest normalizer $N_P(Q)$ out of all of its conjugates, and in Theorem 4.12 we saw that this same condition implied that the subgroup was fully automized as well. This motivates the following definition.

Definition 4.13 Let $\mathcal{F}$ be a fusion system on a finite p-group P.

(i) A subgroup Q is said to be *fully $\mathcal{F}$-centralized* if, whenever R is $\mathcal{F}$-conjugate to Q, we have that

$$|C_P(Q)| \geq |C_P(R)|.$$

(ii) A subgroup Q is said to be *fully $\mathcal{F}$-normalized* if, whenever R is $\mathcal{F}$-conjugate to Q, we have that

$$|N_P(Q)| \geq |N_P(R)|.$$

(If it is clear which fusion system is meant we will simply talk about fully normalized and fully centralized subgroups.)

Thus for fusion systems of finite groups, a subgroup is receptive if it is fully normalized. It is true, for any fusion system, that a subgroup that is both fully centralized and fully automized must be fully normalized, since

$$|N_P(Q)| = |C_P(Q)| \cdot |\operatorname{Aut}_P(Q)|.$$

No other set of implications involving fully normalized, fully centralized, and fully automized subgroups is true for arbitrary fusion systems, as the following example shows.

Example 4.14 By E_{p^n} we mean the elementary abelian group of order p^n. Let Q denote a Sylow 2-subgroup of the finite group $E_8 \rtimes \mathrm{GL}_3(2)$, where $\mathrm{GL}_3(2)$ acts on E_8 as it does on the 3-dimensional vector space over $\mathbb{F}_2$. Let R denote the finite 2-group $E_{16} \rtimes V_4$, where the V_4 acts like the permutation group $\langle (1,2),(3,4) \rangle$ on a basis for the normal subgroup E_{16}. Finally, let $P = Q \times R$.

Let A be the obvious normal subgroup of Q isomorphic with E_8, and let B denote the subgroup of the normal E_{16} in R generated by the first three basis elements. Let $\mathcal{F}$ be a fusion system on P in which A and B are $\mathcal{F}$-conjugate. It is clear that A is normal in P, and that $\mathrm{Aut}_P(A) = D_8$. Therefore A is certainly fully normalized and fully automized, since $A \trianglelefteq P$ and $\mathrm{Aut}_P(A)$ is the maximum possible. However, A is not fully centralized; its centralizer has index 8, where $\mathrm{C}_P(B)$ has index 4.

In fact, B need not be fully centralized (for example, the centre of P is isomorphic with E_8, and there are many classes of E_8 subgroup with centralizer of index 2), but B cannot possibly be fully automized since $|P/\mathrm{C}_P(B)| = 4$ (and $\mathrm{Aut}_{\mathcal{F}}(B) \cong \mathrm{Aut}_{\mathcal{F}}(A) \cong D_8$), and is not fully normalized either, since B is not normalized by Q. Hence A is fully normalized and fully automized, but not fully centralized, and B can be fully centralized, but is neither fully automized nor fully normalized.

The only other remaining combination is a subgroup that is both fully normalized and fully centralized: in the group $Q \times E_8$, let A be as before, and let C be the direct E_8 factor. Let $\mathcal{F}$ be any fusion system in which A and C are $\mathcal{F}$-conjugate; then C is both fully normalized and fully centralized, but is not fully automized, since it has trivial automizer. (Another example is the centre of a finite p-group, which is clearly fully normalized and fully centralized, but not fully automized if it is given an automorphism of order p; for example, the Klein four-group given an automorphism swapping two non-trivial elements and fixing the other.)

This example shows that a fusion system itself really can be a very loose structure. In a saturated fusion system, some of the situations in this example cannot occur.

We will now examine the interplay between the notions of fully normalized, fully centralized, fully automized, and receptive subgroups, for *saturated* fusion systems. Saturated fusion systems will be the focus of the vast majority of the results in this text. We will show that, in a saturated fusion system, being receptive is the same thing as being fully centralized, and being fully automized and receptive is the same thing as being fully normalized. One implication from each of these equivalences, Proposition 4.15 and Corollary 4.16, go through without needing saturation, but the other direction requires it.

Proposition 4.15 *For any fusion system, every receptive subgroup is fully centralized.*

Proof Let $\mathcal{F}$ be a fusion system on a p-group P. Suppose that Q is receptive, and let R be a subgroup of P that is $\mathcal{F}$-isomorphic with Q such that R is fully centralized. Let $\phi : R \to Q$ be an isomorphism in $\mathcal{F}$; since Q is receptive, the map ϕ extends to an injective map $\bar{\phi} : N_\phi \to P$. Since $R\,\mathrm{C}_P(R) \leq N_\phi$, there is a map $\theta = \bar{\phi}|_{R\,\mathrm{C}_P(R)}$ mapping $R\,\mathrm{C}_P(R)$ into P, with the property that $R\theta = Q$. Clearly, therefore, $\mathrm{C}_P(R)\theta \leq \mathrm{C}_P(Q)$, and so Q is fully centralized, as claimed. $\square$

This yields the following corollary.

Corollary 4.16 *For any fusion system, every subgroup that is receptive and fully automized is fully normalized.*

Proof If a subgroup is both fully centralized and fully automized then it is fully normalized, as we have remarked earlier, and since all receptive subgroups are fully centralized, we get the result. $\square$

Proposition 4.15 and Corollary 4.16 yield one direction of the two equivalences that we mentioned earlier, with no requirement on saturation. To do the opposite directions, we do need a saturation condition. Define an $\mathcal{F}$-conjugacy class of subgroups to be *saturated* if it contains a member that is both receptive and fully automized; then a fusion system is saturated if and only if all of its conjugacy classes are saturated.

We begin with an interesting result, that essentially says that if we have two $\mathcal{F}$-conjugate subgroups, with one fully automized, then we can choose a map ϕ from the one into the other, such that the subgroup N_ϕ is the whole of the normalizer.

Proposition 4.17 *Let $\mathcal{F}$ be a fusion system on a finite p-group P. Let Q and R be $\mathcal{F}$-isomorphic subgroups of P such that R is fully automized. There is an isomorphism $\phi : Q \to R$ such that $N_\phi = N_P(Q)$.*

Proof Let ϕ be any isomorphism $\phi : Q \to R$. The group $\mathrm{Aut}_P(Q)^\phi$ is a p-subgroup of $\mathrm{Aut}_\mathcal{F}(R)$, and since R is fully automized, $\mathrm{Aut}_P(R)$ is a Sylow p-subgroup of $\mathrm{Aut}_\mathcal{F}(R)$, so there is some $\alpha \in \mathrm{Aut}_\mathcal{F}(R)$ such that $\mathrm{Aut}_P(Q)^{\phi\alpha}$ is contained within $\mathrm{Aut}_P(R)$. Set $\psi = \phi\alpha$; then N_ψ is the preimage of $\mathrm{Aut}_P(Q) \cap \mathrm{Aut}_P(R)^{\psi^{-1}} = \mathrm{Aut}_P(Q)$ since $\mathrm{Aut}_P(Q)^\psi \leq \mathrm{Aut}_P(R)$, and so $N_\psi = \mathrm{N}_P(Q)$. $\square$

Before we continue, we need to get a more precise handle on how extensions of $\phi : Q \to R$ to subgroups of $\mathrm{N}_P(Q)$ and the induced map $c_\phi :$ $\mathrm{Aut}(Q) \to \mathrm{Aut}(R)$ are related. This is given in the proof of Proposition 4.8, but here we make it explicit.

Lemma 4.18 *Let $\mathcal{F}$ be a fusion system on a finite p-group P, and let Q and R be subgroups with an isomorphism $\phi : Q \to R$. Suppose that $\bar\phi : S \to \mathrm{N}_P(R)$ is an extension of ϕ with $S \leq \mathrm{N}_P(Q)$. If $g \in S$, then, as elements of $\mathrm{Aut}(R)$,*

$$c_{g\bar\phi} = (c_g)^\phi.$$

Consequently, if R is receptive, and $\bar\phi : N_\phi \to R$ is an extension of ϕ, then

$$\left((N_\phi)\bar\phi\right) c_R = (N_\phi c_Q)^\phi.$$

This control of N_ϕ is crucial in the next result, where we prove the other half of one of the equivalences.

Proposition 4.19 (Stancu) *Let $\mathcal{F}$ be a fusion system on a finite p-group P, and let R be a fully centralized subgroup of P. If the $\mathcal{F}$-conjugacy class containing R is saturated, then R is receptive.*

Proof Let Q be any subgroup in the $\mathcal{F}$-conjugacy class of R, and let $\phi : Q \to R$ be an isomorphism. Since the class containing R is saturated, there is a receptive, fully automized subgroup S and an isomorphism $\psi :$ $R \to S$, which may be chosen so that $N_\psi = \mathrm{N}_P(R)$ by Proposition 4.17. We claim that $N_\phi \leq N_{\phi\psi}$: to see this, notice that $\mathrm{Aut}_P(R)^\psi \leq \mathrm{Aut}_P(S)$ since $N_\psi = \mathrm{N}_P(R)$, and so

$$\mathrm{Aut}_P(Q) \cap \mathrm{Aut}_P(R)^{\phi^{-1}} \leq \mathrm{Aut}_P(Q) \cap \left(\mathrm{Aut}_P(S)^{\psi^{-1}}\right)^{\phi^{-1}}$$
$$= \mathrm{Aut}_P(Q) \cap \mathrm{Aut}_P(S)^{(\phi\psi)^{-1}},$$

and taking preimages under the map c_Q gives $N_\phi \leq N_{\phi\psi}$.

Since ψ has an extension to $\bar\psi : \mathrm{N}_P(R) \to \mathrm{N}_P(S)$ and $\phi\psi$ has an extension $\theta : N_\phi \to \mathrm{N}_P(S)$ (the images being contained within $\mathrm{N}_P(S)$ by Proposition 4.8), we need to show that $\mathrm{im}\,\theta \leq \mathrm{im}\,\bar\psi$; if this is true, then $\theta \circ \bar\psi^{-1}|_{N_\phi\theta}$ is a map from N_ϕ to $\mathrm{N}_P(R)$ extending $\phi = (\phi\psi)\psi^{-1}$, as needed.

Since both R and S are fully centralized (by Proposition 4.15) we have that $\bar{\psi} : C_P(R) \to C_P(S)$ is an isomorphism, and so $\operatorname{im} \bar{\psi}$ is the full preimage in $N_P(S)$ of $\operatorname{Aut}_P(R)^\psi$ by Lemma 4.18. Finally, $\operatorname{im} \theta$ is a preimage in $N_P(S)$ of $(N_\phi)\theta c_R = (N_\phi c_Q)^{\phi\psi} \leq \operatorname{Aut}_P(R)^\psi$ since $(N_\phi c_Q)^\phi \leq \operatorname{Aut}_P(R)$, so that $\operatorname{im} \theta \leq \operatorname{im} \bar{\psi}$, as needed. □

The final piece in the puzzle is to prove that, in a saturated $\mathcal{F}$-conjugacy class, every fully normalized subgroup is receptive and fully automized.

Proposition 4.20 *Let $\mathcal{F}$ be a fusion system on a finite p-group P. Any fully normalized subgroup in a saturated $\mathcal{F}$-conjugacy class is fully automized and receptive.*

Proof Let $\mathcal{F}$ be a fusion system on a p-group P. Suppose that Q is fully normalized, and let R be a subgroup of P that is $\mathcal{F}$-isomorphic with Q such that R is fully centralized (which is the same as receptive) and fully automized. Since R is fully normalized as well, by Proposition 4.16, we see that $|N_P(Q)| = |N_P(R)|$, whence

$$|N_P(Q)| = |\operatorname{Aut}_P(Q)| \cdot |C_P(Q)| = |\operatorname{Aut}_P(R)| \cdot |C_P(R)| = |N_P(R)|.$$

Since R is both fully centralized and fully automized, we have that $|C_P(R)| \geq |C_P(Q)|$ and $|\operatorname{Aut}_P(R)| \geq |\operatorname{Aut}_P(Q)|$ (since $\operatorname{Aut}_P(Q)$ is a Sylow p-subgroup of $\operatorname{Aut}_\mathcal{F}(Q)$); clearly, therefore, we must have equality for both centralizers and automizers, and so Q is both fully centralized and fully automized, as needed. □

Collating Propositions 4.15, 4.19, and 4.20, and Corollary 4.16, we get the following theorem.

Theorem 4.21 *Let $\mathcal{F}$ be a saturated fusion system on a finite p-group P.*

(i) *A subgroup is fully centralized if and only if it is receptive.*

(ii) *A subgroup is fully normalized if and only if it is both receptive and fully automized.*

4.3 The equivalent definitions

In this section, we use the work of the previous section to prove the equivalence of the four definitions most commonly used in the literature. The first is as described above, which is that of Roberts and Shpectorov

[RoS09]. We will refer to each as a definition, meaning that we will define the concept of saturation multiple times. We will remark in each case how the definition is equivalent to the original definition, Definition 4.11.

We start with the definition from [Sta06].

Definition 4.22   Let $\mathcal{F}$ be a fusion system on a finite p-group P. We say that $\mathcal{F}$ is *saturated* if

(i) P is fully automized, and
(ii) every fully normalized subgroup is receptive.

Suppose that our original definition holds; then P is fully automized (since P lies in a singleton $\mathcal{F}$-conjugacy class), and a subgroup is fully normalized if and only if it is receptive and fully automized, and so the conditions for Definition 4.22 hold.

Conversely, suppose that Definition 4.22 holds, and let Q be a fully normalized subgroup of P; then Q is receptive, and we need to show that every fully normalized, receptive subgroup is fully automized. We will assume that P is fully automized, and choose a fully normalized, receptive subgroup Q of largest order in P such that Q is not fully automized. Since all fully normalized, receptive subgroups of larger index are fully automized, all $\mathcal{F}$-conjugacy classes of subgroups of larger order than $|Q|$ are saturated. The next lemma proves that Q is indeed fully automized, and so all $\mathcal{F}$-conjugacy classes are saturated.

Lemma 4.23 (Stancu)   *Let $\mathcal{F}$ be a fusion system on a finite p-group P. Suppose that Q is a fully normalized, receptive subgroup. If P is fully automized, and every conjugacy class of subgroups containing an $\mathcal{F}$-conjugate of Q is saturated, then Q is fully automized.*

Proof   Suppose for a contradiction that Q is fully normalized and receptive, but that $\mathrm{Aut}_P(Q)$ is not a Sylow p-subgroup of $\mathrm{Aut}_{\mathcal{F}}(Q)$. Choose Q to be of maximal order with this property; certainly Q is not equal to P by assumption. Choose an automorphism ϕ of p-power order in $\mathrm{Aut}_{\mathcal{F}}(Q) \setminus \mathrm{Aut}_P(Q)$ such that $\langle \phi \rangle$ normalizes $\mathrm{Aut}_P(Q)$, which exists since $\mathrm{Aut}_P(Q)$ is not a Sylow p-subgroup of $\mathrm{Aut}_{\mathcal{F}}(Q)$.

Since ϕ normalizes $\mathrm{Aut}_P(Q)$, for every $x \in \mathrm{N}_P(Q)$, there is some $y \in \mathrm{N}_P(Q)$ such that $(g^x)\phi = (g\phi)^y$ for all $g \in Q$. Therefore $N_\phi = \mathrm{N}_P(Q)$ and, since Q is receptive, there is an extension $\bar{\phi}$ of ϕ to the whole of $\mathrm{N}_P(Q)$. Since ϕ has p-power order, we may assume that $\bar{\phi}$ has p-power order (by raising $\bar{\phi}$ to a suitable power).

Let ψ be a map in $\mathcal{F}$ from $\mathrm{N}_P(Q)$ such that its image, R, is fully normalized. We see that $(\bar{\phi})^\psi$ is a p-element of $\mathrm{Aut}_{\mathcal{F}}(R)$. Since the

$\mathcal{F}$-conjugacy class containing R is saturated, we see that $\mathrm{Aut}_P(R)$ is a Sylow p-subgroup of $\mathrm{Aut}_{\mathcal{F}}(R)$, and so $(\bar{\phi})^{\psi}$ may be conjugated into $\mathrm{Aut}_P(R)$; hence we may choose ψ so that $(\bar{\phi})^{\psi} \in \mathrm{Aut}_P(R)$. Thus there is some $g \in \mathrm{N}_P(R)$ such that $x(\bar{\phi})^{\psi} = x^g$ for all $x \in R$.

Since $\bar{\phi}|_Q = \phi$, we see that $Q\psi$ is invariant under $(\bar{\phi})^{\psi}$, and so g normalizes $Q\psi$. However, Q is fully normalized, and so $\mathrm{N}_P(Q)\psi$ contains $\mathrm{N}_P(Q\psi)$. Therefore $g \in \mathrm{im}\,\psi$, and so if h denotes the preimage of g, we have that $x\bar{\phi} = x^h$ for all $x \in Q$.

Now we see a contradiction: in fact, ϕ may be defined by conjugation, and so lies in $\mathrm{Aut}_P(Q)$, whereas it was chosen not to. Hence $\mathrm{Aut}_P(Q)$ is a Sylow p-subgroup of $\mathrm{Aut}_{\mathcal{F}}(Q)$, as required. $\square$

Next, we have the definition from [BLO04].

Definition 4.24 Let $\mathcal{F}$ be a fusion system on a finite p-group P. We say that $\mathcal{F}$ is *saturated* if

(i) every fully normalized subgroup Q is fully centralized and fully automized, and
(ii) every fully centralized subgroup is receptive.

Suppose that Definition 4.24 holds; therefore every fully normalized subgroup is both fully automized and receptive, whence Definition 4.22 holds, and therefore so does our original definition.

Conversely, suppose that our original definition holds; then Theorem 4.21 yields the two requirements in Definition 4.24, as needed.

Finally, we have another suggested definition from [RoS09], which slightly simplifies both our original definition and Definition 4.22.

Definition 4.25 Let $\mathcal{F}$ be a fusion system on a finite p-group P. We say that $\mathcal{F}$ is *saturated* if

(i) P is fully automized;
(ii) every $\mathcal{F}$-conjugacy class of subgroups contains a fully normalized, receptive member.

It is immediate that Definition 4.22 implies Definition 4.25, and so it is implied by our original definition. Suppose that P is fully automized and every $\mathcal{F}$-conjugacy class of subgroups contains a fully normalized, receptive member, and choose such a subgroup Q of smallest index in P subject to not being fully automized. Lemma 4.23 provides a contradiction here, since Q is fully automized.

This proves the equivalence of all of the definitions we have given here.

4.4 Local subsystems

In local finite group theory, the idea of normalizers is the fundamental notion. When working with fusion systems, similar arguments can be used, and in order to do this, we need the concepts of centralizers and normalizers. In group theory, if $g \in C_G(Q)$, then conjugation by g acts trivially on Q and, if $g \in N_G(Q)$, then conjugation by g acts as an automorphism of Q. There are analogous notions for fusion systems, due to Puig [Pui06].

Definition 4.26 Let $\mathcal{F}$ be a fusion system on a finite p-group P, and let Q be a subgroup of P.

(i) The category $C_{\mathcal{F}}(Q)$ has as objects all subgroups of $C_P(Q)$, and has morphisms $\mathrm{Hom}_{C_{\mathcal{F}}(Q)}(R, S)$ given by all $\phi \in \mathrm{Hom}_{\mathcal{F}}(R, S)$ such that ϕ extends to $\bar{\phi} \in \mathrm{Hom}_{\mathcal{F}}(QR, QS)$ with $\bar{\phi}|_Q = 1$. This category is the *centralizer* of Q in $\mathcal{F}$.

(ii) The category $N_{\mathcal{F}}(Q)$ has as objects all subgroups of $N_P(Q)$, and has morphisms $\mathrm{Hom}_{N_{\mathcal{F}}(Q)}(R, S)$ given by all $\phi \in \mathrm{Hom}_{\mathcal{F}}(R, S)$ such that ϕ extends to $\bar{\phi} \in \mathrm{Hom}_{\mathcal{F}}(QR, QS)$ with $\bar{\phi}|_Q \in \mathrm{Aut}_{\mathcal{F}}(Q)$. This category is the *normalizer* of Q in $\mathcal{F}$. (Note that we need only that $\bar{\phi}|_Q \in \mathrm{Aut}(Q)$, since if this is true then it must be an automorphism in $\mathcal{F}$ by Proposition 4.4.)

These are two special cases of a more general concept, that of K-*normalizers*, which we will talk about later. We delay the proof that normalizers and centralizers are in fact fusion systems in their own right and deal with K-normalizers. However, first we should make sure that these definitions are the 'right' ones, and for that we should check that they are correct for groups.

Theorem 4.27 *Let G be a finite group with Sylow p-subgroup P, and let Q be a subgroup of P.*

(i) *If Q is fully normalized, then $\mathcal{F}_{N_P(Q)}(N_G(Q)) = N_{\mathcal{F}_P(G)}(Q)$.*
(ii) *If Q is fully centralized, then $\mathcal{F}_{C_P(Q)}(C_G(Q)) = C_{\mathcal{F}_P(G)}(Q)$.*

Proof We will prove (i), and note that the proof of (ii) is very similar. Firstly, write $\mathcal{F} = \mathcal{F}_P(G)$. Suppose that Q is fully normalized; by Proposition 1.38, $N_P(Q)$ is a Sylow p-subgroup of $N_G(Q)$, and so the left-hand side, $\mathcal{F}_{N_P(Q)}(N_G(Q))$, makes sense. Let $\phi : A \to B$ be a morphism in $N_{\mathcal{F}}(Q)$; then it extends to some $\bar{\phi} : QA \to QB$, such that $\bar{\phi}|_Q \in \mathrm{Aut}_{\mathcal{F}}(Q)$. Since $\mathcal{F} = \mathcal{F}_P(G)$, there is $g \in G$ such that $\bar{\phi} = c_g$ on

QA whence, since $c_g|_Q \in \mathrm{Aut}_{\mathcal{F}}(Q)$, we see that $g \in \mathrm{N}_G(Q)$. Therefore $c_g : A \to B$ is also a morphism in $\mathcal{F}_{\mathrm{N}_P(Q)}(\mathrm{N}_G(Q))$.

Now suppose that $\phi : A \to B$ is a morphism in $\mathcal{F}_{\mathrm{N}_P(Q)}(\mathrm{N}_G(Q))$. Since this is a group fusion system, there is $g \in \mathrm{N}_G(Q)$ such that $\phi = c_g$ on A. Certainly ϕ is a morphism in $\mathcal{F}$, and so it remains to show that ϕ extends to a morphism $\bar{\phi} : QA \to QB$ such that $\bar{\phi}|_Q \in \mathrm{Aut}_{\mathcal{F}}(Q)$. However, since c_g normalizes Q, it is clear that $(QA)^g = QB$, and that $c_g : QA \to QB$ both extends ϕ and acts as an automorphism of Q. Therefore ϕ is also in the normalizer of Q in $\mathcal{F}$, as needed. $\qquad\square$

Thus, for group fusion systems, our guide to defining the corresponding concepts for fusion systems in general, everything is correct. We now embark on a proof of the following theorem.

Theorem 4.28 (Puig [Pui06]) *Let $\mathcal{F}$ be a fusion system on a finite p-group P, and let Q be a subgroup of P.*

(i) *The categories $\mathrm{C}_{\mathcal{F}}(Q)$ and $\mathrm{N}_{\mathcal{F}}(Q)$ are fusion systems on $\mathrm{C}_P(Q)$ and $\mathrm{N}_P(Q)$ respectively.*

(ii) *If $\mathcal{F}$ is saturated, then $\mathrm{C}_{\mathcal{F}}(Q)$ is saturated whenever Q is fully centralized, and $\mathrm{N}_{\mathcal{F}}(Q)$ is saturated whenever Q is fully normalized.*

The proof of the first part of this theorem is fairly elementary, but the second requires considerable work, and will occupy much of the rest of this section. The reader should not feel obligated to study this proof in depth at a first reading, since it is long and complicated. However, in this case it would be helpful to read the definitions in this section since they will be used later.

We first note that the need for something like fully normalized is required in the condition for $\mathrm{N}_{\mathcal{F}}(Q)$ to be saturated, since at least fully automized is necessary.

Proposition 4.29 *Let $\mathcal{F}$ be a fusion system on a finite p-group P, and let Q be a subgroup of P. The $\mathrm{N}_{\mathcal{F}}(Q)$-conjugacy class containing Q is $\{Q\}$, and if Q is not fully $\mathcal{F}$-automized then $\mathrm{N}_{\mathcal{F}}(Q)$ is not saturated.*

Proof Write $\mathcal{N}$ for the subsystem $\mathrm{N}_{\mathcal{F}}(Q)$. Every morphism ϕ of $\mathcal{N}$ extends to a morphism $\bar{\phi}$, which acts as an automorphism of Q. Therefore, if Q is the domain of a morphism ϕ, then ϕ must be an automorphism of Q, i.e., $\{Q\}$ is the $\mathcal{N}$-conjugacy class containing Q, proving our first claim. Therefore, if $\mathcal{N}$ is saturated, certainly Q must be fully $\mathcal{N}$-automized and receptive.

By definition $\operatorname{Aut}_{\mathcal{N}}(Q) = \operatorname{Aut}_{\mathcal{F}}(Q)$, and $\operatorname{Aut}_{\mathrm{N}_P(Q)}(Q) = \operatorname{Aut}_P(Q)$, and hence Q is fully $\mathcal{N}$-automized if and only if it is fully $\mathcal{F}$-automized. $\qquad\square$

We move on to a proof of Theorem 4.28: we would like to deal with the centralizer and normalizer subsystems together, and indeed there are two other related subsystems that we will mention after proving this theorem. As we said before, this will go via the concept of K-normalizers. Broadly speaking, for a subgroup K of $\operatorname{Aut}(Q)$, the K-normalizer is the set of morphisms that extend to include Q in the domain, and act like an element of K on Q.

We begin by defining the analogues of the subgroup $\mathrm{N}_P(Q)$ and fully normalized subgroups.

Definition 4.30   Let $\mathcal{F}$ be a fusion system on a finite p-group P, and let Q be a subgroup of P. Let K be a subgroup of $\operatorname{Aut}(Q)$.

(i) The *K-normalizer* of Q is the set
$$\mathrm{N}_P^K(Q) = \{g \in \mathrm{N}_P(Q) \mid c_g|_P \in K\}.$$

(ii) We say that Q is *fully K-normalized* if, for all subgroups R that are $\mathcal{F}$-conjugate to Q, and all isomorphisms $\phi : Q \to R$, we have that
$$|\mathrm{N}_P^K(Q)| \geq |\mathrm{N}_P^{K^\phi}(R)|.$$

(iii) Denote by $\operatorname{Aut}_{\mathcal{F}}^K(Q)$ the set $K \cap \operatorname{Aut}_{\mathcal{F}}(Q)$, and by $\operatorname{Aut}_P^K(Q)$ the set $K \cap \operatorname{Aut}_P(Q)$.

The following is an obvious fact.

Lemma 4.31   *In any fusion system, a subgroup Q is fully $\operatorname{Aut}(Q)$-normalized if and only if it is fully normalized, and it is fully $\{1\}$-normalized if and only if it is fully centralized.*

Two other natural subgroups of $\operatorname{Aut}(Q)$ other than $\operatorname{Aut}(Q)$ itself and $\{1\}$ are $\operatorname{Aut}_Q(Q)$ and $\operatorname{Aut}_P(Q)$. These will correspond to the other two subsystems that we mentioned before.

Attached to each subgroup $K \leq \operatorname{Aut}(Q)$, there is a concept of a K-normalizer and fully K-normalized. We now need the concept of the K-normalizer subsystem as well.

Definition 4.32   Let $\mathcal{F}$ be a fusion system on a finite p-group P, and let Q be a subgroup of P. The category $\mathrm{N}_{\mathcal{F}}^K(Q)$ has as objects all subgroups of $\mathrm{N}_P^K(Q)$, and has morphisms $\operatorname{Hom}_{\mathrm{N}_{\mathcal{F}}^K(Q)}(R, S)$ given by
$$\{\phi \in \operatorname{Hom}_{\mathcal{F}}(R, S) : \phi \text{ extends to } \bar{\phi} \in \operatorname{Hom}_{\mathcal{F}}(QR, QS) \text{ with } \bar{\phi}|_Q \in K\}.$$

This category is the *K-normalizer* of Q in $\mathcal{F}$.

Again, the normalizer subsystem of Q in $\mathcal{F}$ is the $\mathrm{Aut}(Q)$-normalizer of Q in $\mathcal{F}$, and the centralizer of Q in $\mathcal{F}$ is the $\{1\}$-normalizer of Q in $\mathcal{F}$. The general theorem that needs to be proved is the following.

Theorem 4.33 (Puig [Pui06]) *Let $\mathcal{F}$ be a fusion system on a finite p-group P, let Q be a subgroup of P, and let K be a subgroup of $\mathrm{Aut}(Q)$. The K-normalizer $\mathrm{N}_{\mathcal{F}}^{K}(Q)$ is a subsystem of $\mathcal{F}$, and if $\mathcal{F}$ is saturated and Q is fully K-normalized then $\mathrm{N}_{\mathcal{F}}^{K}(Q)$ is saturated.*

We begin with the easy part.

Proposition 4.34 *Let $\mathcal{F}$ be a fusion system on a finite p-group P, and let Q be a subgroup of P. If K is a subgroup of $\mathrm{Aut}(Q)$, then the K-normalizer $\mathrm{N}_{\mathcal{F}}^{K}(Q)$ is a subsystem of $\mathcal{F}$ on $\mathrm{N}_{P}^{K}(Q)$.*

Proof We will check the axioms for a fusion system, noting that the objects in the category are correct. Thus let R and S be subgroups of $\mathrm{N}_{P}^{K}(Q)$. For $g \in \mathrm{N}_{P}^{K}(Q)$, if $c_g : R \to S$ is a conjugation map it clearly extends to a map $QR \to QS$ that acts as an automorphism in K on Q, from the fact that $g \in \mathrm{N}_{P}^{K}(Q)$. Thus the first axiom of a fusion system is satisfied by $\mathrm{N}_{\mathcal{F}}^{K}(Q)$.

If $\phi : R \to S$ is a map in $\mathrm{N}_{\mathcal{F}}^{K}(Q)$, then it extends to a map $\bar{\phi} : QR \to QS$ that acts as an automorphism in K on Q, and so clearly the isomorphism map $R \to R\phi$ also has this condition; thus $\mathrm{N}_{\mathcal{F}}^{K}(Q)$ satisfies the second axiom of a fusion system.

Suppose that $\phi : R \to S$ is an isomorphism in $\mathrm{N}_{\mathcal{F}}^{K}(Q)$; then it extends to a map $\bar{\phi} : QR \to QS$ that acts as an automorphism in K on Q. If $s \in S$, then $s = r\phi$ for some $r \in R$, and so $(qr)\phi = (q\phi)s$, which ranges over all elements of QS as q and r range over all elements of Q and R respectively: in particular, $\bar{\phi} \in \mathcal{F}$ is an isomorphism. Thus its inverse lies in $\mathcal{F}$, and so the map ϕ^{-1} has an extension $\bar{\phi}^{-1} : QS \to QR$ with the necessary properties. (This last part requires K to be closed under inverses.)

Finally, we need to note that this is a category, in the sense that the composition of two morphisms in $\mathrm{N}_{\mathcal{F}}^{K}(Q)$ also lies in $\mathrm{N}_{\mathcal{F}}^{K}(Q)$; this comes from the fact that the composition of two elements of K is also in K. $\square$

Next, we need some preliminary lemmas about K-normalizers. We begin with a generalization of the statement that a subgroup is fully normalized if and only if it is fully centralized and fully automized.

Lemma 4.35 *Let $\mathcal{F}$ be a saturated fusion system on a finite p-group P, and suppose that Q is a subgroup of P. If K is a subgroup of $\mathrm{Aut}(Q)$, then Q is fully K-normalized if and only if it is fully centralized and $\mathrm{Aut}_P^K(Q)$ is a Sylow p-subgroup of $\mathrm{Aut}_{\mathcal{F}}^K(Q)$. Consequently, if $K \leq \mathrm{Aut}_P(Q)$, then Q is fully K-normalized whenever it is fully centralized.*

Proof Suppose that Q is fully K-normalized, and let R be a fully normalized subgroup $\mathcal{F}$-conjugate to Q. By Proposition 4.17, there is a morphism $\phi : Q \to R$ such that ϕ extends to a morphism $\bar{\phi} : \mathrm{N}_P(Q) \to \mathrm{N}_P(R)$. (The image is in $\mathrm{N}_P(R)$ by Proposition 4.8.) Write $L = K^\phi$, so that $\bar{\phi}$ maps $\mathrm{N}_P^K(Q)$ to $\mathrm{N}_P^L(R)$. We know that Q is fully K-normalized, and so $\bar{\phi}$ induces an isomorphism $\theta : \mathrm{N}_P^K(Q) \to \mathrm{N}_P^L(R)$. This map θ, restricted to $\mathrm{C}_P(Q)$, must also be an isomorphism, and since R is fully normalized it is fully centralized. We see, therefore, that Q itself is also fully centralized.

We now prove that $\mathrm{Aut}_P^K(Q)$ is a Sylow p-subgroup of $\mathrm{Aut}_{\mathcal{F}}^K(Q)$. Since R is fully normalized it is fully automized, and so $\mathrm{Aut}_P(R)$ is a Sylow p-subgroup of $\mathrm{Aut}_{\mathcal{F}}(R)$. We have that $\mathrm{Aut}_P^K(Q)$ is a p-subgroup of $\mathrm{Aut}_{\mathcal{F}}^K(Q)$, and so $\bar{\phi}$ induces a map $\bar{\phi} : \mathrm{Aut}_P^K(Q) \to \mathrm{Aut}_P^L(R)$. Since any p-subgroup may be conjugated into a Sylow p-subgroup of a group, there is some conjugate of $\mathrm{Aut}_P(R)$ that intersects $\mathrm{Aut}_{\mathcal{F}}^L(R)$ in a Sylow p-subgroup, and this may be chosen to contain the image of $\mathrm{Aut}_P^K(Q)$ under $\bar{\phi}$. Therefore, there is an element $\psi \in \mathrm{Aut}_{\mathcal{F}}(R)$ such that

$$\mathrm{Aut}_P^K(Q)\bar{\phi} \leq \mathrm{Aut}_P(R)^\psi \cap \mathrm{Aut}_{\mathcal{F}}^L(R) = S,$$

where S is a Sylow p-subgroup of $\mathrm{Aut}_{\mathcal{F}}^L(R)$. Therefore,

$$S^{\psi^{-1}} = \mathrm{Aut}_P^{\psi K^\phi \psi^{-1}}(R)$$

is a Sylow p-subgroup of $\mathrm{Aut}_{\mathcal{F}}^{\psi L \psi^{-1}}(R)$.

Note that Q and R have the same centralizer orders in P, and Q is fully K-normalized; therefore, since $\phi \psi^{-1}$ is an isomorphism $Q \to R$, we have that

$$|\mathrm{Aut}_P^K(Q)| \cdot |\mathrm{C}_P(Q)| = |\mathrm{N}_P^K(Q)| \geq |\mathrm{N}_P^{\psi L \psi^{-1}}(R)|$$
$$= |\mathrm{Aut}_P^{\psi L \psi^{-1}}(R)| \cdot |\mathrm{C}_P(R)|,$$

and so $|\mathrm{Aut}_P^K(Q)| \geq |\mathrm{Aut}_P^{\psi L \psi^{-1}}(R)|$, with the latter being a Sylow p-subgroup of $\mathrm{Aut}_{\mathcal{F}}^{\psi L \psi^{-1}}(R)$. Therefore, $\mathrm{Aut}_P^K(Q)$ is a Sylow p-subgroup of $\mathrm{Aut}_{\mathcal{F}}^K(Q)$, as claimed.

Now suppose that Q is fully centralized, and that $\operatorname{Aut}_P^K(Q)$ is a Sylow p-subgroup of $\operatorname{Aut}_{\mathcal{F}}^K(Q)$. Let $\psi : Q \to R$ be any $\mathcal{F}$-isomorphism, and again write L for K^ψ. Since Q is fully centralized, we have that $|C_P(Q)| \geq |C_P(R)|$, and since $\operatorname{Aut}_{\mathcal{F}}^K(Q)$ and $\operatorname{Aut}_{\mathcal{F}}^L(R)$ are isomorphic – this follows easily from Proposition 4.6 – we see that $|\operatorname{Aut}_P^K(Q)| \geq |\operatorname{Aut}_P^L(R)|$. We get that

$$|N_P^K(Q)| = |C_P(Q)| \cdot |\operatorname{Aut}_P^K(Q)| \geq |C_P(R)| \cdot |\operatorname{Aut}_P^L(R)| = |N_P^L(R)|,$$

and so Q is fully K-normalized, as claimed.

Finally, suppose that Q is fully centralized, so that its centralizer has maximal order across all $\mathcal{F}$-conjugates of Q. The fact that $|N_P^K(Q)| = |C_P(Q)| \cdot |K|$ (since K is a subgroup of $\operatorname{Aut}_P(Q)$) yields that Q is fully K-normalized, as required by the final statement. $\qquad\square$

Using this, we give a technical lemma, needed in the proof to follow.

Lemma 4.36 *Let $\mathcal{F}$ be a saturated fusion system on the finite p-group P, let Q be a subgroup of P, and let K be a subgroup of $\operatorname{Aut}(Q)$. Suppose that $\phi : Q \to R$ is an isomorphism with R fully K^ϕ-normalized. There is a morphism $\psi : Q \, N_P^K(Q) \to P$ in $\mathcal{F}$ and $\alpha \in K$ such that $\psi|_Q = \alpha\phi$, and $\beta \in K^\phi$ such that $(\psi|_Q)\beta = \phi$.*

Proof Write $X = (Q \, N_P^K(Q))c_Q$. The map $\phi : Q \to R$ induces an isomorphism $\phi : \operatorname{Aut}_{\mathcal{F}}(Q) \to \operatorname{Aut}_{\mathcal{F}}(R)$; we need to show that some conjugate of X^ϕ lies inside $\operatorname{Aut}_P(R)$. Certainly X is a p-group; the subgroup $Qc_Q = \operatorname{Inn}(Q)$ is a normal p-subgroup of $\operatorname{Aut}(Q)$, invariant under any isomorphism, and so $(Qc_Q)^\phi = \operatorname{Inn}(R)$ is a normal p-subgroup of $\operatorname{Aut}(R)$ contained in $\operatorname{Aut}_{\mathcal{F}}(R)$.

Note that $(N_P^K(Q)c_Q)^\phi = \operatorname{Aut}_P^K(Q)^\phi \leq \operatorname{Aut}_P^{K^\phi}(R)$ since R is fully K-normalized, and also $\operatorname{Aut}_P^K(R)$ is a Sylow p-subgroup of $\operatorname{Aut}_{\mathcal{F}}(R)$. Since $\operatorname{Aut}_P^K(Q)^\phi$ is contained in $\operatorname{Aut}_{\mathcal{F}}^{K^\phi}(R)$, there is some $\lambda \in \operatorname{Aut}_{\mathcal{F}}^{K^\phi}(R)$ that conjugates $\operatorname{Aut}_P^K(Q)^\phi$ into $\operatorname{Aut}_P^{K^\phi}(R) \leq \operatorname{Aut}_P(R)$. Therefore $(X^\phi)^\lambda \leq \operatorname{Aut}_P(R)$ (as $\operatorname{Inn}(R)^\lambda = \operatorname{Inn}(R)$), and so $N_{\phi\lambda}$ contains $Q \, N_P^K(Q)$. Let ψ be an extension of $\phi\lambda$ to $Q \, N_P^K(Q)$, and let $\alpha = \phi\lambda\phi^{-1}$; the maps ψ and α are as needed.

For the final statement, let $\beta = (\psi^{-1}\alpha\psi)^{-1}$, and notice that $(\psi|_Q)\beta = \alpha^{-1}(\psi|_Q)$. $\qquad\square$

We now prove Theorem 4.33: this proof follows that of Oliver in [AKO11]. Let Q be a subgroup of P and let K be a subgroup of $\operatorname{Aut}(Q)$. For brevity, we write $N = N_P^K(Q)$ and $\mathcal{N} = N_{\mathcal{F}}^K(Q)$. For any subgroup

R of N, we will find a subgroup $\mathcal{N}$-conjugate to R that is both fully $\mathcal{N}$-automized and receptive in $\mathcal{N}$. In order to find such a subgroup, for any subgroup $R \leq N$ we define K_R to be the subgroup of $\mathrm{Aut}(QR)$ consisting of those automorphisms ϕ with $\phi|_R$ an automorphism and $\phi|_Q$ an automorphism in K.

Step 1: *If QR is fully K_R-normalized in $\mathcal{F}$ then R is fully $\mathcal{N}$-automized and receptive in $\mathcal{N}$.* Let R be a subgroup of N, and write $L = K_R$. If QR is fully L-normalized in $\mathcal{F}$ then by Lemma 4.35 we see that $\mathrm{Aut}_P^L(QR)$ is a Sylow p-subgroup of $\mathrm{Aut}_{\mathcal{F}}^L(QR)$. Notice that every element of $\mathrm{Aut}_N(R)$ extends to an element of $\mathrm{Aut}_P^L(QR)$, by definition of N, and so restriction to R gives a surjective map from $\mathrm{Aut}_P^L(QR)$ to $\mathrm{Aut}_N(R)$. Similarly, every automorphism $\phi \in \mathrm{Aut}_{\mathcal{N}}(R)$ extends to an automorphism $\bar{\phi}$ of QR in $\mathcal{F}$ such that $\bar{\phi}|_Q \in K$, by the definition of $\mathcal{N}$. Hence restriction to R gives a surjective map from $\mathrm{Aut}_{\mathcal{F}}^L(QR)$ to $\mathrm{Aut}_{\mathcal{N}}(R)$. As $\mathrm{Aut}_P^L(QR)$ is a Sylow p-subgroup of $\mathrm{Aut}_{\mathcal{F}}^L(QR)$, its image under the restriction map, $\mathrm{Aut}_N(R)$, is a Sylow p-subgroup of $\mathrm{Aut}_{\mathcal{N}}(R)$. Hence R is fully $\mathcal{N}$-automized.

It remains to show that R is receptive. In order to prove this, we need to restrict the possible actions of automorphisms on R; write

$$M = \{\alpha \in \mathrm{Aut}_{\mathcal{F}}(QR) \mid \alpha|_Q \in K, \ \alpha|_R \in \mathrm{Aut}_N(R)\} \leq L.$$

We claim that QR is also fully M-normalized; notice that $\mathrm{Aut}_P^L(QR) = \mathrm{Aut}_P^M(QR)$ (as if $g \in \mathrm{N}_P^L(QR)$ then $c_g|_Q \in K$, and so $g \in N$), and so since QR is fully L-normalized it must be fully M-normalized (using Lemma 4.35).

Let S be any other subgroup of N and let $\phi : S \to R$ be an isomorphism in $\mathcal{N}$. Since ϕ lies in $\mathcal{N}$, we may extend ϕ to an isomorphism $\bar{\phi} : QS \to QR$ acting as an automorphism of Q lying in K. Write $M' = M^{\phi^{-1}}$; since QR is fully M-normalized in $\mathcal{F}$, Lemma 4.36 implies that there exist maps $\beta \in M$ and $\psi \in \mathrm{Hom}_{\mathcal{F}}(QS\,\mathrm{N}_P^{M'}(QS), P)$ such that $\psi = \bar{\phi}\beta$ on QS. Since the domain of ψ is $QS\,\mathrm{N}_P^{M'}(QS)$, we see that the image of ψ is contained in $QR\,\mathrm{N}_P^M(QR)$.

By the definition of M, we have that $\beta|_R = c_g$ for some $g \in N$, and so $\psi c_g^{-1} = \phi$ on S. As $N_\phi \leq S\,\mathrm{N}_P^{M'}(QS)$, we replace ψ by $\psi|_{QN_\phi}$, so $\psi c_g^{-1}|_{N_\phi} \in \mathrm{Hom}_{\mathcal{N}}(N_\phi, N)$. Hence ψ is an extension of ϕ to N_ϕ in $\mathcal{N}$, and so R is receptive in $\mathcal{N}$, as claimed.

Step 2: *Every subgroup of N is $\mathcal{N}$-conjugate to a subgroup S such that QS is fully K_S-normalized.* Let $\phi : QR \to P$ be a morphism in $\mathcal{F}$ such that $(QR)\phi$ is fully K_R^ϕ-normalized. Write $\bar{Q} = Q\phi$ and $\bar{R} = R\phi$. To avoid too much clutter, we write $L = K_R^\phi$. As Q is fully K-normalized, we may apply Lemma 4.35 to the map $\phi^{-1} : \bar{Q} \to Q$, to get a morphism $\psi : \bar{Q} \mathrm{N}_P^{K^\phi}(\bar{Q}) \to P$ and $\beta \in K$ such that $\psi\beta = \phi^{-1}$ on $\bar{Q}$.

As $R \leq N = \mathrm{N}_P^K(Q)$, we see that $R\phi \leq \mathrm{N}_P^{K^\phi}(\bar{Q})$; consider the map $\phi\psi : QR \to P$. Notice that, by construction, $(\phi\psi)|_Q = \beta^{-1} \in K$, and so $(\phi\psi)|_R$ lies in $\mathcal{N}$. Write $S = R\phi\psi$; we claim that QS is fully K_S-normalized, completing the proof of Theorem 4.33.

Notice that we have the equalities

$$\mathrm{N}_P^L(\bar{Q}\bar{R}) = \mathrm{N}_P^{K^\phi}(\bar{Q}) \cap \mathrm{N}_P(\bar{R}) \quad \text{and} \quad \mathrm{N}_P^{K_S}(QS) = \mathrm{N}_P^K(Q) \cap \mathrm{N}_P(S).$$

Since ψ maps $\mathrm{N}_P^{K^\phi}(\bar{Q})$ to $\mathrm{N}_P^{K^{\phi\psi}}(Q) = \mathrm{N}_P^K(Q)$, it maps the normalizer of R inside the domain into the normalizer of S in the codomain. We see therefore that $\left(\mathrm{N}_P^L(\bar{Q}\bar{R})\right)\psi \leq \mathrm{N}_P^{K_S}(QS)$. The final piece of the proof is to notice that since $\bar{Q}\bar{R}$ is fully K_R^ϕ-normalized and ψ maps the K_R^ϕ-normalizer of $\bar{Q}\bar{R}$ inside the K_S-normalizer of QS, we must have that QS is fully K_S-normalized, as required.

Having finally proved Theorem 4.33 completely, we introduce two more subsystems, in between the centralizer and normalizer.

Definition 4.37 Let $\mathcal{F}$ be a fusion system on a finite p-group P, and let Q be a subgroup of P.

(i) By $Q\mathrm{C}_\mathcal{F}(Q)$, we mean the $\mathrm{Aut}_Q(Q)$-normalizer subsystem. Thus, $Q\mathrm{C}_\mathcal{F}(Q)$ is the subcategory of $\mathrm{N}_\mathcal{F}(Q)$ on $Q\mathrm{C}_P(Q)$, having as objects all subgroups of $Q\mathrm{C}_P(Q)$, and with morphisms $\mathrm{Hom}_{Q\mathrm{C}_\mathcal{F}(Q)}(R, S)$ given by all morphisms $\phi \in \mathrm{Hom}_\mathcal{F}(R, S)$ such that ϕ extends to $\bar{\phi} \in \mathrm{Hom}_\mathcal{F}(QR, QS)$ with $\bar{\phi}|_Q \in \mathrm{Aut}_Q(Q)$.

(ii) By $\mathrm{N}_P(Q)\mathrm{C}_\mathcal{F}(Q)$, we mean the $\mathrm{Aut}_P(Q)$-normalizer subsystem. In this case, $\mathrm{N}_P(Q)\mathrm{C}_\mathcal{F}(Q)$ is the subcategory of $\mathrm{N}_\mathcal{F}(Q)$ on $\mathrm{N}_P(Q)$, having as objects all subgroups of $\mathrm{N}_P(Q)$, and having as morphisms $\mathrm{Hom}_{\mathrm{N}_P(Q)\mathrm{C}_\mathcal{F}(Q)}(R, S)$ given by all $\phi \in \mathrm{Hom}_\mathcal{F}(R, S)$ such that ϕ extends to $\bar{\phi} \in \mathrm{Hom}_\mathcal{F}(QR, QS)$ with $\bar{\phi}|_Q \in \mathrm{Aut}_P(Q)$.

Theorem 4.33 and Lemma 4.35 imply the following result.

Corollary 4.38 *Let $\mathcal{F}$ be a saturated fusion system on a finite p-group P, and let Q be a subgroup of P. If Q is fully centralized then $\mathrm{C}_\mathcal{F}(Q)$, $Q\mathrm{C}_\mathcal{F}(Q)$, and $\mathrm{N}_P(Q)\mathrm{C}_\mathcal{F}(Q)$ are saturated subsystems of $\mathcal{F}$. If, in addition, Q is fully normalized, then $\mathrm{N}_\mathcal{F}(Q)$ is saturated.*

We introduce a final piece of notation, and prove a result about normalizers, which will help embed our understanding of them.

Definition 4.39 Let $\mathcal{F}$ be a fusion system on a finite p-group P, and let Q and R be subgroups of P such that $Q \leq \mathrm{N}_P(R)$ and $R \leq \mathrm{N}_P(Q)$. The *double normalizer* of Q and R in $\mathcal{F}$ is the subsystem $\mathrm{N}_{\mathrm{N}_{\mathcal{F}(Q)}}(R)$, denoted by $\mathrm{N}_{\mathcal{F}}(Q,R)$. The subgroup $\mathrm{N}_P(Q) \cap \mathrm{N}_P(R)$ is denoted by $\mathrm{N}_P(Q,R)$.

The double normalizer $\mathrm{N}_{\mathcal{F}}(Q,R)$ is a subsystem of P on $\mathrm{N}_P(Q,R)$. Clearly $\mathrm{N}_P(Q,R) = \mathrm{N}_P(R,Q)$, and indeed this holds for the normalizer subsystem as well.

Proposition 4.40 *Let $\mathcal{F}$ be a fusion system on a finite p-group P, and let Q and R be subgroups of P such that $QR \leq \mathrm{N}_P(Q,R)$, i.e., such that $Q \leq \mathrm{N}_P(R)$ and $R \leq \mathrm{N}_P(Q)$.*

(i) *The subsystem $\mathrm{N}_{\mathcal{F}}(Q,R)$ is the subsystem on $\mathrm{N}_P(Q,R)$ consisting of exactly those morphisms $\phi : A \to B$ with $A, B \leq \mathrm{N}_P(Q,R)$ such that there exists a morphism $\psi : AQR \to BQR$ in $\mathcal{F}$ with $\psi|_Q$ and $\psi|_R$ automorphisms and $\psi|_A = \phi$.*

(ii) *The subsystems $\mathrm{N}_{\mathcal{F}}(Q,R)$ and $\mathrm{N}_{\mathcal{F}}(R,Q)$ are equal.*

(iii) *If Q is fully $\mathcal{F}$-normalized and R is fully $\mathrm{N}_{\mathcal{F}}(R)$-normalized then $\mathrm{N}_{\mathcal{F}}(Q,R)$ is saturated.*

Proof Suppose that $\psi : AQR \to BQR$ is a morphism in $\mathcal{F}$ such that $\psi|_Q$ and $\psi|_R$ are automorphisms. Clearly ψ lies in $\mathrm{N}_{\mathcal{F}}(Q)$ and indeed also in $\mathrm{N}_{\mathrm{N}_{\mathcal{F}(Q)}}(R) = \mathrm{N}_{\mathcal{F}}(Q,R)$. Hence ψ is a morphism in $\mathrm{N}_{\mathcal{F}}(Q,R)$, and thus so is $\phi = \psi|_A$. Conversely, if $\phi : A \to B$ is a morphism in $\mathrm{N}_{\mathcal{F}}(Q,R)$, there is a morphism $\bar{\phi} : AR \to BR$ in $\mathrm{N}_{\mathcal{F}}(Q)$ extending ϕ and such that $\bar{\phi}|_R \in \mathrm{Aut}(R)$. Therefore there is a morphism $\psi : AQR \to BQR$ in $\mathcal{F}$ extending $\bar{\phi}$ such that $\psi|_Q \in \mathrm{Aut}(Q)$, completing the proof of (i).

The proof of (ii) is clear, since the equivalent condition to morphisms lying in $\mathrm{N}_{\mathcal{F}}(Q,R)$ given in (i) is symmetric in Q and R. To see (iii), we notice that $\mathcal{E} = \mathrm{N}_{\mathcal{F}}(Q)$ is saturated since $\mathcal{F}$ is saturated and Q is fully $\mathcal{F}$-normalized, and $\mathrm{N}_{\mathcal{E}}(R) = \mathrm{N}_{\mathcal{F}}(Q,R)$ is saturated since $\mathcal{E}$ is saturated and R is fully $\mathcal{E}$-normalized. This completes the proof. $\square$

We can of course extend this definition to any collection $Q_1, \ldots, Q_n$ of subgroups such that $Q_i \leq \mathrm{N}_P(Q_j)$ for all i and j, and form the subsystem $\mathrm{N}_{\mathcal{F}}(Q_1, Q_2, \ldots, Q_n)$. However, we will only need the case $n = 2$ here. For K-normalizers, there is a suitable analogue; see Exercise 4.2.

4.5 Centric and radical subgroups

The goal of the next section is to prove a version of Alperin's fusion theorem for fusion systems. The set of subgroups that control conjugation in a fusion system, in the sense that all morphisms in $\mathcal{F}$ are compositions of (restrictions of) automorphisms of elements of this set, is a small subset of all subgroups. To define the types of subgroups we are interested in, we need the new concepts of centric, radical, and essential subgroups. Being centric is a statement about the centralizer of the subgroup, whereas being radical or essential is a statement about the automorphisms of the subgroup.

We begin with centric subgroups.

Definition 4.41 Let $\mathcal{F}$ be a fusion system on a finite p-group P. A subgroup Q is $\mathcal{F}$-*centric* if, whenever R is $\mathcal{F}$-isomorphic to Q, then R contains its centralizer (or equivalently, $C_P(R) = Z(R)$).

Clearly, if Q is $\mathcal{F}$-centric, then so are all subgroups $\mathcal{F}$-conjugate to Q. Also, any $\mathcal{F}$-centric subgroup is fully centralized, since the centralizers of all $\mathcal{F}$-conjugate subgroups have the same order. Not every fully centralized subgroup is centric, but we do have something along those lines.

Lemma 4.42 *Let $\mathcal{F}$ be a saturated fusion system on a finite p-group P. If Q is a fully centralized subgroup of P then $Q \, C_P(Q)$ is $\mathcal{F}$-centric.*

Proof Suppose that Q is fully $\mathcal{F}$-centralized. Write $\bar{Q} = Q \, C_P(Q)$, and notice that clearly $C_P(\bar{Q}) \leq \bar{Q}$. Let $\phi : \bar{Q} \to R$ be an $\mathcal{F}$-isomorphism: notice that $R \leq (Q\phi) \, C_P(Q\phi)$, and since Q is fully centralized, $R = (Q\phi) \, C_P(Q\phi)$, so that $C_P(R) \leq R$ for the same reason that $C_P(\bar{Q}) \leq \bar{Q}$. Thus $\bar{Q}$ is $\mathcal{F}$-centric. $\square$

For finite groups we have a notion of centric, and it is relatively easy to define.

Proposition 4.43 *Let G be a finite group, and let P be a Sylow p-subgroup of G. A subgroup Q of P is $\mathcal{F}_P(G)$-centric if and only if*

$$C_G(Q) = Z(Q) \times O_{p'}(C_G(Q)).$$

Proof Suppose that Q is $\mathcal{F}$-centric; then Q is fully centralized, $C_P(Q) = Z(Q)$, and $C_{\mathcal{F}}(Q) = \mathcal{F}_{Z(Q)}(Z(Q))$. To see this last equality, if ϕ is a morphism in $C_{\mathcal{F}}(Q)$, then ϕ extends to a morphism that acts trivially on Q, and hence ϕ acts trivially on the subgroup of $Z(Q)$ that is its domain. By Frobenius's normal p-complement theorem, this implies that $C_G(Q)$ has a normal p-complement.

It remains to show that $Z(Q) \trianglelefteq C_G(Q)$. However, $Z(Q)$ is clearly a central subgroup of $C_G(Q)$, hence normal in $C_G(Q)$.

Now suppose that $C_G(Q) = Z(Q) \times R$, where $R = O_{p'}(C_G(Q))$; if g is an element of G such that $Q^g \leq P$, then $Z(Q)^g$ is a Sylow p-subgroup of $C_G(Q)^g = C_G(Q^g)$, it must be equal to $C_P(Q^g)$. This proves that Q is $\mathcal{F}$-centric. $\qquad\square$

The set of all $\mathcal{F}$-centric subgroups is also closed under inclusion.

Lemma 4.44 *Let $\mathcal{F}$ be a fusion system on a finite p-group P, and let Q and R be subgroups of P with $Q \leq R$. If Q is $\mathcal{F}$-centric then so is R, and $Z(Q) \leq Z(R)$.*

Proof Let $\phi : R \to S$ be an isomorphism in $\mathcal{F}$; since Q is $\mathcal{F}$-centric, $C_P(R\phi) \leq C_P(Q\phi) \leq Q\phi \leq R\phi$, and so R is $\mathcal{F}$-centric. Also, $Z(R\phi) = C_P(R\phi) \leq C_P(Q\phi) = Z(Q\phi)$, and letting $\phi = 1$ gives us the second statement. $\qquad\square$

Thus the centric subgroups are at the top of the group, and once you find a centric subgroup, all of the subgroups above it are centric. Centric subgroups are important in Alperin's fusion theorem, as is the next notion.

Let $\mathcal{F}$ be a fusion system on a finite p-group P, and let Q be a subgroup of P. Notice that $\mathrm{Aut}_Q(Q)$ is a p-group, and in fact $\mathrm{Aut}_Q(Q) = \mathrm{Inn}(Q)$ is a normal p-subgroup of $\mathrm{Aut}_{\mathcal{F}}(Q)$.

Definition 4.45 Let $\mathcal{F}$ be a fusion system on a finite p-group P. A subgroup Q is said to be *radical* if

$$O_p(\mathrm{Aut}_{\mathcal{F}}(Q)) = \mathrm{Inn}(Q).$$

For fusion systems of finite groups, we do not really have a very nice interpretation of radical subgroups. The only way to describe them is the obvious one: if Q is a subgroup of a Sylow p-subgroup P of a finite group G, then Q is $\mathcal{F}_P(G)$-radical if and only if $N_G(Q)/Q\,C_G(Q)$ has no non-trivial normal p-subgroups.

Proposition 4.46 *Let $\mathcal{F}$ be a fusion system on a finite p-group P, and let Q be a subgroup of P. If $\mathcal{F} = N_{\mathcal{F}}(Q)$, then Q is contained within every centric, radical subgroup of P.*

Proof Let R be a subgroup of P, and suppose that R is radical and centric. We claim that the image of Q in $\mathrm{Aut}_{\mathcal{F}}(R)$ is, in fact, a normal subgroup of $\mathrm{Aut}_{\mathcal{F}}(R)$. If this is true, then it is contained in the image $\mathrm{Inn}(R) = O_p(\mathrm{Aut}_{\mathcal{F}}(R))$ of R in $\mathrm{Aut}_{\mathcal{F}}(R)$. Thus $Q \leq R\,\mathrm{C}_P(R)$, and since R is centric, $R\,\mathrm{C}_P(R) = R$, yielding the result.

It remains to prove the claim. Let ϕ be an automorphism in $\mathrm{Aut}_{\mathcal{F}}(R)$, and extend ϕ to an automorphism of QR. Note that both Q and R are ϕ-invariant, and so $\mathrm{N}_Q(R) = Q \cap \mathrm{N}_{QR}(R)$ is ϕ-invariant.

If $g \in Q$, then g normalizes R if and only if $c_g \in \mathrm{Aut}_Q(R)$, and so it suffices to show that $\phi^{-1} c_g \phi = c_{g\phi}$, for $c_g \in \mathrm{Aut}_Q(R)$. This calculation is well known:

$$x(\phi^{-1} c_g \phi) = (x\phi^{-1}) c_g \phi = (g^{-1} x\phi^{-1} g)\phi = (g\phi)^{-1} x(g\phi),$$

as claimed. $\qquad\qquad\qquad\qquad\qquad\qquad\qquad\qquad\qquad\qquad\qquad\square$

Fully normalized, centric, radical subgroups are interesting in fusion systems, as their automorphisms in some sense 'generate' the fusion system: we will see this in the next section. We can restrict still further the collection of subgroups that we need for this generation statement, to so-called essential subgroups.

Recall from Chapter 1 the following definition. Let G be a finite group with $p \mid |G|$, and let M be a subgroup of G. We say that M is *strongly p-embedded* in G if M contains a Sylow p-subgroup of G and $M \cap M^g$ is a p'-group for all $g \in G \setminus M$. Note that if G has a strongly p-embedded subgroup then $O_p(G) = 1$.

The reason we are bringing it up again here is that this is the property that we want for the outer automorphisms of a subgroup in a fusion system.

Definition 4.47 Let $\mathcal{F}$ be a fusion system on a finite p-group P. A subgroup Q of P is *$\mathcal{F}$-essential* if Q is $\mathcal{F}$-centric and $\mathrm{Out}_{\mathcal{F}}(Q) = \mathrm{Aut}_{\mathcal{F}}(Q)/\mathrm{Inn}(Q)$ contains a strongly p-embedded subgroup.

Notice that every essential subgroup is centric radical, so the set of fully normalized, essential subgroups is contained within the fully normalized, centric, radical subgroups. The interpretation for finite groups is obvious: a subgroup Q is essential if and only if $\mathrm{N}_G(Q)/Q\,\mathrm{C}_G(Q)$ has a strongly p-embedded subgroup.

Groups with a strongly p-embedded subgroup can be characterized in terms of their Quillen complex. Rather than deal with the whole Quillen complex, we simply consider the partially ordered set of all non-identity

p-subgroups of a finite group, which we will turn into an undirected graph via inclusion in the obvious way, and denote this by $A_p(G)$.

Proposition 4.48 *Let G be a finite group such that $p \mid |G|$. Then G has a strongly p-embedded subgroup if and only if the graph $A_p(G)$ is disconnected.*

Proof Suppose that G has a strongly p-embedded subgroup, M, containing a Sylow p-subgroup P. Let g be an element of $G \setminus M$, and consider P^g. We claim that P^g and P lie in different components of $A_p(G)$. Since $M \cap M^g$ is a p'-group, we see that $P \cap P^g = 1$. Suppose that $Q = Q_0, Q_1, \ldots, Q_n = Q^g$ is a path of minimal length linking $Q \le P$ and $Q^g \le P^g$, as we range over all subgroups of P and all paths. Since $Q \le P$ and $Q \cap Q_1 \ne 1$, we must have that Q_1 is contained within P, contradicting the minimal length claim. Thus P and P^g lie in different components, as claimed.

Now suppose that $A_p(G)$ is disconnected, and let P be a Sylow p-subgroup of G. Since $A_p(G)$ is disconnected, this splits $\mathrm{Syl}_p(G)$ into (at least two) components (else all p-subgroups, which are contained in Sylow p-subgroups, would be connected to each other), and let S denote the subset of $\mathrm{Syl}_p(G)$ lying in the same component as P. Let M denote the set of all $g \in G$ such that $P^g \in S$. The claim is that M is a strongly p-embedded subgroup of G. Firstly, M is clearly a subgroup, and contains a Sylow p-subgroup of G. Furthermore, if $g \notin M$, then for any (non-trivial) p-subgroup Q of M, we have that Q and Q^g are not connected in $A_p(G)$, so certainly $Q \cap Q^g = 1$. Hence $M \cap M^g$ is a p'-group, as required. □

In the case where $p = 2$, the presence of a strongly 2-embedded subgroup (often abbreviated simply to 'strongly embedded') in a finite group G implies a considerable restriction on the structure of G, because of a theorem of Bender.

Theorem 4.49 (Bender [Ben71]) *Let G be a finite group with a strongly embedded subgroup. Either*

(i) *G has cyclic or generalized quaternion Sylow 2-subgroups, or*
(ii) *$G / O_{2'}(G)$ has a normal subgroup of odd index, isomorphic to one of the groups $\mathrm{PSL}_2(2^n)$, $\mathrm{PSU}_3(2^n)$ or $\mathrm{Sz}(2^n)$, for some $n \ge 2$.*

This substantially restricts the possible choices for the outer automorphism groups of essential subgroups, at least for the prime 2.

Example 4.50 Let P be a finite 2-group, and let Q be an elementary abelian subgroup of P of order 2^n; then $\mathrm{Aut}(Q) = \mathrm{Out}(Q) = \mathrm{GL}_n(2)$, which has a strongly embedded subgroup if and only if $n = 2$. Thus the full automorphism group of the Klein four-group has a strongly embedded subgroup, for example.

If $Q \cong V_4$, then $\mathrm{Aut}(Q) \cong S_3$, and there is only one subgroup of S_3 with a strongly embedded subgroup, namely S_3 itself. Thus $\mathrm{Out}_{\mathcal{F}}(Q)$ must be isomorphic with $\mathrm{Aut}(Q)$ and so, if Q is fully normalized, then $\mathrm{N}_P(Q)/\mathrm{C}_P(Q)$ has order 2. For Q to be essential, it must also be centric, and since it is abelian this means that $Q = \mathrm{C}_P(Q)$, and so $\mathrm{N}_P(Q) \cong D_8$. (It is not difficult to prove that if a finite 2-group P has a self-centralizing Klein four-subgroup, then P is either dihedral or semidihedral (see Lemma 5.81).)

For the larger elementary abelian groups, it might still be possible that they are essential, but in this case it cannot be that $\mathcal{F}$ induces the full automorphism group on the subgroup.

For odd primes p, the classification of finite groups with a strongly p-embedded subgroup is a result of Gorenstein–Lyons [GL83] and Aschbacher [Asc93], based on the classification of the finite simple groups. The structure of finite groups with a strongly p-embedded subgroup is very restricted.

4.6 Alperin's fusion theorem

In this section we will provide a proof of Alperin's fusion theorem for fusion systems. In its original statement, it essentially ran as follows: any $\mathcal{F}$-isomorphism may be 'factored' into restrictions of automorphisms of fully normalized, centric, radical subgroups of the ambient p-group P. In the refined version that we give here, due to Puig, the class of subgroups needed to factor an automorphism is restricted still further, with the loss of granularity being an automorphism of P itself.

Theorem 4.51 (Alperin's fusion theorem) *Let $\mathcal{F}$ be a saturated fusion system on a finite p-group P, let $\mathscr{S}$ denote the set of all fully normalized, essential subgroups of P, and let Q and R denote two subgroups of P, with $\phi : Q \to R$ an $\mathcal{F}$-isomorphism. There exist*

(i) *a sequence of $\mathcal{F}$-isomorphic subgroups $Q = Q_0, Q_1, \ldots, Q_{n+1} = R$,*
(ii) *a sequence $S_1, S_2, \ldots, S_n$ of elements of $\mathscr{S}$, with $Q_{i-1}, Q_i \leq S_i$,*

(iii) *a sequence of $\mathcal{F}$-automorphisms ϕ_i of S_i such that $Q_{i-1}\phi_i = Q_i$, and*

(iv) *an $\mathcal{F}$-automorphism ψ of P (mapping Q_n to Q_{n+1}), such that*
$$(\phi_1\phi_2\ldots\phi_n\psi)|_Q = \phi.$$

Proof We begin by showing that if θ is a $\mathcal{F}$-automorphism of P, and ρ is an $\mathcal{F}$-automorphism of some fully normalized, essential subgroup E, then there exists an $\mathcal{F}$-automorphism ρ' of some other fully normalized, essential subgroup $\tilde{E}$ such that $\theta\rho = \rho'\theta$. Notice that

$$u\theta\rho = u(\theta\rho\theta^{-1})\theta$$

for all $u \in E\theta^{-1}$. We need to show that $E\theta^{-1}$ is a fully normalized, essential subgroup of P, for then $\rho' = \theta\rho\theta^{-1}$ is an automorphism of it, and we have proved our claim. However, $\theta \in \mathrm{Aut}_{\mathcal{F}}(P)$, and so $\mathrm{N}_P(E)\theta^{-1} = \mathrm{N}_P(E\theta^{-1})$. Since E is fully normalized, $|\mathrm{N}_P(E)|$ is maximal amongst subgroups $\mathcal{F}$-isomorphic to E, and so therefore $E\theta^{-1}$ is fully normalized as well. The property of being essential is clearly transported by θ^{-1} and so the claim holds.

This proves that the product of two $\mathcal{F}$-isomorphisms that possess a decomposition of the required form also possesses a decomposition of the required form, as does the inverse of such an $\mathcal{F}$-isomorphism. This will be invaluable in what follows.

We proceed by induction on $|P : Q|$; if $Q = P$ then ϕ is an automorphism of P, and so $n = 0$ and the theorem is true. Thus we may assume that $Q < P$. The proof will proceed in stages.

Suppose firstly that R is fully normalized. By Proposition 4.17, there is a map ϕ from Q to R such that $N_\phi = \mathrm{N}_P(Q)$. Since any two isomorphisms between two subgroups differ by an automorphism of R, there exists $\chi \in \mathrm{Aut}_{\mathcal{F}}(R)$ such that $\phi\chi = \phi$. Thus there is a morphism $\overline{\phi\chi} : \mathrm{N}_P(Q) \to P$ extending $\phi\chi$, and since $Q < \mathrm{N}_P(Q)$, we may apply the inductive hypothesis to $\overline{\phi\chi}$, to get that this morphism, and hence $\phi\chi$, has such a decomposition.

It remains to show that χ has such a decomposition, since then $\phi = (\phi\chi)\chi^{-1}$ has a decomposition of the required form. Thus let χ be an element of $\mathrm{Aut}_{\mathcal{F}}(R)$, where R is fully normalized. If R is not centric then $R\,\mathrm{C}_P(R) > R$, and since N_χ contains $R\,\mathrm{C}_P(R)$, we may decompose $\bar{\chi}$ (which extends χ to $R\,\mathrm{C}_P(R)$), so we may decompose χ, as claimed.

Since R is fully normalized, it cannot be essential, since otherwise χ would be of the required form.

By Proposition 4.48, there are two sequences of subgroups, $\mathrm{Aut}_P(R) = A_1, A_2, \ldots, A_n = \mathrm{Aut}_P(R)^\chi$ and $B_1, \ldots, B_{n-1}$ such that

(i) $B_i, \leq A_i, A_{i+1}$ for $i < n$, and

(ii) $\mathrm{Aut}_R(R) < B_i$ for all i.

Replacing the A_i with Sylow p-subgroups of $\mathrm{Aut}_{\mathcal{F}}(R)$ containing each A_i, for each i we have an element $\chi_i \in \mathrm{Aut}_{\mathcal{F}}(R)$ such that $A_1^{\chi_i} = A_i$ (taking χ_1 to be the identity) as all Sylow p-subgroups of $\mathrm{Aut}_{\mathcal{F}}(R)$ are conjugate inside $\mathrm{Aut}_{\mathcal{F}}(R)$, with $\chi_{n-1} = \chi$. Writing $\theta_i = \chi_{i+1}\chi_i^{-1}$, we have that $\chi_{i+1} = \theta_i\theta_{i-1}\ldots\theta_1$ for all $1 \leq i < n-1$.

We want to show that $N_{\theta_i} \leq N_P(R)$ properly contains R, for then each θ_i extends to an overgroup of R, and hence has a decomposition of the required form, whence θ_i does. Finally, the composition of the θ_i is χ, and so that has a decomposition of the required form.

The subgroup $N_{\theta_i}/Z(R)$ properly contains $\mathrm{Aut}_R(R)$ because

$$(N_{\theta_i}/Z(R))^{\chi_{i+1}} = \left(\mathrm{Aut}_P(R) \cap \mathrm{Aut}_P(R)^{\theta_i^{-1}}\right)^{\chi_{i+1}}$$

$$= A_{i+1} \cap A_i \geq B_i > \mathrm{Aut}_R(R).$$

Hence $N_\theta > R$, as required.

The last step is to remove the assumption that R is fully normalized. Let $\nu : R \to S$ be an $\mathcal{F}$-isomorphism such that S is fully normalized. Both ν and $\phi\nu$ have decompositions of the required form, since they are $\mathcal{F}$-isomorphisms mapping onto a fully normalized subgroup of P. Therefore ϕ has such a decomposition, by the conclusion of the first paragraph. □

A weaker form of Alperin's fusion theorem is also useful, and in most cases is all that is needed for applications.

Theorem 4.52 *Let $\mathcal{F}$ be a saturated fusion system on a finite p-group P, and let $\phi : Q \to R$ be an isomorphism. There exist*

(i) *a sequence of $\mathcal{F}$-isomorphic subgroups $Q = Q_0, Q_1, \ldots, Q_n = R$,*

(ii) *a sequence $S_1, S_2, \ldots, S_n$ of fully normalized, $\mathcal{F}$-radical, $\mathcal{F}$-centric subgroups, with $Q_{i-1}, Q_i \leq S_i$, and*

(iii) *a sequence of $\mathcal{F}$-automorphisms ϕ_i of S_i with $Q_{i-1}\phi_i = Q_i$, such that*

$$(\phi_1\phi_2\ldots\phi_n)|_Q = \phi.$$

Proof   Since every essential subgroup is radical and centric, we have expanded the collection of subgroups for which we may consider automorphisms. In particular, the whole group P is fully normalized, centric, and radical (since $\operatorname{Aut}_P(P)$ is a Sylow p-subgroup of $\operatorname{Aut}_{\mathcal{F}}(P)$), and so an $\mathcal{F}$-automorphism of P, as required by Theorem 4.51, is allowed as one of the ϕ_i. Thus Theorem 4.51 implies this weaker version, as claimed. $\square$

The question of whether a fusion system $\mathcal{F}$ on a finite p-group P has any fully normalized, essential subgroups is an interesting one. If a given finite p-group has the property that any saturated fusion system $\mathcal{F}$ on P has no essential subgroups, then P is called *resistant*; in Section 7.1 we will explore this concept further. One may turn the question on its head, and ask whether a particular p-group may be an essential subgroup of some overgroup. Indeed, can the automorphism group of a p-group contain a strongly p-embedded subgroup at all?

If the automorphism group of a p-group is itself a p-group, then this group cannot be an essential subgroup in any fusion system.

Proposition 4.53   *Let P be a dihedral 2-group of order at least 8, a generalized quaternion 2-group of order at least 16, or a semidihedral 2-group of order at least 16. Then $\operatorname{Aut}(P)$ is a 2-group, and consequently P cannot be an essential subgroup in any fusion system.*

Proof   Notice that, in all cases, $P/\Phi(P)$ has order 4. Any non-trivial odd automorphism of a 2-group induces a non-trivial automorphism of the quotient $P/\Phi(P)$ (Burnside's theorem [Gor80, Theorem 5.1.4]). The only non-trivial odd automorphisms of V_4 have order 3, and permute transitively the three non-identity elements; these correspond to the three maximal subgroups of P under the map $P \to P/\Phi(P)$. Therefore, in order to have a non-trivial odd automorphism, all maximal subgroups of P must be isomorphic.

Also, each of the groups above has a cyclic subgroup of index 2 and so all maximal subgroups of P are cyclic. If P is dihedral or semidihedral of order at least 8, then P possesses Klein four-subgroups, which must be contained within a non-cyclic subgroup of P. If P is generalized quaternion, then P is generated by a and b such that $a^q = b^4 = 1$ and $a^b = a^{-1}$, and $\langle a^2, b \rangle$ is a quaternion subgroup of index 2. This gives the result. $\square$

Therefore of the subgroups of the 2-groups of maximal class, only the quaternion group of order 8 and the Klein four-group can be essential subgroups. (The automorphism group of a cyclic 2-group is itself also a 2-group.) One may use this result to prove the following theorem.

Theorem 4.54 *Let P be a dihedral 2-group of order at least 8, let Q be a generalized quaternion group of order at least 16, and let R be a semidihedral 2-group, of order at least 16. There are exactly three non-isomorphic saturated fusion systems on P and Q, and exactly four non-isomorphic saturated fusion systems on R. All of these fusion systems are realizable by finite groups.*

Proof Firstly, let P be a dihedral 2-group of order $4q$, presented as $\langle a, b \mid a^{2q} = b^2 = 1, a^b = a^{-1} \rangle$. There are three conjugacy classes of involutions: those of the form $a^{2i}b$, those of the form $a^{2i+1}b$, and the central involution a^q. A Klein four-subgroup of P has elements 1, a^q, ab^j, and ab^{q+j}, and hence there are q of these, separated into two conjugacy classes, with representatives $A_1 = \langle a^q, b \rangle$ and $A_2 = \langle a^q, ab \rangle$. Notice that $\mathrm{Aut}_P(A_i)$ has order 2, given by the element swapping b and $a^q b$ in A_1, and ab and $a^{q+1}b$ in A_2. Since all subgroups of P are dihedral, cyclic, or V_4, and dihedral and cyclic 2-groups have outer automorphism groups of order 2 (see Corollary 1.14 and Proposition 4.53), the only two possible $\mathcal{F}$-conjugacy classes of essential subgroups are those of A_1 and A_2, and so there are three isomorphism types of saturated fusion system, depending on whether both A_1 and A_2, exactly one of A_1 or A_2, or neither A_1 nor A_2, has an automorphism of order 3 in $\mathcal{F}$. (Alperin's fusion theorem proves that the two groups $\mathrm{Aut}_{\mathcal{F}}(A_i)$ determine $\mathcal{F}$.) If exactly one of the A_i has an $\mathcal{F}$-automorphism of order 3, then it does not matter which one, because P has an outer automorphism of order 2 swapping A_1 and A_2. (This is the automorphism fixing a and swapping b with ab.)

If neither A_1 nor A_2 has a non-trivial automorphism, then this is simply the fusion system $\mathcal{F} = \mathcal{F}_P(P)$, and $\mathcal{F}$ has three conjugacy classes of involutions. If A_1 has an $\mathcal{F}$-automorphism of order 3, but not A_2, then $\mathcal{F}$ has two conjugacy classes of involutions, since b and a^q are $\mathcal{F}$-conjugate, and $\mathcal{F} = \mathcal{F}_P(\mathrm{PGL}_2(p^d))$, for some suitable choice of p^d so that the order of the Sylow 2-subgroup is $4q$. If both of the A_i have $\mathcal{F}$-automorphisms of order 3, then $\mathcal{F}$ has a single conjugacy class of involutions, and $\mathcal{F} = \mathcal{F}_P(\mathrm{PSL}_2(p^d))$, for some suitable choice of p^d so that the order of the Sylow 2-subgroup is $4q$.

Now let Q be a quaternion group of order $4q$, and note that the only possible essential subgroups are those isomorphic with Q_8 (as all subgroups are quaternion or cyclic). Using the fact that $Q/Z(Q)$ is a dihedral group, we see that there are two conjugacy classes of subgroups of Q isomorphic with Q_8, with representatives B_1 and B_2. Again, there are three possibilities, according to whether neither B_1 nor B_2, exactly one of B_1 and B_2, or both B_1 and B_2, have an $\mathcal{F}$-automorphism of order 3. The first and last possibilities are realized by Q itself and $\mathrm{SL}_2(q)$ respectively, for suitable choices of q. The intermediate fusion system, when exactly one of B_i has an $\mathcal{F}$-automorphism of order 3, is the fusion system of a group with $\mathrm{SL}_2(q)$ of index 2; see Exercise 4.4.

Finally, we consider the semidihedral groups R, and assume that $|R| \geq 32$. These have three maximal subgroups, isomorphic with a cyclic group, a dihedral group S_1, and a generalized quaternion group S_2, and $\mathrm{Out}_R(S_i)$ has order 2. This outer automorphism swaps the two conjugacy classes of Klein four-subgroups in S_1, and the two conjugacy classes of Q_8 subgroups in S_2, and so there is exactly one class of Klein four-subgroups (with representative D_1) and one class of quaternion subgroups (with representative D_2) in R. Thus we get four possibilities for fusion systems on R, depending on whether neither, just D_1, just D_2, or both, have $\mathcal{F}$-automorphisms of order 3.

If $|R| = 16$, then we get the same result as before, that there is one conjugacy class each of Klein four- and Q_8 subgroups, but this time there is only one subgroup in total of R isomorphic with Q_8.

The semidihedral groups have two conjugacy classes of involutions, which are fused in $\mathcal{F}$ if and only if D_1 has an automorphism of order 3. They also have two conjugacy classes of elements of order 4, which are fused in $\mathcal{F}$ if and only if D_2 has an automorphism of order 3. Hence one may determine which fusion system a finite group with semidihedral Sylow 2-subgroups has by examining the conjugacy classes of involutions and elements of order 4.

We include a table here, displaying the number of conjugacy classes of elements of orders 2 and 4, along with the group realizing the fusion system $\mathcal{F}$.

Order 2	Order 4	Group
2	2	SD_{2^n}
1	2	$\mathrm{PSL}_2(q^2) \cdot \langle \sigma\tau \rangle$
2	1	$\mathrm{GL}_2(q)$
1	1	$\mathrm{PSL}_3(q)$

In this table, the second group down is the group obtained by extending the group $PSL_2(q^2)$ by the diagonal-field automorphism, which is a non-split extension of $PSL_2(q^2)$ by C_2. It has semidihedral Sylow 2-subgroups, and since it is a non-split extension, it has a single conjugacy class of involutions. Thus it must have more than one class of elements of order 4, since some lie outside $PSL_2(q^2)$. □

By a theorem of Martin [Mar86], it is known that, for almost all p-groups, the automorphism group (and hence the outer automorphism group) is itself a p-group. Hence almost no isomorphism types of p-group can be found as essential subgroups of saturated fusion systems.

4.7 Weak and strong closure

In finite group theory, a subgroup H is said to be *weakly closed* in K with respect to G if, whenever $g \in G$ has the property that $H^g \leq K$, then $H = H^g$; in other words, no G-conjugate of H other than H itself lies inside K. In an extension of this notion, H is said to be *strongly closed* in K with respect to G if, whenever $\bar{H}$ is a subgroup of H and $\bar{H}^g \leq K$, then $\bar{H}^g \leq H$; in other words, if any subgroup of H is conjugated to a subgroup inside K, then this subgroup lies inside H. Clearly, if a subgroup is strongly closed in K then it is weakly closed in K.

The corresponding notions for fusion systems are easy to define.

Definition 4.55 Let $\mathcal{F}$ be a fusion system on a finite p-group P, and let Q be a subgroup of P.

(i) Q is said to be *weakly $\mathcal{F}$-closed* if, whenever $\phi : Q \to P$ is a morphism in $\mathcal{F}$, the image of ϕ is Q (or equivalently, Q is only $\mathcal{F}$-conjugate to itself).

(ii) Q is said to be *strongly $\mathcal{F}$-closed* if, whenever R is a subgroup of Q and $\phi : R \to P$ is a morphism in $\mathcal{F}$, then $R\phi \leq Q$.

The following verification is easy, and left to the reader.

Proposition 4.56 *Let G be a finite group, and let P be a Sylow p-subgroup of G. Let Q be a subgroup of P.*

(i) *Q is weakly $\mathcal{F}_P(G)$-closed if and only if it is weakly closed in P with respect to G.*

(ii) *Q is strongly $\mathcal{F}_P(G)$-closed if and only if it is strongly closed in P with respect to G.*

There is a sometimes useful relation between normalizers and strongly closed subgroups.

Lemma 4.57 *Let $\mathcal{F}$ be a fusion system on a finite p-group P. If Q is a subgroup of P, then Q is strongly $N_{\mathcal{F}}(Q)$-closed, and if K is a normal subgroup of $\mathrm{Aut}_{\mathcal{F}}(Q)$, then $N_P^K(Q)$ is strongly $N_{\mathcal{F}}(Q)$-closed. In particular, $C_P(Q)$ and $Q\,C_P(Q)$ are strongly $N_{\mathcal{F}}(Q)$-closed.*

Proof Let $\phi : R \to N_P(Q)$ be a morphism in $N_{\mathcal{F}}(Q)$, where $R \leq Q$. Since ϕ extends to a morphism $\bar{\phi} : QR \to N_P(Q)$ such that $\bar{\phi}|_Q$ is an automorphism, and $QR = Q$, we see that $\bar{\phi} \in \mathrm{Aut}(Q)$; therefore $\mathrm{im}\,\phi \leq Q$, as claimed.

For the proof of the second claim, suppose that $R \leq N_P^K(Q)$ and let $\phi : R \to N_P(Q)$ be a morphism in $N_{\mathcal{F}}(Q)$; as ϕ is a morphism in $N_{\mathcal{F}}(Q)$, ϕ extends to $\bar{\phi} : QR \to N_P(Q)$. Writing $\psi = \bar{\phi}|_Q$, by Lemma 4.18 we have that $\mathrm{Aut}_{R\bar{\phi}}(Q) = \mathrm{Aut}_R(Q)^\psi \leq K$, since $\mathrm{Aut}_R(Q) \leq K \trianglelefteq \mathrm{Aut}_{\mathcal{F}}(Q)$. Since $N_P^K(Q)$ is the full preimage of K under the map $c_Q : N_P(Q) \to \mathrm{Aut}(Q)$, we must have that $R\bar{\phi} \leq N_P^K(Q)$, as claimed. $\qquad\square$

Another important proposition is the following, although the proof will use results from the next chapter; see Exercise 5.3.

Proposition 4.58 *Let $\mathcal{F}$ be a saturated fusion system on a finite p-group P. If Q is a strongly $\mathcal{F}$-closed subgroup of P then so is $Q\,C_P(Q)$.*

The intersection of two strongly closed subgroups is strongly closed, as is the product of two weakly closed subgroups. These two assertions are easy to prove, and we do these now.

Lemma 4.59 *Let $\mathcal{F}$ be a fusion system on a finite p-group P, and let Q and R be subgroups of P.*

(i) *If Q and R are weakly $\mathcal{F}$-closed, then so is QR.*
(ii) *If Q and R are strongly $\mathcal{F}$-closed, then so is $Q \cap R$.*

Proof Suppose firstly that Q and R are weakly $\mathcal{F}$-closed, and let $\phi : QR \to P$ be a morphism in $\mathcal{F}$. We have that $\psi = \phi|_Q$ and $\theta = \phi|_R$ are morphisms in $\mathcal{F}$, and so $Q\psi = Q$ and $R\theta = R$; hence $(QR)\phi = (Q\phi)(R\phi) = QR$, as needed for (i).

Now suppose that Q and R are strongly $\mathcal{F}$-closed, and let S be a subgroup of $Q \cap R$. Let $\phi : S \to P$ be a morphism in $\mathcal{F}$; then, since Q is strongly $\mathcal{F}$-closed, $S\phi \leq Q$. Similarly, $S\phi \leq R$, and so $S\phi \leq Q \cap R$, proving that $Q \cap R$ is strongly $\mathcal{F}$-closed. $\qquad\square$

It is easy to construct examples of weakly $\mathcal{F}$-closed subgroups Q and R such that $Q \cap R$ is not weakly $\mathcal{F}$-closed. For example, let P be the dihedral group of order 8, and let Q and R denote the two subgroups isomorphic with V_4. If $\mathcal{F} = \mathcal{F}_P(S_4)$, then the centre of P is not weakly $\mathcal{F}$-closed, but both Q and R are weakly $\mathcal{F}$-closed. Thus Lemma 4.59 cannot be extended in that direction. The only other direction is to ask whether the product of two strongly $\mathcal{F}$-closed subgroups is strongly $\mathcal{F}$-closed.

Theorem 4.60 (Aschbacher [Asc11]) *In any saturated fusion system $\mathcal{F}$, the product of two strongly $\mathcal{F}$-closed subgroups is strongly $\mathcal{F}$-closed.*

The proof in [Asc11, Theorem 2] is a series of reductions. In Section 5.1, we will provide an elementary proof (Theorem 5.22), but to do so we need the concept of factor systems and morphisms of fusion systems.

To end this chapter, we *nearly* prove a theorem of Stancu [Sta06] giving an equivalent condition for a subgroup Q to have the property that $\mathcal{F} = N_{\mathcal{F}}(Q)$; the sticking point, as with the result claimed above, is that we do not yet have the notions of factor systems and morphisms. However, we will prove a few more results that will be of use later. We begin by finishing off a loose end with Proposition 4.46, which said that if $\mathcal{F} = N_{\mathcal{F}}(Q)$, then Q is contained within every centric, radical subgroup.

Proposition 4.61 *Let $\mathcal{F}$ be a saturated fusion system on a finite p-group P, and let Q be a strongly $\mathcal{F}$-closed subgroup of P. The following are equivalent:*

(i) $\mathcal{F} = N_{\mathcal{F}}(Q)$;
(ii) *Q is contained in every fully normalized, $\mathcal{F}$-centric, $\mathcal{F}$-radical subgroup of P;*
(iii) *Q is contained in every fully normalized, $\mathcal{F}$-essential subgroup of P.*

Proof By Proposition 4.46, if $\mathcal{F} = N_{\mathcal{F}}(Q)$, then Q is contained in every $\mathcal{F}$-radical, $\mathcal{F}$-centric subgroup of P, so (i) implies (ii). Clearly (ii) implies (iii); therefore, suppose that Q is contained in every fully normalized, $\mathcal{F}$-essential subgroup of P, and let $\phi : R \to S$ be a morphism in $\mathcal{F}$. By Alperin's fusion theorem, ϕ may be decomposed as the restriction of $\phi_1 \phi_2 \dots \phi_n$, where $\phi_i \in \mathrm{Aut}_{\mathcal{F}}(U_i)$ for fully normalized, $\mathcal{F}$-essential subgroups U_i. Let $R = R_0$ and $R_i = R_{i-1}\phi_i$, so that R_{i-1} is $\mathcal{F}$-conjugate to R_i via ϕ_i for all $1 \leq i \leq n$. Since Q is contained in each U_i, we have restrictions $\psi_i : QR_{i-1} \to QR_i$ of each ϕ_i, and the composition of the

ψ_i gives $\psi : QR \to QS$, extending ϕ. Since Q is strongly $\mathcal{F}$-closed, $\psi|_Q \in \mathrm{Aut}_{\mathcal{F}}(Q)$, proving that (iii) implies (i). $\qquad\square$

This gives an equivalent condition (technically two, very similar conditions) to a subgroup having the property that $\mathcal{F} = \mathrm{N}_{\mathcal{F}}(Q)$. In the remainder, we will prove one more and state another. We begin with a result of Aschbacher.

Proposition 4.62 (Aschbacher [AKO11, Lemma I.4.6]) *Let $\mathcal{F}$ be a saturated fusion system on a finite p-group P, and let Q be a subgroup of P. The following are equivalent:*

(i) *$\mathcal{F} = \mathrm{N}_{\mathcal{F}}(Q)$;*
(ii) *there exists a central series*

$$1 = Q_0 \leq Q_1 \leq \cdots \leq Q_n = Q$$

for Q, all of whose terms are strongly $\mathcal{F}$-closed.

Proof Suppose that $\mathcal{F} = \mathrm{N}_{\mathcal{F}}(Q)$, and let $Q_i = \mathrm{Z}_i(Q)$. We need to show that every Q_i is strongly $\mathcal{F}$-closed, and we are done. Let $R \leq Q_i$, and $\phi \in \mathrm{Hom}_{\mathcal{F}}(R, P)$; since Q is a normal subgroup of $\mathcal{F}$, and R is contained in Q, ϕ extends to a map $\bar{\phi} \in \mathrm{Hom}_{\mathcal{F}}(Q, P)$ that acts like an automorphism on Q (since Q is strongly $\mathcal{F}$-closed). By assumption Q_i is characteristic in Q (which we write $Q_i \,\mathrm{char}\, Q$), so we see that $\mathrm{im}\,\phi \leq Q_i$, as needed.

Now suppose that (ii) holds, and let T be any fully $\mathcal{F}$-normalized, $\mathcal{F}$-radical, $\mathcal{F}$-centric subgroup of P. If we can show that in this case $Q \leq T$, then we are done, since then Q is contained in every $\mathcal{F}$-radical, $\mathcal{F}$-centric subgroup of P, and so $\mathcal{F} = \mathrm{N}_{\mathcal{F}}(Q)$ by Proposition 4.61. Choose i maximal such that $Q_i \leq T$, so that $Q_{i+1} \not\leq T$. (For a contradiction, assume that Q is not a subgroup of T.) Set $R = Q_{i+1} \cap T$ and $S = \mathrm{N}_{Q_{i+1}}(T)$; as $Q \not\leq T$, we have that $R < S$. Since S normalizes T and $Q_{i+1} \trianglelefteq P$, we have that $[T, S] \leq R$. As Q_{i+1}/Q_i is central in Q/Q_i, we see that S centralizes R/Q_i, and since each Q_{j+1}/Q_j is central in Q/Q_j, we see that S centralizes Q_{j+1}/Q_j for all $j < i$.

Each Q_i is strongly $\mathcal{F}$-closed, and therefore $\mathrm{Aut}_{\mathcal{F}}(T)$ acts on $R = Q_{i+1} \cap T$ and Q_j for all $j \leq i$. Hence there is an $\mathrm{Aut}_{\mathcal{F}}(T)$-invariant series

$$1 = Q_0 \leq \cdots \leq Q_i \leq R \leq T,$$

with S centralizing each factor. The set of all such automorphisms of T is clearly a normal subgroup of $\mathrm{Aut}_{\mathcal{F}}(T)$, and is a p-subgroup by

Exercise 4.5. Therefore, $\mathrm{Aut}_S(T)$ is contained in a normal p-subgroup of $\mathrm{Aut}_{\mathcal{F}}(T)$, and so in particular $\mathrm{Aut}_S(T) \leq \mathrm{Inn}(T)$ since T is $\mathcal{F}$-radical. Thus $S = \mathrm{N}_{Q_{i+1}}(T) \leq T\,\mathrm{C}_P(T) = T$ since T is $\mathcal{F}$-centric. Therefore, $Q_{i+1} \cap T \geq S > R = Q_{i+1} \cap T$, a contradiction. Hence Q is contained in every $\mathcal{F}$-centric, $\mathcal{F}$-radical subgroup of P, and so $\mathcal{F} = \mathrm{N}_{\mathcal{F}}(Q)$, as claimed. $\qquad\square$

Stancu's theorem is that, in the proposition above, the Q_i (apart from Q_n) need only be *weakly* closed, rather than strongly closed. In order to prove this condition in the next chapter, we need a preliminary lemma on weak and strong closure, which will be placed here.

Lemma 4.63 *Let $\mathcal{F}$ be a saturated fusion system on a finite p-group P, and let Q be a strongly $\mathcal{F}$-closed subgroup of P. If Z is a weakly $\mathcal{F}$-closed, central subgroup of Q, then Z is strongly $\mathcal{F}$-closed.*

Proof Let $\phi : X \to Y$ be an $\mathcal{F}$-isomorphism with either X or Y contained in Z, and Y fully normalized. Since Q is strongly $\mathcal{F}$-closed, both X and Y lie inside Q. The subgroup Z is central in Q, and so $Z \leq \mathrm{C}_P(X)$; as Y is fully normalized and $\mathcal{F}$ is saturated, ϕ extends to a morphism $\psi : ZX \to Q$ which, as Z is weakly $\mathcal{F}$-closed, restricts to an automorphism $\psi|_Z : Z \to Z$. Therefore if either X or Y lies in Z, then the other also does.

To see how this proves our result, suppose that X is a subgroup of Z, and that $\phi : X \to Y$ is an isomorphism with Y fully normalized; then Y is contained in Z by the argument above. Now let $\psi : Y \to W$ be any $\mathcal{F}$-isomorphism. By the argument above applied to the *inverse* of ψ, we see that W lies in Z as well, proving our result. $\qquad\square$

Exercise 4.6 gives a much quicker way of proving this lemma, using Alperin's fusion theorem. Note that, for example, the fusion system of S_4 on D_8 is an example of a weakly closed abelian subgroup that is not strongly closed, so one cannot extend the last lemma in that direction.

Strongly closed subgroups in particular will be important in most of the next chapter, and are a fundamental notion in the rest of the text.

Exercises

4.1   Give an example of a fusion system $\mathcal{F}$ such that every $\mathcal{F}$-conjugacy class of subgroups contains a fully automized member and a receptive member, but is not saturated.

4.2 Let $\mathcal{F}$ be a fusion system on a finite p-group P, and let Q and R be subgroups of P. Let K be a subgroup of $\mathrm{Aut}(Q)$ and L be a subgroup of $\mathrm{Aut}(R)$. Suppose that $Q \leq \mathrm{N}_P^L(R)$ and $R \leq \mathrm{N}_P^K(Q)$. Prove that

$$\mathrm{N}_{\mathrm{N}_{\mathcal{F}}^L(R)}^K(Q) = \mathrm{N}_{\mathrm{N}_{\mathcal{F}}^K(Q)}^L(R).$$

4.3 Let $\mathcal{F}$ be a fusion system on a finite p-group P, and let Q be a subgroup of P. Suppose that $\mathcal{E}$ is a subsystem of $\mathcal{F}$ on Q, and let R be a fully $\mathcal{E}$-normalized subgroup of Q. If $\phi \in \mathrm{Hom}_{\mathcal{F}}(Q, P)$, prove that $\mathrm{N}_{\mathcal{E}}(R)^\phi = \mathrm{N}_{\mathcal{E}^\phi}(R\phi)$, where if $\mathcal{D}$ is a subsystem of $\mathcal{F}$, by $\mathcal{D}^\phi$ we mean the subsystem consisting of all maps $\psi^\phi = \phi^{-1}\psi\phi$ where ψ lies in $\mathcal{D}$.

4.4 In Theorem 4.54, we claimed that all three saturated fusion systems on a generalized quaternion 2-group were given by group fusion systems, and gave examples for two of them. Give an example of a finite group possessing the third fusion system described there.

4.5 Suppose that A is a p'-group of automorphisms acting on a p-group P, and suppose that P possesses a series

$$1 = P_0 \leq P_1 \leq P_2 \leq \cdots \leq P_n = P$$

of normal, A-invariant subgroups such that the induced action of A on P_i/P_{i-1} is trivial for all i. Show that $A = 1$.

4.6 Prove Lemma 4.63 via Alperin's fusion theorem.

4.7 (Grün's second theorem) A group is said to be p-*normal* if $Z(P)$ is weakly closed in P with respect to G. Grün's second theorem says that if G is p-normal then $\mathrm{N}_G(Z(P))$ controls G-fusion in P. Prove the generalization that if a central subgroup Z is weakly $\mathcal{F}$-closed for a saturated fusion system $\mathcal{F}$, then $\mathcal{F} = \mathrm{N}_{\mathcal{F}}(Z)$.

4.8 Let P be the 2-group $C_{2^a} \times C_{2^b}$.

   (i) Prove that $\mathrm{Aut}(P)$ is a 2-group unless $a = b$.
   (ii) Suppose that $P \cong C_{2^a} \times C_{2^a}$, and let A be a subgroup of $\mathrm{Aut}(P)$ such that $O_2(A) = 1$ and $|A|$ is even. Prove that $A \cong S_3$.
   (iii) Let ϕ be an element of order 3 in A, let x be an element of order 2^a in P, and write $y = x\phi$. Prove that $y\phi = x^{-1}y^{-1}$. (Notice that $P = \langle x, y \rangle$.)
   (iv) Let ψ be an element of order 2 in $A \cong S_3$. By considering the action of A on $\Omega_1(P)$, show that there are generators x and y such that ψ swaps x and y.

4.9 (Frobenius's normal p-complement theorem) Let $\mathcal{F}$ be a saturated fusion system on a finite p-group P. Prove that the following are equivalent:

(i) $\mathcal{F} = \mathcal{F}_P(P)$.
(ii) $\mathrm{Aut}_{\mathcal{F}}(Q)$ is a p-group for all $Q \leq P$.
(iii) $\mathrm{N}_{\mathcal{F}}(Q) = \mathcal{F}_{\mathrm{N}_P(Q)}(\mathrm{N}_P(Q))$ for all $Q \leq P$.

5

Weakly normal subsystems, quotients, and morphisms

A morphism Φ of a fusion system is a natural notion: since a fusion system is a category, this map Φ should be a functor on the category. It should also be a group homomorphism on the underlying group P. It turns out that, whenever Q is a strongly $\mathcal{F}$-closed subgroup of P, there is a surjective morphism of fusion systems with kernel Q, and this morphism of fusion systems is determined uniquely by the group homomorphism (and hence by Q). The theory of morphisms of fusion systems is very satisfactory, and might be said to be complete, in a quite reasonable sense.

However, there is one aspect of the theory that is distinctly less appealing: while the kernel of a morphism is a subgroup, there is no nice and obvious way to get a *subsystem* on this kernel. There is an obvious way to get a 'kernel subsystem', but this subsystem does not have the 'nice' properties that we want such a subsystem to have. We clearly want this subsystem to be 'normal' in some sense; we will introduce the concept of weak normality here, and normality in Chapter 8.

In this chapter, we introduce another important notion: a normal subgroup. A normal subgroup is simply a subgroup Q such that $\mathcal{F} = N_{\mathcal{F}}(Q)$: we have seen some results about this type of subgroup already, and will see more in this chapter and Chapter 7. The concept of a weakly normal subsystem, as alluded to before, is slightly more complicated, but essentially it should be a subsystem in which one may 'conjugate' a morphism by another and stay within the subsystem.

This definition of a weakly normal subsystem, while natural, does not furnish us with the tools to do local finite group theory with fusion systems, and so in [Asc08a] Aschbacher introduced what we call here a *normal* subsystem. (In [Asc08a], normal subsystems are referred to as 'normal', and weakly normal subsystems are referred to as 'invariant', saturated systems.) The definition is slightly less natural, but is in some respects a better notion. For example, if $\mathcal{E}_1$ and $\mathcal{E}_2$ are two weakly normal subsystems on subgroups that intersect trivially and commute, we should have that $\mathcal{E}_1 \times \mathcal{E}_2$, the direct product (which we shall see later) should be a subsystem of the fusion system. This is not true, however, but *is* true when weak normality is replaced by normality. The theory of normal subsystems is considerably more complicated than that of weakly normal subsystems and so it will be delayed until Chapter 8.

In Section 5.6, we consider simple fusion systems, and in Section 5.7 we consider the centre and hypercentre, and prove results concerning the relationship between a fusion system and its images. We end with a definition of soluble fusion systems, and prove that they are constrained, so come from finite groups. This chapter, together with the previous one, contains most of the major definitions in the basic theory of fusion systems.

5.1 Morphisms of fusion systems

The material is this section is mostly due to Puig, but the exposition here is new, with a different (but equivalent), simpler definition of a morphism of fusion systems than is normally used. The rest of the exposition broadly follows that of the author in [Cra10a].

If P and Q are finite p-groups with a group homomorphism $\phi : P \to Q$, and A and B are subgroups of P with a homomorphism $\psi : A \to B$, then we can 'transfer' ψ to a homomorphism ψ^ϕ between $A\phi$ and $B\phi$, which are of course subgroups of Q. The natural way to do this is to define, for $a \in A$,

$$(a\phi)\psi^\phi = (a\psi)\phi.$$

It is easy to see that ψ^ϕ is a homomorphism from $A\phi$ to $B\phi$. This definition will be used to define a morphism of fusion systems.

Definition 5.1 Let $\mathcal{F}$ and $\mathcal{E}$ be fusion systems on the finite p-groups P and Q respectively. A *morphism* $\Phi : \mathcal{F} \to \mathcal{E}$ of fusion systems is a group homomorphism $\phi : P \to Q$ such that, for all morphisms ψ in $\mathcal{F}$, $\psi\Phi = \psi^\phi$ is a morphism in $\mathcal{E}$. In this case, we say that ϕ *induces* a morphism of fusion systems, and that ϕ is the *underlying* group homomorphism of Φ.

Notions of kernels, injectivity and surjectivity are standard.

Definition 5.2 Let $\mathcal{F}$ and $\mathcal{E}$ be fusion systems on the finite p-groups P and Q respectively, and let $\Phi : \mathcal{F} \to \mathcal{E}$ be a morphism.

(i) The *kernel* of Φ is the kernel of the underlying group homomorphism $\phi : P \to Q$, necessarily a normal subgroup of P.
(ii) The map Φ is *injective* if $\ker \Phi = 1$.
(iii) The map Φ is *surjective* if the underlying homomorphism ϕ is surjective, and for any two subgroups $\bar{R}$ and $\bar{S}$ of Q and $\bar{\psi} : \bar{R} \to \bar{S}$ in $\mathcal{E}$, there are subgroups R and S of P and a map $\psi : R \to S$ in $\mathcal{F}$ such that $\psi\Phi = \bar{\psi}$.

If we are to get an understanding of morphisms of fusion systems, an important first step is knowing when the map ψ^ϕ is injective. We will do this now.

Proposition 5.3 *Let $\mathcal{F}$ be a fusion system on a finite p-group P, and let $\phi : P \to Q$ be a homomorphism. The kernel T of ϕ is strongly $\mathcal{F}$-closed if and only if, whenever ψ is a map in $\mathcal{F}$, the image ψ^ϕ is an injection.*

Proof Suppose that T is strongly $\mathcal{F}$-closed, and let $\psi : A \to B$ be a morphism in $\mathcal{F}$, so that in particular it is injective. Let $a \in A$, and suppose that $(a\phi)\psi^\phi = 1$. We have $(a\phi)\psi^\phi = (a\psi)\phi$, so that $a\psi \in \ker \phi = T$: since T is strongly $\mathcal{F}$-closed, and $a\psi \in T$, we must have that $a \in T$, so that $a\phi = 1$. Hence ψ^ϕ is an injection.

Conversely, suppose that ψ^ϕ is an injection for all ψ in $\mathcal{F}$, and let B be a subgroup of T. Let $\psi : A \to B$ be an isomorphism in $\mathcal{F}$. Since $B \leq T$, the image of ψ^ϕ is trivial, so its domain must be trivial as well; hence $A \leq T$, and so T is strongly $\mathcal{F}$-closed. $\qquad\square$

This yields the following trivial corollary.

Corollary 5.4 *Let $\mathcal{F}$ and $\mathcal{E}$ be fusion systems on the finite p-groups P and Q respectively. If $\Phi : \mathcal{F} \to \mathcal{E}$ is a morphism then $\ker \Phi$ is strongly $\mathcal{F}$-closed.*

We have seen that a (surjective) morphism of fusion systems has a strongly $\mathcal{F}$-closed kernel: it would be nice if, to every group homomorphism with strongly $\mathcal{F}$-closed kernel, there were a surjective morphism of fusion systems. This is not true in general, but for *saturated* fusion systems we can get this. We begin with a definition.

Definition 5.5 Let $\mathcal{F}$ be a fusion system on a finite p-group P, and let Q be a strongly $\mathcal{F}$-closed subgroup of P. Let $\phi : P \to P/Q$ denote the natural homomorphism. We will denote by $\bar{\mathcal{F}}_Q$ the subobject of the universal fusion system $\mathcal{U}(P/Q)$, containing all objects in $\mathcal{U}(P/Q)$, and with morphisms $\mathrm{Hom}_{\bar{\mathcal{F}}_Q}(R/Q, S/Q)$ consisting of those morphisms ψ^ϕ for $\psi \in \mathrm{Hom}_\mathcal{F}(\tilde{R}, \tilde{S})$, as $\tilde{R}$ and $\tilde{S}$ range over all subgroups of P such that $Q\tilde{R} = R$ and $Q\tilde{S} = S$. (Since Q is strongly $\mathcal{F}$-closed, any such morphism $\psi : \tilde{R} \to \tilde{S}$ gives rise to a morphism $\psi^\phi : R/Q \to S/Q$ in $\mathcal{U}(P/Q)$.)

By $\langle \bar{\mathcal{F}}_Q \rangle$ we denote the subcategory of $\mathcal{U}(P/Q)$ on P/Q consisting of all finite compositions of morphisms from $\bar{\mathcal{F}}_Q$.

As suggested by the introduction of $\langle \bar{\mathcal{F}}_Q \rangle$, the set $\bar{\mathcal{F}}_Q$ does not in general form a category. This is because one might have two morphisms, $\phi : A \to B$ and $\psi : C \to D$, with $QB = QC$ for some strongly $\mathcal{F}$-closed subgroup Q; in this case, the images of ϕ and ψ in $\bar{\mathcal{F}}_Q$ are composable, but the composition might well not be the image of a map in $\mathcal{F}$. This deficiency is repaired by including compositions, i.e., considering $\langle \bar{\mathcal{F}}_Q \rangle$: it turns out that $\langle \bar{\mathcal{F}}_Q \rangle$ is a fusion system.

Lemma 5.6 *Let $\mathcal{F}$ be a fusion system on a finite p-group P, and let Q be a subgroup of P. If Q is strongly $\mathcal{F}$-closed, then $\langle \bar{\mathcal{F}}_Q \rangle$ is a fusion system on P/Q.*

Proof That $\langle \bar{\mathcal{F}}_Q \rangle$ is a category is obvious, since we are guaranteed compositions of morphisms by definition. The first axiom of a fusion system is satisfied, since conjugation by $Qg \in P/Q$ is induced from conjugation by $g \in P$, taken modulo Q. If we prove the final two axioms for the subset $\bar{\mathcal{F}}_Q$, then since $\langle \bar{\mathcal{F}}_Q \rangle$ is obtained from $\bar{\mathcal{F}}_Q$ by compositions of morphisms, those axioms would also hold there. If $\psi : R/Q \to S/Q$ is a morphism in $\bar{\mathcal{F}}_Q$, then there is some morphism $\tilde{\psi} : \tilde{R} \to \tilde{S}$ in $\mathcal{F}$ inducing ψ, and the corresponding isomorphism $\tilde{\theta} : \tilde{R} \to \tilde{R}\tilde{\psi}$ induces an isomorphism in $\bar{\mathcal{F}}_Q$ corresponding to ψ. Finally, if $\psi : R/Q \to S/Q$ is an isomorphism in $\bar{\mathcal{F}}_Q$, then it comes from some isomorphism $\tilde{\psi} : \tilde{R} \to \tilde{S}$ in $\mathcal{F}$, and the inverse of $\tilde{\psi}$ induces the inverse of ψ. Hence $\langle \bar{\mathcal{F}}_Q \rangle$ is a fusion system. $\square$

Proposition 5.7 *Let $\mathcal{F}$ be a fusion system on a finite p-group P, and let Q be a strongly $\mathcal{F}$-closed subgroup of P.*

(i) *The natural map $\Phi_Q : \mathcal{F} \to \langle \bar{\mathcal{F}}_Q \rangle$ is a morphism of fusion systems.*
(ii) *The map Φ_Q is surjective if and only if $\bar{\mathcal{F}}_Q = \langle \bar{\mathcal{F}}_Q \rangle$.*

Proof Let $\phi : P \to P/Q$ denote the natural homomorphism. By definition, ψ^ϕ lies in $\langle \bar{\mathcal{F}}_Q \rangle$ for all ψ in $\mathcal{F}$, so that this is a morphism of fusion systems, proving (i). Also, the condition for Φ_Q being surjective is that ϕ is surjective, which it is, and that every morphism in $\langle \bar{\mathcal{F}}_Q \rangle$ is the image ψ^ϕ of some morphism ψ in $\mathcal{F}$, which it is if and only if $\bar{\mathcal{F}}_Q = \langle \bar{\mathcal{F}}_Q \rangle$, proving (ii). □

This proposition makes the question of when $\bar{\mathcal{F}}_Q$ and $\langle \bar{\mathcal{F}}_Q \rangle$ coincide important.

Proposition 5.8 *Let $\mathcal{F}$ be a fusion system on a finite p-group P, and suppose that Q is a subgroup of P. If $\mathcal{F} = \mathrm{N}_\mathcal{F}(Q)$, then $\bar{\mathcal{F}}_Q = \langle \bar{\mathcal{F}}_Q \rangle$, and hence the natural map $\mathcal{F} \to \bar{\mathcal{F}}_Q$ is a surjective morphism of fusion systems.*

Proof Any morphism $\phi : R \to S$ extends to a morphism $\psi : QR \to QS$ that acts as an automorphism on Q. Certainly, $\bar{\phi} = \bar{\psi}$ in $\bar{\mathcal{F}}_Q$, since the action of ϕ and ψ on QR/Q is the same. We need to show that the composition of two morphisms in $\bar{\mathcal{F}}_Q$ also lies in $\bar{\mathcal{F}}_Q$.

Suppose that $\phi : R \to S$ and $\psi : T \to U$ are two morphisms in $\mathcal{F}$ whose images in $\bar{\mathcal{F}}_Q$ are composable, so that $QS \leq QT$. Write $\hat{\phi}$ for an extension of ϕ to QR with $\hat{\phi}|_Q \in \mathrm{Aut}(Q)$, and similarly with $\hat{\psi}$. The composition $\hat{\phi}\hat{\psi} : QR \to QU$ is a morphism in $\mathcal{F}$ that acts as an automorphism of Q; the images of ϕ and $\hat{\phi}$ in $\bar{\mathcal{F}}_Q$ are clearly the same, and so the composition of the images of ϕ and ψ is in $\bar{\mathcal{F}}_Q$, since it is the image of $\hat{\phi}\hat{\psi}$. □

The same statement – that $\bar{\mathcal{F}}_Q = \langle \bar{\mathcal{F}}_Q \rangle$ – is true when $\mathcal{F}$ is saturated, but the proof is not nearly as easy. In order to do it, we will introduce another fusion system, this time without reference to morphisms of fusion systems; this is a construction of Puig, given in [Pui06], but see also [Lin07].

Definition 5.9 Let $\mathcal{F}$ be a fusion system on a finite p-group P, and let Q be a normal subgroup of P. Let $\phi : P \to P/Q$ be the natural homomorphism. The *factor system* $\mathcal{F}/Q$ is the category with objects all subgroups of P/Q, and for any two subgroups R and S containing Q, we have that $\mathrm{Hom}_{\mathcal{F}/Q}(R/Q, S/Q)$ is the set of homomorphisms ψ^ϕ, where $\psi \in \mathrm{Hom}_\mathcal{F}(R, S)$ is a morphism such that $Q\phi = Q$.

Traditionally, in the definition above, the subgroup Q is strongly $\mathcal{F}$-closed, but this is not necessary for the following two results. The first is easy, and its proof is Exercise 5.1.

Proposition 5.10 *Let $\mathcal{F}$ be a fusion system on a finite p-group P. If Q is a normal subgroup of P, then the category $\mathcal{F}/Q$ is a fusion system on P/Q.*

Our proof of the next proposition follows [Lin07, Theorem 6.2], and although our hypotheses are weaker, the method of proof is the same.

Proposition 5.11 *Let $\mathcal{F}$ be a saturated fusion system on a finite p-group P. If Q is a weakly $\mathcal{F}$-closed subgroup of P then the fusion system $\mathcal{F}/Q$ is saturated.*

Proof All automorphisms in $\mathrm{Aut}_{\mathcal{F}/Q}(P/Q)$ are induced from automorphisms in $\mathrm{Aut}_{\mathcal{F}}(P)$, and so the obvious homomorphism $\mathrm{Aut}_{\mathcal{F}}(P) \to \mathrm{Aut}_{\mathcal{F}/Q}(P/Q)$ is surjective. The image of $\mathrm{Aut}_P(P)$ in $\mathrm{Aut}_{\mathcal{F}/Q}(P/Q)$ is clearly $\mathrm{Aut}_{P/Q}(P/Q)$, so that it satisfies the first axiom of saturation.

Suppose that $\phi \in \mathrm{Hom}_{\mathcal{F}/Q}(R/Q, S/Q)$ is an isomorphism such that S/Q is fully $\mathcal{F}/Q$-normalized. We claim that S is also fully $\mathcal{F}$-normalized. Since $Q \leqslant R$, and Q is weakly $\mathcal{F}$-closed, for all T that are $\mathcal{F}$-isomorphic to R we have that $Q \leqslant T \leqslant \mathrm{N}_P(T)$. Also, $\mathrm{N}_P(R)/Q = \mathrm{N}_{P/Q}(R/Q)$, and therefore

$$|\mathrm{N}_P(T)| = |\mathrm{N}_{P/Q}(T/Q)| \cdot |Q| \leqslant |\mathrm{N}_{P/Q}(S/Q)| \cdot |Q| = |\mathrm{N}_P(S)|;$$

hence S is fully $\mathcal{F}$-normalized.

Now let ϕ be an automorphism of a fully $\mathcal{F}/Q$-normalized subgroup R/Q, and let ψ be an $\mathcal{F}$-automorphism of R with image ϕ in $\mathcal{F}/Q$. At this point we would like to prove that $N_\psi/Q = N_\phi$, but it is not necessarily true. However, there is *some* ψ with image ϕ for which $N_\psi/Q = N_\phi$, as we shall demonstrate now. Notice that $N_\psi/Q \leqslant N_\phi$ trivially.

Let K be the kernel of the natural map $\mathrm{Aut}_{\mathcal{F}}(R) \to \mathrm{Aut}(R/Q)$, a normal subgroup of $\mathrm{Aut}_{\mathcal{F}}(R)$; then K consists of all elements of $\mathrm{Aut}_{\mathcal{F}}(R)$ that act trivially on R/Q, and hence are sent to the identity automorphism of R/Q under the map $\mathcal{F} \to \mathcal{F}/Q$. The idea is that if $\chi \in K$, then $\chi\psi$ and ψ both have the image ϕ in $\mathcal{F}/Q$, so one may 'ignore' elements in K. We will prove that there are morphisms $\chi \in K$ and $\theta : R \to R$ such that $\psi = \chi\theta$ and θ has the property that $N_\theta/Q = N_\phi$. Since θ and ψ define the same image ϕ in $\mathcal{F}/Q$, we prove that ϕ extends to N_ϕ.

Since K is a normal subgroup of $\mathrm{Aut}_{\mathcal{F}}(R)$ and $\mathrm{Aut}_P(R)$ is a Sylow p-subgroup of $\mathrm{Aut}_{\mathcal{F}}(R)$ (as R is fully normalized), we have that

$\operatorname{Aut}_P^K(R) = K \cap \operatorname{Aut}_P(R)$ is a Sylow p-subgroup of K, and by the Frattini argument

$$\operatorname{Aut}_{\mathcal{F}}(R) = K \, \mathrm{N}_{\operatorname{Aut}_{\mathcal{F}}(R)}(\operatorname{Aut}_P^K(R)).$$

As $\operatorname{Aut}_P(R)$ normalizes $\operatorname{Aut}_P^K(R)$, we see that $\operatorname{Aut}_P(R)$ is a Sylow p-subgroup of $\mathrm{N}_{\operatorname{Aut}_{\mathcal{F}}(R)}(\operatorname{Aut}_P^K(R))$. Writing $X = \mathrm{N}_{\operatorname{Aut}_{\mathcal{F}}(R)}(\operatorname{Aut}_P^K(R))$, we have

$$X/(X \cap K) \cong KX/K = \operatorname{Aut}_{\mathcal{F}}(R)/K \cong \operatorname{Aut}_{\mathcal{F}/Q}(R/Q),$$

by the second isomorphism theorem and the definition of K. Since $X = \mathrm{N}_{\operatorname{Aut}_{\mathcal{F}}(R)}(\operatorname{Aut}_P^K(R))$, we may form the quotient group $X/\operatorname{Aut}_P^K(R)$, and as $S = \operatorname{Aut}_P^K(R)$ is a Sylow p-subgroup of K, it must be a (normal) Sylow p-subgroup of $X \cap K$; hence $(X \cap K)/S$ is a p'-group. Quotienting out by this Sylow p-subgroup, we see that

$$X/(X \cap K) \cong (X/S)/\big((X \cap K)/S\big) \cong \operatorname{Aut}_{\mathcal{F}/Q}(R/Q).$$

Notice that the subgroup N_ϕ is the preimage in P/Q of the intersection $\operatorname{Aut}_{P/Q}(R/Q) \cap \operatorname{Aut}_{P/Q}(R/Q)^{\phi^{-1}}$ of two Sylow p-subgroups of $\operatorname{Aut}_{\mathcal{F}/Q}(R/Q)$. Since $(X \cap K)/S$ is a p'-group, we see that there are two Sylow p-subgroups, A/S and B/S, of X/S, projecting onto the subgroups $\operatorname{Aut}_{P/Q}(R/Q)$ and $\operatorname{Aut}_{P/Q}(R/Q)^{\phi^{-1}}$ of $\operatorname{Aut}_{\mathcal{F}/Q}(R/Q)$ respectively, and an element $Sg \in X/S$, such that $A/S \cap (B/S)^{Sg^{-1}}$ projects onto $\operatorname{Aut}_{P/Q}(R/Q) \cap \operatorname{Aut}_{P/Q}(R/Q)^{\phi^{-1}}$ and Sg projects onto an element inducing the morphism ϕ, by Exercise 5.2.

Since S is a normal p-subgroup of $X \cap K$, we may take preimages in X, and so there is some element $\theta \in X = \mathrm{N}_{\operatorname{Aut}_{\mathcal{F}}(R)}(\operatorname{Aut}_P^K(R))$ such that $\operatorname{Aut}_P(R) \cap \operatorname{Aut}_P(R)^{\theta^{-1}}$ maps onto $\operatorname{Aut}_{P/Q}(R/Q) \cap (\operatorname{Aut}_{P/Q}(R/Q))^{\phi^{-1}}$. Therefore the map $N_\theta \to N_\phi$ is surjective, and so $N_\theta/Q = N_\phi$. As the map θ extends to N_θ, the map ϕ extends to N_ϕ.

It remains to deal with any map $\phi : S/Q \to R/Q$ in $\mathcal{F}/Q$, where R/Q is fully $\mathcal{F}/Q$-normalized. This lifts to a map $\psi : S \to R$ in $\mathcal{F}$ with R fully $\mathcal{F}$-normalized. By Proposition 4.17, there is some map $\theta : S \to R$ with $N_\theta = N_P(S)$, and so ϕ extends to N_ϕ if and only if both the image χ of θ in $\mathcal{F}/Q$ extends to $N_\chi = \mathrm{N}_{P/Q}(S/Q)$ and $\chi^{-1}\phi$ extends to $N_{\chi^{-1}\phi} = N_\phi$. The first of these claims is obvious, and the second has just been proved above (since $\chi\theta^{-1}$ is an automorphism of R/Q), concluding the proof. $\qquad\square$

5.2 The isomorphism theorems

We now have two competing notions for a quotient of a fusion system by a strongly $\mathcal{F}$-closed subgroup: $\mathcal{F}/Q$ and $\langle \bar{\mathcal{F}}_Q \rangle$. All morphisms in $\mathcal{F}/Q$ are also in $\bar{\mathcal{F}}_Q$, and hence

$$\mathcal{F}/Q \subseteq \bar{\mathcal{F}}_Q \subseteq \langle \bar{\mathcal{F}}_Q \rangle.$$

If $\mathcal{F}/Q = \bar{\mathcal{F}}_Q$, then we also have that $\bar{\mathcal{F}}_Q = \langle \bar{\mathcal{F}}_Q \rangle$, and the natural morphism Φ_Q is surjective. Certainly this does not always happen since, as we have already said, $\bar{\mathcal{F}}_Q$ is not always a category, whereas $\mathcal{F}/Q$ is always a category. We begin with an example to see how bad things can get.

Example 5.12 Let $P = \langle a, b, c, d \rangle$ be elementary abelian of order 16, and let $\mathcal{F}$ be the fusion system generated by $\mathcal{F}_P(P)$ and the two morphisms $\langle ab \rangle \to \langle c \rangle$ and $\langle ac \rangle \to \langle d \rangle$. The subgroup $A = \langle a \rangle$ is strongly $\mathcal{F}$-closed, and so we may form the object $\bar{\mathcal{F}}_A$. Here, the cosets Ab and Ac are $\bar{\mathcal{F}}_A$-conjugate, as are the cosets Ac and Ad. However, there is no map sending Ab to Ad, and so $\bar{\mathcal{F}}_A$ is not a fusion system.

Note also that there are no non-trivial morphisms on overgroups of A, and so $\mathcal{F}/A = \mathcal{F}_{P/A}(P/A)$.

This fusion system is far from being saturated, so we should expect things to be bad; we have seen in the past that general fusion systems are quite loose objects, and so expecting anything nice from them might be ill advised.

Proposition 5.13 *Let $\mathcal{F}$ be a fusion system on a finite p-group P, and suppose that Q is a subgroup of P. If $\mathcal{F} = N_{\mathcal{F}}(Q)$, then $\mathcal{F}/Q = \bar{\mathcal{F}}_Q$.*

Proof Any morphism $\phi : R \to S$ extends to a morphism $\psi : QR \to QS$ that acts as an automorphism on Q. Certainly, $\bar{\phi} = \bar{\psi}$ in $\bar{\mathcal{F}}_Q$, since the action of ϕ and ψ on QR/Q is the same. Also, $\bar{\psi} \in \mathcal{F}/Q$, and since $\mathcal{F}/Q \subseteq \bar{\mathcal{F}}_Q$ we must have equality. $\square$

Other than the case where $\mathcal{F} = N_{\mathcal{F}}(Q)$, there is another case in which $\mathcal{F}/Q = \bar{\mathcal{F}}_Q$, and that is when $\mathcal{F}$ is saturated. The proof of this theorem follows [Pui06, Proposition 6.3], although simplifications have been made.

Theorem 5.14 (Puig [Pui06]) *Let $\mathcal{F}$ be a saturated fusion system on a finite p-group P, and let Q be a subgroup of P. If Q is strongly $\mathcal{F}$-closed then $\mathcal{F}/Q = \bar{\mathcal{F}}_Q$.*

Proof If $\phi : R \to S$ is a map in $\mathcal{F}$, write $\bar{\phi}$ for the image of this map in $\bar{\mathcal{F}}_Q$, i.e., write $\bar{\phi}$ for the induced map $\bar{\phi} : QR/Q \to QS/Q$. Firstly, notice that both $\mathcal{F}/Q$ and $\bar{\mathcal{F}}_Q$ are defined on the same group, namely P/Q. Certainly, $\mathcal{F}/Q$ is contained in $\bar{\mathcal{F}}_Q$, so we need to prove the converse; in other words, given a morphism $\phi : R \to S$ in $\mathcal{F}$, we need to show that there is some morphism $\psi : QR \to P$ such that $\bar{\phi} = \bar{\psi}$, for then $\bar{\phi} \in \mathcal{F}/Q$, as needed.

We proceed by induction on $n = |P : R|$, noting that if R contains Q then we are done trivially; in particular, this implies that $n > 1$. By Alperin's fusion theorem, any morphism may be factored as (restrictions of) a sequence of automorphisms ϕ_i of fully normalized, $\mathcal{F}$-centric, $\mathcal{F}$-radical subgroups U_i. Suppose that the images $\bar{\phi}_i$ of each of the ϕ_i lie in $\mathcal{F}/Q$: since $\mathcal{F}/Q$ is a fusion system and $\mathcal{F} \to \langle \bar{\mathcal{F}}_Q \rangle$ is a morphism, we have that $\bar{\phi}$, the product of (restrictions of) the $\bar{\phi}_i$, also lies in $\mathcal{F}/Q$.

Therefore we may assume that one of the $\bar{\phi}_i$ lies in $\bar{\mathcal{F}}_Q$ but not in $\mathcal{F}/Q$. By our inductive hypothesis, we see that $|U_i| = |R|$, and so we may replace R and ϕ by U_i and ϕ_i; therefore R is now a fully normalized, $\mathcal{F}$-centric, $\mathcal{F}$-radical subgroup of P, and ϕ is an automorphism of R. Also, $Q \not\leq R$ as we saw above.

Similar to the proof of Proposition 5.11, let K be the kernel of the natural map $\mathrm{Aut}_{\mathcal{F}}(R) \to \mathrm{Aut}(QR/Q)$, a normal subgroup of $A = \mathrm{Aut}_{\mathcal{F}}(R)$; then K consists of all elements of A that act trivially on QR/Q, and hence are sent to the identity automorphism of QR/Q under the map $\mathcal{F} \to \mathcal{F}/Q$.

Let $T = \mathrm{N}_P^K(R)$; since R is fully normalized, $\mathrm{Aut}_P(R)$ is a Sylow p-subgroup of A, and so $K \cap \mathrm{Aut}_P(R) = \mathrm{Aut}_T(R)$ is a Sylow p-subgroup of K. Therefore, by the Frattini argument,

$$A = K \, \mathrm{N}_A(\mathrm{Aut}_T(R)).$$

Step 1: *We have $Q \cap T = \mathrm{N}_Q^K(R) = \mathrm{N}_Q(R)$, and $R \mathrm{N}_Q(R) > R$.* The first equality is obvious. Let $g \in \mathrm{N}_Q(R)$, and $x \in R$. It is easy to see that $Qx^g = Qx$, so that g acts trivially on QR/Q. Hence the automorphism determined by g is in K, and so our first claim is proved. To see the second part, notice that $Q \cap R < Q$, and so $\mathrm{N}_Q(R) = \mathrm{N}_Q(Q \cap R) > Q \cap R$.

Step 2: *If $\psi \in \mathrm{N}_A(\mathrm{Aut}_T(R))$, then N_ψ contains T.* Since N_ψ is the inverse image under c_R of the subgroup $\mathrm{Aut}_P(R) \cap \mathrm{Aut}_P(R)^{\psi^{-1}}$ in A, we need to show that $\mathrm{Aut}_T(R)$ is contained in both terms of the intersection. That it is contained in the first is clear, and, for the second, since $\psi \in N_A(\mathrm{Aut}_T(R))$, we have that $(\mathrm{Aut}_T(R))^\psi = \mathrm{Aut}_T(R)$. Thus it is contained in both terms, and so our claim is proved.

Now we may prove the result: since $A = K N_A(\mathrm{Aut}_T(R))$, the morphism ϕ may be written as $\phi = \chi\psi$, where $\chi \in K$ and $\psi \in N_A(\mathrm{Aut}_T(R))$. Since χ acts trivially on QR/Q, we see that $\bar\phi = \bar\psi$ in $\bar{\mathcal{F}}_Q$. Furthermore, by Step 2 we see that N_ψ contains T. However, if $N_\psi > R$, then ψ extends to ψ', on an overgroup of R, and so by induction $\bar{\psi}'$ lies in $\mathcal{F}/Q$. Since $\mathcal{F}/Q$ is a fusion system, this would imply that $\bar\psi$ is in $\mathcal{F}/Q$. Therefore $T \le R$, and hence $Q \cap T = N_Q(R) \le R$. However, by Step 1, $R N_Q(R) > R$, a contradiction, proving the theorem. □

This can be reinterpreted as the following result.

Corollary 5.15 (First isomorphism theorem) *Let $\mathcal{F}$ and $\mathcal{E}$ be saturated fusion systems on the finite p-groups P and Q respectively. If $\Phi : \mathcal{F} \to \mathcal{E}$ is a morphism of fusion systems then $\mathcal{F}/\ker\Phi \cong \mathrm{im}\,\Phi$.*

We will now prove the other two isomorphism theorems for fusion systems. Getting a second isomorphism theorem might seem unlikely considering that we do not have either products or intersections of subsystems, but nevertheless we will make sense of the statement. The next result is from [Cra10a, Proposition 5.11].

Proposition 5.16 (Second isomorphism theorem, Craven [Cra10a]) *Let $\mathcal{F}$ be a saturated fusion system on a finite p-group P, let Q be a strongly $\mathcal{F}$-closed subgroup of P, and let $\mathcal{E}$ be a saturated subsystem of $\mathcal{F}$ on a subgroup R of P. Writing $\mathcal{E}Q/Q$ for the image of $\mathcal{E}$ in $\mathcal{F}/Q$, we have*

$$\mathcal{E}Q/Q \cong \mathcal{E}/(Q \cap R).$$

Proof The isomorphism $QR/Q \to R/(Q \cap R)$ induces an isomorphism $\Phi : \mathcal{U}(QR/Q) \to \mathcal{U}(R/(Q \cap R))$ of the universal fusion systems, so we need to prove that the image of $\mathcal{E}Q/Q$ in $\mathcal{U}(R/(Q \cap R))$ lies inside $\mathcal{E}/(Q \cap R)$ and vice versa.

Let S/Q and T/Q be subgroups of QR/Q. A morphism $\phi : S/Q \to T/Q$ lies in $\mathcal{E}Q/Q$ if and only if there exist subgroups $\tilde{S}$ and $\tilde{T}$ of R with $Q\tilde{S} = S$ and $Q\tilde{T} = T$, and a morphism $\psi \in \mathrm{Hom}_\mathcal{E}(\tilde{S}, \tilde{T})$ such that the image of ψ in $\mathcal{F}/Q$ is ϕ. The image of ψ in $\mathcal{E}/(Q \cap R)$ is clearly $\phi\Phi$, and so the image of $\mathcal{E}Q/Q$ under Φ is contained in $\mathcal{E}/(Q \cap R)$. Conversely, if $\theta : S/(Q \cap R) \to T/(Q \cap R)$ is a morphism in $\mathcal{E}/(Q \cap R)$, then there is a morphism $\chi : S \to T$ in $\mathcal{E}$ with image θ in $\mathcal{E}/(Q \cap R)$, and the image

$\bar{\chi}$ of χ in $\mathcal{E}Q/Q$ also satisfies $\bar{\chi}\Phi = \theta$, and so Φ induces an isomorphism $\mathcal{E}Q/Q \to \mathcal{E}/(Q \cap R)$, as needed. $\qquad\square$

We get the following corollary, which is more difficult to prove directly.

Corollary 5.17 *Let $\mathcal{F}$ be a saturated fusion system on a finite p-group P, and let Q be a strongly $\mathcal{F}$-closed subgroup of P. The image of any saturated subsystem of $\mathcal{F}$ in $\mathcal{F}/Q$ is saturated.*

We now consider the third isomorphism theorem for fusion systems; this result is [Cra10a, Proposition 5.13].

Proposition 5.18 (Third isomorphism theorem, Craven [Cra10a])   *Let $\mathcal{F}$ be a saturated fusion system on a finite p-group P, and let Q and R be strongly $\mathcal{F}$-closed subgroups of P with $Q \leq R$. Then*

$$(\mathcal{F}/Q)/(R/Q) \cong \mathcal{F}/R.$$

Proof By the third isomorphism theorem for groups, the two fusion systems $\mathcal{F}/R$ and $\mathcal{E} = (\mathcal{F}/Q)/(R/Q)$ are on the same group. Suppose that $\bar{\phi} : S/R \to T/R$ is a morphism in $\mathcal{F}/R$. There is some morphism $\phi \in \mathrm{Hom}_{\mathcal{F}}(S, T)$ with image $\bar{\phi}$. Furthermore, the image $\phi' : S/Q \to T/Q$ of ϕ in $\mathcal{F}/Q$ has image $\phi'' : S/R \to T/R$ in $\mathcal{E}$, and since both $\bar{\phi}$ and ϕ'' are derived from ϕ, they must be the same morphism. The converse is a similar calculation, and is safely omitted. $\qquad\square$

To summarize, we have morphisms of fusion systems whenever we have a group homomorphism with strongly closed kernel, and any morphism arises in this way. We can also construct quotients $\mathcal{F}/Q$ in a slightly different way, which are saturated whenever $\mathcal{F}$ itself is saturated and Q is strongly $\mathcal{F}$-closed. Finally, if either $\mathcal{F} = \mathrm{N}_{\mathcal{F}}(Q)$ or $\mathcal{F}$ is saturated, then $\mathcal{F}/Q = \bar{\mathcal{F}}_Q$, and so when arguing with quotients, we may use either definition. From now on we will denote this quotient system by $\mathcal{F}/Q$ unless $\mathcal{F}$ is not saturated (and hence $\mathcal{F}/Q \neq \bar{\mathcal{F}}_Q$ in general).

A final point should be noted here: the fusion system $\mathcal{F}/Q$ is exactly the same as the fusion system $\mathrm{N}_{\mathcal{F}}(Q)/Q$, since the only morphisms in $\mathcal{F}/Q$ are those $R/Q \to S/Q$ that come from morphisms that act as an automorphism of Q, and so lie in $\mathrm{N}_{\mathcal{F}}(Q)$. Therefore, even though the map $\mathcal{F} \to \mathcal{F}/Q$ is a morphism of fusion systems, the image is determined by $\mathrm{N}_{\mathcal{F}}(Q)$. This local control of quotients of fusion systems should have important consequences for the structure of fusion systems. We give this now as a formal proposition.

Proposition 5.19 *Let $\mathcal{F}$ be a saturated fusion system on a finite p-group P. If Q is a strongly $\mathcal{F}$-closed subgroup of P, then*

$$\mathcal{F}/Q = \mathrm{N}_{\mathcal{F}}(Q)/Q.$$

Having determined the relationship between $\mathcal{F}/Q$ and $\bar{\mathcal{F}}_Q$, we turn to proving results about the factor systems themselves. We begin with the situation for finite groups: it should be that when one takes a finite group, the quotient morphism constructed above is the same as taking quotients modulo a normal p-subgroup.

Theorem 5.20 (Stancu [Sta03]) *Let G be a finite group with Sylow p-subgroup P. If H is a normal subgroup of G and $Q = P \cap H$, then*

$$\mathcal{F}_P(G)/Q = \mathcal{F}_{P/Q}(G/H).$$

Proof Both $\mathcal{F}_P(G)/Q$ and $\mathcal{E} = \mathcal{F}_{P/Q}(G/H)$ are fusion systems on P/Q, so the statement makes sense. The fusion system $\mathcal{F}_P(G)/Q$ contains all morphisms of P/Q induced from conjugation by elements of G. Therefore, $\mathcal{F}_P(G)/Q$ contains $\mathcal{E}$. To prove that $\mathcal{E}$ contains $\mathcal{F}_P(G)/Q$, we need to prove that if R and S are subgroups of P containing Q, and $\phi : R \to S$ is a morphism in $\mathcal{F}_P(G)$, then the induced map $R/Q \to S/Q$ is also in $\mathcal{F}_{P/Q}(G/H)$; in other words, we must show that those morphisms induced by H on P/Q are trivial, for then all morphisms induced by G on P/Q are also induced by G/H.

Let h be an element of H such that $R^h = S$ in P. For all $r \in R$ we have that $[h, r] \in H$, since $H \trianglelefteq G$, and $[h, r]$ also lies in P, since

$$h^{-1}r^{-1}hr = \left(h^{-1}r^{-1}h\right)r \in SR \le P.$$

Thus $[h, r] \in H \cap P = Q$, and so h acts trivially on R/Q, as needed. $\square$

If Q is a strongly $\mathcal{F}$-closed subgroup, then one may study the relationship between subgroups and subsystems of $\mathcal{F}$, and those of $\mathcal{F}/Q$. Here we will compare weak and strong closure. This theorem comes from [Cra10a], and generalizes results of Stancu [Sta06, Lemma 4.7] and Aschbacher [Asc08a, Lemma 8.9].

Theorem 5.21 (Craven [Cra10a]) *Let $\mathcal{F}$ be a fusion system on a finite p-group P, let Q be a strongly $\mathcal{F}$-closed subgroup of P, and let R be a subgroup of P.*

(i) *If R is weakly $\mathcal{F}$-closed then the image of R in $\mathcal{F}/Q$ is weakly $\mathcal{F}/Q$-closed.*

(ii) *The map $P \to P/Q$ induces a bijection between the weakly $\mathcal{F}$-closed subgroups of P containing Q and the weakly $\mathcal{F}/Q$-closed subgroups of P/Q.*

(iii) *If R is strongly $\mathcal{F}$-closed then the image of R in $\mathcal{F}/Q$ is strongly $\mathcal{F}/Q$-closed.*

(iv) *The map $P \to P/Q$ induces a bijection between the strongly $\mathcal{F}$-closed subgroups of P containing Q and the strongly $\langle \bar{\mathcal{F}}_Q \rangle /Q$-closed subgroups of P/Q.*

Proof Suppose that R is a subgroup of P that contains Q, and that R is weakly $\mathcal{F}$-closed. Any morphism of $\mathcal{F}$ starting in R ends in R, and so the same must be true of R/Q in $\mathcal{F}/Q$, since all morphisms in $\mathcal{F}/Q$ starting in R/Q must come from morphisms in $\mathcal{F}$ starting in R. This proves one direction of (ii). To prove the second half of (ii), if $\phi : R \to S$ is a morphism in $\mathcal{F}$ with $R \neq S$, then the image $\bar{\phi} : R/Q \to S/Q$ in $\mathcal{F}/Q$ has $R/Q \neq S/Q$, and the contrapositive is what is required; thus (ii) holds.

Now suppose that R is a weakly $\mathcal{F}$-closed subgroup of P, not necessarily containing Q. By Lemma 4.59(i), QR is weakly $\mathcal{F}$-closed, and so by (ii) above QR/Q is weakly $\mathcal{F}/Q$-closed, proving (i).

Let R be any strongly $\mathcal{F}$-closed subgroup of P, let S be any subgroup of R, and let $\phi \in \mathrm{Hom}_\mathcal{F}(QS, P)$. Since Q is strongly $\mathcal{F}$-closed, $\phi|_Q$ is an automorphism of Q, and since R is strongly $\mathcal{F}$-closed, $\phi|_S$ maps S inside R. Therefore, $(QS)\phi \leq QR$, and so the morphism's image $\bar{\phi} : QS/Q \to P/Q$ in $\mathcal{F}/Q$ would have image inside R/Q, proving that QR/Q is strongly $\mathcal{F}/Q$-closed. If every subgroup of QR/Q is of the form QS/Q for some $S \leq R$, then QR/Q would be strongly $\mathcal{F}/Q$-closed, proving (iii). However, this follows from the modular law, since if $Q \leq A \leq QR$ then $A = A \cap QR = Q(A \cap R)$, so writing $S = A \cap R$ we have that $A/Q = QS/Q$, as needed.

Finally, we prove (iv). Let R be a subgroup containing Q, and let $\phi : S \to T$ be a morphism in $\mathcal{F}$ with $S \leq R$; then $T \leq R$ if and only if $QT/Q \leq R/Q$, and so $\mathrm{im}\,\phi \leq R$ if and only if the image of the corresponding morphism $\bar{\phi} : QS/Q \to QT/Q$ in $\langle \bar{\mathcal{F}}_Q \rangle$ is contained in R/Q. Therefore R is strongly $\mathcal{F}$-closed if and only if R/Q is strongly $\langle \bar{\mathcal{F}}_Q \rangle$-closed, since every morphism with domain inside R/Q in $\langle \bar{\mathcal{F}}_Q \rangle$ arises as a composition of such morphisms.

Thus all parts are proved. $\square$

An important point to make is that there is no requirement in this theorem that the fusion system be saturated. If you add the condition that the fusion system is saturated, you can get the following theorem. (This theorem was mentioned in the previous chapter as Theorem 4.60.) This theorem is due to Aschbacher, but the proof given here is based upon work of the author in [Cra10a].

Theorem 5.22 (Aschbacher [Asc11]) *Let $\mathcal{F}$ be a saturated fusion system on a finite p-group P. The product of two strongly $\mathcal{F}$-closed subgroups is strongly $\mathcal{F}$-closed.*

Proof Let Q and R be strongly $\mathcal{F}$-closed subgroups of P. Since $\mathcal{F}$ is saturated, $\mathcal{F}/Q = \langle \bar{\mathcal{F}}_Q \rangle$. By Theorem 5.21(iii), QR/Q is strongly $\mathcal{F}/Q$-closed, and by (iv) of that theorem, this implies that QR is strongly $\mathcal{F}$-closed, as required. $\square$

In Theorem 5.21, various statements were made; each of these is the best that can be done, in the sense that there are counterexamples to the obvious attempts to extend them.

Example 5.23 Let P be elementary abelian of order 8, generated by a, b, and c. Let $\mathcal{F}$ be a fusion system on P generated by the map $\langle ab \rangle \to \langle c \rangle$; write $A = \langle a \rangle$, and so on. We see that A and B are strongly $\mathcal{F}$-closed, so in particular weakly $\mathcal{F}$-closed.

Let $\mathcal{E} = \langle \bar{\mathcal{F}}_A \rangle$; this has a morphism sending AB/A to AC/A, so AB/A is not even weakly $\mathcal{E}$-closed, never mind strongly $\mathcal{E}$-closed, and so one may not replace $\mathcal{F}/Q$ by $\langle \bar{\mathcal{F}}_Q \rangle$ in (i) and (iii) of Theorem 5.21. It also shows that, since AB is not strongly $\mathcal{F}$-closed, that Theorem 5.22 requires that the fusion system be saturated.

In addition, AB is weakly $\mathcal{F}$-closed, but AB/A is not weakly $\langle \bar{\mathcal{F}}_A \rangle$-closed, so (ii) of the theorem can not be extended to this case. (The opposite direction is true, however, since $\langle \bar{\mathcal{F}}_Q \rangle$ contains $\mathcal{F}/Q$, and so if a subgroup is weakly $\langle \bar{\mathcal{F}}_Q \rangle$-closed then it is weakly $\mathcal{F}/Q$-closed.)

To see that (iv) will not extend, again we have the same caveat: if R is strongly $\mathcal{F}$-closed then R/Q is strongly $\langle \bar{\mathcal{F}}_Q \rangle$-closed, and so in particular strongly $\mathcal{F}/Q$-closed. However, $\mathcal{F}/A$ above is the fusion system $\mathcal{F}_{P/A}(P/A)$, so AB/A is strongly $\mathcal{F}/A$-closed, but AB is not strongly $\mathcal{F}$-closed.

We end with a definition that will be needed throughout this chapter.

Definition 5.24 Let $\mathcal{F}$ be a fusion system on a finite p-group P. An *automorphism* of $\mathcal{F}$ is an isomorphism $\Phi : \mathcal{F} \to \mathcal{F}$. The group of all automorphisms of $\mathcal{F}$ is denoted, as usual, by $\operatorname{Aut}(\mathcal{F})$.

Since a morphism of fusion systems is determined by the underlying group homomorphism, we see that $\mathrm{Aut}(\mathcal{F})$ is a subgroup of $\mathrm{Aut}(P)$; in general, of course, the two need not be the same. Indeed, from the definition of a morphism, it is easy to see that an automorphism $\alpha :$ $P \to P$ yields an automorphism of $\mathcal{F}$ if and only if the induced map $\mathrm{Hom}_{\mathcal{F}}(Q, R) \to \mathrm{Hom}_{\mathcal{F}}(Q\alpha, R\alpha)$ given by $\phi \mapsto \alpha^{-1}\phi\alpha$ is an isomorphism for all Q and R subgroups of P.

If $\phi \in \mathrm{Aut}_{\mathcal{F}}(P)$, then conjugation by ϕ induces an automorphism of the group, and in fact it also induces an automorphism of the fusion system, since the induced map

$$\mathrm{Hom}_{\mathcal{F}}(Q, R) \to \mathrm{Hom}_{\mathcal{F}}(Q\phi, R\phi)$$

is an injection (as $\phi^{-1}\psi\phi$ is in $\mathcal{F}$ for all $\psi \in \mathrm{Hom}_{\mathcal{F}}(Q, R)$) and hence a bijection, since ϕ^{-1} is also in $\mathrm{Aut}_{\mathcal{F}}(P)$. Therefore we have the chain of inclusions

$$\mathrm{Aut}_{\mathcal{F}}(P) \leq \mathrm{Aut}(\mathcal{F}) \leq \mathrm{Aut}(P).$$

For example, if $\mathcal{F}$ is a fusion system on P, where P is the dihedral group of order 8, and one of the two Klein four subgroups has an automorphism of order 3 but the other does not, so that $\mathcal{F} = \mathcal{F}_P(S_4)$, then $\mathrm{Aut}(\mathcal{F}) = \mathrm{Aut}_{\mathcal{F}}(P) = \mathrm{Inn}(P)$. If both or neither of the Klein four subgroups has an automorphism of order 3 (for example, in the fusion system of $\mathrm{PSL}_2(7)$), then $\mathrm{Aut}(\mathcal{F}) = \mathrm{Aut}(P)$.

5.3 Normal subgroups

In this section, we build upon the concept of a strongly $\mathcal{F}$-closed subgroup to produce the concept of a subgroup that is normal in a fusion system (note not a subsystem that is normal).

Definition 5.25   Let $\mathcal{F}$ be a fusion system on a finite p-group P. A subgroup Q of P is *normal* in $\mathcal{F}$ (denoted $Q \trianglelefteq \mathcal{F}$) if $\mathcal{F} = \mathrm{N}_{\mathcal{F}}(Q)$, i.e., if, for each subgroup $R \leq P$ and $\phi \in \mathrm{Hom}_{\mathcal{F}}(R, P)$, the map ϕ may be extended to a map $\bar{\phi} \in \mathrm{Hom}_{\mathcal{F}}(QR, P)$ such that $\bar{\phi}|_Q$ is an automorphism of Q.

Note that, by Lemma 4.57, a subgroup Q is strongly $\mathrm{N}_{\mathcal{F}}(Q)$-closed, and hence a subgroup normal in a fusion system is strongly $\mathcal{F}$-closed.

The following lemma is easy, and its proof is left to the reader (Exercise 5.4).

Lemma 5.26 *Let G be a finite group and let P be a Sylow p-subgroup of G. If Q is a normal p-subgroup of G then $Q \trianglelefteq \mathcal{F}_P(G)$.*

The intersection and product of normal subgroups is again normal, mirroring the situation for normal subgroups in groups.

Proposition 5.27 *Let $\mathcal{F}$ be a fusion system on a finite p-group P, and let Q and R be subgroups of P.*

(i) *If $Q \trianglelefteq \mathcal{F}$ and R is a strongly $\mathcal{F}$-closed subgroup of P, then $Q \cap R \trianglelefteq \mathcal{F}$.*
(ii) *If Q and R are both normal subgroups, then $Q \cap R \trianglelefteq \mathcal{F}$.*
(iii) *If Q and R are normal in $\mathcal{F}$, then QR is normal in $\mathcal{F}$.*

Proof Lemma 4.59 states that the intersection of two strongly $\mathcal{F}$-closed subgroups is strongly $\mathcal{F}$-closed, and so $Q \cap R$ is strongly $\mathcal{F}$-closed. Since $Q \cap R \leqslant Q$ and Q is contained in every fully $\mathcal{F}$-normalized, $\mathcal{F}$-centric, $\mathcal{F}$-radical subgroup of P by Proposition 4.61, so therefore is $Q \cap R$, and hence $Q \cap R \trianglelefteq \mathcal{F}$, proving (i); then (ii) follows immediately as a special case of (i).

We can prove (iii) in one of two ways: firstly, by Theorem 5.22, the product of two strongly $\mathcal{F}$-closed subgroups is strongly $\mathcal{F}$-closed, and by Proposition 4.61, both Q and R are contained within every fully normalized, centric, radical subgroup of P. Hence QR is contained within a fully normalized, centric, radical subgroup of P and strongly $\mathcal{F}$-closed, and so by Proposition 4.61 again, we see that $QR \trianglelefteq \mathcal{F}$.

Alternatively, we may prove this directly: Since $N_\mathcal{F}(Q) = \mathcal{F}$, any morphism $\phi : S \to P$ extends to a morphism $\psi : QS \to P$ in $\mathcal{F}$, which then extends to a morphism $\theta : (QR)S \to P$ in $\mathcal{F}$ since $N_\mathcal{F}(R)$. Furthermore, $\theta|_Q = \psi|_Q \in \mathrm{Aut}_\mathcal{F}(Q)$ and $\theta|_R \in \mathrm{Aut}(R)$, so $\theta|_{QR}$ is an automorphism in $\mathrm{Aut}_\mathcal{F}(QR)$. Hence ϕ lies in $N_\mathcal{F}(QR)$, and so $\mathcal{F} = N_\mathcal{F}(QR)$, as claimed. $\square$

As a trivial consequence to this proposition, we see that there is a largest subgroup that is normal in a fusion system.

Definition 5.28 Let $\mathcal{F}$ be a fusion system on a finite p-group P. The largest subgroup of P that is normal in $\mathcal{F}$ is denoted by $O_p(\mathcal{F})$. Denote by $\mathcal{O}_p(\mathcal{F})$ the subsystem $\mathcal{F}_{O_p(\mathcal{F})}(O_p(\mathcal{F}))$.

The subgroup $O_p(\mathcal{F})$ is of significant interest – for example, it is the basis of the definition of soluble fusion systems in Section 5.7 – and we will meet it often.

As a definition, the statement that $\mathcal{F} = N_\mathcal{F}(Q)$ might not be easy to confirm directly. In the previous chapter we saw (Propositions 4.61

and 4.62) two equivalent conditions to the statement that $\mathcal{F} = N_{\mathcal{F}}(Q)$, and there are several others. We are now in a position to prove Stancu's result, as mentioned in Section 4.7.

Theorem 5.29 (Stancu [Sta06])   *Let $\mathcal{F}$ be a saturated fusion system on a finite p-group P, and let Q be a strongly $\mathcal{F}$-closed subgroup of P. The following are equivalent:*

(i) $\mathcal{F} = N_{\mathcal{F}}(Q)$;

(ii) Q *possesses a central series*

$$1 = Q_0 \leq Q_1 \leq \cdots \leq Q_r = Q,$$

for which each Q_i is weakly $\mathcal{F}$-closed.

Proof   If $Q \trianglelefteq \mathcal{F}$, then by Proposition 4.62 we have that Q possesses a central series of strongly $\mathcal{F}$-closed subgroups, all of which are clearly weakly $\mathcal{F}$-closed, proving that (i) implies (ii).

To prove the other direction, suppose that Q possesses a central series of weakly $\mathcal{F}$-closed subgroups as in (ii) and proceed by induction on r; by Lemma 4.63, Q_1, which is central, is also strongly $\mathcal{F}$-closed. The subgroups Q_i/Q_1 form a weakly $\mathcal{F}/Q_1$-closed central series for Q/Q_1 by Theorem 5.21, and so by induction Q/Q_1 possesses a central series of strongly $\mathcal{F}/Q_1$-closed subgroups which have preimages that (together with Q_1) form a central series for Q consisting of strongly $\mathcal{F}$-closed subgroups, again by Theorem 5.21. Therefore $Q \trianglelefteq \mathcal{F}$ by another application of Proposition 4.62.   □

Once we have the definition of weakly normal subsystems in the next section, we will be able to prove the last equivalent condition that we state in this chapter, and we will collect all of the equivalent conditions together.

5.4 Weakly normal subsystems

The majority of the interesting results in the theory of fusion systems concern *saturated* fusion systems. In particular, if one wants to prove some version of the classification of finite simple groups – maybe a classification of simple fusion systems – then as well as a definition of a normal subsystem, you will probably want to restrict yourself to saturated fusion systems. However, there are two competing notions for a normal subsystem, which here we will refer to as weakly normal and normal subsystems. The first, due essentially to Puig (although he didn't require the subsystem to be saturated) will be given now, and the second is in Chapter 8.

Definition 5.30 Let $\mathcal{F}$ be a fusion system on a finite p-group P, and let Q be a strongly $\mathcal{F}$-closed subgroup of P. A subsystem $\mathcal{E}$ of $\mathcal{F}$ on Q is $\mathcal{F}$-invariant if, for each $R \leq S \leq Q$, $\phi \in \mathrm{Hom}_{\mathcal{E}}(R, S)$, and $\psi \in \mathrm{Hom}_{\mathcal{F}}(S, P)$ (which is $\mathrm{Hom}_{\mathcal{F}}(S, Q)$ since Q is strongly $\mathcal{F}$-closed), $\psi^{-1}\phi\psi$ is a morphism in $\mathrm{Hom}_{\mathcal{E}}(R\psi, Q)$. If in addition $\mathcal{E}$ is saturated, $\mathcal{E}$ is *weakly normal* in $\mathcal{F}$. We denote weak normality by $\mathcal{E} \prec \mathcal{F}$.

It is easy to see the following equivalent condition to being invariant, and so we will not supply a proof.

Lemma 5.31 *Let $\mathcal{F}$ be a fusion system on a finite p-group P. A subsystem $\mathcal{E}$ on a strongly $\mathcal{F}$-closed subgroup Q is $\mathcal{F}$-invariant if and only if, for all $R \leq S \leq P$ and $\phi : S \to P$, the map ϕ induces a bijection $\mathrm{Hom}_{\mathcal{E}}(R, S) \to \mathrm{Hom}_{\mathcal{E}}(R\phi, S\phi)$ by conjugation by ϕ.*

We can think of being invariant as like being normal for groups. There are two ways in which the two are connected. The first is that normal subgroups yield weakly normal subsystems, as you would hope.

Lemma 5.32 *Let G be a finite group with a Sylow p-subgroup P, let H be a normal subgroup of G, and write $Q = P \cap H$. The subsystem $\mathcal{F}_Q(H)$ is weakly normal in $\mathcal{F}_P(G)$.*

This result is Exercise 5.5. The second result is on the automorphism groups of subgroups.

Lemma 5.33 *Let $\mathcal{F}$ be a fusion system on a finite p-group P, and let $\mathcal{E}$ be an $\mathcal{F}$-invariant subsystem of $\mathcal{F}$, on a subgroup Q of P. If R is a subgroup of Q, then $\mathrm{Aut}_{\mathcal{E}}(R) \trianglelefteq \mathrm{Aut}_{\mathcal{F}}(R)$.*

Proof Let ϕ be an $\mathcal{F}$-automorphism of R, and let $\alpha \in \mathrm{Aut}_{\mathcal{E}}(R)$. By definition $\phi^{-1}\alpha\phi \in \mathrm{Aut}_{\mathcal{E}}(R)$ and so $\mathrm{Aut}_{\mathcal{E}}(R) \trianglelefteq \mathrm{Aut}_{\mathcal{F}}(R)$, as claimed. $\qquad\square$

The intersection of two subsystems is defined in the previous chapter. One big problem with the intersection operation is that it does not preserve saturation.

Example 5.34 Let $P = D_8 \times C_2$, with the D_8 factor generated by an element x of order 4 and y of order 2, and the C_2 factor being generated by z. Let $Q = \langle x, y \rangle$, and $R = \langle x, yz \rangle$; then $S = Q \cap R = \langle x \rangle$ is a normal cyclic subgroup of P, and $\mathrm{Aut}_Q(S) = \mathrm{Aut}_R(S)$ contains the identity and the map inverting x. Thus $\mathcal{E} = \mathcal{F}_Q(Q) \cap \mathcal{F}_R(R)$ contains an automorphism of order 2 on S, and so cannot be saturated, as $\mathrm{Aut}_S(S)$ is not a Sylow 2-subgroup of $\mathrm{Aut}_{\mathcal{E}}(S)$.

It is true that the intersection of two invariant subsystems is invariant, but since saturation cannot be guaranteed, this fact is not very helpful in practice.

Proposition 5.35 *Let $\mathcal{F}$ be a fusion system on a finite p-group P. Let $\mathcal{E}$ and $\mathcal{E}'$ be subsystems on the subgroups Q and R respectively.*

(i) *If $\mathcal{E}$ is $\mathcal{F}$-invariant, then $\mathcal{E} \cap \mathcal{E}'$ is $\mathcal{E}'$-invariant.*
(ii) *If both $\mathcal{E}$ and $\mathcal{E}'$ are $\mathcal{F}$-invariant, then so is $\mathcal{E} \cap \mathcal{E}'$.*

Proof Firstly assume that $\mathcal{E}$ is $\mathcal{F}$-invariant. If Q is strongly $\mathcal{F}$-closed, then it is easy to see that $Q \cap R$ is strongly $\mathcal{E}'$-closed: if $\phi : S \to T$ is a morphism in $\mathcal{E}'$ with $S \leq Q \cap R$ then $S\phi \leq Q$ and $S\phi \leq T \leq R$, so $S\phi \leq Q \cap R$, and $Q \cap R$ is strongly $\mathcal{E}'$-closed.

Now let S and T be subgroups of $Q \cap R$, such that $S \leq T$. Suppose that $\phi \in \mathrm{Hom}_{\mathcal{E} \cap \mathcal{E}'}(S, T)$ and that $\psi \in \mathrm{Hom}_{\mathcal{E}'}(S, R)$; then $\psi^{-1}\phi\psi$ is in $\mathcal{E}'$ since its components are in $\mathcal{E}'$. Also, since $\mathcal{E}$ is $\mathcal{F}$-invariant, then $\psi^{-1}\phi\psi$ is in $\mathcal{E}$, and so it is in $\mathcal{E} \cap \mathcal{E}'$. Thus $\mathcal{E} \cap \mathcal{E}'$ is $\mathcal{E}'$-invariant, proving (i).

The proof of (ii) is similar, and left to the reader. $\square$

The intersection of two weakly normal subsystems need not be weakly normal, but it *contains* a weakly normal subsystem on the same subgroup as the intersection. In Chapter 8 we shall meet the concept of a normal subsystem, and Theorem 8.21 provides exactly this statement for normal subsystems. Using this result we can prove the same result for weakly normal subsystems, Theorem 8.27.

For any given strongly $\mathcal{F}$-closed subgroup Q of P, there is an obvious subsystem $\mathcal{F}_Q(Q)$ of $\mathcal{F}$. It might be that this subsystem is weakly normal (as it is always saturated). The condition as to when it is will seem familiar, since $\mathcal{F}_Q(Q) \prec \mathcal{F}$ if and only if $Q \trianglelefteq \mathcal{F}$. However, we begin by proving a lemma, needed in the proof of this statement.

Lemma 5.36 *Let $\mathcal{F}$ be a saturated fusion system on a finite p-group P, and let Q be a subgroup of P. If $\mathcal{F}_Q(Q) \prec \mathcal{F}$ then every characteristic subgroup of Q is weakly $\mathcal{F}$-closed. In particular, Q has a central series each of whose terms is weakly $\mathcal{F}$-closed.*

Proof Let R be a characteristic subgroup of Q, and note that, since $\mathcal{F}_Q(Q) \prec \mathcal{F}$, we must have that Q is strongly $\mathcal{F}$-closed. Let S be any fully $\mathcal{F}$-normalized subgroup that is $\mathcal{F}$-conjugate to R; by Proposition 4.17, we may choose $\phi : R \to S$ such that $N_\phi = N_P(R) \geq Q$. Since $\mathcal{F}$ is saturated, ϕ has an extension to $\bar{\phi} : Q \to P$, which must be an automorphism of Q since Q is strongly $\mathcal{F}$-closed. However, R is characteristic, so is invariant under the action of $\mathrm{Aut}_{\mathcal{F}}(Q)$. Thus $R = R\phi$, and so R is fully $\mathcal{F}$-normalized. The same argument with S any subgroup $\mathcal{F}$-conjugate to R gives the result.

The second assertion follows since the subgroups $Z_i(G)$ are characteristic. $\qquad\square$

In fact, it is shown in Exercise 5.6 that all characteristic subgroups of Q are strongly $\mathcal{F}$-closed when $\mathcal{F}_Q(Q) \prec \mathcal{F}$, but we do not need this because we will apply Theorem 5.29.

Theorem 5.37 (Stancu [Sta06, Proposition 6.2]) *Let $\mathcal{F}$ be a saturated fusion system on a finite p-group P. If Q is a subgroup of P, then $Q \trianglelefteq \mathcal{F}$ if and only if $\mathcal{F}_Q(Q) \prec \mathcal{F}$.*

Proof Suppose that $\mathcal{F} = N_{\mathcal{F}}(Q)$; then any morphism $\phi : R \to S$ lifts to a morphism $\bar{\phi} : QR \to QS$ that acts as an automorphism on Q and so Q is strongly $\mathcal{F}$-closed. Write $\mathcal{E} = \mathcal{F}_Q(Q)$; if $R \leq S \leq Q$ and $\psi : S \to P$ is a map in $\mathcal{F}$, then we should show that, for all $\phi = c_g$ with $g \in Q$, the map $\psi^{-1}\phi\psi$ is also in $\mathcal{E}$. Since $\psi^{-1}c_g\psi = c_{g\psi}$, and Q is strongly $\mathcal{F}$-closed, we see that the conjugate of c_g by ψ is also an element of $\mathcal{E}$.

The other half of the proof is Lemma 5.36 and Theorem 5.29 combined. $\qquad\square$

In the case of abelian subgroups, when $\mathcal{F}_Q(Q)$ is weakly normal in $\mathcal{F}$ is very easy to understand, using Theorems 4.62 and 5.37; this theorem extends a result of Glauberman [Gla72, Theorem 6.1] to fusion systems.

Corollary 5.38 *Let $\mathcal{F}$ be a fusion system on a finite p-group P, and let Q be a subgroup of P. If Q is abelian then $Q \trianglelefteq \mathcal{F}$ if and only if Q is strongly $\mathcal{F}$-closed.*

We now collect all of the equivalent definitions that we have found for a subgroup Q to be normal in $\mathcal{F}$, an amalgamation of Propositions 4.61 and 4.62, and Theorems 5.29 and 5.37.

Theorem 5.39 *Let $\mathcal{F}$ be a saturated fusion system on a finite p-group P, and let Q be a strongly $\mathcal{F}$-closed subgroup of P. The following are equivalent:*

(i) $\mathcal{F} = \mathrm{N}_{\mathcal{F}}(Q)$;

(ii) $\mathcal{F}_Q(Q) \prec \mathcal{F}$;

(iii) Q possesses a central series all of whose terms are strongly $\mathcal{F}$-closed;

(iv) Q possesses a central series all of whose terms are weakly $\mathcal{F}$-closed;

(v) Q is contained in every fully normalized, $\mathcal{F}$-centric, $\mathcal{F}$-radical subgroup of P.

Having produced a reasonable collection of equivalent conditions for a subgroup to be normal, we now want conditions for a subsystem to be weakly normal. The one that we will consider in this section is the analogue of the Frattini argument for groups. Before we define the concept of an $\mathcal{F}$-Frattini subsystem, we prove a lemma about automorphisms of subsystems. Let Q be a strongly $\mathcal{F}$-closed subgroup, and let $\mathcal{E}$ be an $\mathcal{F}$-invariant subsystem on Q. The automorphisms $\alpha \in \mathrm{Aut}_{\mathcal{F}}(Q)$ in fact induce automorphisms of $\mathcal{E}$, as we will now show.

Lemma 5.40   *Let $\mathcal{F}$ be a fusion system on a finite p-group P, let Q be a strongly $\mathcal{F}$-closed subgroup of P, and let $\mathcal{E}$ be a subsystem on Q. If $\mathcal{E}$ is $\mathcal{F}$-invariant then $\mathrm{Aut}_{\mathcal{F}}(Q) \leq \mathrm{Aut}(\mathcal{E})$.*

Proof   Let α be an automorphism of Q. Since $\mathcal{E}$ is $\mathcal{F}$-invariant, the induced map (also called α), given by

$$\alpha : \phi \mapsto \alpha^{-1}\phi\alpha,$$

where ϕ is a morphism in $\mathcal{E}$, is a bijection $\alpha : \mathrm{Hom}_{\mathcal{E}}(R, S) \to \mathrm{Hom}_{\mathcal{E}}(R, S)$ for all $R, S \leq Q$. Hence $\alpha \in \mathrm{Aut}(\mathcal{E})$, as claimed. $\square$

We now define an analogue of the Frattini argument, as mentioned before. Recall that if H is a normal subgroup of a finite group G, and Q is a Sylow p-subgroup of H, then $G = \mathrm{N}_G(Q)H$. Thinking about the elements of g inducing maps in $\mathcal{F}_P(G)$ on the subgroup Q, this says that c_g may be decomposed as the product of a G-automorphism of Q followed by a morphism in $\mathcal{F}_Q(H)$.

Definition 5.41   Let $\mathcal{F}$ be a fusion system on a finite p-group P, and let $\mathcal{E}$ be a subsystem of $\mathcal{F}$, on the subgroup Q of P. We say that $\mathcal{E}$ is *$\mathcal{F}$-Frattini* if, whenever $R \leq Q$ and $\phi : R \to P$ is a morphism in $\mathcal{F}$, there exist morphisms $\alpha \in \mathrm{Aut}_{\mathcal{F}}(Q)$ and $\beta \in \mathrm{Hom}_{\mathcal{E}}(R\alpha, Q)$ such that $\phi = \alpha\beta$.

It is not true in general that if R and S are two $\mathcal{F}$-conjugate subgroups of Q, and $\mathcal{E}$ is an $\mathcal{F}$-invariant subsystem on Q, then R and S are $\mathcal{E}$-conjugate; this is not true in groups, for example. There, we see that two subgroups that are conjugate in a finite group G may not be conjugate in a normal subgroup H, but that there is an outer automorphism, induced from conjugation of G on H, that sends one of the subgroups to another.

The Frattini argument for groups translates into the following result for saturated fusion systems.

Lemma 5.42 *Let $\mathcal{F}$ be a saturated fusion system on a finite p-group P. If $\mathcal{E}$ is an $\mathcal{F}$-invariant subsystem on a subgroup Q of P, then $\mathcal{E}$ is $\mathcal{F}$-Frattini.*

Proof Let R be a subgroup of Q. A morphism $\phi : R \to S$ in $\mathcal{F}$ has an $\alpha\beta$-*decomposition* if there exists $\alpha \in \mathrm{Aut}_{\mathcal{F}}(Q)$ and $\beta \in \mathrm{Hom}_{\mathcal{E}}(R\alpha, S)$ such that $\phi = \alpha\beta$ on R. The aim is to prove that every morphism $\phi \in \mathrm{Hom}_{\mathcal{F}}(R, S)$, where R and S are arbitrary subgroups of Q, has an $\alpha\beta$-decomposition.

Step 1: *The composition, inverse, and restriction, of morphisms in $\mathcal{F}$ that have an $\alpha\beta$-decomposition also have $\alpha\beta$-decompositions.* For restrictions this is obvious, and for compositions and inverse this follows from the fact that, if α and β are of the form above, then (since $\mathcal{E}$ is an $\mathcal{F}$-invariant subsystem) we have that there is some β', a morphism in $\mathcal{E}$, such that $\alpha^{-1}\beta\alpha = \beta'$, and so

$$\beta\alpha = \alpha\beta'.$$

Hence we may move the $\mathcal{F}$-automorphisms of Q to the left, and so our claim is proved.

For each $R \leq Q$, let $\mathcal{H}(R)$ denote the subset of $\mathrm{Hom}_{\mathcal{F}}(R, Q)$ that have an $\alpha\beta$-decomposition, and choose R maximal in Q subject to $\mathcal{H}(R) \neq \mathrm{Hom}_{\mathcal{F}}(R, Q)$; let ϕ be a morphism in $\mathrm{Hom}_{\mathcal{F}}(R, Q) \setminus \mathcal{H}(R)$.

Step 2: *The map ϕ can be chosen to be an automorphism.* By Alperin's fusion theorem, there is a sequence $R = R_0, R_1, \ldots, R_n = R\phi$ of $\mathcal{F}$-conjugate subgroups of Q, fully $\mathcal{F}$-normalized subgroups $T_1, T_2, \ldots, T_n$, and automorphisms $\phi_i \in \mathrm{Aut}_{\mathcal{F}}(T_i)$ with $R_{i-1}\phi = R_i$ and $\phi = \phi_1\phi_2\ldots\phi_n$ on R: write $Q_i = T_i \cap Q$. Since Q is strongly $\mathcal{F}$-closed, we see that $\phi_i|_{Q_i}$ must be an automorphism. If $|Q_i| > |R|$, then $\phi_i|_{Q_i}$ lies in $\mathcal{H}(Q_i)$, and hence $\phi_i|_{R_{i-1}}$ is in $\mathcal{H}(R_{i-1})$. If all of the $\phi_i|_{R_{i-1}}$ lie in $\mathcal{H}(R_{i-1})$ then ϕ lies in $\mathcal{H}(R)$ by Step 1; therefore there is some i such that $|Q_i| = |R|$. Replace R by Q_i and ϕ by ϕ_i.

Step 3: *If $\psi : R \to S$ is an $\mathcal{F}$-isomorphism with S fully $\mathcal{F}$-normalized, then $\psi \in X_R$.* Recall that, for $U \leq P$, the map $c_U : \mathrm{N}_P(U) \to \mathrm{Aut}_P(U)$ is the natural map: we have that $N_\psi c_R = \operatorname{Aut}_P(R) \cap \operatorname{Aut}_P(S)^{\psi^{-1}}$. Since $\mathcal{E}$ is $\mathcal{F}$-invariant, we have that $\operatorname{Aut}_\mathcal{E}(R) \trianglelefteq \operatorname{Aut}_\mathcal{F}(R)$ by Lemma 5.33. Since S is fully $\mathcal{F}$-automized, $\operatorname{Aut}_P(S)$ is a Sylow p-subgroup of $\operatorname{Aut}_\mathcal{F}(S) \cong \operatorname{Aut}_\mathcal{F}(R)$, and therefore $\operatorname{Aut}_\mathcal{F}(S)^{\psi^{-1}}$ is a Sylow p-subgroup of $\operatorname{Aut}_\mathcal{F}(R)$. Hence $\operatorname{Aut}_\mathcal{E}(R) \cap \operatorname{Aut}_\mathcal{F}(S)^{\psi^{-1}}$ is a Sylow p-subgroup of $\operatorname{Aut}_\mathcal{E}(R)$, and so, by Sylow's theorem, there is some $\nu \in \operatorname{Aut}_\mathcal{E}(R)$ such that

$$\operatorname{Aut}_Q(R)^\nu \leq \operatorname{Aut}_\mathcal{F}(S)^{\psi^{-1}},$$

and hence $\operatorname{Aut}_Q(R) \leq \operatorname{Aut}_\mathcal{F}(S)^{(\nu\psi)^{-1}}$. Thus $N_{\nu\psi}$ contains $\operatorname{N}_Q(R)$, and so $\nu\psi$ has an $\alpha\beta$-decomposition by maximal choice of R. Since ν has one trivially, we see that $\nu^{-1}(\nu\psi)$ lies in $\mathcal{H}(R)$ by Step 1, as needed.

Step 4: *Completion of the proof.* Let S be a fully $\mathcal{F}$-normalized subgroup that is $\mathcal{F}$-conjugate to R via $\psi : R \to S$. Both ψ and $\phi\psi$ are isomorphisms whose image is fully $\mathcal{F}$-normalized, and so by Step 3 they have $\alpha\beta$-decompositions, whence ϕ has an $\alpha\beta$-decomposition. Thus $\phi \in \mathcal{H}(R)$, a contradiction that completes the proof. $\square$

What Lemma 5.42 tells us is that although two $\mathcal{F}$-conjugate subgroups R and S of Q might not be $\mathcal{E}$-conjugate, there is an automorphism $\alpha \in \operatorname{Aut}_\mathcal{F}(Q)$ mapping the $\mathcal{E}$-conjugacy class containing R to that of S. In fact, Lemmas 5.40 and 5.42 together have a converse, a result which is very useful, in the next few pages for example.

Theorem 5.43 (Puig [Pui06, Proposition 6.6])   *Let $\mathcal{F}$ be a saturated fusion system on a finite p-group P, and let Q be a strongly $\mathcal{F}$-closed subgroup of P. A subsystem $\mathcal{E}$ of $\mathcal{F}$ on Q is $\mathcal{F}$-invariant if and only if $\operatorname{Aut}_\mathcal{F}(Q) \leq \operatorname{Aut}(\mathcal{E})$ and $\mathcal{E}$ is $\mathcal{F}$-Frattini.*

Proof   Suppose that $\mathcal{E}$ is $\mathcal{F}$-Frattini and that $\operatorname{Aut}_\mathcal{F}(Q) \leq \operatorname{Aut}(\mathcal{E})$. Let $R \leq S \leq Q$ be subgroups and $\phi : S \to P$ be a morphism in $\mathcal{F}$. We need to show that ϕ induces a bijection $\operatorname{Hom}_\mathcal{E}(R, S) \to \operatorname{Hom}_\mathcal{E}(R\phi, S\phi)$. Since $\mathcal{E}$ is $\mathcal{F}$-Frattini, there exist morphisms $\alpha \in \operatorname{Aut}_\mathcal{F}(Q)$ and $\beta \in \operatorname{Hom}_\mathcal{E}(R\alpha, Q)$ with $\phi = \alpha\beta$. Also, $\alpha \in \operatorname{Aut}_\mathcal{F}(Q) \leq \operatorname{Aut}(\mathcal{E})$, so that α induces a bijection $\operatorname{Hom}_\mathcal{E}(R, S) \to \operatorname{Hom}_\mathcal{E}(R\alpha, S\alpha)$. Also, since β is a morphism in $\mathcal{E}$, it too induces a bijection $\operatorname{Hom}_\mathcal{E}(R\alpha, S\alpha) \to \operatorname{Hom}_\mathcal{E}(R\phi, S\phi)$ and so, as $\phi = \alpha\beta$, we see that ϕ induces the required bijection.

The converse is given by Lemmas 5.40 and 5.42. $\square$

In group theory, if K and H are subgroups of G, and K char $H \trianglelefteq G$, then $K \trianglelefteq G$. The same result is true for fusion systems, but we have not defined characteristic subsystems yet. More generally, if $K \trianglelefteq H \trianglelefteq G$

and all automorphisms of H induced by conjugations by elements of G fix K, then $K \trianglelefteq G$ as well. This second formulation can also be proved for fusion systems.

Proposition 5.44 (Aschbacher [Asc11, 7.4]) *Let $\mathcal{F}$ be a saturated fusion system on a finite p-group P, and let Q and R be subgroups of P with $Q \leq R \leq P$. Suppose that $\mathcal{E}$ and $\mathcal{E}'$ are saturated subsystems of $\mathcal{F}$ on Q and R respectively. If $\mathcal{E} \prec \mathcal{E}' \prec \mathcal{F}$, and all elements of $\mathrm{Aut}_{\mathcal{F}}(R)$ induce automorphisms on $\mathcal{E}$, then $\mathcal{E} \prec \mathcal{F}$.*

Proof If $S \leq Q$ and $\phi : S \to P$ is any map in $\mathcal{F}$, then since R is strongly $\mathcal{F}$-closed, we have that $\mathrm{im}\, \phi \leq R$. Since $\mathcal{E}' \prec \mathcal{F}$, it is $\mathcal{F}$-Frattini by Theorem 5.43, and so $\phi = \alpha\beta$, where $\alpha \in \mathrm{Aut}_{\mathcal{F}}(R)$ and β is a morphism in $\mathcal{E}'$. All $\mathcal{F}$-automorphisms of R induce automorphisms on $\mathcal{E}$, so we see that $S\alpha \leq Q$, and as Q is strongly $\mathcal{E}'$-closed, $S\phi = (S\alpha)\beta \leq Q$. In particular, Q is strongly $\mathcal{F}$-closed as $S\phi \leq Q$.

We will use Theorem 5.43 again: since $\mathcal{E} \prec \mathcal{E}'$, $\mathcal{E}$ is $\mathcal{E}'$-Frattini, so $\beta = \gamma\delta$, where $\gamma \in \mathrm{Aut}_{\mathcal{E}'}(Q)$ and δ is a morphism in $\mathcal{E}$. Notice that $\phi = \alpha\gamma\delta$. We claim that this decomposition proves that $\mathcal{E}$ is $\mathcal{F}$-Frattini; since $\alpha \in \mathrm{Aut}_{\mathcal{F}}(R)$ and Q is strongly $\mathcal{F}$-closed, $\alpha|_Q$ is an automorphism, and hence $\alpha\gamma \in \mathrm{Aut}_{\mathcal{F}}(Q)$. Since $\delta \in \mathrm{Hom}_{\mathcal{E}}(S\alpha\gamma, Q)$, this gives the correct decomposition of ϕ.

If $S = Q$, then we again use the fact that $\mathcal{E}'$ is $\mathcal{F}$-Frattini (and that Q is strongly $\mathcal{F}$-closed) to get that $\phi \in \mathrm{Aut}_{\mathcal{F}}(Q)$ may be written as $\phi = \alpha\beta$, where $\alpha \in \mathrm{Aut}_{\mathcal{F}}(R)$ (and so $\alpha \in \mathrm{Aut}_{\mathcal{F}}(Q)$) and $\beta \in \mathrm{Aut}_{\mathcal{E}'}(Q)$. Because all elements of $\mathrm{Aut}_{\mathcal{F}}(R)$ induce automorphisms on $\mathcal{E}$ by hypothesis (so $\alpha \in \mathrm{Aut}(\mathcal{E})$), and $\mathrm{Aut}_{\mathcal{E}'}(Q) \leq \mathrm{Aut}(\mathcal{E})$ by Theorem 5.43 (so $\beta \in \mathrm{Aut}(\mathcal{E})$), we see that $\phi \in \mathrm{Aut}(\mathcal{E})$, so that $\mathrm{Aut}_{\mathcal{F}}(Q) \leq \mathrm{Aut}(\mathcal{E})$. Hence $\mathcal{E}$ is weakly normal in $\mathcal{F}$ by Theorem 5.43. $\square$

Proposition 5.44 can be used to easily prove a theorem of Linckelmann.

Theorem 5.45 (Linckelmann [Lin07, Theorem 9.1]) *Let $\mathcal{F}$ be a saturated fusion system on a finite p-group P, and let $\mathcal{E}$ be a weakly normal subsystem of $\mathcal{F}$, on a subgroup Q of P. Suppose that $\mathcal{E} = \mathrm{N}_{\mathcal{E}}(R)$ for some subgroup R of Q. If S is the subgroup of Q generated by all $\mathcal{F}$-conjugates of R, then $\mathcal{F} = \mathrm{N}_{\mathcal{F}}(S)$.*

Proof Suppose that $\mathcal{E} = \mathrm{N}_{\mathcal{E}}(R)$, and write S for the product of all $\mathcal{F}$-conjugates of R (all of which are subgroups of Q, since Q is strongly $\mathcal{F}$-closed). Then $\mathcal{E} = \mathrm{N}_{\mathcal{E}}(S)$ since the product of subgroups normal in $\mathcal{E}$

is also normal in $\mathcal{E}$, and if $\phi \in \mathrm{Aut}_{\mathcal{F}}(Q)$, then ϕ leaves S invariant, so induces an automorphism on $\mathcal{D} = \mathcal{F}_S(S)$; i.e., $\mathrm{Aut}_{\mathcal{F}}(Q) \leq \mathrm{Aut}(\mathcal{D})$. (By Theorem 5.37, $\mathcal{F}_S(S) \prec \mathcal{E}$ if and only if $\mathcal{E} = N_{\mathcal{F}}(S)$.) Therefore, since $\mathcal{D} \prec \mathcal{E} \prec \mathcal{F}$, by Proposition 5.44, $\mathcal{D} \prec \mathcal{F}$. Again, by Theorem 5.37, we get that $\mathcal{F} = N_{\mathcal{F}}(S)$, as claimed. $\qquad\square$

This theorem relates normal subgroups and weakly normal subsystems, and says that if $R \leq \mathrm{O}_p(\mathcal{E})$ then $R \leq \mathrm{O}_p(\mathcal{F})$ for some weakly normal subsystem $\mathcal{E}$ of $\mathcal{F}$.

Lemma 5.46 *Let $\mathcal{F}$ be a saturated fusion system on a finite p-group P, and let Q be a subgroup of P. If $\mathcal{E}$ is a saturated subsystem of $\mathcal{F}$ on Q, then*

$$\mathrm{O}_p(\mathcal{E}) \geq \mathrm{O}_p(\mathcal{F}) \cap Q.$$

Proof Let $R = \mathrm{O}_p(\mathcal{F})$; by Proposition 4.62, R possesses a central series

$$1 = R_0 \leq R_1 \leq \cdots \leq R_d = R,$$

such that each R_i is strongly $\mathcal{F}$-closed. We claim that $Q_i = Q \cap R_i$ is strongly $\mathcal{E}$-closed; in this case,

$$1 = Q_0 \leq Q_1 \leq \cdots \leq Q_d = Q \cap R$$

is a central series for $Q \cap R$, each of whose terms is strongly $\mathcal{E}$-closed, yielding that $Q \cap R \leq \mathrm{O}_p(\mathcal{E})$, as required. It remains to show that Q_i is strongly $\mathcal{E}$-closed; however, any morphism in $\mathcal{E}$ that originates inside $Q_i = Q \cap R_i$ must have image inside Q since $\mathcal{E}$ lies on Q, and must also lie in R_i since it is strongly $\mathcal{F}$-closed, and so Q_i is strongly $\mathcal{E}$-closed. $\quad\square$

In the case where $\mathcal{E} \prec \mathcal{F}$, we can say more.

Proposition 5.47 *Let $\mathcal{F}$ be a saturated fusion system on a finite p-group P. If $\mathcal{E}$ is a weakly normal subsystem on a subgroup Q of P, then $\mathrm{O}_p(\mathcal{F}) \cap Q = \mathrm{O}_p(\mathcal{E})$. In particular, if $\mathrm{O}_p(\mathcal{E}) \neq 1$ for some $\mathcal{E} \prec \mathcal{F}$, then $\mathrm{O}_p(\mathcal{F}) \neq 1$.*

Proof Because of Lemma 5.46, we have that $\mathrm{O}_p(\mathcal{E}) \geq Q \cap \mathrm{O}_p(\mathcal{F})$; on the other hand, Theorem 5.45 tells us that $\mathrm{O}_p(\mathcal{E}) \leq \mathrm{O}_p(\mathcal{F})$, and so we get equality. $\qquad\square$

In group theory, characteristic subgroups are an integral part of the theory. The definition is natural, and because of this, there is a natural analogue in the theory of fusion systems. Recall that α is an automorphism of $\mathcal{F}$ if and only if, for all A and B subgroups of P, we have that the induced map $\alpha : \mathrm{Hom}_{\mathcal{F}}(A, B) \to \mathrm{Hom}_{\mathcal{F}}(A\alpha, B\alpha)$ is an isomorphism, where the induced map is $\phi \mapsto \phi^{\alpha}$.

Definition 5.48 Let $\mathcal{F}$ be a saturated fusion system on a finite p-group P. A subsystem $\mathcal{E}$ of $\mathcal{F}$ is *weakly characteristic* in $\mathcal{F}$ if it is weakly normal and for any automorphism α of $\mathcal{F}$, we have that $\mathcal{E}\alpha = \mathcal{E}$.

For groups, if a subgroup is invariant under all automorphisms then it is in particular normal, since conjugation maps form automorphisms of the group. For fusion systems, this no longer holds, and so to be weakly characteristic it must be *both* invariant under automorphisms and weakly normal.

We would like to know how automorphisms of fusion systems move the various special types of subgroups and subsystems around.

Lemma 5.49 *Let $\mathcal{F}$ be a fusion system, and let Q and R be subgroups of P. If α is an automorphism of $\mathcal{F}$, then Q and R are $\mathcal{F}$-conjugate if and only if $Q\alpha$ and $R\alpha$ are $\mathcal{F}$-conjugate. Furthermore, $\mathrm{Aut}_{\mathcal{F}}(Q) \cong \mathrm{Aut}_{\mathcal{F}}(Q\alpha)$.*

Proof Let Q and R be of the same order; then $\mathrm{Hom}_{\mathcal{F}}(Q, R)$ is non-empty if and only if Q and R are $\mathcal{F}$-conjugate, and clearly $\mathrm{Hom}_{\mathcal{F}}(Q, R)$ is non-empty if and only if $\mathrm{Hom}_{\mathcal{F}}(Q\alpha, R\alpha)$ is non-empty. The second part follows from setting $Q = R$, and noting that the map α induces an isomorphism $\mathrm{Aut}_{\mathcal{F}}(Q) \to \mathrm{Aut}_{\mathcal{F}}(Q\alpha)$. Hence the lemma is proved. □

Using this lemma, we may easily get the following two results, whose proof we shall leave to the reader.

Lemma 5.50 *Let $\mathcal{F}$ be a fusion system on a finite p-group P, and let α be an automorphism of $\mathcal{F}$. Let Q be a subgroup of P. If Q has a property $\mathscr{P}$, then so does $Q\alpha$, where $\mathscr{P}$ comes from the following list: fully normalized, fully centralized, fully automized, receptive, $\mathcal{F}$-centric, radical, essential, strongly $\mathcal{F}$-closed, weakly $\mathcal{F}$-closed, and normal.*

Lemma 5.51 *Let $\mathcal{F}$ be a fusion system, and let α be an automorphism of $\mathcal{F}$. Let $\mathcal{E}$ be a subsystem of $\mathcal{F}$, on a subgroup Q of P. If $\mathcal{E}$ has a property $\mathscr{P}$, then so does $\mathcal{E}\alpha$, where $\mathscr{P}$ comes from the following list: saturated, invariant, weakly normal.*

Since the image of a normal subgroup under an automorphism is a normal subgroup, we notice that the subgroup $O_p(\mathcal{F})$ is $\mathrm{Aut}(\mathcal{F})$-invariant, and hence $\mathcal{O}_p(\mathcal{F})$, the subsystem on $O_p(\mathcal{F})$, is a weakly characteristic subsystem of $\mathcal{F}$. In Section 5.5 we will meet another, the centre.

To finish this brief discussion on weakly characteristic subsystems, Proposition 5.44 has the following corollary.

Corollary 5.52   *Let $\mathcal{F}$ be a saturated fusion system on a finite p-group P, and let $\mathcal{E}$ and $\mathcal{E}'$ be subsystems of $\mathcal{F}$. If $\mathcal{E}'$ is weakly characteristic in $\mathcal{E}$ and $\mathcal{E}$ is weakly normal in $\mathcal{F}$, then $\mathcal{E}'$ is weakly normal in $\mathcal{F}$, and if $\mathcal{E}$ is also weakly characteristic in $\mathcal{F}$ then $\mathcal{E}'$ is weakly characteristic in $\mathcal{F}$.*

By defining a weakly normal subsystem to be one that is both invariant and saturated, we managed to get round one of the most irksome aspects of invariant subsystems: that they need not be saturated. The theory of weakly normal subsystems seems reasonably robust for our purposes. Indeed, it is interesting enough for the following definition to lead to non-trivial results.

Definition 5.53   A saturated fusion system $\mathcal{F}$ is *simple* if the only weakly normal subsystems of $\mathcal{F}$ are $\{1\}$ and $\mathcal{F}$.

5.5 Correspondences for quotients

In group theory, if one has a group G and a normal subgroup H, then there is a one-to-one correspondence between the subgroups of G containing H and the subgroups of G/H, and this bijection preserves normality. For fusion systems, this does not hold in general, because, while the image of a saturated or weakly normal subsystem is always saturated or weakly normal respectively, one does not get any such result about preimages, as the following example shows.

Example 5.54   Let $Q = \langle a, b \rangle$ be an elementary abelian group of order 9, and let x and y be automorphisms of Q, both fixing b, with x inverting a, and y mapping a to ab. (Note that x has order 2 and y has order 3.) The group $\langle x, y \rangle$ is isomorphic with S_3, and we form the semidirect product $G = Q \rtimes \langle x, y \rangle$. Let K_i denote the subgroup $\langle a, b, xy^i \rangle$. Write P for the Sylow 3-subgroup of G, and let $\mathcal{F} = \mathcal{F}_P(G)$ and $\mathcal{E}_i = \mathcal{F}_Q(K_i)$.

Clearly, if $B = \langle b \rangle$, then $\mathcal{F} = C_{\mathcal{F}}(B)$, so we may form the quotient system $\mathcal{F}/B$. Notice that none of the $\mathcal{E}_i$ are equal, but the subsystems $\mathcal{E}_i/B$ are equal. Furthermore, $\mathcal{E}_i/B$ is $\mathcal{F}/B$-invariant, so weakly normal in $\mathcal{F}/B$, but the $\mathcal{E}_i$ are not $\mathcal{F}$-invariant.

Hence there is no bijection between saturated subsystems of $\mathcal{F}$ on subgroups containing B with saturated subsystems of $\mathcal{F}/B$, and if the image of a saturated subsystem is $\mathcal{F}/B$-invariant then this need not mean that the original subsystem is $\mathcal{F}$-invariant.

In this section we will investigate what one can say about the quotient given knowledge about the original fusion system, and vice versa. While one may take quotients with any strongly closed subgroup as the kernel, we will find that what one can say depends on the structure of $\mathcal{F}$; for example, if $\mathcal{F}$ is a saturated fusion system on P, and Q is a strongly $\mathcal{F}$-closed subgroup of P, then the behaviour of, for example, centric radical subsystems varies greatly depending on whether $\mathcal{F} = P\,\mathrm{C}_{\mathcal{F}}(Q)$ or not.

Since the centre of a group is all elements whose centralizer is the whole group, we have a clear analogue for fusion system.

Definition 5.55 Let $\mathcal{F}$ be a fusion system on a finite p-group P. The *centre* of $\mathcal{F}$ is the subgroup

$$\mathrm{Z}(\mathcal{F}) = \{z \in P \mid \mathrm{C}_{\mathcal{F}}(\langle z \rangle) = \mathcal{F}\}.$$

Writing $Z = \mathrm{Z}(\mathcal{F})$, by $\mathcal{Z}(\mathcal{F})$, we mean the subgroup $\mathcal{F}_Z(Z)$.

The subgroup B from Example 5.54 is an example of a central subgroup of a fusion system. The next result is trivial, and is left to the reader.

Lemma 5.56 *Let $\mathcal{F}$ be a fusion system on a finite p-group P. If $\phi : X \to Y$ is a morphism with $X, Y \leq \mathrm{Z}(\mathcal{F})$ then $X \leq Y$ and ϕ is the inclusion map.*

Notice that, since for $z \in \mathrm{Z}(\mathcal{F})$ every morphism extends to a morphism acting trivially on z, every morphism in $\mathcal{F}$ extends to a morphism acting trivially on the whole of $\mathrm{Z}(\mathcal{F})$ *simultaneously*, and so $\mathcal{F} = \mathrm{C}_{\mathcal{F}}(\mathrm{Z}(\mathcal{F}))$. Clearly $\mathrm{Z}(\mathcal{F})$ is the largest subgroup for which this is true, and $\mathcal{Z}(\mathcal{F})$ is a weakly characteristic subsystem of $\mathcal{F}$.

This has a useful corollary.

Corollary 5.57 *Let $\mathcal{F}$ be a fusion system on a finite p-group P. If $X \leq \mathrm{Z}(\mathcal{F})$ and $\mathcal{E}$ is a subsystem of $\mathcal{F}$, based on X, then $\mathcal{E} = \mathcal{F}_X(X)$.*

Having introduced the centre, we go back to discussing our original problem of images and preimages under quotient maps. We have five possibilities for the subgroup Q, when we take a quotient $\mathcal{F}/Q$:

(i) $\mathcal{F} = \mathrm{C}_{\mathcal{F}}(Q)$;
(ii) $\mathcal{F} = Q\,\mathrm{C}_{\mathcal{F}}(Q)$ but $\mathcal{F} \neq \mathrm{C}_{\mathcal{F}}(Q)$;
(iii) $\mathcal{F} = P\,\mathrm{C}_{\mathcal{F}}(Q)$ but $\mathcal{F} \neq Q\,\mathrm{C}_{\mathcal{F}}(Q)$;

(iv) $\mathcal{F} = \mathrm{N}_{\mathcal{F}}(Q)$ but $\mathcal{F} \neq P\,\mathrm{C}_{\mathcal{F}}(Q)$;

(v) Q is strongly $\mathcal{F}$-closed but $\mathcal{F} \neq \mathrm{N}_{\mathcal{F}}(Q)$.

Which of these is the correct one to consider depends on the property we want to pull back from the quotient to the original fusion system. We concentrate on saturated fusion systems since for these systems there is a unique notion for the quotient. If the reader is interested, he may attempt to remove saturation and deal with the systems $\mathcal{F}/Q$ and $\langle \bar{\mathcal{F}}_Q \rangle$.

We begin by understanding what happens to the images of the various types of subgroup and subsystem.

Proposition 5.58 *Let $\mathcal{F}$ be a fusion system on a finite p-group P, and let Q be a strongly $\mathcal{F}$-closed subgroup of P. Let R and S be subgroups of P containing Q, and let $\alpha : \mathrm{Aut}_{\mathcal{F}}(R) \to \mathrm{Aut}_{\mathcal{F}/Q}(R/Q)$ be the natural map.*

(i) *R and S are $\mathcal{F}$-conjugate if and only if R/Q and S/Q are $\mathcal{F}/Q$-conjugate.*

(ii) *For normalizers we have $\mathrm{N}_{P/Q}(R/Q) = \mathrm{N}_P(R)/Q$, and for automizers we have $\mathrm{Aut}_P(R)\alpha = \mathrm{Aut}_{P/Q}(R/Q)$ and $\mathrm{Inn}(R)\alpha = \mathrm{Inn}(R/Q)$.*

(iii) *R is fully $\mathcal{F}$-normalized if and only if R/Q is fully $\mathcal{F}/Q$-normalized.*

(iv) *If R/Q is $\mathcal{F}/Q$-centric then R is $\mathcal{F}$-centric.*

Proof If R and S are $\mathcal{F}$-conjugate then clearly R/Q and S/Q are $\mathcal{F}/Q$-conjugate, and if R/Q and S/Q are $\mathcal{F}/Q$-conjugate via $\bar{\phi}$, then there is some morphism $\phi : R \to S$ in $\mathcal{F}$, and so R and S are $\mathcal{F}$-conjugate, proving (i).

Let Qg be a coset in P/Q; we have that Qg normalizes R/Q if and only if $Q(g^{-1}xg) \in R/Q$ for all $x \in R$, and this is true if and only if $g^{-1}xg \in R$ for all $x \in R$, so $Qg \in \mathrm{N}_{P/Q}(R/Q)$ if and only if $Qg \in \mathrm{N}_P(R)/Q$, proving that $\mathrm{N}_{P/Q}(R/Q) = \mathrm{N}_P(R)/Q$. For the next statement, notice that, if $c_g : x \mapsto g^{-1}xg$ is a map in $\mathrm{Aut}_P(R)$, then $c_g\alpha : Qx \to Q(g^{-1}xg)$, and so $c_g\alpha = c_{Qg}$. Thus $\mathrm{Aut}_P(R)\alpha = \mathrm{Aut}_{P/Q}(R/Q)$, and $\mathrm{Aut}_R(R)\alpha = \mathrm{Aut}_{R/Q}(R/Q)$.

Since $\mathrm{N}_{P/Q}(R/Q) = \mathrm{N}_P(R)/Q$, we must have that $|\mathrm{N}_{P/Q}(R/Q)| = |\mathrm{N}_P(R)|/|Q|$; by (i), this proves that $|\mathrm{N}_P(R)|$ is maximal amongst $\mathcal{F}$-conjugates of R if and only if $|\mathrm{N}_{P/Q}(R/Q)|$ is maximal amongst $\mathcal{F}/Q$-conjugates of R/Q. Thus R is fully $\mathcal{F}$-normalized if and only if R/Q is fully $\mathcal{F}/Q$-normalized. Thus (iii) is proved.

Finally, we prove (iv). Notice that $Q\,\mathrm{C}_P(S)/Q \leq \mathrm{C}_{P/Q}(S/Q)$, so that if $\mathrm{C}_{P/Q}(S/Q) \leq S/Q$ then $Q\,\mathrm{C}_P(S) \leq S$. Since R/Q is $\mathcal{F}/Q$-centric and being $\mathcal{F}/Q$-conjugate is the same as being $\mathcal{F}$-conjugate for the preimages, we see that R is $\mathcal{F}$-centric. $\qquad\square$

The converse of (iv) is false in general, in that there are centric subgroups whose image is not centric. For example, the subgroup Q from Example 5.54 is $\mathcal{F}$-centric, but Q/B is not $\mathcal{F}/B$-centric, so this does not hold, even when quotienting out by a central subgroup.

We can get a similar statement to Proposition 5.58 for images of subsystems.

Lemma 5.59 *Let $\mathcal{F}$ be a saturated fusion system on a finite p-group P, and let Q be a strongly $\mathcal{F}$-closed subgroup of P. If $\mathcal{E}$ is an $\mathcal{F}$-invariant subsystem of $\mathcal{F}$ on the subgroup R of P, then the image $\bar{\mathcal{E}}$ of $\mathcal{E}$ in $\mathcal{F}/Q$ is $\mathcal{F}/Q$-invariant, and so the image of a weakly normal subsystem is weakly normal.*

Proof Let $\bar{\phi} \in \mathrm{Hom}_{\bar{\mathcal{E}}}(A/Q, B/Q)$, where A and B are subgroups of QR with $A \le B$, and let $\bar{\psi} \in \mathrm{Hom}_{\mathcal{F}/Q}(B/Q, QR/Q)$. We need to show that $\bar{\phi}^{\bar{\psi}}$ is a morphism in $\bar{\mathcal{E}}$.

Let $\phi \in \mathrm{Hom}_{\mathcal{E}}(\tilde{A}, \tilde{B})$ be a map such that $Q\tilde{A} = A$, $Q\tilde{B} = B$ and ϕ has image $\bar{\phi}$ in $\mathcal{F}/Q$, and let $\psi \in \mathrm{Hom}_{\mathcal{F}}(B, QR)$ be a map such that ψ has image $\bar{\psi}$. Notice that $\psi|_{\tilde{B}} \in \mathrm{Hom}_{\mathcal{F}}(\tilde{B}, R)$, and so the map ϕ^{ψ} is in $\mathcal{E}$, since $\mathcal{E}$ is $\mathcal{F}$-invariant. It remains to show that the image of ϕ^{ψ} in $\mathcal{F}/Q$ is actually $\bar{\phi}^{\bar{\psi}}$; this is true since the map $\mathcal{F} \to \mathcal{F}/Q$ is a morphism of fusion systems. Thus $\bar{\mathcal{E}}$ is $\mathcal{F}/Q$-invariant, as required. Since the image of a saturated subsystem is saturated by Corollary 5.17, we get that the image of a weakly normal subsystem is weakly normal, completing the proof. $\qquad\square$

We now want to introduce the condition that $\mathrm{Aut}_{\mathcal{F}}(Q)$ is a p-group. We will see that this condition (and the associated condition $\mathcal{F} = P\,\mathrm{C}_{\mathcal{F}}(Q)$) plays a considerable role in what follows. The first two parts of the next proposition are due to Kessar and Linckelmann, in [KL08, Proposition 3.1].

Proposition 5.60 (Kessar–Linckelmann [KL08]) *Let $\mathcal{F}$ be a saturated fusion system on a finite p-group P, and let Q be a strongly $\mathcal{F}$-closed subgroup such that $\mathrm{Aut}_{\mathcal{F}}(Q)$ is a p-group (i.e., $\mathrm{N}_{\mathcal{F}}(Q) = P\,\mathrm{C}_{\mathcal{F}}(Q)$). Let R be a subgroup of P containing Q.*

(i) *If R is $\mathcal{F}$-radical then R/Q is $\mathcal{F}/Q$-radical.*
(ii) *R is $\mathcal{F}$-centric and $\mathcal{F}$-radical if and only if R/Q is $\mathcal{F}/Q$-centric and $\mathcal{F}/Q$-radical.*

(iii) R is $\mathcal{F}$-essential if and only if R/Q is $\mathcal{F}/Q$-essential.

In particular, if $\mathcal{F} = P\,\mathrm{C}_{\mathcal{F}}(Q)$ then there is a one-to-one correspondence between the $\mathcal{F}$-centric $\mathcal{F}$-radical subgroups of P and the $\mathcal{F}/Q$-centric, $\mathcal{F}/Q$-radical subgroups of P/Q, and similarly for essential subgroups.

Proof Let α denote the natural map $\mathrm{Aut}_{\mathcal{F}}(R) \to \mathrm{Aut}_{\mathcal{F}/Q}(R/Q)$. The kernel K of α is the collection of all maps that act trivially on R/Q. Let ϕ be a p'-automorphism of $\mathrm{Aut}_{\mathcal{F}}(R)$ that acts trivially on R/Q. Since $\mathrm{Aut}_{\mathcal{F}}(Q)$ is a p-group and $\phi|_Q$ is a p'-automorphism, it must also be trivial; Exercise 4.5 proves that $\phi = 1$, so that K is a p-group.

If R is $\mathcal{F}$-radical then $\mathrm{O}_p(\mathrm{Aut}_{\mathcal{F}}(R)) = \mathrm{Inn}(R)$, and so as K is a normal p-subgroup of $\mathrm{Aut}_{\mathcal{F}}(R)$, we see that $K \leq \mathrm{Inn}(R)$. By Proposition 5.58(ii), $\mathrm{Inn}(R/Q) = \mathrm{O}_p(\mathrm{Aut}_{\mathcal{F}/Q}(R/Q))$, so that R/Q is $\mathcal{F}/Q$-radical, proving (i).

Since $K \leq \mathrm{Inn}(R)$, the preimage C of K under the map $c_R : R \to \mathrm{Aut}(R)$ must be contained within $R\,\mathrm{C}_P(R)$, and hence $\mathrm{C}_{P/Q}(R/Q)$, which is contained in QC/Q, is contained within $R\,\mathrm{C}_P(R)/Q$. If R is $\mathcal{F}$-centric then $R\,\mathrm{C}_P(R) = R$, so $\mathrm{C}_{P/Q}(R/Q) \leq R/Q$. Since this applies for any $\mathcal{F}$-centric, $\mathcal{F}$-radical subgroup that is $\mathcal{F}$-conjugate to R, we see that R/Q is $\mathcal{F}/Q$-centric and $\mathcal{F}/Q$-radical as well.

Conversely, suppose that R/Q is $\mathcal{F}/Q$-centric, so that $\mathrm{C}_{P/Q}(R/Q) \leq R/Q$. Therefore, if $g \in P$ induces the trivial automorphism on R/Q, we have that $g \in R\,\mathrm{C}_P(R) = R$ (as R is $\mathcal{F}$-centric by Proposition 5.58), and so K, the set of all such automorphisms, must lie in $\mathrm{Inn}(R)$. Therefore, since $\mathrm{Inn}(R)\alpha = \mathrm{O}_p(\mathrm{Aut}_{\mathcal{F}/Q}(R/Q))$, and $\ker \alpha \leq \mathrm{Inn}(R)$, we have that $\mathrm{Inn}(R) = \mathrm{O}_p(\mathrm{Aut}_{\mathcal{F}}(R))$, so that R is $\mathcal{F}$-radical, as needed for (ii).

If R is $\mathcal{F}$-centric and $\mathcal{F}$-radical, then $\mathrm{Out}_{\mathcal{F}}(R) = \mathrm{Out}_{\mathcal{F}/Q}(R/Q)$, and so R is $\mathcal{F}$-essential if and only if R/Q is $\mathcal{F}/Q$-essential. This proves (iii). The final remark follows since if $\mathcal{F} = \mathrm{N}_{\mathcal{F}}(Q)$ then all $\mathcal{F}$-essential (and $\mathcal{F}$-radical $\mathcal{F}$-centric subgroups more generally) contain Q, by Proposition 4.61. $\qquad\square$

In the case where $\mathcal{F} = \mathrm{N}_{\mathcal{F}}(Q)$ but $\mathcal{F} \neq P\,\mathrm{C}_{\mathcal{F}}(Q)$, we do not get this correspondence. For example, let $H = S_4$ with Sylow 2-subgroup Q and normal subgroup $R \cong V_4$, and let $G = H \times C_2$, with P the Sylow 2-subgroup of G, and S the direct product of R with the C_2 factor. In this case, R is a strongly $\mathcal{F}$-closed subgroup of P, where $\mathcal{F} = \mathcal{F}_P(G)$. Notice that S is $\mathcal{F}$-centric and $\mathcal{F}$-radical (with automorphism group S_3), indeed even $\mathcal{F}$-essential. (It is the only such subgroup.) However, P/R is abelian and $\mathcal{F}/R$ is simply $\mathcal{F}_{P/R}(P/R)$, and so S/R is not $\mathcal{F}/R$-centric, and not $\mathcal{F}/R$-radical either since there are no $\mathcal{F}/R$-radical subgroups of P/R. Thus the image of radical, radical centric, and essential subgroups need not be radical, radical centric, or essential, even if the quotient is by a normal subgroup of the fusion system.

Suppose that $\mathcal{F} = P\mathrm{C}_{\mathcal{F}}(Q)$. Proposition 5.60 gives us a one-to-one correspondence between the $\mathcal{F}$-essential subgroups of P and the $\mathcal{F}/Q$-essential subgroups of P/Q, such that it preserves the outer $\mathcal{F}$-automorphism subgroups. We can use this to prove the next result. In the special case of normalizer subsystems, this was proved by Kessar and Linckelmann in [KL08, Proposition 3.4].

Lemma 5.61 *Let $\mathcal{F}$ be a saturated fusion system on a finite p-group P, and suppose that Q is a strongly $\mathcal{F}$-closed subgroup such that $\mathcal{F} = P\mathrm{C}_{\mathcal{F}}(Q)$. If $\mathcal{E}_1$ and $\mathcal{E}_2$ are saturated subsystems of $\mathcal{F}$ on a subgroup R containing Q with $\mathcal{E}_1 \leq \mathcal{E}_2$, then $\mathcal{E}_1/Q = \mathcal{E}_2/Q$ if and only if $\mathcal{E}_1 = \mathcal{E}_2$.*

Proof Let $\mathscr{S}_i$ denote the set of $\mathcal{E}_i$-centric, $\mathcal{E}_i$-radical subgroups of R, and let $\bar{\mathscr{S}}_i$ denote the set of $\mathcal{E}_i/Q$-centric $\mathcal{E}_i/Q$-radical subgroups of R/Q.

Since $\mathcal{F} = P\mathrm{C}_{\mathcal{F}}(Q)$, we have that $\mathrm{Aut}_{\mathcal{F}}(Q)$ is a p-group and $\mathcal{F} = \mathrm{N}_{\mathcal{F}}(Q)$. By Lemma 5.46, $Q \leq \mathrm{O}_p(\mathcal{E}_i)$, and therefore $\mathcal{E}_i = \mathrm{N}_{\mathcal{E}_i}(Q)$; since $\mathrm{Aut}_{\mathcal{E}_i}(Q)$ is also a p-group, we have that $\mathcal{E}_i = R\mathrm{C}_{\mathcal{E}_i}(Q)$. Thus there is a one-to-one correspondence between the elements of $\mathscr{S}_i$ and the elements of $\bar{\mathscr{S}}_i$, such that $\mathrm{Out}_{\mathcal{E}_i}(S)$ and $\mathrm{Out}_{\mathcal{E}_i/Q}(S/Q)$ are equal (where $S \in \mathscr{S}_i$).

However, since $\mathcal{E}_1/Q = \mathcal{E}_2/Q$, this means that $\bar{\mathscr{S}}_1 = \bar{\mathscr{S}}_2$, and so $\mathscr{S}_1 = \mathscr{S}_2$, and $\mathrm{Out}_{\mathcal{E}_1}(S) = \mathrm{Out}_{\mathcal{E}_2}(S)$ for all $S \in \mathscr{S}_1$. Therefore $|\mathrm{Aut}_{\mathcal{E}_1}(S)| = |\mathrm{Aut}_{\mathcal{E}_2}(S)|$ and, since $\mathcal{E}_1 \leq \mathcal{E}_2$, we therefore have that $\mathrm{Aut}_{\mathcal{E}_1}(S) = \mathrm{Aut}_{\mathcal{E}_2}(S)$ for all $S \in \mathscr{S}_1$. By Alperin's fusion theorem, $\mathcal{E}_1 = \mathcal{E}_2$, as required. $\qquad\square$

The importance of $P\mathrm{C}_{\mathcal{F}}(Q)$ is such that we will characterize it now.

Definition 5.62 Let $\mathcal{F}$ be a saturated fusion system on a finite p-group P. Define $\mathrm{Z}_1(\mathcal{F}) = \mathrm{Z}(\mathcal{F})$ and $\mathrm{Z}_i(\mathcal{F})$ inductively by $\mathrm{Z}_i(\mathcal{F})$ being the preimage in P of $\mathrm{Z}(\mathcal{F}/\mathrm{Z}_{i-1}(\mathcal{F}))$. The largest term of this ascending sequence is the *hypercentre* of $\mathcal{F}$, and is denoted by $\mathrm{Z}_{\infty}(\mathcal{F})$.

The main theorem about the subgroup $P\mathrm{C}_{\mathcal{F}}(Q)$ is the following, proved by the author in [Cra10b].

Theorem 5.63 (Hypercentral subgroup theorem [Cra10b]) *Let $\mathcal{F}$ be a saturated fusion system on a finite p-group P, and let Q and R be subgroups of P.*

(i) *If* $\mathcal{F} = P\,C_{\mathcal{F}}(Q)$ *and* $\mathcal{F} = P\,C_{\mathcal{F}}(R)$ *then* $\mathcal{F} = P\,C_{\mathcal{F}}(QR)$.

Let $X_{\mathcal{F}}$ *denote the largest (strongly $\mathcal{F}$-closed) subgroup of* P *such that* $\mathcal{F} = P\,C_{\mathcal{F}}(X_{\mathcal{F}})$.

(ii) *If* Q *is a normal subgroup of* P *contained in* $X_{\mathcal{F}}$ *then* Q *is strongly $\mathcal{F}$-closed, and* $X_{\mathcal{F}/Q} = X_{\mathcal{F}}/Q$ *and* $O_p(\mathcal{F})/Q = O_p(\mathcal{F}/Q)$.

(iii) $X_{\mathcal{F}} = Z_{\infty}(\mathcal{F})$.

Proof   Certainly $\mathcal{F} = N_{\mathcal{F}}(QR)$ since both Q and R are contained within $O_p(\mathcal{F})$, so to prove (i) it remains to show that $\text{Aut}_{\mathcal{F}}(QR)$ is a p-group. As $\mathcal{F} = N_{\mathcal{F}}(Q) = N_{\mathcal{F}}(R)$, both Q and R are strongly $\mathcal{F}$-closed. If ϕ is a p'-automorphism in $\text{Aut}_{\mathcal{F}}(QR)$, then the restrictions to Q and R are both p'-automorphisms, and must therefore be trivial; thus $\phi = 1$. Hence $\mathcal{F} = P\,C_{\mathcal{F}}(QR)$, proving (i).

If Q is a normal subgroup of P contained in $X_{\mathcal{F}}$, and $\phi : A \to B$ is a morphism with $A \leq Q$, then ϕ extends to an automorphism of $X_{\mathcal{F}}$, which must be a conjugation map c_g for some $g \in P$, and this fixes Q, so that $B \leq Q$, proving that Q is strongly $\mathcal{F}$-closed.

Let R/Q be a subgroup of P/Q such that $\mathcal{F}/Q = N_{\mathcal{F}/Q}(R/Q)$. Let $\mathcal{E} = N_{\mathcal{F}}(R)$, and note that $\mathcal{E}$ is a saturated subsystem of $\mathcal{F}$ and

$$\mathcal{E}/Q = N_{\mathcal{F}}(R)/Q = N_{\mathcal{F}/Q}(R/Q) = \mathcal{F}/Q,$$

by Exercise 5.11. Lemma 5.61 implies that $\mathcal{E} = \mathcal{F}$, so that $\mathcal{F} = N_{\mathcal{F}}(R)$. Therefore $O_p(\mathcal{F})/Q = O_p(\mathcal{F}/Q)$.

Next, suppose in addition that $\text{Aut}_{\mathcal{F}/Q}(R/Q)$ is a p-group (i.e., $R/Q \leq X_{\mathcal{F}/Q}$). Since $\text{Aut}_{\mathcal{F}}(Q)$ is a p-group, any p'-automorphism in $\text{Aut}_{\mathcal{F}}(R)$ must act trivially on both Q and R/Q, and is therefore trivial by Exercise 4.5; thus $\mathcal{F} = P\,C_{\mathcal{F}}(R)$, and so $X_{\mathcal{F}}/Q = X_{\mathcal{F}/Q}$, proving (ii).

We now show that $X_{\mathcal{F}} = 1$ if and only if $Z(\mathcal{F}) = 1$. The one direction is clear, so suppose that $X_{\mathcal{F}} \neq 1$; write $Z = Z(P) \cap X_{\mathcal{F}}$, a non-trivial central subgroup of P since $X_{\mathcal{F}} \trianglelefteq P$. We claim that Z is strongly $\mathcal{F}$-closed, and in fact $Z \leq Z(\mathcal{F})$. If $\phi : A \to B$ is a morphism, then ϕ extends to a morphism $\psi : AX_{\mathcal{F}} \to BX_{\mathcal{F}}$ and $\psi|_{X_{\mathcal{F}}} = c_g$ for some $g \in P$ (since $\mathcal{F} = P\,C_{\mathcal{F}}(X_{\mathcal{F}})$). However, c_g acts trivially on $Z(P)$, so must act trivially on Z. Hence $\psi' = \psi|_{AZ}$ is a morphism extending ϕ and acting trivially on Z, so that $\mathcal{F} = C_{\mathcal{F}}(Z)$, as claimed. Thus $X_{\mathcal{F}} = 1$ if and only if $Z(\mathcal{F}) = 1$.

Induction and (ii) of this theorem imply that $Z_{\infty}(\mathcal{F}) \leq X_{\mathcal{F}}$. If we can show that $X_{\mathcal{F}/Z_{\infty}(\mathcal{F})} = 1$ then we are done by (ii) again. However, certainly $Z(\mathcal{F}/Z_{\infty}(\mathcal{F})) = 1$, and hence $X_{\mathcal{F}/Z_{\infty}(\mathcal{F})} = 1$, as needed. $\square$

Identifying the subgroups $Z_i(\mathcal{F})$ in the fusion system $\mathcal{F}$, once you know $Z_\infty(\mathcal{F})$, is very easy, as the next lemma shows. Before we can prove it, we need to understand what being in the ith centre of $\mathcal{F}$ means. Write $Z = Z_{i-1}(\mathcal{F})$. If $x \in Z_i(\mathcal{F})$ then any morphism $A/Z \to B/Z$ in $\mathcal{F}/Z$ extends to a map $A\langle x \rangle/Z \to B\langle x \rangle/Z$ acting trivially on $Z\langle x \rangle/Z$. Taking preimages, we need that any morphism $A \to B$ in $\mathcal{F}$ extends to a map $A\langle x \rangle \to B\langle x \rangle$ acting trivially on $Z\langle x \rangle/Z$.

We claim that $Z\langle x \rangle$ is a normal subgroup of P. To see this, notice that $Z \trianglelefteq P$; if $x^g \in Z\langle x \rangle$ for all $g \in P$ then $Z\langle x \rangle$ is indeed a normal subgroup of P. However, since $Z\langle x \rangle$ lies in the centre of $\mathcal{F}/Z$, $xc_g \in (Zx)c_g = Zx$, so that $Z\langle x \rangle \trianglelefteq P$. Hence by Theorem 5.63, we see that $Z\langle x \rangle$ is strongly $\mathcal{F}$-closed. Because of this, we don't have to examine all maps $A \to B$, but just maps $\langle x \rangle \to P$. If every map $\phi : \langle x \rangle \to P$ acts trivially on $Z\langle x \rangle/Z$ (i.e., $x\phi = zx$ for some $z \in Z$) then x lies in $Z_i(\mathcal{F})$.

Lemma 5.64 *Let $\mathcal{F}$ be a saturated fusion system on a finite p-group P. Then*

$$Z_i(\mathcal{F}) = Z_\infty(\mathcal{F}) \cap Z_i(P).$$

Proof Clearly $Z_i(\mathcal{F}) \leq Z_i(P) \cap Z_\infty(\mathcal{F})$, so we need to show the converse. Let x be an element in $Z_\infty(\mathcal{F}) \cap Z_i(P)$, and let $\phi : \langle x \rangle \to P$ be any morphism in $\mathcal{F}$. We need to show that ϕ acts trivially on $\langle x \rangle Z_{i-1}(\mathcal{F})/Z_{i-1}(\mathcal{F})$, i.e., that $x\phi = zx$ for some $z \in Z_{i-1}(\mathcal{F})$. Thus we need to evaluate $x\phi x^{-1}$.

Since $Z_\infty(\mathcal{F})$ is a normal subgroup of $\mathcal{F}$, ϕ extends to an automorphism ψ of $Z_\infty(\mathcal{F})$, and by Theorem 5.63, $\psi = c_g$ for some $g \in P$. Since $x\phi = x^g$, we see that $x\phi x^{-1} = [g, x^{-1}] \in Z_{i-1}(P)$ (as $x \in Z_i(P)$). Also $x\phi x^{-1}$ clearly lies in $Z_\infty(\mathcal{F})$, so by induction $x\phi x^{-1}$ lies in $Z_{i-1}(P) \cap Z_\infty(\mathcal{F}) = Z_{i-1}(\mathcal{F})$, as needed. $\qquad\square$

The centre of a fusion system and the centre of a finite group are connected, but this result is incredibly difficult. This follows the general pattern that theorems that relate properties of the fusion system to properties of the finite group are difficult to prove.

Theorem 5.65 *Let G be a finite group with $O_{p'}(G) = 1$ and let P be a Sylow p-subgroup of G. If $x \in P$ is an element of order p such that the only G-conjugate of x lying in P is x itself, then $x \in Z(G)$.*

For the prime 2, this theorem is Glauberman's Z^*-theorem, and was proved by Glauberman in [Gla66], but for odd primes Theorem 5.65 requires the classification of the finite simple groups (for now, at least).

We can use the results above to translate this theorem into the language of fusion systems.

Proposition 5.66   *Let G be a finite group with a Sylow p-subgroup P. If $O_{p'}(G) = 1$ then $Z_i(G) = Z_i(\mathcal{F})$, where $\mathcal{F} = \mathcal{F}_P(G)$.*

Proof   By Theorem 5.65, if x is an element of order p in P such that $x^G \cap P = \{x\}$ then $x \in Z(G)$. The condition that $x^G \cap P = \{x\}$ is the same as $x \in Z(\mathcal{F})$. Therefore, $Z(G) = 1$ if and only if $Z(\mathcal{F}) = 1$, and so $Z_\infty(G) = Z_\infty(\mathcal{F})$. (Notice that $Z_\infty(G)$ is a p-subgroup because it is a nilpotent normal subgroup of G, so is the direct product of its Sylow subgroups, and since $O_{p'}(G) = 1$ there can be only one of these, so that $Z_\infty(G)$ is a p-group.)

By Lemma 5.64, $Z_i(\mathcal{F}) = Z_i(P) \cap Z_\infty(\mathcal{F})$. We must similarly prove that $Z_i(G) = Z_i(P) \cap Z_\infty(G)$; to see this, again we only have to prove that $Z(G) = Z(P) \cap Z_\infty(G)$. Certainly $Z(G) \leq Z(P) \cap Z_\infty(G)$, so let x be an element of $Z(P) \cap Z_\infty(G)$. Any p'-element g of G must centralize $Z_\infty(G)$, since it acts trivially on $Z_i(G)/Z_{i-1}(G)$ for all i. Since P centralizes x as well, $C_G(x)$ contains all p'-elements of G and a Sylow p-subgroup of G, so that $C_G(x) = G$, as claimed.

Therefore $Z_i(G) = Z_i(P) \cap Z_\infty(G) = Z_i(P) \cap Z_\infty(\mathcal{F}) = Z_i(\mathcal{F})$, as claimed. $\square$

We end this section by considering perfect fusion systems. A perfect group is one with no non-trivial abelian quotients. Since we are interested in p-subgroups and p-quotients, it makes sense to make the following definitions.

Definition 5.67   A finite group G is *p-perfect* if, whenever $\phi : G \to A$ is a surjective homomorphism with A an abelian p-group, then $A = 1$. A fusion system $\mathcal{F}$ is *perfect* if, whenever $\Phi : \mathcal{F} \to \mathcal{F}_A(A)$ is a surjective morphism with A is an abelian p-group, then $A = 1$.

We make these definitions because p-perfect groups and perfect fusion systems have an interesting property with respect to the hypercentre.

Proposition 5.68   *Let G be a finite group such that $O_{p'}(G) = 1$.*

(i)   *If $x \in Z_2(G)$ then the map $\lambda_x : g \to [x, g]$ is a homomorphism into $Z(G)$.*

(ii)   *If G is p-perfect, then $Z_\infty(G) = Z(G)$.*

Proof Let x be an element of $Z_2(G)$; we will prove that λ_x is a group homomorphism: notice that $[x, g] \in Z(G)$ for all $g \in G$, and hence

$$(gh)\lambda_x = [x, gh] = [x, h][x, g]^h = [x, g][x, h] = (g\lambda_x)(h\lambda_x).$$

This shows that λ_x is a homomorphism, proving (i).

To prove (ii), we start by noticing that, since G is p-perfect, λ_x must be a trivial homomorphism, as $Z(G)$ is a p-group. Therefore $g\lambda_x = 1$ for all g, so that $x \in Z(G)$. As we chose x arbitrarily in $Z_2(G)$, $Z_2(G) = Z(G)$, as claimed. □

Using this homomorphism, we can prove the corresponding result about fusion systems.

Proposition 5.69 *Let $\mathcal{F}$ be a saturated fusion system on a finite p-group P. If $\mathcal{F}$ is perfect then $Z_\infty(\mathcal{F}) = Z(\mathcal{F})$.*

Proof Write $Z = Z(\mathcal{F})$, let x be an element of $Z_2(\mathcal{F})$, and for $g \in P$ let λ_x be the map defined by $g\lambda_x = [x, g]$. By Proposition 5.68(i) this a group homomorphism into $Z(P)$, an abelian p-group. Writing A for the image of λ_x, we wish to show that λ_x induces a surjective morphism of fusion systems with image $\mathcal{F}_A(A)$. Since $\mathcal{F}$ is perfect, $A = 1$, so that $x \in Z(P)$.

We begin by showing that the kernel of λ_x is a strongly $\mathcal{F}$-closed subgroup of P, for then there is a surjective morphism of fusion systems with kernel $\ker \lambda_x$. Let X denote the kernel of λ_x, and notice that $Z \leq X$. Let Q and R be subgroups of P with $Q \leq X$, and let $\phi : Q \to R$ be an $\mathcal{F}$-isomorphism. Since $x \in Z_2(\mathcal{F})$, by the discussion before Lemma 5.64 we see that $Z\langle x \rangle$ is a strongly $\mathcal{F}$-closed subgroup of $Z_\infty(\mathcal{F})$. Hence we may extend $\phi : Q \to R$ to a morphism $\psi : Q(Z\langle x \rangle) \to R(Z\langle x \rangle)$; such a morphism must act as an automorphism of $Z\langle x \rangle$, and act trivially on Z, and also on $Z\langle x \rangle/Z$. Write $z = x(x\psi^{-1})$, an element of Z, so that $(zx)\psi = x$. As $[x, Q] = 1$, we have

$$[x, Q\psi] = [zx, Q]\psi \leq ([z, Q]^x[x, Q])\psi = 1.$$

Therefore $R \leq X$, so that X is strongly $\mathcal{F}$-closed. Hence there is a surjective morphism $\Phi : \mathcal{F} \to \mathcal{E}$ of fusion systems with kernel X, where $\mathcal{E}$ is a fusion system on A.

Since A is an abelian p-group, every automorphism in $\mathcal{E}$ extends to an $\mathcal{E}$-automorphism of A. Therefore, if we can show that any p'-automorphism α of P satisfies $\alpha\Phi = \mathrm{id}_A$ then $\mathcal{E} = \mathcal{F}_A(A)$. Let g be an element in P; we need to show that $g^{-1}(g\alpha)$ lies in X, i.e., that

$[x, g^{-1}(g\alpha)] = 1$. Since α is a p'-automorphism, and $\mathcal{F} = P\,\mathrm{C}_{\mathcal{F}}(Z\langle x\rangle)$, we must have that α acts as the identity on $Z\langle x\rangle$ by Exercise 4.5 (as it acts trivially on Z and $Z\langle x\rangle/Z$); clearly the subgroup $Z\langle x\rangle$ contains both x and $[x, g]$. Hence (since $[x, g]$ and $[x, g^{-1}]$ are central in P)

$$[x, g^{-1}g\alpha] = [x, g\alpha][x, g^{-1}]^{g\alpha} = ([x, g]\alpha)[x, g^{-1}]^g$$
$$= [x, g][x, g^{-1}]^g = [x, g^{-1}g] = 1.$$

Therefore $\mathcal{E}$ is the fusion system $\mathcal{F}_A(A)$ for some abelian p-group A. Since $\mathcal{F}$ is perfect, we must have $A = 1$, so that $X = P$. Therefore $x \in Z(P)$, which means that $x \in Z(P) \cap Z_\infty(\mathcal{F}) = Z(\mathcal{F})$ by Lemma 5.64; therefore $Z_2(\mathcal{F}) = Z(\mathcal{F})$, as required.   □

In Chapter 9, we will need to consider a certain perfect fusion system, and so we need a little background on central extensions. For finite groups G, the Schur multiplier $M(G)$ is the cohomology group $H^2(G, \mathbb{C}^\times)$. If G has a Sylow p-subgroup P, then one may think of $M(G)$ as the G-stable elements of $H^2(P, \mathbb{C}^\times)$, since G acts on this group. Let G be a finite perfect group. One obvious central extension of G is to simply take $G \times A$ where A is an abelian group. We will only consider the case where the central extensions themselves are perfect. In this case there is a universal perfect central extension $\hat{G}$, a central extension of $M(G)$ by G, and any other central extension $\bar{G}$ of G such that $\bar{G}$ is perfect is a quotient of $\hat{G}$.

For fusion systems, we get a similar setup, as described in [Lin06]. If $\mathcal{F}$ is a saturated fusion system on a finite p-group P, define the *Schur multiplier* of $\mathcal{F}$, denoted $M(\mathcal{F})$, to be the $\mathcal{F}$-stable elements of $H^2(P, \mathbb{C}^\times)$. If $\mathcal{F} = \mathcal{F}_P(G)$ then $M(\mathcal{F})$ is the Sylow p-subgroup of $M(G)$. If $\mathcal{F}$ is a perfect fusion system then there is a universal perfect central extension $\hat{\mathcal{F}}$ of $\mathcal{F}$ by $M(\mathcal{F})$, and any other perfect central extension of $\mathcal{F}$ is a quotient of $\hat{\mathcal{F}}$. In particular, if $M(\mathcal{F})$ is trivial then any central extension of $\mathcal{F}$ is simply $\mathcal{F} \times \mathcal{F}_A(A)$ for some abelian p-group A.

Using this theory, the following result is clear.

Proposition 5.70 (Linckelmann [Lin06, Corollary 4.4])   *Let G be a finite perfect group and let P be a Sylow p-subgroup of G. Suppose that $M(G)$ is a p'-group. If $\mathcal{F}$ is a saturated fusion system on P containing $\mathcal{F}_P(G)$ then $M(\mathcal{F}) = 1$, and so any central extension $\hat{\mathcal{F}}$ of $\mathcal{F}$ is of the form $\mathcal{F} \times \mathcal{F}_A(A)$ for some abelian p-group A.*

The point here is that the $\mathcal{F}$-stable elements of $H^2(P, \mathbb{C}^\times)$ are a subset of the $\mathcal{F}_P(G)$-stable elements of $H^2(P, \mathbb{C}^\times)$, with the latter set a singleton.

5.6 Simple fusion systems

This section begins with some general statements about detecting weakly normal subsystems on the same subgroup as the fusion system itself, and on simple fusion systems, before performing a classification of all simple fusion systems on 2-groups of 2-rank 2; this latter task uses many techniques introduced so far.

The theory of simple fusion systems is subtly different from that of finite groups, in that a fusion system on a simple group need not be simple. For example, we have seen that if $\mathcal{F}$ is a saturated fusion system based on an abelian finite p-group P, then $O_{\mathcal{F}}(P) = P$ and so $\mathcal{F}_P(P) \prec \mathcal{F}$. Thus if G is a finite simple group with abelian Sylow p-subgroup (say $\mathrm{PSL}_2(p^n)$) then the fusion system at the prime p is not simple. However, if a simple fusion system comes from a finite group, then we can say quite a lot.

Theorem 5.71 *Let $\mathcal{F} = \mathcal{F}_P(G)$ be a fusion system on a finite p-group P, suppose that $O_{p'}(G) = 1$, and suppose that $\mathcal{F}_P(G) \neq \mathcal{F}_P(H)$ for any proper subgroup H of G containing P. If $\mathcal{F}$ is simple then G is simple.*

Proof Let N be a normal subgroup of G; then $\mathcal{F}_{P \cap N}(N)$ is a weakly normal subsystem of $\mathcal{F}$, and so $\mathcal{F} = \mathcal{F}_{P \cap N}(N)$ or $\mathcal{F}_{P \cap N}(N) = 1$. In the second case, $N \leq O_{p'}(G) = 1$, and in the first case $N = G$ since $\mathcal{F}_P(G) \neq \mathcal{F}_P(H)$ for any $P \leq H \leq G$. Therefore G is simple, as claimed. $\square$

Thus if a simple fusion system comes from a finite group, it comes from a simple group. There are two questions to ask at this point: the first is to understand which simple groups yield simple fusion systems, and the second is to understand simple fusion systems that do not arise from any finite group.

The first question in some sense is much easier than the second, because we already have a classification of finite simple groups, as a result of many years of work. Furthermore, much is known about the local structure of these groups, which means that we have a good chance of being able to answer this question.

The second question is much more difficult, and at this stage there is growing evidence that there are many simple fusion systems for odd primes that do not arise from groups. However, for the prime 2 there is only one series of such fusion systems known, and they arose during the proof of the classification of the finite simple groups; these are the Solomon fusion systems, which have been mentioned before. We will not say anything about them here, and delay a discussion until Chapter 9.

One might be able to classify simple fusion systems, at least for the prime 2, without appeal to the classification of the finite simple groups; it might even be true that the theory of fusion systems might help improve some of the proof of the classification theorem itself, by classifying all 'local structures' (fusion systems) first, and then deciding which simple groups can demonstrate the fusion systems. For instance, if P is a dihedral 2-group of order at least 8, we can quite easily classify all saturated fusion systems over P, and just one of them is simple. (This is the fusion system of any simple group with dihedral Sylow 2-subgroups, which are simply $\mathrm{PSL}_2(q)$ for various q and A_7, by the classification of such simple groups in [GW65].) At the moment, such a programme is still very far off.

In order for there to be a weakly normal subsystem on a proper subgroup of P, we need the presence of a strongly $\mathcal{F}$-closed subgroup; these might well be easy to detect, so we are interested in the case where there is a weakly normal subsystem $\mathcal{E} \prec \mathcal{F}$ on P itself. (We saw such an example when P is abelian and $\mathcal{F}_P(P) \prec \mathcal{F}$.)

Lemma 5.72   *Let $\mathcal{F}$ be a saturated fusion system on a finite p-group P, and suppose that $\mathcal{E}$ is a weakly normal subsystem of $\mathcal{F}$ on P. For every subgroup Q of P, the index of $\mathrm{Aut}_{\mathcal{E}}(Q)$ in $\mathrm{Aut}_{\mathcal{F}}(Q)$ is prime to p.*

Proof   Let Q be any subgroup of P, and let R be a fully normalized subgroup $\mathcal{F}$-isomorphic to Q via an isomorphism ϕ. Since R is fully normalized, $\mathrm{Aut}_P(R)$ is a Sylow p-subgroup of both $\mathrm{Aut}_{\mathcal{E}}(R)$ and $\mathrm{Aut}_{\mathcal{F}}(R)$ (Proposition 4.20), confirming the result for fully normalized subgroups. As

$$\mathrm{Aut}_P(Q)^{\phi^{-1}} \leq \mathrm{Aut}_{\mathcal{E}}(R),$$

and since $\mathcal{E}$ is weakly normal in $\mathcal{F}$, we see that

$$\mathrm{Aut}_P(R)^{\phi} \leq \mathrm{Aut}_{\mathcal{E}}(Q),$$

and this is a Sylow p-subgroup of $\mathrm{Aut}_{\mathcal{F}}(Q)$. Hence the index $|\mathrm{Aut}_{\mathcal{F}}(Q) : \mathrm{Aut}_{\mathcal{E}}(Q)|$ is prime to p.   □

In the same vein, we have a result of Oliver, which gives a useful criterion for the existence of a weakly normal subsystem on the same subgroup as the fusion system itself.

Proposition 5.73 (Oliver)   *Let $\mathcal{F}$ be a saturated fusion system on a finite p-group P. If $\mathcal{E}$ is a weakly normal subsystem of $\mathcal{F}$ on P, then $\mathcal{F} = \mathcal{E}$ if and only if $\mathrm{Aut}_{\mathcal{F}}(P) = \mathrm{Aut}_{\mathcal{E}}(P)$.*

Proof Let Q be a subgroup of P of smallest index such that $\operatorname{Aut}_{\mathcal{E}}(P) < \operatorname{Aut}_{\mathcal{F}}(P)$: by hypothesis, $Q \neq P$. Firstly assume that Q is fully normalized; then $\operatorname{Aut}_P(Q)$ is a Sylow p-subgroup of $\operatorname{Aut}_{\mathcal{F}}(Q)$ by Proposition 4.20. Since $\mathcal{E}$ is weakly normal in $\mathcal{F}$, $\operatorname{Aut}_{\mathcal{E}}(Q)$ is a normal subgroup of $\operatorname{Aut}_{\mathcal{F}}(Q)$ containing a Sylow p-subgroup of $\operatorname{Aut}_{\mathcal{F}}(Q)$, and so we may apply the Frattini argument; this yields

$$\operatorname{Aut}_{\mathcal{F}}(Q) = \operatorname{Aut}_{\mathcal{E}}(Q) \, \mathrm{N}_{\operatorname{Aut}_{\mathcal{F}}(Q)}(\operatorname{Aut}_P(Q)).$$

Every automorphism of Q in $\mathrm{N}_{\operatorname{Aut}_{\mathcal{F}}(Q)}(\operatorname{Aut}_P(Q))$ extends to an automorphism of $\mathrm{N}_P(Q)$ in $\mathcal{F}$, because if $\phi \in \mathrm{N}_{\operatorname{Aut}_{\mathcal{F}}(Q)}(\operatorname{Aut}_P(Q))$ then $N_\phi = \mathrm{N}_P(Q)$, and Q is fully normalized so ϕ extends to N_ϕ. Since $\mathrm{N}_P(Q) > Q$, it must be true that this extended automorphism also lies in $\mathcal{E}$. Thus

$$\mathrm{N}_{\operatorname{Aut}_{\mathcal{F}}(Q)}(\operatorname{Aut}_P(Q)) \leq \operatorname{Aut}_{\mathcal{E}}(Q),$$

and so therefore $\operatorname{Aut}_{\mathcal{F}}(Q) = \operatorname{Aut}_{\mathcal{E}}(Q)$. If all automorphism groups coincide then the fusion systems coincide, by Alperin's fusion theorem. It remains to remove the hypothesis that Q is fully normalized. Let R be a fully $\mathcal{F}$-normalized subgroup $\mathcal{F}$-conjugate to Q via ϕ. By Alperin's fusion theorem, ϕ may be chosen to lie in $\mathcal{E}$, and hence

$$\operatorname{Aut}_{\mathcal{F}}(Q) \cong \operatorname{Aut}_{\mathcal{F}}(R) \cong \operatorname{Aut}_{\mathcal{E}}(R) \cong \operatorname{Aut}_{\mathcal{E}}(Q).$$

This proves the result. $\square$

In Corollary 8.24, we extend this result, and prove that if $\mathcal{E}_1$ and $\mathcal{E}_2$ are weakly normal subsystems of $\mathcal{F}$ on the same subgroup Q, and $\operatorname{Aut}_{\mathcal{E}_1}(Q) = \operatorname{Aut}_{\mathcal{E}_2}(Q)$, then $\mathcal{E}_1 = \mathcal{E}_2$. With Proposition 5.73, we can now get the following sufficient condition for simplicity.

Corollary 5.74 *Let $\mathcal{F}$ be a saturated fusion system on a finite p-group P. Assume that $\operatorname{Aut}_{\mathcal{F}}(P)$ is a p-group. If P has no proper, non-trivial, strongly $\mathcal{F}$-closed subgroups, then $\mathcal{F}$ is simple.*

Proof Since $\mathcal{F}$ has no strongly $\mathcal{F}$-closed subgroups, any weakly normal subsystem $\mathcal{E}$ must be on P, and by Proposition 5.73, since $\operatorname{Aut}_P(P) = \operatorname{Aut}_{\mathcal{E}}(P) = \operatorname{Aut}_{\mathcal{F}}(P)$, we see that $\mathcal{F} = \mathcal{E}$. Hence $\mathcal{F}$ is simple, as required. $\square$

While this result is helpful for $p = 2$, its usefulness in odd primes is stymied by Corollary 7.82, which states that there are *no* simple fusion systems $\mathcal{F}$ on p-groups P with $\operatorname{Aut}_{\mathcal{F}}(P)$ a p-group if $p \geq 5$.

We have a more restrictive corollary, which is still enough for a lot of purposes.

Corollary 5.75 *Let $\mathcal{F}$ be a fusion system on a finite p-group P, and suppose that P is generated by its elements of order p. Suppose that $\mathrm{Aut}_{\mathcal{F}}(P)$ is a p-group (in particular, if $\mathrm{Aut}(P)$ is a p-group). If all elements of order p are $\mathcal{F}$-conjugate, then $\mathcal{F}$ is simple.*

Proof Let Q be a strongly $\mathcal{F}$-closed subgroup of P. If $Q \neq 1$, then Q contains an element of order p, whence Q contains all elements of order p. Thus $Q = P$, and Corollary 5.74 proves that $\mathcal{F}$ is simple.   □

The fusion systems $\mathcal{F}_P(P)$ should not be simple unless P is cyclic of prime order, since they are meant to represent finite p-groups. The next two simple lemmas deal with this case, and the case where P is abelian.

Lemma 5.76 *Let P be a p-group. The fusion system $\mathcal{F}_P(P)$ is simple if and only if P is cyclic of order p.*

Proof This follows immediately from Theorem 5.37 and Lemma 5.26.
    □

Lemma 5.77 *Let $\mathcal{F}$ be a saturated fusion system on the finite abelian p-group P. If $\mathcal{F}$ is simple then $\mathcal{F} = \mathcal{F}_P(P)$ and $P = C_p$.*

Proof Since P is abelian, we have that $\mathcal{F} = \mathrm{N}_{\mathcal{F}}(P)$ by Exercise 1.8, and so $\mathcal{F}_P(P)$ is a weakly normal subsystem of $\mathcal{F}$. Hence $\mathcal{F} = \mathcal{F}_P(P)$ and, by Lemma 5.76, P is cyclic of order p, as claimed.   □

We will now prove our first result explicitly classifying simple fusion systems, by classifying all those over 2-groups of rank 2. The finite simple groups of 2-rank 2 were classified by Alperin, Brauer, Gorenstein, Lyons, and Walter, in several papers ([ABG70], [ABG73], [Bra71], [GW65], and [Lyo72]) during the 1960s and early 1970s. We adapt the fusion-theoretic analysis of Alperin given in [ABG73], but see also [GLS05, Section 3.2].

The general strategy is to let $\mathcal{F}$ be a simple fusion system on a 2-group P of 2-rank 2, and prove that there must be an $\mathcal{F}$-centric, $\mathcal{F}$-radical subgroup Q of P that is isomorphic with either the abelian group $C_{2^n} \times C_{2^n}$ or the Sylow 2-subgroup of $\mathrm{PSU}_3(4)$. Once this is done, we then classify those fusion systems that contain such a subgroup as an $\mathcal{F}$-centric, $\mathcal{F}$-radical subgroup.

It turns out that it is not much more difficult to classify a slightly larger collection of fusion systems. We let $\mathcal{F}$ be a centre-free saturated fusion system on a 2-group P of 2-rank 2, and later specialize to the case where $\mathcal{F}$ is simple. Let z be an involution in the centre of P. Since $\mathcal{F}$ is centre-free, there must be an $\mathcal{F}$-morphism that moves z, and by Alperin's fusion theorem we may assume that this morphism is actually an automorphism ϕ of a fully $\mathcal{F}$-normalized, $\mathcal{F}$-centric, $\mathcal{F}$-radical subgroup Q of P. Furthermore, we may assume that ϕ has odd order, since all elements of $\mathrm{Aut}_P(Q)$ act trivially on z, and $\mathrm{Aut}_{\mathcal{F}}(Q)$ is generated by $\mathrm{Aut}_P(Q)$ and the $\mathcal{F}$-automorphisms of odd order. Thus we have an automorphism ϕ of odd order in $\mathrm{Aut}_{\mathcal{F}}(Q)$ such that $z\phi \neq z$. In particular, since $z\phi \in \mathrm{Z}(Q)$, the centre of Q must be non-cyclic.

We pause briefly to introduce a definition.

Definition 5.78   A finite group G is *homocyclic* if G is isomorphic to the direct product of cyclic groups, each of the same order.

Suppose that Q is abelian; as P has 2-rank 2, Q has two generators and possesses an odd-order automorphism, and so Q is homocyclic by Exercise 4.8. This proves our claim in the case where Q is abelian. When Q is non-abelian, we need some more information. Notice that, as Q has 2-rank 2 and the centre is non-cyclic, Q has exactly three involutions, all central, and ϕ permutes these involutions transitively. A non-abelian group with these properties is called a Suzuki 2-group (see Section 7.1 for more information). These have been studied extensively, and in this case we have the following result.

Proposition 5.79   (See [GLS05, Lemma 2.2.6])   *Let Q be a finite 2-group with three involutions, and a cyclic group of automorphisms that permutes these automorphisms transitively. We have that Q is isomorphic to a Sylow 2-subgroup of* $\mathrm{PSU}_3(4)$.

Collecting these two results together, we have proved the following lemma.

Lemma 5.80   *Let $\mathcal{F}$ be a saturated fusion system on a finite p-group P of 2-rank 2. If $\mathcal{F}$ is centre-free then there exists an $\mathcal{F}$-centric, $\mathcal{F}$-radical subgroup Q of P, and an odd-order automorphism $\phi \in \mathrm{Aut}_{\mathcal{F}}(Q)$ such that ϕ does not centralize $\mathrm{Z}(P)$. Furthermore, Q is either homocyclic or the Sylow 2-subgroup of* $\mathrm{PSU}_3(4)$.

The next step in the classification is to determine the possible 2-groups P that possess one of these subgroups Q as an $\mathcal{F}$-centric, $\mathcal{F}$-radical subgroup, for some saturated fusion system $\mathcal{F}$ on P.

We deal with the easier, homocyclic, case first. This splits into two cases, depending on whether the group is V_4 or not.

Lemma 5.81   *Let P be a finite 2-group. If P possesses a subgroup $Q \cong V_4$ such that $Q = \mathrm{C}_P(Q)$, then P is either dihedral or semidihedral.*

Proof   We will first prove that P has maximal class. Since $Q = \mathrm{C}_P(Q)$, $\mathrm{Z}(P) \leq Q$, and therefore $Z = \mathrm{Z}(P)$ has order 2. If P/Z also contains a subgroup $R \cong V_4$ such that $\mathrm{C}_{P/Z}(R) = R$, then by induction P/Z, and so P itself, have maximal class.

Since $\mathrm{Aut}(Q)$ has order 6, we see that $R = \mathrm{N}_P(Q)$ has order 8 (and contains Z), and so R/Z has order 4. Also, since clearly $R \cong D_8$, we have that $R/Z \cong V_4$. It remains to show that $\mathrm{C}_{P/Z}(R/Z) = R/Z$; suppose that Zg centralizes R/Z; then g centralizes the cosets Z and $Q \setminus Z$, and so normalizes Q, whence $g \in R$, as claimed.

Finally, the fact that there are only three types of 2-groups of maximal class – dihedral, semidihedral, and quaternion (see for example [Gor80, Theorem 5.4.5]) – and there is no such subgroup Q in the quaternion groups gives us the result. $\square$

In the case where Q is abelian but of larger order, we will meet another type of 2-group. Recall that a 2-group is *wreathed* if it is isomorphic with the group $C_{2^n} \wr C_2$ for some $n \geq 2$. It may be presented as

$$P = \langle\, x, y, z \,:\, x^{2^n} = y^{2^n} = z^2 = 1, xy = yx, x^z = y \,\rangle;$$

some basic results about wreathed groups, which we will not prove here, are collected in Exercise 5.7.

Lemma 5.82   *Let $\mathcal{F}$ be a saturated fusion system on a finite 2-group P of 2-rank 2. Suppose that there is an abelian $\mathcal{F}$-centric, $\mathcal{F}$-radical subgroup Q of P. If $|Q| = 4$ then P is either dihedral or semidihedral (or V_4), and if $|Q| > 4$ then P is either wreathed or abelian (and $Q = P$ in this case).*

Proof   If $Q = P$, then P is abelian, and so we are done. Thus we may assume that $Q < P$, and so $Q < \mathrm{N}_P(Q) = R$. As Q is abelian, it is also self-centralizing. If $|Q| = 4$, then by Lemma 5.81, P is either dihedral or semidihedral, as needed, so suppose that $|Q| > 4$. Since Q has an automorphism of odd order, and has 2-rank 2, it is of the form $C_{2^n} \times C_{2^n}$, with an automorphism ϕ of order 3.

By Exercise 4.8(ii), $\mathrm{Aut}_{\mathcal{F}}(Q) \cong S_3$. Let x and y be generators of Q, and let t be an element of $R \setminus Q$; by Exercise 4.8(iv), we may choose x and y such that t swaps x and y. As $|R : Q| = 2$, we have that $t^2 \in Q$, and since t must centralize t^2 we have $t^2 = (xy)^i$. Then

$$(x^{-i}t)^2 = x^{-i}t^2(t^{-1}x^{-i}t) = x^{-i}x^iy^iy^{-i} = 1,$$

so that R is wreathed. In particular, Q char $R \trianglelefteq N_P(R)$ (since Q is the unique abelian subgroup of index 2), and hence $Q \trianglelefteq N_P(R)$, proving that $R = P$ since $R = N_P(Q)$. Hence P is wreathed, as claimed. □

We turn our attention to the Sylow 2-subgroup of $\mathrm{PSU}_3(4)$, a group Q of order 64.

Lemma 5.83 (See [GLS05, Lemma 2.2.9]) *Let Q be the Sylow 2-subgroup of $\mathrm{PSU}_3(4)$, and let $A = \mathrm{Aut}(Q)$. Writing $\bar{A} = A / O_2(A)$, we have that $\bar{A}$ is a group of order 60, with a normal subgroup $B \cong C_3 \times C_5$, and a cyclic group of order 4 acting faithfully on the C_5 factor and trivially on the C_3 factor. In particular, all involutions in $\bar{A}$ centralize the elements of order 3.*

Again, we will not prove this lemma here, but we refer instead to the relevant part of [GLS05]. This lemma gives the structure of $\mathrm{Out}(Q)$, so that we can pin down the structure of P when Q is of this type. Any odd-order automorphism of Q that does not fix the (central) involutions has order 3, and so we may assume that $3 \mid \mid \mathrm{Aut}_{\mathcal{F}}(Q) \mid$.

Lemma 5.84 *Let $\mathcal{F}$ be a saturated fusion system on a finite 2-group P of 2-rank 2. Suppose that there is an $\mathcal{F}$-radical, $\mathcal{F}$-centric subgroup Q of P isomorphic with the Sylow 2-subgroup of $\mathrm{PSU}_3(4)$. If $3 \mid \mid \mathrm{Aut}_{\mathcal{F}}(Q) \mid$ then $Q = P$.*

Proof Suppose that $Q < P$, and let R be a subgroup of P such that $|P : R| = 2$; we have that $\mathrm{Aut}_R(Q)/\mathrm{Inn}(Q)$ is of order 2, as Q is $\mathcal{F}$-centric. Since Q is $\mathcal{F}$-radical, $\mathrm{Out}_R(Q)$ projects onto $\mathrm{Aut}(Q)/O_2(\mathrm{Aut}(Q))$, and hence commutes with the subgroup of order 3 in $\mathrm{Out}_{\mathcal{F}}(Q)$, by Lemma 5.83. Writing ϕ for an automorphism of order 3 in $\mathrm{Aut}_{\mathcal{F}}(Q)$, we see therefore that $\mathrm{Aut}_R(Q)^\phi = \mathrm{Aut}_R(Q)$, and so ϕ extends to $\psi \in \mathrm{Aut}_{\mathcal{F}}(R)$, as Q is fully $\mathcal{F}$-centralized, hence receptive. If all involutions of R lie in Q then R also possesses a cyclic group of automorphisms permuting the three involutions of R, and so, by Proposition 5.79, $R = Q$, a contradiction. Hence there is an involution t in $R \setminus Q$.

Finally, since ϕ and c_t commute as elements of $\mathrm{Out}(Q)$, the element ϕc_t has order 6; its action on $Z(Q) \cong V_4$ must have order a multiple of 3, and since $\mathrm{Aut}(V_4)$ has no elements of order 6, c_t must act trivially on $Z(Q)$. Thus $\langle Q, t \rangle$ is elementary abelian of rank 3, contradicting the fact that P has 2-rank 2. Thus $Q = P$, as claimed. □

Combining Lemmas 5.82 and 5.84, we arrive at the following theorem.

Theorem 5.85   *Let $\mathcal{F}$ be a saturated fusion system on a finite 2-group P, and suppose that P has 2-rank 2. If $\mathcal{F}$ is centre-free then P is abelian, dihedral, semidihedral, wreathed, or isomorphic with the Sylow 2-subgroup of $\mathrm{PSU}_3(4)$.*

We should now describe the possible fusion systems on these groups. Theorem 4.54 deals with the case where P is dihedral or semidihedral: here, it is easy to see that both non-trivial fusion systems on D_{2^n} are centre-free and the fusion system of $\mathrm{PSL}_2(q)$ is simple, and for semidihedral groups, the two fusion systems with odd-order automorphisms on V_4 subgroups are centre-free, and the one of those with automorphisms of the Q_8 subgroups is simple.

If P is either an abelian 2-group or the Sylow 2-subgroup of $\mathrm{PSU}_3(4)$, we have shown that $\mathcal{F} = \mathrm{N}_{\mathcal{F}}(P)$, so that no saturated fusion system over P is simple. For it to be centre-free, P must have an $\mathcal{F}$-automorphism of order 3.

The remaining 2-groups are the wreathed 2-groups. Let P denote the wreathed group $C_{2^n} \wr C_2$, and let U denote the homocyclic subgroup of index 2. We use the properties of wreathed groups given in Exercise 5.7. We want to determine all fusion systems on P; to do this we must determine the possible centric radical subgroups of P. We prove that there are four saturated fusion systems on P, each with a different focal subgroup (see Exercise 5.9). This analysis was originally suggested in [ABG70, Section II.1] (but left mostly as an exercise), but see also [BW71] for an alternative viewpoint. Here we will broadly follow the plan of [ABG70], but with some simplifications.

Let $\mathcal{F}$ be a saturated fusion system on P, and let Q denote an $\mathcal{F}$-centric, $\mathcal{F}$-radical subgroup of P. If Q is abelian, then Q must be homocyclic abelian of rank 2, and contain $\mathrm{Z}(P)$, a cyclic subgroup of order 2^n by Exercise 5.7(i). Hence $Q \cong C_{2^n} \times C_{2^n}$, and so $Q = U$.

Now suppose that Q is non-abelian. Since Q is $\mathcal{F}$-centric, it contains $Z = \mathrm{Z}(P)$, cyclic of order 2^n. The subgroup Q possesses an odd-order automorphism, which must act trivially on Q/Z since cyclic 2-groups have no odd-order automorphisms (and using Exercise 4.5). Hence Q/Z is a subgroup of $P/Z \cong D_{2^n}$ (via Exercise 5.7(iii)) with an odd-order automorphism. Thus $Q/Z \cong V_4$, and there are two P-conjugacy classes of subgroups of this form. To specify them, we keep the notation of Exercise 5.7, so that U is generated by x and y, and z is an involution in $P \setminus U$ with $x^z = y$ (and $y^z = x$).

Representatives from the two conjugacy classes of Klein four-groups are generated by the central element of $P/Z(P)$, namely $Zx^{2^{n-1}}$, and either Zz or Zxz, and so the two subgroups that we are interested in are

$$V = \langle xy, x^{2^{n-1}}, z \rangle \text{ and } W = \langle xy, x^{2^{n-1}}, xz \rangle = \langle x^{2^{n-1}}, xz \rangle.$$

Let us consider the three maximal subgroups of W: two of these are $\langle xy, x^{2^{n-1}} \rangle$, which is not cyclic, and $\langle xy, xz \rangle$, which is (since xz has order 2^{n+1}), and so W cannot support an automorphism of order 3. Therefore V (or any conjugate of V) is the only candidate for a subgroup with an automorphism of odd order acting on it.

Write $m = 2^{n-2}$, and let $a = x^{2m}z$, $b = x^m y^{-m}$, and $c = x^m y^m z$. Then $ab = c$, $bc = a$ and $ca = b$, and each of a, b, and c squares to the central element $x^{2m}y^{2m}$, and so $X = \langle a, b \rangle$ is a quaternion group of order 8, which affords an automorphism of order 3. It is not difficult to see that V is the central product of Z and X, and so the odd-order automorphism of X lifts to an automorphism of V.

We have proved the following result.

Proposition 5.86 *Let P, U, and V be as above. Then U and V are representatives of the only two conjugacy classes of subgroups of P that are self-centralizing and with automorphism group not a 2-group.*

Thus there are at most four saturated fusion systems on P, depending on whether or not U and V have odd-order automorphisms acting on them. If just U has an odd-order automorphism in $\mathcal{F}$, then $x^{2m}y^{2m}$ and x^{2m} are $\mathcal{F}$-conjugate, and clearly U is a strongly $\mathcal{F}$-closed subgroup of P, so there are two $\mathcal{F}$-conjugacy classes of involutions, namely the one with x^{2m} in it, and the one with z in it. (See Exercise 5.7(ii).)

If only V has an odd-order automorphism, then Z is strongly $\mathcal{F}$-closed since it is central and fixed under this automorphism, and so $x^{2m}y^{2m}$ lies in the centre of $\mathcal{F}$. Furthermore, letting ϕ be the automorphism that cycles a, b, and c in that order, we have

$$z\phi^{-1} = [(x^{-m}y^{-m})c]\phi^{-1} = (x^{-m}y^{-m})(c\phi^{-1}) = y^{2m},$$

and so z and y^{2m} are $\mathcal{F}$-conjugate, proving that in this case there are two $\mathcal{F}$-conjugacy classes of involutions.

Finally, assume that $\mathcal{F}$ has odd-order automorphisms of both U and V; by the arguments above, there is now a single conjugacy class of involutions. There are no strongly $\mathcal{F}$-closed subgroups of P apart from 1 and P by Exercise 5.8, and together with Exercise 5.7(iv) and Corollary 5.74, we see that $\mathcal{F}$ is simple.

Proposition 5.87   *Let P be a wreathed 2-group, and let $\mathcal{F}$ be a saturated fusion system on P. One of the following holds:*

(i) $\mathcal{F} = \mathcal{F}_P(P)$ *and there are three $\mathcal{F}$-conjugacy classes of involutions;*

(ii)   *there are two conjugacy classes of involutions (one central), and the subgroup $\Omega_1(\mathrm{Z}(P))$ is strongly $\mathcal{F}$-closed;*

(iii)   *there are two conjugacy classes of involutions (neither central), and the abelian subgroup of P of index 2 is strongly $\mathcal{F}$-closed;*

(iv)   *all involutions are $\mathcal{F}$-conjugate and there are no strongly $\mathcal{F}$-closed subgroups apart from 1 and P.*

(In (i), neither U nor V has an odd-order automorphism, in (ii) V does, in (iii) U does, and in (iv) both do.) In each case, $\mathcal{F}$ is isomorphic to some group fusion system, with only (iv) being simple.

Proof   In the remarks preceding this proposition all apart from the construction of finite groups was performed. This is performed in [ABG70, Section II.2], and we merely state their results here, with brief justification.

In case (i), this is realized by $\mathcal{F}_P(P)$, and in case (iv), this is realized by $\mathrm{PSL}_3(q)$ for $q \equiv 1 \bmod 4$ and $\mathrm{PSU}_3(q)$ for $q \equiv 3 \bmod 4$. In case (ii), this is realized by the centralizer of an involution in $\mathrm{PSL}_3(q)$, and so we make take $\mathrm{GL}_2(q)$, which only leaves (iii). Indeed, in this case the subgroup U is centric and normal, and so there must be a finite group modelling it: the group $(C_{2^n} \times C_{2^n}) \rtimes S_3$ – if x and y generate the base group, then the S_3 of automorphisms acts by permutations on the elements x, y, and $x^{-1}y^{-1}$ – will perform this function.   □

In Exercise 5.9, we construct the focal subgroup for each of the four fusion systems above; they are all different, giving us another way of characterizing the different fusion systems. In Exercise 5.8, we prove that if both U and V have odd-order $\mathcal{F}$-automorphisms then P has no strongly $\mathcal{F}$-closed subgroups, so is simple as $\mathrm{Aut}(P)$ is a 2-group, by Corollary 5.74.

Explicitly, we have the following theorem, determining the simple fusion systems of 2-rank 2.

Theorem 5.88   *Let $\mathcal{F}$ be a simple fusion system on a finite 2-group P of rank 2. Then P is dihedral, semidihedral, or wreathed, and $\mathcal{F}$ is the fusion system of either $\mathrm{PSL}_2(q)$ or $\mathrm{PSL}_3(q)$.*

While we have determine the centre-free saturated fusion systems on 2-groups of 2-rank 2, it would obviously be of interest to try to understand *all* saturated fusion systems on 2-groups of 2-rank 2. The associated problem for odd primes has been solved by Díaz, Ruiz, and Viruel in [DRV07], but for $p = 2$ this is still open. Recently, the 2-groups of 2-rank 2 with an odd-order automorphism were classified completely by the author and Glesser in [CG10], an obvious first step in the determination of all saturated fusion systems on 2-groups of 2-rank 2. (See Exercise 5.15 for a start to this classification.)

5.7 Soluble fusion systems

The theory of soluble fusion systems was initiated by Aschbacher; he proves an analogue of the Jordan–Hölder theorem in [Asc11], getting a finite collection of (simple) composition factors for a given (saturated) fusion system $\mathcal{F}$, and defines a fusion system to be soluble if all of its composition factors are the fusion system $\mathcal{F}_P(P)$ for a cyclic group of order p.

However, here we will proceed as in the paper [Cra10a] of the author, and define a soluble fusion system as follows.

Definition 5.89 Let $\mathcal{F}$ be a saturated fusion system on a finite p-group P. We say that $\mathcal{F}$ is *soluble* if there exists a chain of strongly $\mathcal{F}$-closed subgroups

$$1 = P_0 \leq P_1 \leq \cdots \leq P_n = P,$$

such that $P_i/P_{i-1} \leq O_p(\mathcal{F}/P_{i-1})$ for all $1 \leq i \leq n$. If $\mathcal{F}$ is soluble, then the length n of a smallest such chain above is the *p-length* of $\mathcal{F}$.

Define $O_p^{(0)}(\mathcal{F}) = 1$, and the ith term by

$$O_p^{(i)}(\mathcal{F})/O_p^{(i-1)}(\mathcal{F}) = O_p\left(\mathcal{F}/O_p^{(i-1)}(\mathcal{F})\right).$$

The next lemma will prove the same simple facts for soluble fusion systems that exist for p-soluble groups – namely, that all subsystems and quotients of soluble systems are soluble and that the class of soluble systems is extension closed – and give a characterization of the p-length of a soluble system.

Lemma 5.90 *Let $\mathcal{F}$ be a saturated fusion system on a finite p-group P, let Q be a strongly $\mathcal{F}$-closed subgroup of P and let $\mathcal{E}$ be a weakly normal subsystem of $\mathcal{F}$ on the subgroup T of P.*

(i) If $\mathcal{F}$ is soluble then all saturated subsystems and quotients $\mathcal{F}/Q$ are soluble.

(ii) If $\mathcal{E}$ and $\mathcal{F}/T$ are soluble then so is $\mathcal{F}$.

(iii) $\mathcal{F}$ is soluble if and only $O_p^{(n)}(\mathcal{F}) = P$ for some n, and the smallest such n is the p-length of $\mathcal{F}$.

Proof Let $\mathcal{F}$ be a soluble fusion system on P, and let Q be a strongly $\mathcal{F}$-closed subgroup such that $\mathcal{F}/Q$ is not soluble, and choose P of minimal order subject to this. Write $R = O_p(\mathcal{F}) \neq 1$; we claim that $QR/Q \leq O_p(\mathcal{F}/Q)$. To see this, notice that $\mathcal{F}_R(R) \prec \mathcal{F}$, and so by Lemma 5.59 the image of $\mathcal{F}_R(R)$, which is clearly $\mathcal{F}_{QR/Q}(QR/Q)$, is also weakly normal. Hence $QR/Q \leq O_p(\mathcal{F}/Q)$.

Since $\mathcal{F}/Q$ is not soluble, neither is (using the third isomorphism theorem)

$$(\mathcal{F}/Q)/(QR/Q) \cong \mathcal{F}/QR.$$

However, clearly $\mathcal{F}/O_p(\mathcal{F})$ is soluble, and P/R is a p-group of smaller order; thus all quotients of $\mathcal{F}/R$ – including $\mathcal{F}/QR$ – are soluble, a contradiction.

Now let $\mathcal{E}$ be a saturated subsystem of $\mathcal{F}$, and suppose that $\mathcal{E}$ is not soluble but $\mathcal{F}$ is. Choose P of minimal order with this property. Again, write $R = O_p(\mathcal{F}) \neq 1$. By Lemma 5.46, $O_p(\mathcal{E}) \geq R \cap T$, and by the second isomorphism theorem,

$$\mathcal{E}/(R \cap T) \cong \mathcal{E}R/R,$$

where $\mathcal{E}R/R$ is the image of $\mathcal{E}$ in $\mathcal{F}/R$. Since $\mathcal{F}/R$ is soluble, and P/R has smaller order than P, every saturated subsystem of $\mathcal{F}/R$, including $\mathcal{E}R/R$, is soluble, a contradiction. This proves (i).

By Theorem 5.21, the preimages of $O_p^{(j)}(\mathcal{F}/R)$ are strongly $\mathcal{F}$-closed. If $\mathcal{E}$ is soluble, then $O_p(\mathcal{E})$ is weakly characteristic in $\mathcal{E}$, which is weakly normal in $\mathcal{F}$, so that $O_p(\mathcal{E})$ is strongly $\mathcal{F}$-closed, and by an obvious induction (and Theorem 5.21 again) this applies to all $O_p^{(i)}(\mathcal{E})$. The concatenation of the two series $O_p^{(i)}(\mathcal{E})$ and the preimages of $O_p^{(j)}(\mathcal{F}/R)$ satisfy the requirement for $\mathcal{F}$ to be soluble, proving (ii).

Finally, we prove (iii). Again write $R = O_p(\mathcal{F})$. One direction of the claim is clear. We firstly prove that if S is a strongly $\mathcal{F}$-closed subgroup such that $S \leq O_p(\mathcal{F})$, then the preimage of $O_p(\mathcal{F}/S)$ in P is contained within $O_p^{(2)}(\mathcal{F})$. Firstly, $O_p(\mathcal{F}/S)$ is contained within R/S, as we saw earlier in this proof. Applying this to $O_p(\mathcal{F}/S)$, we see that

$O_p(\mathcal{F}/S)/(R/S)$ is contained within $O_p(\mathcal{F}/R)$, and so taking preimages we get $O_p(\mathcal{F}/S) \leq O_p^{(2)}(\mathcal{F})$, as claimed.

Suppose that $\mathcal{F}$ has a series (Q_i) such that $Q_i/Q_{i-1} \leq O_p(\mathcal{F}/Q_{i-1})$ for all i. Induction, together with the previous statement, proves that this series grows at most as slowly as the $O_p^{(i)}(\mathcal{F})$ series, proving (iii). $\square$

We come to the main result of the section, the proof that soluble fusion systems are constrained. This was proved by Aschbacher [Asc11] in terms of the generalized Fitting subsystem of a fusion system, a concept that we will meet in Chapter 8. The proof here follows that of the author in [Cra10a].

Theorem 5.91 (Aschbacher [Asc11]) *Let $\mathcal{F}$ be a saturated fusion system on a finite p-group P. If $\mathcal{F}$ is soluble then $C_P(O_p(\mathcal{F})) \leq O_p(\mathcal{F})$, so that $\mathcal{F}$ is constrained.*

Proof Let $Q = O_p(\mathcal{F})$, and let $\mathcal{E} = C_\mathcal{F}(Q)$. Since $\mathcal{F} = N_\mathcal{F}(Q)$, we see that $\mathcal{E} \prec \mathcal{F}$ by Exercise 5.10. Therefore, by Proposition 5.47,

$$O_p(\mathcal{E}) = O_p(\mathcal{F}) \cap C_P(Q) = Q \cap C_P(Q) = Z(Q).$$

However, since $\mathcal{E} = C_\mathcal{F}(Q)$, we notice that every morphism in $\mathcal{E}$ centralizes $Z(Q) = Q \cap C_P(Q)$, so that $Z(Q) \leq Z(\mathcal{E})$. Since $Z(\mathcal{E}) \leq O_p(\mathcal{E})$ obviously, we see that $O_p(\mathcal{E}) = Z(\mathcal{E})$. By the hypercentral subgroup theorem (Theorem 5.63), we see that $O_p(\mathcal{E}/O_p(\mathcal{E})) = 1$, and since $\mathcal{E}$ is soluble (as $\mathcal{F}$ is) this means that

$$Z(Q) = Z(\mathcal{E}) = O_p(\mathcal{E}) = C_P(Q);$$

hence $C_P(Q) \leq Q$, as required. $\square$

The fact that soluble fusion systems are constrained means that they are fusion systems of finite groups, by Theorem 3.70. We now characterize those fusion systems that come from *p-soluble* groups. Since every p-soluble group has a soluble fusion system, it is a necessary condition that they be constrained, so we start from here.

We recall some notation and results from finite group theory. A *component* of a finite group G is a subnormal quasisimple subgroup of G. The *layer* of G, $E(G)$, is the (characteristic) subgroup generated by all components of G, and the *generalized Fitting subgroup*, $F^*(G)$, is the subgroup generated by the *Fitting subgroup*, $F(G)$, the largest normal nilpotent subgroup of G, and $E(G)$. The subgroup $F^*(G)$ is the central product of $F(G)$ and $E(G)$.

Proposition 5.92   *Let G be a finite group and let p be a prime dividing $|G|$.*

(i) $C_G(F^*(G)) \le F^*(G)$.

(ii) *If $O_{p'}(G) = 1$, then G is p-constrained if and only if G has no components, so $F^*(G) = O_p(G)$.*

Proof   The first part of this proof is well known, and we refer to [Asc00, 31.13] for a proof. For the second part, if G is p-constrained then $F(G) = O_p(G)$, and since $C_G(O_p(G)) \le O_p(G)$, and $F^*(G)$ is the central product of $E(G)$ and $O_p(G)$, we must have $E(G) = 1$. Conversely, if $E(G) = 1$ and $O_{p'}(G) = 1$, then $F^*(G) = O_p(G)$; we use (i) of this proposition to prove that G is p-constrained.   □

It is well known (see, for example, [Asc00, 33.12]) that if G is a simple p'-group, then any quasisimple group with quotient G is also a p'-group. This gives us the following, originally proved by Hall and Higman [HH56].

Lemma 5.93   *If G is a p-soluble group with $O_{p'}(G) = 1$, then*

$$C_G(O_p(G)) \le O_p(G).$$

Proof   Since G is p-soluble, any subnormal quasisimple group is a p'-group, and therefore $E(G) \le O_{p'}(G) = 1$. Therefore $F^*(G) = O_p(G)$, proving the claim.   □

Using this lemma, we give our first characterization of fusion systems of p-soluble groups. In the rest of this section, for a constrained fusion system $\mathcal{F}$, we denote the unique finite group given in Theorem 3.70 by $L^{\mathcal{F}}$.

Proposition 5.94   *Let $\mathcal{F}$ be a saturated fusion system on a finite p-group P. There is a p-soluble group G such that $\mathcal{F} = \mathcal{F}_P(G)$ if and only if $\mathcal{F}$ is constrained and $L^{\mathcal{F}}$ is p-soluble.*

Proof   Suppose that $\mathcal{F} = \mathcal{F}_P(G)$ for some p-soluble group G. We may assume that $O_{p'}(G) = 1$, since the fusion systems on G and $G/O_{p'}(G)$ are the same. By Lemma 5.93, $O_p(G)$ contains its centralizer. Therefore, by the uniqueness of the group in Theorem 3.70, $G = L^{\mathcal{F}}$.   □

Finally, this allows us to reach an internal characterization of fusion systems of p-soluble groups, without reference to groups.

Corollary 5.95 (Craven [Cra10a]) *Let $\mathcal{F}$ be a saturated fusion system on a finite p-group P, and write $Q = O_p(\mathcal{F})$. Then $\mathcal{F}$ is the fusion system of a p-soluble group if and only if $\mathcal{F}$ is constrained and $\mathrm{Aut}_{\mathcal{F}}(Q)$ is p-soluble.*

Proof Suppose that $\mathcal{F}$ is constrained and that $\mathrm{Aut}_{\mathcal{F}}(Q)$ is p-soluble. Since $\mathcal{F}$ is constrained, we have that $\mathcal{F} = \mathcal{F}_P(L^{\mathcal{F}})$; then $L^{\mathcal{F}}$ is an extension of $C_{L^{\mathcal{F}}}(Q) = Z(Q)$ by $\mathrm{Aut}_{L^{\mathcal{F}}}(Q) = \mathrm{Aut}_{\mathcal{F}}(Q)$, which is p-soluble. Hence $L^{\mathcal{F}}$ is p-soluble, and so $\mathcal{F}$ is the fusion system of a p-soluble group. The converse is similarly clear. $\square$

Having understood which soluble fusion systems come from p-soluble groups, it remains to understand which finite groups G yield soluble fusion systems. Since the class of soluble fusion systems is extension closed, we will end up only with restrictions on the *composition factors* of G. Firstly notice that, since $O_p(\mathcal{F}) \neq 1$, by Proposition 4.62, G possesses a strongly closed abelian p-subgroup. Thus we need to understand which simple groups possess a strongly closed abelian p-subgroup.

For $p = 2$ this is a famous result of Goldschmidt.

Theorem 5.96 (Goldschmidt [Gol74]) *Let G be a finite simple group, and let P be a Sylow 2-subgroup of G. If P contains an abelian subgroup strongly closed in P with respect to G, then G is isomorphic with one of the following groups:*

  (i) $\mathrm{PSL}_2(q)$ *for $q \equiv 3, 5 \bmod 8$;*
 (ii) $\mathrm{SL}_2(2^n)$ *for $n \geq 2$;*
(iii) $\mathrm{PSU}_3(2^n)$ *for some $n \geq 2$;*
 (iv) $^2G_2(3^{2n+1})$ *for $n \geq 2$;*
  (v) $\mathrm{Sz}(2^{2n+1})$ *for some $n \geq 2$;*
 (vi) *the Janko group J_1.*

Clearly if G has an abelian Sylow 2-subgroup then it has a strongly closed abelian 2-subgroup, and so nearly all of the groups on this list definitely have a soluble 2-fusion system. The remaining groups to check are $\mathrm{PSU}_3(2^n)$ and $\mathrm{Sz}(2^{2n+1})$, and we will see in Corollaries 7.11 and 7.12 that the 2-fusion systems of these groups are soluble as well. The simple groups on the list above will be called *Goldschmidt groups*.

We get the following corollary.

Corollary 5.97 *Let $\mathcal{F}$ be a saturated fusion system on a finite 2-group P. Then $\mathcal{F}$ is soluble if and only if $\mathcal{F} = \mathcal{F}_P(G)$ for some finite group G*

with Sylow 2-subgroup P, with the composition factors of G being either cyclic of prime order or Goldschmidt groups.

Such a result for odd primes is possible but only using the classification of the finite simple groups. The reason for this is that no analogue of Theorem 5.96 is known for odd primes without the classification of the finite simple groups.

Exercises

5.1   Prove Proposition 5.10.

5.2   Let G be a finite group, let H be a normal p'-subgroup of G, and let $\alpha : G \to G/H$ be the quotient map. Let Q be a Sylow p-subgroup of G/H, let P be a Sylow p-subgroup of G such that $P\alpha = Q$, and let h be any element of H. Prove that there exists $g \in G$ with $g\alpha = h$, such that

$$(P \cap P^g)\alpha = Q \cap Q^h.$$

5.3   Let $\mathcal{F}$ be a saturated fusion system on a finite p-group P, and let Q be a strongly $\mathcal{F}$-closed subgroup of P. Prove that $Q\,C_P(Q)$ is strongly $\mathcal{F}$-closed.

5.4   Prove Lemma 5.26.

5.5   Prove Lemma 5.32.

5.6   Suppose that $\mathcal{F}_Q(Q) \prec \mathcal{F}$. Prove that every characteristic subgroup of Q is strongly $\mathcal{F}$-closed.

5.7   Let P be a wreathed 2-group of order 2^{2n+1}. Then P is generated by commuting elements x and y, both of order 2^n, together with an involution z that swaps x and y. Write $U = \langle x, y \rangle$.

  (i)   Show that $Z(P) = \langle xy \rangle$ has order 2^n, $P' = \langle xy^{-1} \rangle$ has order 2^n, and $\Phi(P) = \langle x^2, xy \rangle$ has order 2^{2n-1}.

  (ii) Show that the conjugacy classes of P are as follows: $\{(xy)^i\}$, $\{x^i, y^i\}$, $\{x^iy^j, x^jy^i\}$, and $\{x^ay^bz \mid a + b \equiv i \bmod 2^n\}$ for $1 \le i, j \le 2^n - 1$. In particular, there are three conjugacy classes of involutions.

  (iii) Show that $P/Z(P)$ is dihedral of order 2^{n+1}.

  (iv) Show that U is the unique abelian maximal subgroup, and hence that $\mathrm{Aut}(P)$ is a 2-group.

5.8 Let P, U, and V be as in Proposition 5.87, and let $\mathcal{F}$ be the fusion system in (iv) of that result, so that neither $\text{Aut}_{\mathcal{F}}(U)$ nor $\text{Aut}_{\mathcal{F}}(V)$ is a 2-group. Show that there are no strongly $\mathcal{F}$-closed subgroups of P apart from 1 and P.

5.9 The *focal subgroup* of a fusion system is generated by the elements $x^{-1}x\phi$, as ϕ ranges over all morphisms in $\mathcal{F}$ and x ranges over all elements in the domain (see Sections 1.4 and 7.5 for more on the focal subgroup). Show that, for the four fusion systems on a wreathed 2-group, the focal subgroups are $P' = \langle xy^{-1}\rangle$, $\langle xy^{-1}, x^{2m}z\rangle$, U, and P itself, respectively. (Do not use the realizations of the fusion systems as group fusion systems.)

5.10 Let $\mathcal{F}$ be a saturated fusion system on a finite p-group P. Suppose that Q is a fully normalized subgroup of P, and let K be a normal subgroup of $\text{Aut}_{\mathcal{F}}(Q)$. Prove that $\text{N}_{\mathcal{F}}^{K}(Q)$ is a weakly normal subsystem of $\text{N}_{\mathcal{F}}(Q)$.

In particular, this proves that $\text{C}_{\mathcal{F}}(Q) \prec \text{N}_{\mathcal{F}}(Q)$ and $Q\,\text{C}_{\mathcal{F}}(Q) \prec \text{N}_{\mathcal{F}}(Q)$.

5.11 Let $\mathcal{F}$ be a fusion system on a finite p-group P, and let Q be a strongly $\mathcal{F}$-closed subgroup of $\mathcal{F}$. Prove that, if R is a fully $\mathcal{F}$-normalized subgroup containing Q, then $\text{N}_{\mathcal{F}}(R)/Q = \text{N}_{\mathcal{F}/Q}(R/Q)$.

5.12 Prove that a saturated fusion system is soluble if and only if it possesses no simple subquotients (i.e., a strongly $\mathcal{F}$-closed subgroup Q and a simple subsystem $\mathcal{E}$ of $\mathcal{F}/Q$).

5.13 Let $\mathcal{F}$ be a saturated fusion system on a finite p-group P, and let $\mathcal{E}$ be an $\mathcal{F}$-invariant subsystem of $\mathcal{F}$ on a strongly $\mathcal{F}$-closed subgroup Q of P.

(i) Prove that any fully $\mathcal{F}$-normalized subgroup of Q is fully $\mathcal{E}$-normalized and fully $\mathcal{E}$-centralized.

(ii) If R is a subgroup of Q, prove that R is $\mathcal{E}$-conjugate to a fully $\mathcal{E}$-normalized, fully $\mathcal{E}$-centralized subgroup S, such that $S\alpha$ is fully $\mathcal{F}$-normalized for some $\alpha \in \text{Aut}_{\mathcal{F}}(Q)$.

5.14 ([BCGLO07, Lemma 3.5]) Let $\mathcal{F}$ be a saturated fusion system on a finite p-group P, and let $\mathcal{E}$ be a saturated subsystem of $\mathcal{F}$, on the strongly $\mathcal{F}$-closed subgroup P. Prove that if R is an $\mathcal{F}$-centric, $\mathcal{F}$-radical subgroup of P, then $R \cap Q$ is $\mathcal{E}$-centric.

5.15 Let P be a finite 2-group. Suppose that $Q \cong Q_8$ is a subgroup of P. If $\text{C}_P(Q) = \text{Z}(Q)$, prove that P is semidihedral or generalized quaternion, and hence that any saturated fusion system $\mathcal{F}$ with Q_8 as an $\mathcal{F}$-centric, $\mathcal{F}$-radical subgroup is on one of those 2-groups.

6
Proving saturation

This chapter is concerned with various ways to prove that a fusion system, or a subsystem of a fusion system, is saturated. Finding out whether a given system is saturated is one of the most important, and one of the most difficult, aspects of the theory of fusion systems, and it will be useful to have different equivalent conditions, each tailored to a specific need.

The first section introduces the surjectivity property, a condition on the automorphisms of a given subgroup that, when combined with another condition on extensions of maps between subgroups, yields saturation. The power of this condition lies in its combination with the subsystem being invariant, which we will see later.

The second section introduces the notion of $\mathcal{H}$-saturation, which was considered by Broto, Castellana, Grodal, Levi, and Oliver in [BCGLO05]. This reduces the number of subgroups for which the saturation axioms need to be checked, basically to the class of centric subgroups. In fact, we can combine this result with the surjectivity property of the previous section to get another equivalent condition to saturation.

The third section moves in a different direction, describing the theory of invariant maps, as introduced by Aschbacher in [Asc08a]. Broadly speaking, they associate to each subgroup U of a strongly $\mathcal{F}$-closed subgroup Q a normal subgroup $A(U)$ of $\mathrm{Aut}_{\mathcal{F}}(U)$, in a consistent way. The subsystem generated by the subgroups $A(U)$ will be an $\mathcal{F}$-invariant subsystem $\mathcal{E}$, although in general $\mathrm{Aut}_{\mathcal{E}}(U) \neq A(U)$ for some subgroups U of Q. Even worse is that not every $\mathcal{F}$-invariant subsystem comes from an invariant map, since not all $\mathcal{F}$-invariant subsystems are generated by automorphisms of their subgroups.

The power of invariant maps is when you strengthen them to weakly normal maps, which are invariant maps that have some compatibility conditions. With weakly normal maps, $A(U) = \mathrm{Aut}_{\mathcal{E}}(U)$ for all subgroups U of Q, and the subsystem $\mathcal{E}$ is actually saturated as well. Crucially, every weakly normal subsystem comes from a unique weakly normal map, and every weakly normal subsystem yields a (unique) weakly normal map.

6.1 The surjectivity property

We start this section by slightly refining our definition of receptivity.

Definition 6.1 Let $\mathcal{F}$ be a fusion system on a finite p-group P, and let R be a subgroup of P. A subgroup Q of P is *R-receptive* if all isomorphisms $\phi : R \to Q$ extend to morphisms $\bar{\phi} : N_\phi \to N_P(Q)$.

Clearly a subgroup Q is receptive if and only if it is R-receptive for all $R \leq P$, the condition holding vacuously if Q and R are not $\mathcal{F}$-conjugate.

Proposition 6.2 *Let $\mathcal{F}$ be a fusion system on a finite p-group P. If Q is a subgroup of P such that*

(i) *Q is Q-receptive, and*
(ii) *for any R that is $\mathcal{F}$-conjugate to Q, there is a morphism $R \to Q$ that extends to a morphism $N_P(R) \to N_P(Q)$,*

then Q is receptive. In particular, a fusion system $\mathcal{F}$ is saturated if and only if, for every $\mathcal{F}$-conjugacy class $\mathcal{Q}$ of subgroups of P, there exists a fully automized subgroup $Q \in \mathcal{Q}$ such that the above two conditions hold.

Proof Let $\phi : R \to Q$ be any isomorphism in $\mathcal{F}$, and let $\psi : N_P(R) \to N_P(Q)$ be an $\mathcal{F}$-isomorphism that restricts to a map $R \to Q$. The map $\theta = \psi^{-1}\phi$ is an automorphism of Q, and has an extension $\bar{\theta} : N_\theta \to N_P(Q)$.

The subgroup N_ϕ is the preimage of $\mathrm{Aut}_P(R) \cap \mathrm{Aut}_P(Q)^{\phi^{-1}}$ in R, and $N_\phi\psi$ induces the automorphism group $\mathrm{Aut}_P(R)^\psi \cap \mathrm{Aut}_P(Q)^{\phi^{-1}\psi}$ of Q. Since $\mathrm{Aut}_P(R)^\psi$ is contained in $\mathrm{Aut}_P(Q)$, we see that this intersection is contained in $\mathrm{Aut}_P(Q) \cap \mathrm{Aut}_P(Q)^{\theta^{-1}}$, so that the image of N_ϕ under ψ is contained in N_θ. Hence the composition $\psi\bar{\theta} : N_\phi \to N_P(Q)$ is an extension of ϕ to N_ϕ, proving that Q is receptive, as required. $\square$

We now come to the definition of the surjectivity property. Recall that $N_{\mathcal{F}}(Q, R) = N_{N_{\mathcal{F}}(Q)}(R)$ (see Definition 4.39). Suppose that Q is an $\mathcal{F}$-centric subgroup of P, and that ϕ is an automorphism of Q. If

$Q \leq R \leq \mathrm{N}_P(Q)$, and $\mathrm{Aut}_R(Q)^\phi \leq \mathrm{Aut}_P(Q)$, then $R \leq N_\phi$, and any extension of ϕ to R maps R to the preimage of $\mathrm{Aut}_R(Q)^\phi$; in particular, if $\mathrm{Aut}_R(Q)^\phi = \mathrm{Aut}_R(Q)$ – i.e., $\phi \in \mathrm{N}_{\mathrm{Aut}_{\mathcal{F}}(Q)}(\mathrm{Aut}_R(Q))$ – then ϕ extends to an automorphism of R (if it extends to any map at all).

Definition 6.3 Let $\mathcal{F}$ be a fusion system on a finite p-group P, and let Q be a subgroup of P.

(i) If R is a subgroup of P with $Q \leq R \leq \mathrm{N}_P(Q)$, then by $\mathrm{Aut}_{\mathcal{F}}(Q \leq R)$ we mean the collection of all $\mathcal{F}$-automorphisms of R that restrict to automorphisms of Q, i.e., we have

$$\mathrm{Aut}_{\mathcal{F}}(Q \leq R) = \mathrm{Aut}_{\mathrm{N}_{\mathcal{F}}(Q,R)}(R).$$

Similarly, $\mathrm{Aut}_P(Q \leq R) = \mathrm{Aut}_{\mathrm{N}_P(Q,R)}(R)$.

(ii) The subgroup Q has the *surjectivity property* for $\mathcal{F}$ if, whenever R is a subgroup of P containing $Q\,\mathrm{C}_P(Q)$, the map

$$\mathrm{Aut}_{\mathcal{F}}(Q \leq R) \to \mathrm{N}_{\mathrm{Aut}_{\mathcal{F}}(Q)}(\mathrm{Aut}_R(Q))$$

is surjective.

Therefore if Q has the surjectivity property and $Q\,\mathrm{C}_P(Q) \leq R \leq \mathrm{N}_P(Q)$, then any map $\phi \in \mathrm{Aut}_{\mathcal{F}}(Q)$ that normalizes $\mathrm{Aut}_R(Q)$ extends to some map $\psi \in \mathrm{Aut}_{\mathcal{F}}(R)$. Of course, if Q is receptive then this holds.

We leave the proof of the following easy lemma to the reader; it says that receptivity and the surjectivity property are preserved by isomorphisms of fusion systems (and in particular automorphisms).

Lemma 6.4 *Let P and $\bar{P}$ be finite p-groups, and let $\alpha : P \to \bar{P}$ be an isomorphism. Let $\mathcal{F}$ and $\bar{\mathcal{F}}$ be fusion systems on P and $\bar{P}$ respectively, such that α induces an isomorphism $\mathcal{F} \to \bar{\mathcal{F}}$. Let R be a subgroup of P. A subgroup Q of P is R-receptive if and only if $Q\alpha$ is $R\alpha$-receptive, and Q has the surjectivity property if and only if $Q\alpha$ has the surjectivity property.*

We will often have to deal with fusion systems that are not saturated, or at least not necessarily saturated. To prove that a fusion system is saturated we need to show that each $\mathcal{F}$-conjugacy class contains a fully automized, receptive member, i.e., that it is saturated. We generally choose an $\mathcal{F}$-conjugacy class of subgroups of smallest index subject to not being saturated. Because of this, we make the following definition.

Definition 6.5 Let $\mathcal{F}$ be a fusion system on a finite p-group P, and let Q be a subgroup of P. We say that $\mathcal{F}$ is *inductively saturated with respect* *to* Q if all $\mathcal{F}$-conjugacy classes of subgroups of P properly containing $\mathcal{F}$-conjugates of Q are saturated. Similarly, $\mathcal{F}$ is inductively saturated with respect to $\mathcal{Q}$ for some $\mathcal{F}$-conjugacy class $\mathcal{Q}$ if it is inductively saturated with respect to Q, for some $Q \in \mathcal{Q}$.

As a quick remark on this definition, notice that if R properly contains an $\mathcal{F}$-conjugate of Q, any $\mathcal{F}$-conjugate of R also contains an $\mathcal{F}$-conjugate of Q.

We will keep the second condition of Proposition 6.2 and alter the first, to make being fully automized and Q-receptive into a single condition, namely that it has the surjectivity property. Because this condition only deals with extensions, it will be vacuously true for P itself, and so we will need to assume that P is fully automized.

We first note the following easy lemma.

Lemma 6.6 *Let $\mathcal{F}$ be a saturated fusion system on a finite p-group P. If Q is a fully $\mathcal{F}$-normalized subgroup of P then Q has the surjectivity property, and if R is $\mathcal{F}$-conjugate to Q then there is a morphism $\phi : R \to Q$ that extends to a morphism $\bar{\phi} : \mathrm{N}_P(R) \to \mathrm{N}_P(Q)$.*

We will now start examining the surjectivity property in more detail, in an inductively saturated situation.

Lemma 6.7 *Let $\mathcal{F}$ be a fusion system on a finite p-group P, and suppose that P is fully automized. Let Q be a fully $\mathcal{F}$-normalized subgroup of P, and suppose that $\mathcal{F}$ is inductively saturated with respect to Q. If Q has the surjectivity property, then Q is fully automized.*

Proof If $Q = P$ then there is nothing to prove, so suppose that Q is a proper subgroup of P. Write $N = \mathrm{N}_P(Q)$, and suppose firstly that N is fully normalized, so that it is fully automized (as $\mathcal{F}$ is inductively saturated with respect to Q); then $\mathrm{Aut}_P(N)$ is a Sylow p-subgroup of $\mathrm{Aut}_{\mathcal{F}}(N)$, and since $\mathrm{Aut}_{\mathcal{F}}(Q \leq N)$ is a normal subgroup of $\mathrm{Aut}_{\mathcal{F}}(N)$, we see that $\mathrm{Aut}_P(Q \leq N)$ is a Sylow p-subgroup of $\mathrm{Aut}_{\mathcal{F}}(Q \leq N)$. The map $\mathrm{Aut}_{\mathcal{F}}(Q \leq N) \to \mathrm{N}_{\mathrm{Aut}_{\mathcal{F}}(Q)}(\mathrm{Aut}_N(Q))$ is surjective, and so the image of $\mathrm{Aut}_P(Q \leq N)$, namely $\mathrm{Aut}_P(Q)$, is a Sylow p-subgroup of its normalizer; if a p-group is a Sylow p-subgroup of its normalizer then it is a Sylow p-subgroup of the group, completing the proof.

Now assume that N is not fully normalized, and let $\phi : N \to \bar{N}$ be an $\mathcal{F}$-isomorphism such that $\bar{N}$ is fully normalized; write $\bar{Q} = Q\phi$ (and note that $\bar{N} = \mathrm{N}_P(\bar{Q})$). Since Q is fully normalized and $|N| = |\bar{N}|$, $\bar{Q}$ is also fully normalized. There is an isomorphism $\phi : \mathrm{N}_P(Q) \to$

$N_P(\bar{Q})$ in $\mathcal{F}$ sending Q to $\bar{Q}$, so we see that normalizers, centralizers, and P-automizers are isomorphic for both Q and $\bar{Q}$; hence Q is fully automized if and only if $\bar{Q}$ is fully automized.

It remains to show that $\bar{Q}$ inherits the surjectivity property. We have that $\phi : N_P(Q) \to N_P(\bar{Q})$ is an isomorphism, so that ϕ transports $\operatorname{Aut}_P(Q)$ to $\operatorname{Aut}_P(\bar{Q})$. Let $\bar{R}$ be a subgroup lying between $\bar{Q}\,C_P(\bar{Q})$ and $N_P(\bar{Q})$, and let R be the corresponding subgroup of $N_P(Q)$, so that $R\phi = \bar{R}$. Let $\theta' \in N_{\operatorname{Aut}_\mathcal{F}(\bar{Q})}(\operatorname{Aut}_{\bar{R}}(\bar{Q}))$ so that, in $\operatorname{Aut}_\mathcal{F}(\bar{Q})$, we have $(\operatorname{Aut}_{\bar{R}}(\bar{Q}))^{\theta'} = \operatorname{Aut}_{\bar{R}}(\bar{Q})$. Applying ϕ^{-1} to the map θ', we see that $\theta = \theta'^{\phi^{-1}}$ normalizes $\operatorname{Aut}_R(Q)$, and so has an extension $\hat{\theta}$ to R, by hypothesis. Since R and $\bar{R}$ are $\mathcal{F}$-conjugate, $\hat{\theta}^{\phi}$ is an extension of θ', proving that the map is surjective, as claimed. Since both $\bar{Q}$ and $N_P(\bar{Q})$ are fully normalized, $\bar{Q}$ is fully automized by the first paragraph, and so therefore is Q. $\qquad\square$

We now move to the second part of the proof, showing that if Q is fully normalized then the surjectivity property implies that Q is Q-receptive.

Lemma 6.8   *Let $\mathcal{F}$ be a fusion system on a finite p-group P, and suppose that P is fully automized. Let Q be a fully $\mathcal{F}$-normalized subgroup of P, and suppose that $\mathcal{F}$ is inductively saturated with respect to Q. If Q has the surjectivity property then Q is Q-receptive.*

Proof   By Lemma 6.7, $\operatorname{Aut}_P(Q)$ is a Sylow p-subgroup of $\operatorname{Aut}_\mathcal{F}(Q)$. Let G denote the group $\operatorname{Aut}_\mathcal{F}(Q)$ and S denote the Sylow p-subgroup $\operatorname{Aut}_P(Q)$ of $\operatorname{Aut}_\mathcal{F}(Q)$. Let ϕ be an element of G, and let U denote the subgroup $S \cap S^{\phi^{-1}}$. The subgroup N_ϕ of P is the preimage of U in $N_P(Q)$.

Notice that ϕ sends U to U^ϕ, both of which are subgroups of S. We will use Alperin's fusion theorem for groups, Theorem 1.29; this states that there exist $\chi_1, \ldots, \chi_r$ and θ in G, with $\phi = \chi_1 \ldots \chi_r \theta$, and such that $\chi_i \in N_G(V_i)$ for subgroups $V_i \leq S$, and $\theta \in N_G(S)$. Importantly, writing $U_0 = U$, $U_i = U_0^{\chi_1 \ldots \chi_i}$ for $1 \leq i \leq r$, and $U_{r+1} = V = U_r^\theta$, we have $U_{i-1}, U_i \leq V_i$ for all $1 \leq i \leq r$.

Since θ normalizes S, we have that θ extends to an $\mathcal{F}$-automorphism $\hat{\theta}$ of $N_P(Q)$ as by assumption Q is fully $\mathcal{F}$-normalized. Note that $\chi_i \in N_G(V_i)$; hence χ_i extends to an automorphism $\hat{\chi}_i$ of the preimage W_i of V_i in $N_P(Q)$. As $U_0 \leq V_1$, by taking preimages we obtain $N_\phi \leq W_1$. Finally, $U_{i-1}, U_i \leq V_i$ for all i, so we see that $(\hat{\chi}_1 \ldots \hat{\chi}_r \hat{\theta})|_{N_\phi}$ is a morphism in $\operatorname{Hom}_\mathcal{F}(N_\phi, N_P(Q))$. This completes the proof. $\qquad\square$

We now combine Proposition 6.2 and Lemmas 6.7 and 6.8 to get the following result.

Theorem 6.9 (Puig [Pui06, Corollary 2.14]) *Let $\mathcal{F}$ be a fusion system on a finite p-group P, and suppose that P is fully automized. Then $\mathcal{F}$ is saturated if and only if, for each $\mathcal{F}$-conjugacy class, there is some subgroup Q such that*

(i) *if R is $\mathcal{F}$-conjugate to Q then there is a map $\mathrm{N}_P(R) \to \mathrm{N}_P(Q)$ in $\mathcal{F}$ that restricts to a map $R \to Q$, and*

(ii) *Q has the surjectivity property.*

Proof We need to show that every $\mathcal{F}$-conjugacy class contains a fully automized, receptive member, i.e., that every $\mathcal{F}$-conjugacy class is saturated. Suppose that $\mathcal{F}$ is inductively saturated with respect to Q, where Q is a subgroup of P satisfying the two conditions of the theorem. By Lemmas 6.7 and 6.8, Q is fully automized and Q-receptive, and by Proposition 6.2, Q is receptive, completing the proof of one direction. The converse is Lemma 6.6, and so the proof is complete. $\square$

This theorem is a slightly modified version of that given by Puig in [Pui06] and [Pui09], making use of the more flexible modern definition of a fusion system. In the former of these there is a slight omission in the statement of the theorem, and it is corrected in the latter.

6.2 Reduction to centric subgroups

In this section we will prove a result of Broto, Castellana, Grodal, Levi, and Oliver, which proves that, in order to check saturation, it is only necessary to check that some of the $\mathcal{F}$-conjugacy classes are saturated. For a more precise statement, we make the following definition.

Definition 6.10 Let $\mathcal{F}$ be a fusion system on a finite p-group P. If $\mathcal{H}$ is a union of $\mathcal{F}$-conjugacy classes of subgroup of P, then $\mathcal{F}$ is said to be $\mathcal{H}$-*generated* if $\mathcal{F}$ is generated by $\mathrm{Hom}_{\mathcal{F}}(U, V)$, for all U and V in $\mathcal{H}$.

The main theorem we will prove in this section is the following, from [BCGLO05].

Theorem 6.11 ([BCGLO05, Theorem A]) *Let $\mathcal{F}$ be a fusion system on a finite p-group P. Let $\mathcal{H}$ be a union of $\mathcal{F}$-conjugacy classes of subgroups of P such that $\mathcal{F}$ is $\mathcal{H}$-generated, and suppose that every*

$\mathcal{F}$-conjugacy class in $\mathcal{H}$ is saturated. If, for every conjugacy class of $\mathcal{F}$-centric subgroups $\mathcal{R}$ of P not in $\mathcal{H}$, there is a subgroup $R \in \mathcal{R}$ such that $\mathrm{Out}_P(R) \cap \mathrm{O}_p(\mathrm{Out}_{\mathcal{F}}(R))$ is non-trivial, then $\mathcal{F}$ is saturated.

Although the last condition looks technical, it is necessary. What it says is that the particular subgroup R has the property that the inverse image of $\mathrm{O}_p(\mathrm{Aut}_{\mathcal{F}}(R))$ in $\mathrm{N}_P(R)$ (taken under the map c_R) is strictly larger than R. This condition implies that $\mathcal{H}$ contains all centric radical subgroups. By specifying $\mathcal{H}$ to be all centric subgroups we get the following special case.

Corollary 6.12 (Puig [Pui06, Theorem 3.8]) *Let $\mathcal{F}$ be a fusion system on a finite p-group P. If $\mathcal{F}$ is generated by morphisms between $\mathcal{F}$-centric subgroups of P, and all conjugacy classes of $\mathcal{F}$-centric subgroups of P are saturated, then $\mathcal{F}$ is saturated.*

The strategy to prove Theorem 6.11 is fairly simple: suppose that $\mathcal{F}$ is $\mathcal{H}$-generated, and that every conjugacy class in $\mathcal{H}$ is saturated. Let $\mathcal{Q}$ be a conjugacy class of subgroups of smallest index subject to not being in $\mathcal{H}$, and show that $\mathcal{Q}$ is saturated. In this case, $\mathcal{F}$ is $(\mathcal{H} \cup \mathcal{Q})$-generated, and every conjugacy class in $\mathcal{H} \cup \mathcal{Q}$ is saturated, and so an obvious induction will complete the proof.

To get started, we require the definition of a proper $\mathcal{Q}$-pair, which will only be needed in this section. If $\mathcal{F}$ is a fusion system on P, and $\mathcal{Q}$ is an $\mathcal{F}$-conjugacy class of subgroups of P, then a *proper $\mathcal{Q}$-pair* is a pair (Q, R), where Q is an element of $\mathcal{Q}$ and R is a subgroup with $Q < R \leq \mathrm{N}_P(Q)$. Obviously $\mathcal{F}$ acts on the set of proper $\mathcal{Q}$-pairs, and we say that (Q, R) is *fully normalized* if, whenever $(\tilde{Q}, \tilde{R})$ is $\mathcal{F}$-conjugate to (Q, R), we have

$$|\mathrm{N}_P(Q, R)| \geq |\mathrm{N}_P(\tilde{Q}, \tilde{R})|.$$

These behave quite well, and we can get some nice properties for proper $\mathcal{Q}$-pairs. The first gives some idea of what fully normalized proper pairs look like: we work in an inductively saturated situation.

Proposition 6.13 *Let $\mathcal{F}$ be a fusion system on a finite p-group P, and suppose that $\mathcal{F}$ is inductively saturated with respect to Q; write $\mathcal{Q}$ for the $\mathcal{F}$-conjugacy class of subgroups containing Q. Let (Q, R) be a fully normalized proper $\mathcal{Q}$-pair.*

(i) *The subgroup R is fully centralized, and $\mathrm{Aut}_P(Q \leq R)$ is a Sylow p-subgroup of $\mathrm{Aut}_{\mathcal{F}}(Q \leq R)$.*

(ii) *If $(\tilde{Q}, \tilde{R})$ is another proper Q-pair $\mathcal{F}$-conjugate to (Q, R), then there is a morphism $\phi : \tilde{R} \to R$ with $\tilde{Q}\phi = Q$ that extends to a map*

$$\bar{\phi} \in \mathrm{Hom}_{\mathcal{F}}\left(\mathrm{N}_P(\tilde{Q}, \tilde{R}), \mathrm{N}_P(Q, R)\right).$$

Proof Since $R > Q$, R lies in a saturated $\mathcal{F}$-conjugacy class. If K denotes $\mathrm{Aut}(Q \leq R)$, then R is fully K-normalized in $\mathcal{F}$, and so by Lemma 4.35 (whose proof only requires that all $\mathcal{F}$-conjugacy classes of subgroups containing conjugates of R (not properly) are saturated), R is fully centralized and $\mathrm{Aut}_P(Q \leq R)$ is a Sylow p-subgroup of $\mathrm{Aut}_{\mathcal{F}}(Q \leq R)$, proving (i).

To prove (ii), since $(\tilde{Q}, \tilde{R})$ is $\mathcal{F}$-conjugate to (Q, R), there is an $\mathcal{F}$-isomorphism $\psi : \tilde{R} \to R$ mapping $\tilde{Q}$ to Q. Since (Q, R) is fully normalized, R is fully centralized, hence receptive (as $\mathcal{F}$ is inductively saturated) and $\mathrm{Aut}_P(Q \leq R)$ is a Sylow p-subgroup of $\mathrm{Aut}_{\mathcal{F}}(Q \leq R)$. The image A of $\mathrm{Aut}_P(\tilde{Q} \leq \tilde{R})$ under ψ is a p-subgroup of $\mathrm{Aut}_{\mathcal{F}}(Q \leq R)$, and hence there is an automorphism $\theta \in \mathrm{Aut}_{\mathcal{F}}(Q \leq R)$ such that $\phi = \psi\theta$ maps $\mathrm{Aut}_P(\tilde{Q} \leq \tilde{R})$ inside $\mathrm{Aut}_P(Q \leq R)$. Since R is fully centralized, this means that N_ϕ contains $\mathrm{N}_P(\tilde{Q}, \tilde{R})$, and therefore ϕ extends to a map $\bar{\phi}$ as claimed. □

In an inductively saturated setting, we would like to prove that a particular conjugacy class is saturated. We can use Proposition 6.2, so one thing we need to show is that there is some subgroup $Q \in \mathcal{Q}$ such that, for any $\tilde{Q} \in \mathcal{Q}$, there is a map $\mathrm{N}_P(\tilde{Q}) \to \mathrm{N}_P(Q)$ in $\mathcal{F}$. This is the content of the next result.

Proposition 6.14 *Let $\mathcal{F}$ be a fusion system on a finite p-group P, and let $\mathcal{H}$ be a union of $\mathcal{F}$-conjugacy classes of subgroups of P. Suppose that $\mathcal{F}$ is $\mathcal{H}$-generated, and that every $\mathcal{F}$-conjugacy class in $\mathcal{H}$ is saturated. Let $\mathcal{Q}$ be an $\mathcal{F}$-conjugacy class of subgroups such that all subgroups of P properly containing a member of $\mathcal{Q}$ lie in $\mathcal{H}$ (so that $\mathcal{F}$ is inductively saturated with respect to $\mathcal{Q}$). Let Q be a fully $\mathcal{F}$-normalized subgroup of P in $\mathcal{Q}$.*

(i) *The subgroup Q is fully centralized, and for any other subgroup $\tilde{Q} \in \mathcal{Q}$ there is a morphism $\phi : \mathrm{N}_P(\tilde{Q}) \to \mathrm{N}_P(Q)$ with $\tilde{Q}\phi = Q$.*
(ii) *If $\{Q\}$ is saturated in $\mathrm{N}_{\mathcal{F}}(Q)$, then $\mathcal{Q}$ is saturated in $\mathcal{F}$.*

Proof If $\mathcal{Q} = \{P\}$ then the result is clear, so we may assume that $\mathcal{Q}$ consists of proper subgroups of P. Before we begin, notice that, if there exists a fully normalized subgroup $Q \in \mathcal{Q}$ with the property that for every $\tilde{Q} \in \mathcal{Q}$ we have a map $N_P(\tilde{Q}) \to N_P(Q)$ extending $\tilde{Q} \to Q$, then the result holds, since if $\tilde{Q}$ is fully centralized, the map $N_P(\tilde{Q}) \to N_P(Q)$ must injectively map the centralizer of $\tilde{Q}$ into the centralizer of Q, proving that Q is also fully centralized.

Let R and $\tilde{R}$ be any two subgroups in $\mathcal{Q}$. We will prove that there is another subgroup $S \in \mathcal{Q}$, such that there are maps $\phi : N_P(R) \to N_P(S)$ and $\phi' : N_P(\tilde{R}) \to N_P(S)$ in $\mathcal{F}$ with $R\phi = \tilde{R}\phi' = S$. Choosing R to be fully normalized, we see that ϕ must be an isomorphism, whence $\phi'\phi^{-1}$ gives the result.

Let $\mathscr{X}$ denote the set of all triples, (A, B, C), where A is a sequence $R = A_1, \ldots, A_t = \tilde{R}$, B is a sequence of B_i with $A_i < B_i \leq N_P(A_i)$ (for $1 \leq i < t$), and C is a sequence of ϕ_i (for $1 \leq i < t$), with $\phi_i : B_i \to N_P(A_{i+1})$ in $\mathcal{F}$ and mapping A_i to A_{i+1}. (Hence the composition of the maps ϕ_i is a map from R to $\tilde{R}$.)

Since $\mathcal{F}$ is $\mathcal{H}$-generated, the set $\mathscr{X}$ is non-empty. To see this, let ϕ be any map $R \to \tilde{R}$ in $\mathcal{F}$; since $\mathcal{F}$ is $\mathcal{H}$-generated, ϕ may be decomposed as the composition of maps $\phi_i : U_i \to V_i$ with $R < U_1$. Writing $A_{i+1} = A_i\phi_i$, $B_i = N_{U_i}(A_i)$, and keeping ϕ_i, this gives a triple (A, B, C) in $\mathscr{X}$.

Let $\mathscr{X}'$ be the subset of all triples in $\mathscr{X}$ such that, for all $1 \leq i < t$, either $B_i\phi < N_P(A_{i+1})$ or $B_{i+1} < N_P(A_{i+1})$, i.e., all triples where there is no map ϕ_i that maps B_i onto the entire normalizer, and where ϕ_{i+1} maps the entire normalizer into the next subgroup. If there is such an i, then we may 'collapse' the sequence, by composing ϕ_i with ϕ_{i+1} and removing A_{i+1}, B_{i+1}. Doing this for all i defines a map $\mathscr{X} \to \mathscr{X}'$, and, since $\mathscr{X}$ is non-empty, this proves that $\mathscr{X}'$ is non-empty as well.

Let (A, B, C) be a triple in $\mathscr{X}'$. Let I denote the set of all $1 \leq i < t$ such that both $B_i < N_P(A_i)$ and $B_i\phi_i < N_P(A_{i+1})$. We claim that there is a triple in $\mathscr{X}'$ such that I is empty. Suppose that this is true; then for all i, either $B_i = N_P(A_i)$ or $B_i\phi_i = N_P(A_{i+1})$ or both. However, since the triple lies in $\mathscr{X}'$, not just in $\mathscr{X}$, if $B_i\phi_i = N_P(A_{i+1})$ then we must have $B_{i+1} \neq N_P(A_{i+1})$. What this means is that if there is some $1 \leq j < t$ such that $B_j\phi_j = N_P(A_{j+1})$ then, for all $i \geq j$, we must have $B_i\phi_i = N_P(A_{i+1})$; let j denote the first such integer. We claim that $S = A_j$ fulfils the requirements of the result. To see this, notice that $\phi' = (\phi_j\phi_{j+1}\ldots\phi_{t-1})^{-1}$ is a map $\tilde{R} \to S$, and the inverse of each of these ϕ_i maps $N_P(A_{i+1})$ into $N_P(A_i)$. Hence ϕ' extends to a map $N_P(\tilde{R}) \to N_P(S)$. For R the situation is the same: the composition $\phi = \phi_1\ldots\phi_{j-1}$ is a map $R \to S$ and since $B_i = N_P(A_i)$ for each i, ϕ extends to a map $N_P(R) \to N_P(S)$, as needed.

It remains to show that there is a triple (A, B, C) in $\mathscr{X}'$ such that I is empty. Let μ denote the minimum over all $i \in I$ of $|B_i : A_i|$, which is at least p since $A_i < B_i$. Notice that μ is bounded above by $|\mathrm{N}_P(Q) : Q|$, where Q is a fully normalized member of $\mathcal{Q}$. Given (A, B, C), with I and μ as defined, we will construct another triple $(\bar{A}, \bar{B}, \bar{C})$ in $\mathscr{X}'$ with either $\bar{I} = \emptyset$ or $\bar{\mu} > \mu$. Since μ is bounded above, eventually we must find one with the corresponding I empty, completing the proof of (i).

For each $i \in I$, choose a fully normalized proper $\mathcal{Q}$-pair $(\bar{A}_i, \bar{B}_i)$ that is $\mathcal{F}$-conjugate to (A_i, B_i) (and hence to $(A_{i+1}, B_i \phi_i)$). By Proposition 6.13, there are morphisms

$$\alpha : \mathrm{N}_P(A_i, B_i) \to \mathrm{N}_P(\bar{A}_i, \bar{B}_i) \quad \text{and} \quad \alpha' : \mathrm{N}_P(A_{i+1}, B_i \phi_i) \to \mathrm{N}_P(\bar{A}_i, \bar{B}_i),$$

with $B_i \alpha = \bar{B}_i$ and $A_i \alpha = \bar{A}_i$, and similarly $(B_i \phi_i)\alpha' = \bar{B}_i$ and $A_{i+1}\alpha' = \bar{A}_i$. In the triple (A, B, C), we make modifications to A, B and C:

- for A, we insert $\bar{A}_i$ in between A_i and A_{i+1};
- for B, we replace B_i by $\mathrm{N}_P(A_i, B_i)$ (which contains B_i) and insert $\mathrm{N}_P(A_{i+1}, B_i \phi_i)\alpha'$ (which contains $(B_i \phi_i)\alpha'$);
- for C, we replace ϕ_i by α and insert $(\alpha')^{-1}$.

We perform this for all i, and notice that this is again in $\mathscr{X}$, and that we can compose morphisms as above to get a member of $\mathscr{X}'$; write $(\bar{A}, \bar{B}, \bar{C})$ for this new triple, with the corresponding $\bar{I}$ and $\bar{\mu}$.

It remains to check that $\bar{\mu}$ is greater than μ, or $\bar{I}$ is empty. If $\bar{I}$ is non-empty, it suffices to check that at each i we have increased the index $|B_i : A_i|$, but by construction our new components in B strictly contain the old B_i, and the order of the A_i is always the same, so this index must have increased. This proves (i).

Suppose that $\{Q\}$ is saturated in $\mathrm{N}_{\mathcal{F}}(Q)$. By Exercise 6.1, Q is Q-receptive and fully automized, and so by Proposition 6.2, to show that Q is receptive, we need to show that if $\tilde{Q} \in \mathcal{Q}$ then there is a map $\mathrm{N}_P(\tilde{Q}) \to \mathrm{N}_P(Q)$ in $\mathcal{F}$ where the image of $\tilde{Q}$ is Q; this is guaranteed by the first part of this result, completing the proof of (ii). $\square$

The introduction of the normalizer subsystem $\mathrm{N}_{\mathcal{F}}(Q)$ is important; we will need to know that it is also inductively saturated with respect to Q, and then finally prove that $\{Q\}$ is saturated in $\mathrm{N}_{\mathcal{F}}(Q)$.

Proposition 6.15 *Let $\mathcal{F}$ be a fusion system on a finite p-group P, and let $\mathcal{H}$ be a union of $\mathcal{F}$-conjugacy classes of subgroups of P. Suppose that $\mathcal{F}$ is $\mathcal{H}$-generated, and that every $\mathcal{F}$-conjugacy class in $\mathcal{H}$ is saturated.*

Let Q be an $\mathcal{F}$-conjugacy class of subgroups of P such that all subgroups of P of smaller index than those in Q lie in $\mathcal{H}$. Let Q be a fully $\mathcal{F}$-normalized member of Q.

(i) *The normalizer subsystem $N_{\mathcal{F}}(Q)$ is inductively saturated with respect to Q, and every $\mathcal{F}$-automorphism of Q is the composition of (restrictions of) morphisms between subgroups of $N_P(Q)$ strictly containing Q.*

(ii) *If either Q is not $\mathcal{F}$-centric, or $\mathrm{Out}_P(Q) \cap O_p(\mathrm{Out}_{\mathcal{F}}(Q))$ is non-trivial, then the $N_{\mathcal{F}}(Q)$-conjugacy class $\{Q\}$ is saturated.*

Proof Let R be a subgroup of $N_P(Q)$ properly containing Q, and suppose that R is fully $N_{\mathcal{F}}(Q)$-normalized. We prove that (Q, R) is a fully normalized proper Q-pair. By Proposition 6.14(i), if $\tilde{Q}$ is $\mathcal{F}$-conjugate to Q, then there is a map $\phi \in \mathrm{Hom}_{\mathcal{F}}(N_P(\tilde{Q}), N_P(Q))$. If $(\tilde{Q}, \tilde{R})$ is $\mathcal{F}$-conjugate to (Q, R), then the restriction of ϕ gives an injective map

$$\phi : N_P(\tilde{Q}, \tilde{R}) \to N_P(Q, \tilde{R}\phi).$$

Therefore

$$|N_P(\tilde{Q}, \tilde{R})| \leq |N_P(Q, \tilde{R}\phi)| \leq |N_P(Q, R)|,$$

with the last inequality holding since R is fully normalized in $N_{\mathcal{F}}(Q)$. Hence (Q, R) is a fully normalized proper Q-pair. In particular, by Proposition 6.13(i), $\mathrm{Aut}_{N_P(Q)}(R)$ is a Sylow p-subgroup of $\mathrm{Aut}_{N_{\mathcal{F}}(Q)}(R)$, and so R is fully $N_{\mathcal{F}}(Q)$-automized and R is fully $\mathcal{F}$-centralized, hence $\mathcal{F}$-receptive. Finally, it is clear that any $\mathcal{F}$-receptive subgroup is $N_{\mathcal{F}}(Q)$-receptive, and so R is fully automized and receptive in $N_{\mathcal{F}}(Q)$, proving that all $N_{\mathcal{F}}(Q)$-conjugacy classes of subgroup containing Q are saturated.

Let ϕ be an automorphism of Q in $\mathcal{F}$. By assumption, there are subgroups $R_1, R_2, \ldots, R_t$ of larger order than $|Q|$ and maps $\phi_i : R_i \to R_{i+1}$, and subgroups $Q_1, \ldots, Q_{t+1}$ with $Q_i \phi_i = Q_{i+1}$ and $Q_1, Q_{t+1} = Q$, such that $\phi = \phi_1 \ldots \phi_t$. Since $Q_i < R_i$, we replace R_i by $N_{R_i}(Q_i)$, which still properly contains Q_i, so that each R_i is contained in $N_P(Q_i)$.

Since Q is fully normalized, there are maps $\theta_i : N_P(Q_i) \to N_P(Q)$ with $Q_i \theta_i = Q$, and we may take θ_1 and θ_{t+1} to be the identity. Define ψ_i by $\psi_i = \theta_i^{-1} \phi_i \theta_{i+1}$. This is a map from a subgroup of $N_P(Q)$ to itself, with $Q\psi_i = Q$, and the domain of ψ_i properly contains Q. Finally, it is clear that

$$\phi = \psi_1 \ldots \psi_t = (\theta_1^{-1} \phi_1 \theta_2)(\theta_2^{-1} \phi_2 \theta_3) \ldots (\theta_t^{-1} \phi_t \theta_{t+1}),$$

completing the proof of (i).

For the rest of the proof, we assume that Q is a normal subgroup of $\mathcal{F}$, i.e., that $\mathcal{F} = N_{\mathcal{F}}(Q)$. Let $\bar{Q}$ be the inverse image in P of $\mathrm{Aut}_P(Q) \cap O_p(\mathrm{Aut}_{\mathcal{F}}(Q))$. This is a normal subgroup of P, like Q. By Lemma 4.57, $\bar{Q}$ is strongly $\mathcal{F}$-closed, since $\bar{Q} = N_P^{O_p(\mathrm{Aut}_{\mathcal{F}}(Q))}(Q)$.

Since $\bar{Q} > Q$ and $\bar{Q}$ is strongly $\mathcal{F}$-closed, $\{\bar{Q}\}$ is a saturated $\mathcal{F}$-conjugacy class; in particular, $\bar{Q}$ is receptive. Suppose that every automorphism of Q extends to an automorphism of $\bar{Q}$; we will show that this implies that Q has the surjectivity property, completing the proof. Let R be a subgroup containing $Q\,C_P(Q)$, and let ϕ be an automorphism in $\mathrm{Aut}_{\mathcal{F}}(Q)$ satisfying $\mathrm{Aut}_R(Q)^\phi = \mathrm{Aut}_R(Q)$. Note that any extension of ϕ with domain R must also have image R.

Let ψ be an extension of ϕ to an automorphism of $\bar{Q}$, and consider $\mathrm{Aut}_R(\bar{Q})^\psi$, a p-subgroup of $\mathrm{Aut}_{\mathcal{F}}(\bar{Q})$. Let H denote the normal subgroup of $\mathrm{Aut}_{\mathcal{F}}(\bar{Q})$ consisting of all automorphisms of $\bar{Q}$ that act trivially on Q. Since all elements of $\mathrm{Aut}_R(\bar{Q})^\psi$ act as elements of $\mathrm{Aut}_R(Q)$ upon restriction to Q, we see that $\mathrm{Aut}_R(Q)^\psi \leq \mathrm{Aut}_R(\bar{Q})H \leq \mathrm{Aut}_P(\bar{Q})H$. As $\mathrm{Aut}_R(\bar{Q})^\psi$ is contained in some Sylow p-subgroup of $\mathrm{Aut}_P(\bar{Q})H$, and $\mathrm{Aut}_P(\bar{Q})$ is another Sylow p-subgroup of $\mathrm{Aut}_P(\bar{Q})H$, there exists $\chi h \in \mathrm{Aut}_P(\bar{Q})H$ (with $\chi \in \mathrm{Aut}_P(\bar{Q})$ and $h \in H$), such that $\mathrm{Aut}_R(\bar{Q})^{\psi\chi h} \leq \mathrm{Aut}_P(\bar{Q})$. The element h normalizes $\mathrm{Aut}_P(\bar{Q})$, and so $\mathrm{Aut}_R(\bar{Q})^{\psi\chi} \leq \mathrm{Aut}_P(\bar{Q})$ and $\psi\chi$ extends ϕ. Hence $N_{\psi\chi}$ contains $R\bar{Q}$, so ϕ extends to a morphism with domain $R\bar{Q}$. The restriction of this map to R provides an extension of ϕ to R, which as we mentioned earlier must be an automorphism. This proves that Q has the surjectivity property. This will complete the proof of (ii) via Lemmas 6.7 and 6.8.

It remains to show that every automorphism ϕ extends to $\bar{Q}$. Since every $\mathcal{F}$-automorphism of Q is a composition of morphisms between subgroups properly containing Q, we may assume that $\phi : R \to \tilde{R}$ is an $\mathcal{F}$-map between two overgroups of Q that restricts to an automorphism of Q. If $\tilde{R}$ is not fully normalized we may choose $\psi : \tilde{R} \to \bar{R}$ with $\bar{R}$ fully normalized, and consider the maps $\phi\psi$ and ψ separately. If both of these extend to automorphisms of $\bar{Q}$, then so does ϕ, so we may assume that $\tilde{R}$ is fully normalized. Since $\bar{Q}$ is strongly $\mathcal{F}$-closed, $|R \cap \bar{Q}| = |\tilde{R} \cap \bar{Q}|$. We proceed by induction on the index $n = |\bar{Q} : R \cap \bar{Q}|$. If $n = 1$, then ϕ does indeed extend to $\bar{Q}$.

Let K denote the subgroup of $\mathrm{Aut}_{\mathcal{F}}(\tilde{R})$ of all automorphisms that act trivially on Q, and let L denote the subgroup of $\mathrm{Aut}_{\mathcal{F}}(\tilde{R})$ of all automorphisms that restrict to elements of $O_p(\mathrm{Aut}_{\mathcal{F}}(Q))$ on Q. Since L/K can be embedded inside $O_p(\mathrm{Aut}_{\mathcal{F}}(Q))$, it is a p-group, and so any two Sylow

p-subgroups of L are conjugate by an element of K. Consider the subgroup $\mathrm{Aut}_{\bar{Q}}(R)^\phi$; by definition, $\bar{Q}c_Q \leq O_p(\mathrm{Aut}_{\mathcal{F}}(Q))$, and so $\mathrm{Aut}_{\bar{Q}}(R)^\phi$ is a p-subgroup of K. Since $\tilde{R}$ is fully normalized, it is receptive and fully automized, and so $\mathrm{Aut}_P(\tilde{R})$ is a Sylow p-subgroup of $\mathrm{Aut}_{\mathcal{F}}(\tilde{R})$; intersecting $\mathrm{Aut}_P(\tilde{R})$ with K yields a Sylow p-subgroup of K. Therefore there is some automorphism $\psi \in L$ such that ψ conjugates $\mathrm{Aut}_{\bar{Q}}(R)^\phi$ inside $\mathrm{Aut}_P(\tilde{R})$, and hence $N_{\phi\psi}$ contains the preimage of $\mathrm{Aut}_{\bar{Q}}(R)$ in P. Therefore $\mathrm{N}_{\bar{Q}R}(R) \leq N_{\phi\psi}$. In addition, $\mathrm{N}_{\bar{Q}R}(R)$ is generated by R and $\mathrm{N}_{\bar{Q}}(R)$. (To see this, notice that if G is any finite group and $G = AB$, with B normal, and $C \leq G$ contains A, then $C = A(B \cap C)$; in this case, $\mathrm{N}_{\bar{Q}}(R) = \bar{Q} \cap \mathrm{N}_{\bar{Q}R}(R)$.) As $\mathrm{N}_{\bar{Q}R}(R)$ strictly contains R, we must have that $\mathrm{N}_{\bar{Q}}(R)$ strictly contains $R \cap \bar{Q}$.

Finally, as $\psi \in L$, it acts trivially on Q, and so both ϕ and $\phi\psi$ are extensions of the same automorphism of Q. As $N_{\phi\psi}$ contains $\mathrm{N}_{\bar{Q}}(R) > R \cap \bar{Q}$, there is an extension of $\phi|_Q$ to an automorphism of $\bar{Q}$ by induction, completing the proof.   □

We now summarize the proof of Theorem 6.11: let $\mathcal{F}$ be a fusion system, and let $\mathcal{H}$ be a union of n_0 conjugacy classes such as is given in the hypothesis of the theorem; let $\mathcal{H}'$ be a union of $\mathcal{F}$-conjugacy classes containing $\mathcal{H}$. We proceed by induction on the number n of $\mathcal{F}$-conjugacy classes present in $\mathcal{H}'$. By hypothesis, $\mathcal{F}$ is $\mathcal{H}$-generated, and all $\mathcal{F}$-conjugacy classes in $\mathcal{H}$ are saturated, establishing the case $n = n_0$. Let $\mathcal{Q}$ be an $\mathcal{F}$-conjugacy class of subgroups of P of smallest index subject to not being in $\mathcal{H}'$. By Proposition 6.14, $\mathcal{Q}$ is saturated if, for some fully normalized subgroup $Q \in \mathcal{Q}$, $\{Q\}$ is saturated in $\mathrm{N}_{\mathcal{F}}(Q)$. Lastly, by Proposition 6.15, if either Q is not $\mathcal{F}$-centric, or Q is $\mathcal{F}$-centric and $\mathrm{Out}_P(Q) \cap O_p(\mathrm{Out}_{\mathcal{F}}(Q))$ is non-trivial, then $\{Q\}$ is saturated in $\mathrm{N}_{\mathcal{F}}(Q)$, but by hypothesis on $\mathcal{H}$ and Exercise 6.2 there is some such $Q \in \mathcal{Q}$ that is fully normalized. Hence $\mathcal{Q}$ is saturated, completing the inductive step; thus every $\mathcal{F}$-conjugacy class of subgroups of P is saturated, and so $\mathcal{F}$ is saturated, as claimed.

Combining the saturation theorems from this chapter yields even better results. In particular, we will combine Theorems 6.9 and 6.11, together with Proposition 6.14. In order to get this, we actually have to assume something slightly stronger than $\mathcal{H}$-generation, namely that it is generated by *automorphisms* of subgroups in $\mathcal{H}$. (Of course, by Alperin's fusion theorem, as long as $\mathcal{H}$ contains all centric radical subgroups then this is true for any saturated fusion system.)

Theorem 6.16 (Craven [Cra10b, Theorem 3.6]) *Let $\mathcal{F}$ be a fusion system on a finite p-group P with P fully automized, and let $\mathcal{H}$ be the set of all $\mathcal{F}$-centric subgroups. If $\mathcal{F}$ is generated by $\{\mathrm{Aut}_{\mathcal{F}}(U) : U \in \mathcal{H}\}$, and if in every $\mathcal{F}$-conjugacy class of centric subgroups of P there is a fully normalized subgroup with the surjectivity property, then $\mathcal{F}$ is saturated.*

Proof Let $\mathcal{H}'$ denote the union of all saturated $\mathcal{F}$-conjugacy classes of subgroups of P, and suppose that $\mathcal{Q}$ is a conjugacy class of subgroups of P of smallest index subject to not being in $\mathcal{H}'$. Let $\mathcal{E}$ be the subsystem of $\mathcal{F}$ generated by $\mathrm{Aut}_{\mathcal{F}}(R)$, where R runs over all subgroups of P of larger order than the subgroups in $\mathcal{Q}$. Notice that $\mathrm{Hom}_{\mathcal{F}}(R, S) = \mathrm{Hom}_{\mathcal{E}}(R, S)$ if R has larger order than the subgroups in $\mathcal{Q}$, and so such $\mathcal{E}$-conjugacy classes are identical to the $\mathcal{F}$-conjugacy classes, and hence are also saturated. In addition, $\mathcal{Q}$ forms a single $\mathcal{E}$-conjugacy class; in particular, a subgroup of $\mathcal{Q}$ is $\mathcal{F}$-centric if and only if it is $\mathcal{E}$-centric.

If $\mathcal{Q}$ consists of subgroups that are not $\mathcal{E}$-centric, then by Propositions 6.14(ii) and 6.15(ii), $\mathcal{Q}$ is saturated in $\mathcal{E}$. Thus we may suppose that $\mathcal{Q}$ consists of $\mathcal{F}$-centric subgroups. Let Q be a fully $\mathcal{F}$-normalized subgroup in $\mathcal{Q}$ with the surjectivity property, guaranteed by hypothesis. Proposition 6.14(i), applied to Q and $\mathcal{E}$, is exactly the first requirement for Theorem 6.9, and so, since Q has the surjectivity property, $\mathcal{Q}$ is saturated, as required. □

Of course, we can reduce the subset $\mathcal{H}$ still further, along the lines of Theorem 6.11, but it is Theorem 6.16 that will be most useful, especially when we have an $\mathcal{F}$-invariant subsystem $\mathcal{E}$ and we want to show that $\mathcal{E}$ is saturated. Theorem 6.16 is a vital tool in this chapter: it reduces checking saturation to only considering the automorphism groups $\mathrm{Aut}_{\mathcal{F}}(U)$ of some subgroups U, and their interaction with each other (the surjectivity property).

The next stage is to 'externalize' the property of saturation. We have a condition on saturation that only deals with automorphism groups. In the next section we will consider functions that assign to each subgroup U a subgroup of $\mathrm{Aut}(U)$, which is meant to be $\mathrm{Aut}_{\mathcal{F}}(U)$ for some fusion system $\mathcal{F}$. It is this idea that leads to invariant maps.

6.3 Invariant maps

Our quest for a nice way of checking saturation begins by producing – easily – examples of $\mathcal{F}$-invariant subsystems. Recall that an invariant subsystem $\mathcal{E}$ of $\mathcal{F}$ on a subgroup Q of P is one where the 'conjugate' of a morphism in $\mathcal{E}$ by a morphism in $\mathcal{F}$ remains in $\mathcal{E}$. In Theorem 5.43 we proved that this was equivalent to both $\mathrm{Aut}_{\mathcal{F}}(Q) \leq \mathrm{Aut}(\mathcal{E})$, and every morphism in $\mathcal{F}$ between subgroups of Q being expressible as a product of an element of $\mathrm{Aut}_{\mathcal{F}}(Q)$ and a morphism in $\mathcal{E}$. Because we will eventually be interested only in *saturated* subsystems, we will assume that our subsystem $\mathcal{E}$ is generated by automorphisms. While this leaves out some $\mathcal{F}$-invariant subsystems, we will still get enough of the subsystems we are interested in that we will not worry about the fact that there are $\mathcal{F}$-invariant subsystems that are not generated by automorphisms.

Theorem 6.9 gives an equivalent condition for a subsystem $\mathcal{E}$ to be saturated, one part of which is the surjectivity property. Theorem 6.16 is the important result in this direction; if $\mathcal{E}$ is $\mathcal{F}$-invariant, then we will be able to use this theorem to get a good set of conditions for saturation.

In this section we want to assign to each subgroup U of a strongly $\mathcal{F}$-closed subgroup Q of P a subgroup $A(U)$ of $\mathrm{Aut}_{\mathcal{F}}(U)$, which we will pretend is the group $\mathrm{Aut}_{\mathcal{E}}(U)$, for some $\mathcal{F}$-invariant subsystem $\mathcal{E}$ of a fusion system $\mathcal{F}$. One obvious way to define a subsystem is to take a collection of morphisms, for example automorphisms of various subgroups, and take the fusion system generated by them. For saturated subsystems we know that this can be done, and it is the content of Alperin's fusion theorem. An invariant map is simply a function that takes each subgroup of a given strongly closed subgroup to a collection of automorphisms of it; formally, we have the following definition.

Definition 6.17 (Aschbacher [Asc08a]) Let $\mathcal{F}$ be a saturated fusion system on a finite p-group P, and let Q be a strongly $\mathcal{F}$-closed subgroup of P. An *$\mathcal{F}$-invariant map* on Q is a function $A(-)$ on the set of subgroups R of Q, with $A(R) \leq \mathrm{Aut}_{\mathcal{F}}(R)$, such that

(i) if $R \leq Q$ and ϕ is an $\mathcal{F}$-isomorphism whose domain is R, then $A(R\phi) = A(R)^{\phi}$, and

(ii) if R is fully $\mathcal{F}$-normalized, then $\mathrm{Aut}_Q(R) \leq A(R)$.

If $A(-)$ is an invariant map, the fusion system *generated by* $A(-)$ is the subsystem on Q generated by the subgroups $A(R)$ for all $R \leq Q$.

Notice that, by the first property, $A(R) \trianglelefteq \mathrm{Aut}_{\mathcal{F}}(R)$ for all R. To produce an invariant map, it suffices to define a normal subgroup $A(R)$, containing $\mathrm{Aut}_Q(R)$, for each fully $\mathcal{F}$-normalized subgroup R of Q, and to check that if R and S are fully $\mathcal{F}$-normalized subgroups with $\phi : R \to S$ an isomorphism, then $A(R)^{\phi} = A(S)$. We can then extend the definition to all subgroups of Q in the only way consistent with the first property of the definition. In other words, we have the following result.

Lemma 6.18 *Let $\mathcal{F}$ be a saturated fusion system on a finite p-group P, and let Q be a strongly $\mathcal{F}$-closed subgroup of P. Let $A'(-)$ be a function defined on the set of fully $\mathcal{F}$-normalized subgroups R of Q, with $A'(R) \trianglelefteq \mathrm{Aut}_{\mathcal{F}}(R)$, $\mathrm{Aut}_Q(R) \leq A'(R)$, and if $\phi : R \to S$ is some $\mathcal{F}$-isomorphism between fully $\mathcal{F}$-normalized subgroups then $A'(R)^\phi = A'(S)$. There is a unique invariant map $A(-)$ such that $A(R) = A'(R)$ for all fully $\mathcal{F}$-normalized subgroups $R \leq Q$.*

The reason for invariant maps being called such is the following proposition.

Proposition 6.19 (Aschbacher [Asc08a, Lemma 5.5]) *Let $\mathcal{F}$ be a saturated fusion system on a finite p-group P, and let Q be a strongly $\mathcal{F}$-closed subgroup of P.*

(i) *If $\mathcal{E}$ is an $\mathcal{F}$-invariant subsystem on Q, then the map $A(-)$ sending R to $\mathrm{Aut}_{\mathcal{E}}(R)$, for each $R \leq Q$, is an invariant map.*

(ii) *If $A(-)$ is an invariant map on Q then the subsystem generated by $A(-)$ is an $\mathcal{F}$-invariant subsystem.*

Proof Let $\mathcal{E}$ be an $\mathcal{F}$-invariant subsystem on Q, let R and S be subgroups of Q, and let ϕ be an $\mathcal{F}$-isomorphism from R to S; set $A(R) = \mathrm{Aut}_{\mathcal{E}}(R)$. Since $\mathcal{E}$ is $\mathcal{F}$-Frattini, ϕ may be written as $\alpha\beta$, with $\alpha \in \mathrm{Aut}_{\mathcal{F}}(Q) \leq \mathrm{Aut}(\mathcal{E})$ and β an $\mathcal{E}$-morphism. As α induces an automorphism on $\mathcal{E}$, $\mathrm{Aut}_{\mathcal{E}}(R\alpha) = \mathrm{Aut}_{\mathcal{E}}(R)^\alpha$. Proposition 4.6 now proves that $\mathrm{Aut}_{\mathcal{E}}(R\alpha)^\beta = \mathrm{Aut}_{\mathcal{E}}(R\alpha\beta)$, proving (i).

For the converse, suppose that $A(-)$ is an invariant map, and let $\mathcal{E}$ be the subsystem generated by $A(-)$. If $\phi \in \mathrm{Aut}_{\mathcal{F}}(Q)$, then ϕ permutes the subgroups $A(R)$ for the various subgroups $R \leq Q$, and so must fix $\mathcal{E}$. Therefore $\mathrm{Aut}_{\mathcal{F}}(Q) \leq \mathrm{Aut}(\mathcal{E})$.

We next show that $\mathcal{E}$ is $\mathcal{F}$-Frattini, proving that $\mathcal{E}$ is $\mathcal{F}$-invariant by Theorem 5.43. As in the proof of Lemma 5.42, we say that $\phi \in \mathrm{Hom}_{\mathcal{F}}(R, Q)$ has an $\alpha\beta$-decomposition if $\phi = \alpha\beta$, with $\alpha \in \mathrm{Aut}_{\mathcal{F}}(Q)$ and β a morphism in $\mathcal{E}$. In Step 1 of the proof of Lemma 5.42 we showed that compositions and inverses of morphisms with $\alpha\beta$-decompositions have $\alpha\beta$ decompositions. Let $\mathcal{H}(R)$ denote the subset of $\mathrm{Hom}_{\mathcal{F}}(R, Q)$ of all morphisms with an $\alpha\beta$-decomposition. By the previous statements, $\mathcal{H}(R)$ is the set of all $\phi \in \mathrm{Hom}_{\mathcal{F}}(R, Q)$ such that $\phi = \alpha_1\beta_1 \ldots \alpha_r\beta_r$, with each α_i in $\mathrm{Aut}_{\mathcal{F}}(Q)$ and each β_i in $\mathcal{E}$. Clearly $\mathcal{H}(Q) = \mathrm{Hom}_{\mathcal{F}}(Q, Q)$;

choose R to be of smallest index in Q such that $\mathcal{H}(R) \neq \operatorname{Hom}_{\mathcal{F}}(R, Q)$, and let ϕ be an element of $\operatorname{Hom}_{\mathcal{F}}(R, Q) \setminus \mathcal{H}(R)$. Notice that if $S > R$ and $\phi \in \mathcal{H}(S)$ then $\phi|_R \in \mathcal{H}(R)$.

By Alperin's fusion theorem, ϕ is the composition of restrictions of automorphisms ϕ_i of subgroups U_i. As Q is strongly closed, the restriction of ϕ_i to $U_i \cap Q$ induces an automorphism of $U_i \cap Q$, and therefore we may assume that $U_i \leq Q$. If $|U_i| > |R|$ then by choice of R, $\phi_i \in \mathcal{H}(U_i)$, and so for ϕ to not be in $\mathcal{H}(R)$ for (at least) one of the U_i, $|U_i| = |R|$, so we reduce to the case where $\phi \in \operatorname{Aut}_{\mathcal{F}}(R)$.

Let $\psi : R \to S$ be an isomorphism in $\mathcal{F}$ with S fully $\mathcal{F}$-normalized. We claim that there is $\chi \in A(S)$ such that $N_{\psi\chi}$ contains $\mathrm{N}_Q(R)$. To see this, since S is fully $\mathcal{F}$-automized and $A(S) \trianglelefteq \operatorname{Aut}_{\mathcal{F}}(S)$, we see that $\operatorname{Aut}_P(S) \cap A(S)$ is a Sylow p-subgroup of $A(S)$. Also, $A(R)^\psi = A(S)$, and so there is $\chi \in A(S)$ such that $(\operatorname{Aut}_Q(R)^\psi)^\chi \leq \operatorname{Aut}_P(S)$. Hence $N_{\psi\chi}$ contains $\mathrm{N}_Q(R)$, as claimed.

We will apply this claim also to the map $\phi\psi$, to get $\chi' \in A(S)$ such that $N_{\phi\psi\chi'}$ contains $\mathrm{N}_Q(R)$. Since S is fully $\mathcal{F}$-normalized, both $\psi\chi$ and $\phi\psi\chi'$ extend to $\mathrm{N}_Q(R)$, and so, by our choice of R, we see that $\psi\chi$ and $\phi\psi\chi'$ lie in $\mathcal{H}(R)$, whence so does

$$\phi = (\phi\psi\chi')(\chi'^{-1}\chi)(\psi\chi)^{-1};$$

this proves that in fact $\mathcal{H}(R) = \operatorname{Hom}_{\mathcal{F}}(R, Q)$, contradicting our choice of R. Hence $\mathcal{H}(R) = \operatorname{Hom}_{\mathcal{F}}(R, Q)$ for all $R \leq Q$, so that every morphism in $\operatorname{Hom}_{\mathcal{F}}(R, Q)$ has an $\alpha\beta$-decomposition. Thus $\mathcal{E}$ is $\mathcal{F}$-Frattini, as required. $\square$

Importantly, we are not saying that $\operatorname{Aut}_{\mathcal{E}}(R) = A(R)$, but of course $A(R) \leq \operatorname{Aut}_{\mathcal{E}}(R)$; equality is guaranteed only for Q itself, an easy exercise. However, there are certain invariant maps, like the one attached to any $\mathcal{F}$-invariant subsystem in the previous proposition, for which this is true.

Another point is that not every $\mathcal{F}$-invariant subsystem is the subsystem generated by an invariant map. Clearly there are fusion systems that are not generated by the automorphisms of their subgroups, but there are even $\mathcal{F}$-invariant subsystems $\mathcal{E}$ of *saturated* fusion systems $\mathcal{F}$ such that $\operatorname{Aut}_{\mathcal{E}}(U)$ is a p-group for all subgroups U (so that the subsystem generated by the invariant map $A(U) = \operatorname{Aut}_{\mathcal{E}}(U)$ is simply $\mathcal{F}_Q(Q)$) but $\mathcal{E}$ itself is not $\mathcal{F}_Q(Q)$. For example, consider the group $P \cong V_4$ with fusion system that of the alternating group A_4, and let $\mathcal{E}$ be the subsystem consisting of all morphisms between subgroups of order 2, but no non-identity automorphisms of V_4 itself. We see that $\mathcal{E}$ is clearly $\mathcal{F}$-invariant, but $\operatorname{Aut}_{\mathcal{E}}(U)$ is the trivial group for all $U \leq P$. If an $\mathcal{F}$-invariant subsystem $\mathcal{E}$ is not generated by automorphisms then, if one takes the $\mathcal{F}$-invariant map $U \mapsto \operatorname{Aut}_{\mathcal{E}}(U)$, and then forms the subsystem $\mathcal{E}'$ generated by this, we have $\mathcal{E}' \neq \mathcal{E}$.

Theorem 6.16 proved that there is a way to determine whether a subsystem is saturated just by examining the automorphism groups, and when this is applied to $\mathcal{F}$-invariant subsystems of a saturated fusion system, this will produce a relatively simple criterion for an invariant map to produce a saturated (and hence weakly normal) subsystem.

6.4 Weakly normal maps

In general, $\mathcal{F}$-invariant subsystems are not very interesting: hence invariant maps are not the ideal object of study. We want to impose extra conditions on the subgroups $A(U)$ in order that the subsystem generated by the map is also saturated (and hence a weakly normal subsystem).

Proposition 6.20 *Let $\mathcal{F}$ be a saturated fusion system on a finite p-group P, and let $\mathcal{E}$ be an $\mathcal{F}$-invariant subsystem on a strongly $\mathcal{F}$-closed subgroup Q of P. A subgroup $R \leq Q$ is $\mathcal{E}$-centric if and only if R is $\mathcal{F}$-conjugate to a fully $\mathcal{F}$-normalized subgroup S with $\mathrm{C}_Q(S) \leq S$.*

Proof See Exercise 6.3. $\qquad\qquad\qquad\qquad\qquad\qquad\qquad\qquad\Box$

Theorem 6.16 in fact tells us what we want for an invariant map to be weakly normal. This definition appears in [Cra10b], and is an extension to weakly normal subsystems of a definition of Aschbacher of a normal map, which we shall meet in Chapter 8.

Definition 6.21 Let $\mathcal{F}$ be a saturated fusion system on a finite p-group P, and let Q be a strongly $\mathcal{F}$-closed subgroup of P. A *weakly normal map* on Q is a function $A(-)$ on the set of subgroups U of Q, with $A(U) \leq \operatorname{Aut}_{\mathcal{F}}(U)$, such that

  (i) if $U \leq Q$ and ϕ is an $\mathcal{F}$-isomorphism whose domain is U, then $A(U\phi) = A(U)^{\phi}$;
 (ii) if U is fully $\mathcal{F}$-normalized, then $\operatorname{Aut}_Q(U) \leq A(U)$;
(iii) $\operatorname{Aut}_Q(Q)$ is a Sylow p-subgroup of $A(Q)$;
 (iv) if U is fully $\mathcal{F}$-normalized, then every element of $A(U)$ extends to an element of $A(U\, \mathrm{C}_Q(U))$; and

(v) if U is fully $\mathcal{F}$-normalized and $C_Q(U) \leq U$ then, for any subgroup $U \leq V \leq N_Q(U)$, denoting by $A(U \leq V)$ the set of automorphisms of V in $A(V)$ that restrict to automorphisms of U, the restriction map

$$A(U \leq V) \rightarrow N_{A(U)}(\mathrm{Aut}_V(U))$$

is surjective.

The idea is that if $\mathcal{E}$ is the subsystem generated by the morphisms in $A(U)$ for all $U \leq Q$, then $\mathrm{Aut}_\mathcal{E}(U) = A(U)$. We demonstrated in the previous section that this is not true for general invariant maps, and so we are genuinely improving things here.

We will make some short remarks about this definition now. The first two conditions are simply that $A(-)$ is an invariant map. The third condition is obviously necessary for $\mathcal{E} = \langle A(U) : U \leq Q \rangle$ to be saturated, and the fifth condition is simply the surjectivity property for fully $\mathcal{F}$-normalized, $\mathcal{E}$-centric subgroups.

The fourth condition is there to make sure that $\mathcal{E}$ is actually generated by the automorphisms of centric subgroups: since every automorphism of a fully $\mathcal{F}$-normalized subgroup U extends to an automorphism of the $\mathcal{E}$-centric subgroup $U\,C_Q(U)$, we really do get that $\mathcal{E}$ is generated by automorphisms of $\mathcal{E}$-centric subgroups, the last of the conditions of Theorem 6.16.

As with invariant maps (Lemma 6.18), we have the following reduction.

Lemma 6.22 *Let $\mathcal{F}$ be a saturated fusion system on a finite p-group P, and let Q be a strongly $\mathcal{F}$-closed subgroup of P. Let $A'(-)$ be a function defined on all fully $\mathcal{F}$-normalized subgroups U of Q with $C_Q(U) \leq U$. Suppose that $A'(U)$ is a normal subgroup of $\mathrm{Aut}_\mathcal{F}(U)$, containing $\mathrm{Aut}_T(U)$, such that, if $\phi : U \rightarrow V$ is any isomorphism in $\mathcal{F}$ with V fully $\mathcal{F}$-normalized (and $C_Q(V) \leq V$), then $A'(U)^\phi = A'(V)$. There exists a unique invariant map $A(-)$ such that $A'(U) = A(U)$ for all fully $\mathcal{F}$-normalized subgroups U of Q with $C_Q(U) \leq U$, and such that condition (iv) of being a weakly normal map is satisfied.*

Such a map $A(-)$ satisfies conditions (i), (ii), and (iv) of being a weakly normal map.

This lemma is easy to show: because of condition (iv), we need only define a weakly normal map on fully $\mathcal{F}$-normalized subgroups U with $C_Q(U) \leq U$, and then use (i) to extend to all $\mathcal{F}$-conjugates of such subgroups, and then (iv) to extend to all fully $\mathcal{F}$-normalized subgroups, then (i) again to extend to all subgroups of Q.

Having discussed the reasons behind the definition of a weakly normal map, we now prove the claimed result. This is very similar to the corresponding statement for normal subsystems in [Asc08a], due to Aschbacher, but the proof in [Asc08a] does not extend to weakly normal subsystems, and so in [Cra10b] a new proof was given that does extend.

Theorem 6.23 (Aschbacher, Craven [Cra10b]) *Let $\mathcal{F}$ be a saturated fusion system on a finite p-group P, and let Q be a strongly $\mathcal{F}$-closed subgroup of P. If $A(-)$ is a weakly normal map on Q, then the subsystem $\mathcal{E}$ generated by A is a weakly normal subsystem of $\mathcal{F}$ on Q, with $\mathrm{Aut}_{\mathcal{E}}(R) = A(R)$ for all subgroups R of Q.*

Proof Since $A(-)$ is an invariant map, $\mathcal{E}$ is $\mathcal{F}$-invariant, so we need to prove that $\mathcal{E}$ is saturated. As $\mathrm{Aut}_{\mathcal{E}}(Q) = A(Q)$ for all invariant maps, $\mathrm{Aut}_Q(Q)$ is a Sylow p-subgroup of $\mathrm{Aut}_{\mathcal{E}}(Q)$. Let $\mathcal{H}$ denote the set of $\mathcal{E}$-centric subgroups of Q.

Step 1: *$\mathcal{E}$ is generated by automorphisms of elements of $\mathcal{H}$.* Since $\mathcal{E}$ is generated by the subgroups $A(R)$ for $R \leq Q$, it suffices to show that each element of these is a product of (the restriction of) elements of $A(U)$, for various $\mathcal{E}$-centric subgroups U of Q. If R is fully $\mathcal{F}$-normalized then this is true by condition (iv) of being a weakly normal map. If S is $\mathcal{F}$-conjugate to R via ϕ, then $\phi = \alpha\beta$, with $\alpha \in \mathrm{Aut}_{\mathcal{F}}(Q) \leq \mathrm{Aut}(\mathcal{E})$ and β a morphism in $\mathcal{E}$ (as $\mathcal{E}$ is $\mathcal{F}$-Frattini). Since every element of $A(R)$ is the restriction of an automorphism of some member of $\mathcal{H}$, so is every element of $A(R)^\alpha = A(R\alpha)$. Finally, if β is an automorphism of $R\alpha$ (so that $R\alpha = S$) then we are done; otherwise $R\alpha$ and S are $\mathcal{E}$-conjugate via β, and hence β is a composition of the restriction of automorphisms of subgroups of Q of larger order than $|R|$, which by induction are generated by automorphisms of elements of $\mathcal{H}$. Therefore $A(S) = A(R)^{\beta^{-1}}$ is generated by automorphisms of elements of $\mathcal{H}$, completing the proof of this step.

Let $\mathcal{R}$ be an $\mathcal{E}$-conjugacy class contained in $\mathcal{H}$, and suppose that it consists of subgroups of Q of smallest index subject to the elements R of $\mathcal{R}$ not satisfying $A(R) = \mathrm{Aut}_{\mathcal{E}}(R)$. Since $\mathrm{Aut}_{\mathcal{E}}(Q) = A(Q)$, $\mathcal{R}$ consists of proper subgroups of Q. By condition (iv), all subgroups in $\mathcal{H}$ of larger order than $|R|$ lie in saturated $\mathcal{E}$-conjugacy classes. Applying an automorphism of $\mathcal{E}$ if necessary (and using Lemma 6.4 and Exercise

5.13(ii)) we may assume that $\mathcal{R}$ contains a fully $\mathcal{F}$-normalized subgroup R of P.

Step 2: $A(R) = \mathrm{Aut}_{\mathcal{E}}(R)$. Consider the subsystem $\mathrm{N}_{\mathcal{E}}(R)$, and notice that $\mathrm{Aut}_{\mathcal{E}}(R) = \mathrm{Aut}_{\mathrm{N}_{\mathcal{E}}(R)}(R)$. Since R is fully $\mathcal{F}$-normalized, by Proposition 6.15(i) every element of $\mathrm{Aut}_{\mathcal{E}}(R)$ can be written as the product of the restriction of morphisms between subgroups of $\mathrm{N}_Q(R)$ strictly containing R, and all such $\mathrm{N}_{\mathcal{E}}(R)$-conjugacy classes of subgroups of Q are saturated. We first prove that each of these maps $\phi : S \to T$ may be written as the composition of automorphisms of overgroups of R in $\mathrm{N}_{\mathcal{E}}(R)$, proceeding by induction on $n = |\mathrm{N}_Q(R) : S|$, the case $n = 1$ being clear.

Let $\psi : T \to \tilde{T}$ be an isomorphism with $\tilde{T}$ fully $\mathrm{N}_{\mathcal{E}}(Q)$-normalized. If the result holds for $\phi\psi$ and ψ then it holds for $\phi = \phi\psi(\psi^{-1})$, so we must show it for morphisms whose image is fully $\mathrm{N}_{\mathcal{E}}(Q)$-normalized; hence we assume that T is so. Since the $\mathrm{N}_{\mathcal{E}}(Q)$-conjugacy class containing S is saturated and T is fully normalized, it is fully automized and so there is a map $\theta : S \to T$ in $\mathrm{N}_{\mathcal{E}}(Q)$ that extends to a map $\bar{\theta} : \mathrm{N}_Q(R,S) \to \mathrm{N}_Q(R,T)$. By induction $\bar{\theta}$, and hence θ, is the restriction of automorphisms of elements of $\mathrm{N}_Q(R)$, and so ϕ can be expressed in such a way if and only if $\theta\phi^{-1}$ can, which is an automorphism of a subgroup of $\mathrm{N}_Q(R)$ properly containing R, as claimed.

Now let ϕ be an automorphism in $\mathrm{Aut}_{\mathcal{E}}(R)$. If ϕ does not lie in $A(R)$, then it must be the composition of restrictions of automorphisms ψ_i of subgroups S_i of $\mathrm{N}_Q(R)$ strictly containing R by the previous paragraph; it suffices to check the case where ϕ is the restriction of a single automorphism of some overgroup S. By choice of $\mathcal{R}$ we have that $A(S) = \mathrm{Aut}_{\mathcal{E}}(S)$. However, by condition (v), and the fact that $\psi \in A(R \leq S)$, we see that $\phi \in A(R)$, completing the proof.

Step 3: *In any $\mathcal{E}$-conjugacy class of $\mathcal{E}$-centric subgroups of Q, there is a fully $\mathcal{E}$-normalized subgroup with the surjectivity property.* If R is a fully $\mathcal{F}$-normalized and $\mathcal{E}$-centric subgroup of Q, then condition (v) of being a weakly normal map implies that R has the surjectivity property. By Lemma 6.4, if $\alpha \in \mathrm{Aut}(\mathcal{E})$ then $R\alpha$ also has the surjectivity property, and by Exercise 5.13(ii) there is such a subgroup in every $\mathcal{E}$-conjugacy class of $\mathcal{E}$-centric subgroups of Q.

In particular, we now know that $\mathcal{E}$ is saturated by Theorem 6.16.

Step 4: $\mathrm{Aut}_{\mathcal{E}}(R) = A(R)$ *for all $R \leq Q$.* If R is fully $\mathcal{F}$-normalized then this follows from Step 1 and condition (iv) of being a weakly normal map. Since $\mathcal{E}$ is $\mathcal{F}$-invariant, if $\phi : R \to S$ is an $\mathcal{F}$-isomorphism with S fully $\mathcal{F}$-normalized then $\mathrm{Aut}_{\mathcal{E}}(R)^{\phi} = \mathrm{Aut}_{\mathcal{E}}(S)$, and since $A(R)^{\phi} = A(S)$ as $A(-)$ is an invariant map, we are done. $\qquad\square$

As an example, we show that the condition that $\mathcal{E}$ be generated by automorphisms of centric subgroups is necessary.

Example 6.24 Let P be the group V_4, and let $\mathcal{F}$ be the fusion system of the alternating group A_4. Let $\mathcal{E}$ be the subsystem of $\mathcal{F}$, on P, given by all $\mathcal{F}$-maps between subgroups of order 2, but not their extensions to P. Hence $\mathrm{Aut}_{\mathcal{E}}(Q)$ is trivial for all $Q \leq P$, and so every subgroup Q has the surjectivity property. It is also clear that $\mathcal{E}$ is $\mathcal{F}$-invariant, and that $\mathrm{Aut}_P(P)$ is a Sylow 2-subgroup of $\mathrm{Aut}_{\mathcal{E}}(P)$. However, $\mathcal{E}$ is obviously not saturated, since for example it does not satisfy the conclusion of Alperin's fusion theorem. (Also, all maps in $\mathcal{E}$ should extend to automorphisms of P, since P is abelian.)

Hence there are $\mathcal{F}$-invariant subsystems $\mathcal{E}$ on a subgroup Q of P with $\mathrm{Aut}_Q(Q)$ fully $\mathcal{E}$-automized, and with all subgroups having the surjectivity property, that are not saturated.

We include an application of this theory, proving the existence of a subsystem like the intersection in certain situations.

Theorem 6.25 (Craven [Cra10b, Theorem 8.1]) *Let $\mathcal{F}$ be a saturated fusion system on a finite p-group P, and let $\mathcal{E}_1$ and $\mathcal{E}_2$ be weakly normal subsystems of $\mathcal{F}$, both on P. There exists a weakly normal subsystem $\mathcal{E}$, contained in $\mathcal{E}_1 \cap \mathcal{E}_2$, such that, for any fully $\mathcal{F}$-normalized, $\mathcal{F}$-centric subgroup Q of P, we have*

$$\mathrm{Aut}_{\mathcal{E}}(Q) = \mathrm{Aut}_{\mathcal{E}_1}(Q) \cap \mathrm{Aut}_{\mathcal{E}_2}(Q).$$

Proof Let $A_1(-)$ and $A_2(-)$ be the weakly normal maps corresponding to $\mathcal{E}_1$ and $\mathcal{E}_2$ respectively, and let $A(-)$ be the map given by $A(Q) = A_1(Q) \cap A_2(Q)$ for Q an $\mathcal{F}$-centric subgroup of P, and then extended to all subgroups via Lemma 6.22. We will show that $A(-)$ is again a weakly normal map; clearly the subsystem $\mathcal{E}$ generated by $A(-)$ will satisfy the conclusions of the theorem. We prove conditions (iii) and (v) in Definition 6.21 now, since (i), (ii) and (iv) follow as in Lemma 6.22.

Clearly, since $\mathrm{Aut}_P(P)$ is a Sylow p-subgroup of both $A_1(P)$ and $A_2(P)$, it is a Sylow p-subgroup of $A(P)$, proving the third condition; it remains to prove the fifth condition. Let R be a fully $\mathcal{F}$-normalized, $\mathcal{F}$-centric subgroup of P, and let S be an overgroup of R contained in $N_P(R)$. Let ϕ be an automorphism in $N_{A(R)}(\mathrm{Aut}_S(R))$; since both $A_i(-)$

are weakly normal maps, there are elements ψ_i in $A_i(S)$ that extend ϕ; by Exercise 6.5, $\psi_1 = c_x \psi_2$ for some $x \in Z(R)$, and since $c_x \in A_i(S)$ for $i = 1, 2$, we see that both ψ_i lie in both $A_i(S)$, whence they lie in $A(S)$, as needed. □

In Section 8.3 we will develop a general theory of intersections of weakly normal subsystems along the lines of Theorem 6.25, but it requires the theory of normal subsystems to get this.

To see why we cannot simply take $\mathcal{E}_1 \cap \mathcal{E}_2$ in Theorem 6.25, consider the following example. (The previous example showing that the intersection of two saturated subsystems need not be saturated, Example 5.34, will not work here, because the subsystems did not lie on P itself.)

Example 6.26 Let G be the group $A_4 \times A_4$, and let P be the Sylow 2-subgroup of G, an elementary abelian group of order 16, generated by a and b in the first factor, and c and d in the second. Let x and y be elements of G of order 3, with both sending a to b, and x sending c to d, and y sending d to c. Let H_1 be the group isomorphic with $A_4 \times V_4$ generated by P and x, and H_2 be the group generated by P and y.

Let $\mathcal{F}$ be the fusion system of G on P, and $\mathcal{E}_i = \mathcal{F}_P(H_i) \leq \mathcal{F}$. Since $H_i \trianglelefteq G$, the $\mathcal{E}_i$ are weakly normal in $\mathcal{F}$, but if $\mathcal{E} = \mathcal{E}_1 \cap \mathcal{E}_2$, then $\operatorname{Aut}_{\mathcal{E}}(P)$ is trivial, but there is a non-trivial automorphism on $Q = \langle a, b \rangle$, which therefore cannot extend to P. Hence $\mathcal{E}$ is not saturated.

This example shows why we should define the weakly normal map only on the fully $\mathcal{F}$-normalized, $\mathcal{F}$-centric subgroups, and then extend it in the unique way, rather than try to define it on all subgroups to begin with, as the subgroup Q above was fully $\mathcal{F}$-normalized, and both $\operatorname{Aut}_{\mathcal{E}_i}(Q)$ were the same, but the 'correct' choice for $\operatorname{Aut}_{\mathcal{E}}(Q)$ was the trivial group.

If the subsystems $\mathcal{E}_i$ do not lie on the same subgroup then this construction does not work well; firstly because the $\mathcal{E}$-centric subgroup R need not be $\mathcal{E}_i$-centric, but even in this case things can go wrong, even in a fusion system $\mathcal{F}_P(P)$.

Example 6.27 Let $P = D_8 \times C_2$, with the D_8 factor generated by an element x of order 4 and y of order 2, and the C_2 factor being generated by z. (This is the group from Example 5.34.) Let $Q = \langle x, y \rangle$, $R = \langle x, yz \rangle$ and $S = \langle x \rangle = Q \cap R$. If $\mathcal{F} = \mathcal{F}_P(P)$, and $\mathcal{E}_1$ and $\mathcal{E}_2$ are the subsystems $\mathcal{F}_Q(Q)$ and $\mathcal{F}_R(R)$ respectively, then both $\mathcal{E}_1$ and $\mathcal{E}_2$ are weakly normal in $\mathcal{F}$. We see that $\operatorname{Aut}_{\mathcal{E}_1}(S) = \operatorname{Aut}_{\mathcal{E}_2}(S)$ has order 2. However, the only saturated subsystem of $\mathcal{F}$ on S is $\mathcal{E} = \mathcal{F}_S(S)$, for which $\operatorname{Aut}_{\mathcal{E}}(S)$

is trivial, and so the 'correct' subsystem we want inside $\mathcal{E}_1 \cap \mathcal{E}_2$ is $\mathcal{E}$ itself. Taking the intersection of the $\mathrm{Aut}_{\mathcal{E}_i}(S)$ therefore does not yield a saturated subsystem.

(Notice that, in this case, S is both fully $\mathcal{F}$-normalized (indeed, it is strongly $\mathcal{F}$-closed) and $\mathcal{E}_i$-centric for $i = 1, 2$, so the problem does not lie in not being $\mathcal{E}_i$-centric.)

In order to produce a subsystem inside the intersection, we will therefore have to do something much more subtle than we did in Theorem 6.25: see Section 8.3.

We end with another application of the theory of this chapter, in proving that the direct product of two saturated fusion systems is a saturated fusion system.

Definition 6.28 Let $\mathcal{F}_1$ and $\mathcal{F}_2$ be fusion systems on the finite p-groups P_1 and P_2 respectively. The *direct product* of $\mathcal{F}_1$ and $\mathcal{F}_2$, denoted $\mathcal{F}_1 \times \mathcal{F}_2$, is the fusion system $\mathcal{F}$ on $P_1 \times P_2$ generated by all morphisms $\phi : (Q_1, Q_2) \to (R_1, R_2)$, where $\phi|_{Q_i} \in \mathrm{Hom}_{\mathcal{F}_i}(Q_i, R_i)$ for $i = 1, 2$.

The direct product of two fusion systems, defined as the fusion system generated by a collection of morphisms, is obviously always a fusion system. If the $\mathcal{F}_i$ are saturated, then the system $\mathcal{F}$ is also saturated. In what follows, if $\phi_i : Q_i \to R_i$ are morphisms in $\mathcal{F}_i$, then by (ϕ_1, ϕ_2) we denote the morphism from $Q_1 \times Q_2$ to $R_1 \times R_2$ that acts like ϕ_i on Q_i.

Theorem 6.29 (Broto–Levi–Oliver [BLO03b, Lemma 1.5]) *Let $\mathcal{F}_1$ and $\mathcal{F}_2$ be fusion systems on the finite p-groups P_1 and P_2 respectively. If $\mathcal{F}_1$ and $\mathcal{F}_2$ are saturated then $\mathcal{F}_1 \times \mathcal{F}_2$ is saturated.*

Proof We will prove that $\mathcal{F} = \mathcal{F}_1 \times \mathcal{F}_2$ is saturated using Theorem 6.16. If Q is a subgroup of $P = P_1 \times P_2$, then denote by Q_i the projection of Q along P_i. Notice that $\mathrm{C}_P(Q) \cong \mathrm{C}_{P_1}(Q_1) \times \mathrm{C}_{P_2}(Q_2)$. Since both $\mathcal{F}_i$ are saturated, they are generated by automorphisms of $\mathcal{F}_i$-centric subgroups. As $\mathcal{F}$ is generated by morphisms (ϕ_1, ϕ_2) with ϕ_i morphisms in $\mathcal{F}_i$, and each ϕ_i is the composition of restrictions of automorphisms of $\mathcal{F}_i$-centric subgroups, we see that every morphism in $\mathcal{F}$ is the composition of restrictions of automorphisms of subgroups $S_1 \times S_2$, where S_i is an $\mathcal{F}_i$-centric subgroup; this subgroup is clearly $\mathcal{F}$-centric, and so $\mathcal{F}$ is generated by automorphisms of $\mathcal{F}$-centric subgroups of P.

Let Q be an $\mathcal{F}$-centric subgroup of P such that each Q_i is fully $\mathcal{F}_i$-normalized, and let $\bar{Q} = Q_1 \times Q_2$. As each Q_i is fully $\mathcal{F}_i$-normalized,

$\mathrm{Aut}_{P_i}(Q_i)$ is a Sylow p-subgroup of $\mathrm{Aut}_{\mathcal{F}_i}(Q_i)$. Clearly we have $\mathrm{N}_P(\bar{Q}) = \mathrm{N}_{P_1}(Q_1) \times \mathrm{N}_{P_2}(Q_2)$, and so $\mathrm{Aut}_P(\bar{Q}) = \mathrm{Aut}_{P_1}(Q_1) \times \mathrm{Aut}_{P_2}(Q_2)$. Also, $\mathrm{Aut}_{\mathcal{F}}(\bar{Q}) = \mathrm{Aut}_{\mathcal{F}_1}(Q_1) \times \mathrm{Aut}_{\mathcal{F}_2}(Q_2)$, and hence $\bar{Q}$ is fully $\mathcal{F}$-automized. In particular, this proves that P is fully automized.

As $\mathrm{Aut}_P(\bar{Q})$ is a Sylow p-subgroup of $\mathrm{Aut}_{\mathcal{F}}(\bar{Q})$, by regarding $\mathrm{Aut}_{\mathcal{F}}(Q)$ as a subgroup of $\mathrm{Aut}_{\mathcal{F}}(\bar{Q})$ there is $\phi \in \mathrm{Aut}_{\mathcal{F}}(\bar{Q})$ such that a Sylow p-subgroup of $\mathrm{Aut}_{\mathcal{F}}(Q\phi)$ is contained in $\mathrm{Aut}_P(\bar{Q})$. This choice of ϕ makes $Q\phi$ fully $\mathcal{F}$-automized; to see this, we have

$$\mathrm{Aut}_P(Q\phi) = \mathrm{Aut}_P(\bar{Q}) \cap \mathrm{Aut}_{\mathcal{F}}(Q\phi) \in \mathrm{Syl}_p(\mathrm{Aut}_{\mathcal{F}}(Q\phi)),$$

by choice of ϕ. Replace Q by $Q\phi$, so that $\mathrm{Aut}_P(Q) \leq \mathrm{Aut}_P(\bar{Q})$; notice that the Q_i remain the same.

We will show that Q is fully $\mathcal{F}$-normalized and has the surjectivity property. Let R be a subgroup of $\mathrm{N}_P(Q)$ containing Q, let ϕ be an $\mathcal{F}$-automorphism of Q such that $\mathrm{Aut}_R(Q)^\phi = \mathrm{Aut}_R(Q)$, and let $\phi_i \in \mathrm{Aut}_{\mathcal{F}_i}(Q_i)$ be the projection of ϕ on Q_i. Since ϕ normalizes $\mathrm{Aut}_R(Q)$, ϕ_i normalizes $\mathrm{Aut}_{R_i}(Q_i)$, and so there is an extension $\psi_i \in \mathrm{Aut}_{\mathcal{F}_i}(R_i)$ of ϕ_i (as each Q_i is fully $\mathcal{F}_i$-normalized). The automorphism $\psi = (\psi_1, \psi_2) \in \mathrm{Aut}_{\mathcal{F}}(R_1 \times R_2)$ extends ϕ, and hence the restriction of ψ to R is a map extending ϕ. As ϕ normalizes $\mathrm{Aut}_R(Q)$, $\psi \in \mathrm{Aut}_{\mathcal{F}}(R)$ by Lemma 4.18. Hence Q has the surjectivity property.

It remains to show that Q is fully $\mathcal{F}$-normalized, but since Q is fully centralized and fully automized, it is clearly fully normalized, and hence $\mathcal{F}$ is saturated by Theorem 6.16. $\qquad\square$

In Section 8.5 we will develop the theory of direct and central products much more, a theory that hinges on having the definition of a normal subsystem.

Exercises

6.1 Let $\mathcal{F}$ be a fusion system on a finite p-group P, and let Q be a fully $\mathcal{F}$-normalized subgroup of P. Show that the following are equivalent:

   (i) Q is Q-receptive in $\mathcal{F}$ and fully $\mathcal{F}$-automized;
   (ii) $\{Q\}$ is saturated in $\mathrm{N}_{\mathcal{F}}(Q)$.

6.2 Let $\mathcal{F}$ be a fusion system on a finite p-group P, and suppose that $\mathcal{Q}$ is an $\mathcal{F}$-conjugacy class of subgroups of P. Writing $\mathcal{H}$ for the set of all subgroups of P of smaller index than those in $\mathcal{Q}$, suppose that

$\mathcal{F}$ is $\mathcal{H}$-generated and that $\mathcal{F}$ is inductively saturated with respect to $\mathcal{Q}$. Prove that, if there is a subgroup Q such that either Q is not $\mathcal{F}$-centric or it is and $\mathrm{Out}_P(Q) \cap \mathrm{O}_p(\mathrm{Out}_{\mathcal{F}}(Q))$ is non-trivial, then there is a fully $\mathcal{F}$-normalized such subgroup.

6.3 Prove Proposition 6.20.

6.4 Let U and V be subgroups of a finite p-group P, and suppose that $U \triangleleft V$. If ϕ is an automorphism of V that acts trivially on U, and $C_V(U) \leq U$, prove that ϕ is a p-automorphism.

6.5 Let $\mathcal{F}$ be a fusion system on a finite p-group P, and let Q be an $\mathcal{F}$-centric subgroup of P with an $\mathcal{F}$-automorphism ϕ. Suppose that R is a subgroup of P containing Q, and that ϕ_1 and ϕ_2 are extensions of ϕ in $\mathcal{F}$ to R. Suppose that all subgroups of P containing R lie in saturated $\mathcal{F}$-conjugacy classes. Show that there is some $x \in \mathrm{Z}(Q)$ such that $\phi_1 = c_x \phi_2$.

6.6 Let $\mathcal{F}$ be a saturated fusion system on a finite p-group P, and let $\mathcal{E}$ be an $\mathcal{F}$-invariant subsystem of $\mathcal{F}$, on a strongly $\mathcal{F}$-closed subgroup Q of P.

  (i) Suppose that R is a fully $\mathcal{F}$-normalized subgroup of Q, and that R is fully $\mathcal{E}$-automized. Show that every fully $\mathcal{E}$-normalized subgroup that is $\mathcal{F}$-conjugate to R is fully $\mathcal{E}$-automized and fully $\mathcal{E}$-centralized.

  (ii) Suppose that all fully $\mathcal{F}$-normalized subgroups of Q are fully $\mathcal{E}$-automized. If S is any subgroup of Q, show that there is an $\mathcal{E}$-map $\phi : S \to R$, for some fully $\mathcal{E}$-normalized subgroup R, such that ϕ extends to an $\mathcal{E}$-map $\mathrm{N}_P(S) \to \mathrm{N}_P(R)$. Deduce that, in (i) of Theorem 6.9, we may assume that R is fully $\mathcal{E}$-normalized, fully $\mathcal{E}$-centralized, and fully $\mathcal{E}$-automized.

6.7 Let $\mathcal{F}$ be a saturated fusion system on a finite p-group P, and let $\mathcal{E}$ be a weakly normal subsystem of $\mathcal{F}$, on a strongly $\mathcal{F}$-closed subgroup Q of P. Let U be a fully $\mathcal{F}$-normalized subgroup of Q. Prove that $\mathrm{N}_{\mathcal{E}}(U)$ is weakly normal in $\mathrm{N}_{\mathcal{F}}(U)$.

6.8 Let $\mathcal{F}$ be a saturated fusion system on a finite p-group P, and suppose that P is the direct product of two strongly $\mathcal{F}$-closed subgroups Q and R of P. Prove that the full subcategory of $\mathcal{F}$ on Q (and on R) is a weakly normal subsystem of $\mathcal{F}$.

6.9 Let $\mathcal{F}$ be a saturated fusion system on a finite p-group P, and let $\mathcal{E}$ be a weakly normal subsystem of $\mathcal{F}$, on a strongly $\mathcal{F}$-closed subgroup Q of P. Suppose that there is some subsystem $\mathcal{F}'$ of $\mathcal{F}$, such that $\mathrm{N}_{\mathcal{E}}(U) \leq \mathcal{F}'$ for all $U \leq Q$ that are $\mathcal{E}$-centric and $\mathrm{Aut}_{\mathcal{F}}(Q)$-conjugate to a fully $\mathcal{F}$-normalized subgroup of Q. Prove that $\mathcal{E} \leq \mathcal{F}'$.

7

Control in fusion systems

As we have seen in Chapter 5, the subgroup $O_p(\mathcal{F})$ can have a significant influence on the structure of the fusion system $\mathcal{F}$: in the case where $\mathcal{F}$ is soluble, we saw that $\mathcal{F}$ is the fusion system of a finite group, and indeed the composition factors of such a group are limited. Because of this, it becomes interesting to ask for conditions on $\mathcal{F}$ that force $O_p(\mathcal{F})$ to be non-trivial, i.e., whether there is a subgroup Q such that $\mathcal{F} = N_\mathcal{F}(Q)$.

Let G be a finite group, with a Sylow p-subgroup P. We have said that a subgroup H containing P controls G-fusion in P if $\mathcal{F}_P(H) = \mathcal{F}_P(G)$. Since we are more interested in when local subgroups ($N_G(Q)$ for some $Q \leq P$) control fusion, we make the following definition.

Definition 7.1 Let $\mathcal{F}$ be a saturated fusion system on a finite p-group P. A subgroup $Q \leq P$ *controls fusion* if $\mathcal{F} = N_\mathcal{F}(Q)$, or equivalently Q is a strongly $\mathcal{F}$-closed subgroup contained in $O_p(\mathcal{F})$.

Theorems about whether there are subgroups that control fusion, and if so what they are, are important in the theory of finite groups. We have seen an example in the first chapter, namely Glauberman's ZJ-theorem. In this chapter (among other things) we will see an analogue of the ZJ-theorem for fusion systems.

We begin the chapter with the concept of resistance. A p-group P is resistant if every fusion system on P is just that of $P \rtimes A$, where A is a p'-subgroup of $\mathrm{Aut}(P)$ or, equivalently, P controls fusion in every saturated fusion system on P. We prove a theorem of Stancu, which states that every non-abelian metacyclic p-group is resistant if p is odd; the case where $p = 2$ is slightly more complicated, and we classify all saturated fusion systems on metacyclic 2-groups.

We then move on to positive characteristic p-functors and Glauberman functors. A characteristic p-functor is a function on p-groups that picks out a specific (non-trivial) characteristic subgroup of a given p-group, invariant under automorphisms of the group in the sense that the choice of characteristic subgroup of the p-group does not depend on the way that the p-group is embedded in a finite group. Examples include $Z(P)$, $\Omega_1(P)$, and in particular $Z(J(P))$. A Glauberman functor is a special type of positive characteristic p-functor that we will define in Section 7.2.

After this we prove the analogue of the ZJ-theorem for fusion systems, and in Section 7.4 some analogues of p-complement theorems. Section 7.5 introduces the hyperfocal subsystem $O^p(\mathcal{F})$ and the residual subsystem $O^{p'}(\mathcal{F})$. In the final two sections, bisets and control of transfer will be considered. This chapter completes the basic theory of fusion systems given in this book.

7.1 Resistance

We begin by defining the concept of resistance.

Definition 7.2 A finite p-group P is *resistant* if, for every saturated fusion system $\mathcal{F}$ on P, we have that $O_p(\mathcal{F}) = P$.

In Exercise 1.8, we proved that abelian p-groups are resistant. If P is a resistant p-group, then all saturated fusion systems are fusion systems of finite groups, for groups of the form $G = P \rtimes A$, where A is a p'-group of automorphisms of P; this is Exercise 7.2.

Lemma 7.3 *Let $\mathcal{F}$ be a saturated fusion system on a finite p-group P. There are no $\mathcal{F}$-essential subgroups of P if and only if $\mathcal{F} = \mathrm{N}_{\mathcal{F}}(P)$.*

Proof If there are no essential subgroups of $\mathcal{F}$, then Alperin's fusion theorem says that every morphism is a restriction of an automorphism of P, and hence we have one direction.

Suppose that $\mathcal{F} = \mathrm{N}_{\mathcal{F}}(P)$, and let E be an essential subgroup of P. By Proposition 4.46, P is contained within every $\mathcal{F}$-centric, $\mathcal{F}$-radical subgroup of P, so $P \leq E$, and since $E \neq P$ we have a contradiction. Thus there are no essential subgroups, as claimed. $\square$

What this means is that, to prove that a p-group is resistant, it suffices to show that none of its subgroups can be essential. In this direction, Stancu proves that metacyclic p-groups (for p odd) are resistant in [Sta06]: the proof below is more structural.

Let p be an odd prime. Recall that a group G is *metacyclic* if there is a cyclic normal subgroup H such that G/H is also cyclic. The first result that Stancu proves is the following.

Proposition 7.4 *Let p be an odd prime. A metacyclic p-group Q is realizable as an essential subgroup in some saturated fusion system if and only if Q is a homocyclic abelian p-group, i.e., Q is of the form $C_{p^n} \times C_{p^n}$.*

Proof If $Q = C_{p^n} \times C_{p^n}$, generated by x and y, then $G = Q \rtimes \mathrm{GL}_2(p)$ (with $\mathrm{GL}_2(p)$ acting on $\langle x, y \rangle$ as it acts on $C_p \times C_p$), with Sylow p-subgroup $P > Q$, is an example of a group in which Q is an $\mathcal{F}_P(G)$-essential subgroup.

Suppose that Q is a non-cyclic, metacyclic p-group that is an $\mathcal{F}$-essential subgroup for some saturated fusion system $\mathcal{F}$. By Exercise 7.3, all of the $p+1$ subgroups of Q of index p are isomorphic, and $\mathrm{Aut}_{\mathcal{F}}(Q)$ acts transitively on them. If Q is abelian then the only case where this occurs is when Q is homocyclic. (If $P \cong C_{p^a} \times C_{p^b}$, then Q has maximal subgroups isomorphic to $C_{p^{a-1}} \times C_{p^b}$ and $C_{p^a} \times C_{p^{b-1}}$.)

Thus suppose that Q is non-abelian, and let x be an element of Q with $\langle x \rangle \trianglelefteq Q$ and $Q/\langle x \rangle$ cyclic. Since Q is not cyclic, it has a unique elementary abelian subgroup R of order p^2 by Exercise 7.4. The subgroup R lies inside one of the maximal subgroups of Q, so it must lie inside all of them, since they are all isomorphic. Hence $\mathrm{Aut}_{\mathcal{F}}(Q)$ induces an automorphism group A on Q/R, and since $\mathrm{Aut}_{\mathcal{F}}(Q)$ acts transitively on the maximal subgroups of Q, we have that A acts transitively on the maximal subgroups of Q/R. By induction, $Q/R \cong C_{p^n} \times C_{p^n}$ for some n. In particular, this means that $Q' \le R$, and since $Q' \le \langle x \rangle$ (as its quotient is cyclic) we see that Q' has order p. As Q' is characteristic in Q, we have that $\mathrm{Aut}_{\mathcal{F}}(Q)$ induces automorphisms on Q/Q' that permute the maximal subgroups transitively, and so Q/Q' again must be homocyclic abelian, a contradiction since Q/Q' has order p^{2n+1} so cannot be of the form $C_{p^m} \times C_{p^m}$ for any m. $\qquad\square$

Using this proposition, we can complete the proof that metacyclic p-groups are resistant for odd primes p.

Theorem 7.5 (Stancu [Sta06, Proposition 5.4]) *Let p be an odd prime. If P is a metacyclic p-group then P is resistant.*

Proof Suppose that P is a metacyclic p-group, where p is odd. If P is abelian then P is resistant, so we assume that P is non-abelian. Choose

P with $|P|$ minimal such that there is a saturated fusion system $\mathcal{F}$ for which $P \neq O_p(\mathcal{F})$. Lemma 7.3 implies that there is some fully normalized, essential subgroup Q of P, and by the previous proposition Q is homocyclic (since subgroups of metacyclic groups are metacyclic), say $Q = C_{p^n} \times C_{p^n}$. Exercise 7.3 proves that $\operatorname{Aut}_P(Q)$ has order p (as $\operatorname{Aut}_{\mathcal{F}}(Q)$ is a subgroup of $\operatorname{GL}_2(p)$), and since $Q \triangleleft P$ (by Exercise 7.4) we see that P has order p^{2n+1}.

Firstly, suppose that $n = 1$; then P is a non-abelian group of order p^3 and, by Exercise 7.5, P has exponent p^2 and is resistant. Now suppose that $n > 1$; since there is a unique subgroup R of P of order $C_p \times C_p$ containing all elements of order p, it is strongly $\mathcal{F}$-closed. Notice that P/R is a metacyclic p-group of smaller order, so is resistant. Therefore $\mathcal{F}/R$ has no $\mathcal{F}/R$-essential subgroups, and in particular Q/R is not an $\mathcal{F}/R$-essential subgroup of P/R.

Since $R \leq \Phi(Q)$, any non-trivial p'-automorphism of Q induces a non-trivial p'-automorphism of Q/R, so that the kernel of the map $\operatorname{Aut}_{\mathcal{F}}(Q) \to \operatorname{Aut}_{\mathcal{F}/R}(Q/R)$ is a normal p-subgroup. However, since Q is $\mathcal{F}$-essential (and so $\mathcal{F}$-radical), this kernel must be trivial. Therefore $\operatorname{Aut}_{\mathcal{F}}(Q) \cong \operatorname{Aut}_{\mathcal{F}/R}(Q/R)$, and so $\operatorname{Out}_{\mathcal{F}}(Q/R)$ has a strongly p-embedded subgroup. As $|P/R| = p^{2n-1}$ and $|Q/R| = p^{2n-2}$, if Q/R is not $\mathcal{F}/R$-centric then $Q/R \leq Z(P/R)$, and so P/R is abelian; in particular $|\operatorname{Aut}_{P/R}(Q/R)| = 1$. However, $p \mid |\operatorname{Aut}_{\mathcal{F}/R}(Q/R)|$, and so Q/R is not fully $\mathcal{F}/R$-automized, contradicting the fact that $\mathcal{F}/R$ is saturated. Hence Q/R is $\mathcal{F}/R$-essential, contradicting the fact that P/R is resistant. Thus Q is not $\mathcal{F}$-essential, so P is resistant, as claimed. $\square$

Having dealt with the case where p is odd, we move on to $p = 2$. We have already seen that the exact analogue of the previous result is false, because dihedral 2-groups are metacyclic and they are certainly not resistant. In addition, generalized quaternion and semidihedral 2-groups also are metacyclic, and also not resistant. (See Theorem 4.54 for the description of all saturated fusion systems on these 2-groups.) However, these three groups offer the only roadblock to a metacyclic 2-group being resistant, as we will now show.

We begin in the same vein as for the odd case, finding those metacyclic 2-groups with automorphisms of odd order. There is one more possibility in this case. In the proof of the next proposition, we will use the standard fact that if G is a finite group with a normal subgroup H, and H has exponent n and G/H has exponent m, then G has exponent at most mn; to see this, simply note that, for any $g \in G$, $g^m \in H$ and so $g^{mn} = 1$.

Proposition 7.6 *Let P be a metacyclic 2-group. If $\mathrm{Aut}(P)$ is not a 2-group, then P is homocyclic abelian or Q_8.*

Proof Let ϕ be an automorphism of odd order acting on P. If P is abelian then P is isomorphic with $C_{2^n} \times C_{2^n}$ by Exercise 4.8, so we may assume that P is non-abelian. By Burnside's theorem, ϕ acts non-trivially on $P/\Phi(P)$, so ϕ permutes the three maximal subgroups of P; in particular, the order of ϕ is 3. Since P is metacyclic, the characteristic subgroup $P' = \langle x \rangle$ is cyclic; write $Z = \Omega_1(P')$, a characteristic subgroup of order 2. By induction, P/Z is either Q_8 or $C_{2^n} \times C_{2^n}$.

Assume that P/Z is abelian. By Exercise 4.8, ϕ induces a non-trivial automorphism on $\bar{Y} = \Omega_1(P/Z)$, and hence on the preimage Y in P, which has order 8; hence $Y \cong Q_8$. Since Y is the inverse image of $\Omega_1(P/Z)$, Y contains all involutions of P, and hence P has a unique involution. Finally, it is well known that the 2-groups with a single involution are either cyclic or generalized quaternion (see, for example, [Gor80, Theorem 5.4.10(ii)]). As $Q_{2^n}/Z(Q_{2^n})$ is non-abelian if $n \geq 4$, we see that $P \cong Q_8$.

If $P/Z \cong Q_8$, then either $Z(P) = Z$ or $Z(P) \cong C_4$. If $Z(P) = Z$ then P has maximal class, and there is no maximal class 2-group whose quotient by the centre is Q_8 (as the 2-groups of maximal class are D_{16}, Q_{16} and SD_{16}, and $D_{16}/Z(D_{16})$, $Q_{16}/Z(Q_{16})$ and $SD_{16}/Z(SD_{16})$ are all D_8). In the second case, we must have that Q_8 is the unique quotient of order 8 (as the centre is cyclic), and so $P' = Z(P)$. Since $P/\langle x \rangle$ is a quotient of P/P', we see that x must have order 8, and $Z(P) \leq \langle x \rangle$. The automorphism that y induces on $\langle x \rangle$ must fix x^2, and hence $x \mapsto x^5$ is the only option. In this case however, $[x, y] = x^4 \in Z$, so that $P' = Z$, a contradiction. $\square$

Having identified the potential essential subgroups, we move on to prove the result.

Theorem 7.7 (Craven–Glesser [CG10]) *Let $\mathcal{F}$ be a saturated fusion system on the metacyclic 2-group P. If $\mathcal{F} \neq \mathcal{F}_P(P)$, then P is dihedral, semidihedral, homocyclic abelian, or generalized quaternion.*

Proof By Proposition 7.6, there are three possibilities for a proper $\mathcal{F}$-centric, $\mathcal{F}$-radical subgroup Q: $Q \cong V_4$, $Q \cong C_{2^n} \times C_{2^n}$ for $n \geq 2$, and $Q \cong Q_8$.

If $Q \cong V_4$ or $Q \cong Q_8$, then P is dihedral, semidihedral or generalized quaternion, by Lemma 5.81 and Exercise 5.15.

If $Q \cong C_{2^n} \times C_{2^n}$ for $n > 1$, either $P = Q$, leading to an acceptable case, or $N_P(Q)$ is a wreathed 2-group of order at least 32, by Lemma 5.82, and since wreathed groups are not metacyclic by Exercise 7.7, we see that this case cannot occur. (The class of metacyclic groups is closed under subgroups and quotients.)

This completes the proof of the theorem. □

Finally, we make some remarks about Suzuki 2-groups. We start with their definition, taken from [Hig63].

Definition 7.8 A finite 2-group P is a *Suzuki 2-group* if the following three conditions hold:

(i) P is not abelian;
(ii) P contains at least two involutions;
(iii) P possesses a cyclic group of automorphisms that permutes the involutions of P transitively.

It is clear from the definition that all involutions belong to the centre of P. In [Hig63], Higman proves the following theorem.

Theorem 7.9 (Higman [Hig63]) *If P is a Suzuki 2-group, then*

$$\Omega_1(P) = \mathrm{Z}(P) = \Phi(P) = P',$$

and so P has exponent 4 and class 2.

Let $\mathcal{F}$ be a fusion system over a Suzuki 2-group P; then $\mathrm{Z}(P)$, being elementary abelian and consisting of all involutions in P together with the identity, is strongly $\mathcal{F}$-closed. (We do not even require saturation here.) In addition, $P/\mathrm{Z}(P)$ is abelian, and so by Proposition 4.62 we have the following result.

Proposition 7.10 (Craven–Glesser [CG10]) *All Suzuki 2-groups are resistant.*

In [Hig63], Higman classifies all Suzuki 2-groups, and finds that there are four classes, labelled A to D. In [Suz62], Suzuki proves that the Sylow 2-subgroups of the groups $\mathrm{Sz}(q)$ are Suzuki 2-groups (hence their name), and in fact are of Type A (see also [Col71]). Thus we get the following corollary. (In [Sta06], Stancu notes the case $\mathrm{Sz}(8)$.)

Corollary 7.11 *If P is a Sylow 2-subgroup of a Suzuki simple group $\mathrm{Sz}(q)$ for some $q = 2^{2n+1}$, then P is resistant.*

There are other simple groups whose Sylow 2-subgroups are Suzuki 2-groups (see, for example, [Col72]). In Section 5.5, we said that the

Sylow 2-subgroup of $\mathrm{PSU}_3(4)$ is a Suzuki 2-group. In fact, the same is true for $\mathrm{PSU}_3(2^n)$ in general.

A Sylow 2-subgroup of $\mathrm{PSU}_3(2^n)$ is isomorphic to the group P of matrices of the form

$$\begin{pmatrix} 1 & 0 & 0 \\ x & 1 & 0 \\ y & x^q & 1 \end{pmatrix},$$

where x and y are elements of $\mathbb{F}_{q^2}$, and $y + y^q = x^{1+q}$. It can easily be seen that such a matrix has order 2 if and only if $x = 0$ and $y \in \mathbb{F}_q \setminus \{0\}$, and the group of automorphisms generated by the matrix

$$\begin{pmatrix} \zeta & 0 & 0 \\ 0 & 1 & 0 \\ 0 & 0 & \zeta^{-1} \end{pmatrix}$$

permutes the involutions transitively, meaning that P is a Suzuki 2-group.

Corollary 7.12 *If P is a Sylow 2-subgroup of a simple group $\mathrm{PSU}_3(q)$ for some even q, then P is resistant.*

In [Gol74], Goldschmidt determined all simple groups G that possess a strongly closed abelian 2-subgroup (see Theorem 5.96). By Proposition 4.62, if $\mathcal{F} = \mathcal{F}_P(G)$ is the fusion system at the prime 2 of a finite simple group, then $O_2(\mathcal{F}) \neq 1$ if and only if G is on the list in Theorem 5.96. Since the only groups on this list that do not have abelian Sylow 2-subgroups are $\mathrm{PSU}_3(2^n)$ and $\mathrm{Sz}(q)$, we actually have the following result.

Corollary 7.13 *Let P be a Sylow 2-subgroup of a finite simple group. Either P is resistant or there is a fusion system $\mathcal{F}$ of a simple group such that $O_2(\mathcal{F}) = 1$.*

This corollary emphasizes the dichotomy of the situation: if a 2-group that is a Sylow 2-subgroup of a simple group is not resistant then there is a simple group whose fusion system has no normal subgroups at all.

7.2 Glauberman functors

A positive characteristic p-functor is not a functor in the categorical sense, but rather a map that picks out a certain characteristic subgroup of a p-group.

Definition 7.14   A map W is a *positive characteristic p-functor* if it sends each finite p-group P to a characteristic subgroup $W(P)$ of P, such that

(i) $W(P) > 1$ if $P > 1$, and
(ii) if $\phi : P \to Q$ is an isomorphism of finite p-groups, then $W(P)\phi = W(Q)$.

As examples of positive characteristic p-functors, we have $P \mapsto Z(P)$, $P \mapsto \Omega_1(Z(P))$, $P \mapsto J(P)$, and $P \mapsto Z(J(P))$. The last one of these is special, as it is something called a Glauberman functor, first considered by Glauberman in [Gla71].

Definition 7.15   A positive characteristic p-functor W is a *Glauberman functor* if, whenever P is a Sylow p-subgroup of a p-constrained finite group G not involving the group $Qd(p) = (C_p \times C_p) \rtimes \mathrm{SL}_2(p)$, we have that $W(P) \trianglelefteq G$.

The statement that $W : P \mapsto Z(J(P))$ is a Glauberman functor is precisely the ZJ-theorem. We will denote this functor simply by ZJ.

Theorem 7.16 (Glauberman's ZJ-theorem [Gla68a])   *For any odd prime p, the positive characteristic p-functor ZJ is a Glauberman functor.*

For a proof, see either the original paper [Gla68a], [Gor80, Theorem 8.2.11], or a more modern graph-theoretic proof in [Ste90] and [Ste92] for example.

For the prime 2, the functor ZJ is not a Glauberman functor. Question 16.1 of [Gla71] asks whether there exists a Glauberman functor in the case where $p = 2$, and this was answered in the affirmative by Stellmacher in [Ste96]; however, this proof relied upon the classification of S_4-free simple groups.

Theorem 7.17 (Stellmacher [Ste96])   *There exists a Glauberman functor for $p = 2$.*

In the next section we will prove that a $Qd(p)$-free fusion system (defined later) is a soluble fusion system if a Glauberman functor exists, and so this will prove that S_4-free fusion systems (for the prime 2) are soluble, and hence fusion systems of finite groups. However, in [Asc08b], Aschbacher goes in the other direction and proves directly that S_4-free fusion systems are constrained, yielding the classification of S_4-free simple groups as a corollary.

We will not prove either of Theorems 7.16 or 7.17 here, as it would take us too far away from our real concern.

Apart from ZJ, for odd primes there are another two Glauberman functors, K^∞ and K_∞ (constructed in [Gla70], but see also [Gla71, Section 12]), whose definitions are unfortunately considerably less easy to understand.

Let P be a finite p-group and let Q be a subgroup of P. Define $\mathcal{M}(P; Q)$ to be the set of subgroups R of P normalized by Q and such that $R/Z(R)$ is abelian (i.e., R has class at most 2). The first subset, $\mathcal{M}^*(P; Q)$, is the subset of $\mathcal{M}(P; Q)$ consisting of those subgroups R for which the induced conjugation action of Q on $R/Z(R)$ is trivial. The second subset, $\mathcal{M}_*(P; Q)$, consists of a collection of subgroups R in $\mathcal{M}(P; Q)$ satisfying the following condition: if $S \in \mathcal{M}(P; R)$ is a subgroup of P such that $S \leq Q \cap C_P([Z(R), S])$ and S' centralizes R, then the conjugation action of S induces the trivial action on $R/Z(R)$.

Write $K_{-1}(P) = P$, and define

$$K_i(P) = \begin{cases} \langle \mathcal{M}^*(P; K_{i-1}(P)) \rangle & i \text{ odd} \\ \langle \mathcal{M}_*(P; K_{i-1}(P)) \rangle & i \text{ even.} \end{cases}$$

Definition 7.18 Let P be a finite p-group. Define

$$K^\infty(P) = \bigcap_{i \geq -1,\ \text{odd}} K_i(P) \quad \text{and} \quad K_\infty(P) = \langle K_i(P) \mid i \geq 0,\ \text{even} \rangle.$$

The maps $K_\infty : P \mapsto K_\infty(P)$ and $K^\infty : P \mapsto K^\infty(P)$ are examples of positive characteristic p-functors.

Lemma 7.19 (Glauberman [Gla71, 13.1]) *Let P be a finite p-group, and let W be either of the functors K^∞ or K_∞.*

(i) *$W(P)$ is a characteristic subgroup of P.*
(ii) *$W(P)$ contains $Z(P)$; in particular $W(P) > 1$ if $P \neq 1$.*
(iii) *If $\phi : P \to Q$ is a group isomorphism, then $W(P)\phi = W(Q)$.*

Proof If Q is a characteristic subgroup of P, then $\mathcal{M}(P; Q)$ is a collection of subgroups that is closed under any automorphism of P. If R is a subgroup of P in $\mathcal{M}^*(P; Q)$ and ϕ is an automorphism of P, then $R\phi \in \mathcal{M}^*(P; Q)$. Thus if Q is characteristic in P, then $\langle \mathcal{M}^*(P; Q) \rangle$ is a characteristic subgroup of P. The same is true if $R \in \mathcal{M}_*(P; Q)$, and so $\langle \mathcal{M}_*(P; Q) \rangle$ is also characteristic. Thus each of the subgroups $K_i(P)$ is characteristic, and so therefore are $K^\infty(P)$ and $K_\infty(P)$.

We notice that $Z(P)$ lies in $\mathcal{M}^*(P; Q)$ and $\mathcal{M}_*(P; Q)$ for all $Q \leq P$, since if $R = Z(P)$ then there can only be the trivial action on $R/Z(R)$; thus $Z(P)$ is a subgroup of $K_i(P)$ for all i, and so $Z(P) \leq K_\infty(P)$ and $Z(P) \leq K^\infty(P)$.

Finally, the third part of this statement is clear, proving the lemma.
□

In fact, as well as being positive characteristic p-functors, the functors K^∞ and K_∞ have much more impressive properties. We will see some of these later on in the chapter, but for now we give the first one, remarked on at the very end of [Gla70], but proved in [Gla71].

Theorem 7.20 (Glauberman [Gla71, Theorem 12.9]) *If p is an odd prime then the two functors K^∞ and K_∞ are Glauberman functors.*

We shall not prove this theorem here either, since its proof is very involved.

We have defined Glauberman functors based on their behaviour for $Qd(p)$-free, *p-constrained* groups: we should note what happens in general $Qd(p)$-free groups.

Theorem 7.21 (Glauberman [Gla71, Theorem 6.6]) *Let p be an odd prime, and let G be a $Qd(p)$-free group with Sylow p-subgroup P. If W is a Glauberman functor, then $\mathrm{N}_G(W(P))$ controls G-fusion in P; in other words, $W(P) \leq \mathcal{F}_P(G)$.*

We shall not need this theorem in what follows, and a proof of this via fusion systems will be given in the next section, when we deal with $Qd(p)$-free fusion systems themselves.

We end this section with the concept of being well placed.

Definition 7.22 Let P be a finite p-group, W be a positive characteristic p-functor, and $\mathcal{F}$ be a saturated fusion system on P. For a subgroup $Q \leq P$, define $W_1(Q) = Q$, $P_1(Q) = \mathrm{N}_P(Q)$, and define $W_{i+1}(Q) = W(P_i(Q))$ and $P_{i+1}(Q) = \mathrm{N}_P(W_{i+1}(Q))$. A subgroup Q is $(\mathcal{F}, W)$-*well placed* if $W_i(Q)$ is fully normalized for all i.

Since $W(P_i(Q)) \operatorname{char} P_i(Q)$, we have that $P_{i+1}(Q)$ strictly contains $P_i(Q)$ (if $P_i(Q) \neq P$), since it contains $\mathrm{N}_P(P_i(Q))$. In particular, for all sufficiently large integers i, $P_i(Q) = P$ and $W_i(Q) = W(P)$.

For the next section we need a useful result on the control of fusion by positive characteristic p-functors; to get it, we will analyse the concept of being well placed.

Lemma 7.23 *Let $\mathcal{F}$ be a saturated fusion system on the finite p-group P, and let W be a positive characteristic p-functor. If Q is a subgroup of P, then Q is $\mathcal{F}$-isomorphic to an $(\mathcal{F}, W)$-well-placed subgroup of P.*

Proof Set $W_i = W_i(Q)$ and $P_i = P_i(Q)$ as in the definition of well placed, and notice that $P_i = \mathrm{N}_P(W_i)$ for all i. By Proposition 4.17, we may find a morphism $\phi_1 : \mathrm{N}_P(Q) = P_1 \to P$ such that $W_1\phi_1$ is fully normalized. Replacing Q by W_1, we now have that W_1 is fully normalized.

Suppose that $W_1, \ldots, W_i$ are all fully normalized. Let $\phi_{i+1} : P_{i+1} \to P$ be any map such that $W_{i+1}\phi_{i+1}$ is fully normalized. We claim that $W_j\phi_{i+1}$ is also fully normalized for all $j \leq i$. To see this, notice that $P_j \leq P_{i+1}$, and so ϕ_{i+1} induces a map $\psi : P_j \to P$, whose image must normalize $W_j\phi_{i+1}$. Since W_j is fully normalized, $|P_j| \geq |\mathrm{N}_P(W_j\phi_{i+1})|$, and so we have equality; thus $W_j\phi_{i+1}$ is also fully normalized. By replacing Q with $Q\phi_{i+1}$, we have that $W_1, \ldots, W_{i+1}$ are all fully normalized. Induction, and the fact that $W_i = W_{i+1}$ for sufficiently large i, proves that Q is $\mathcal{F}$-isomorphic to a $(\mathcal{F}, W)$-well-placed subgroup, as claimed. □

In fact, the collection of all $(\mathcal{F}, W)$-well-placed subgroups forms a conjugation family, in a sense very similar to that in Chapter 1.

Definition 7.24 Let $\mathcal{F}$ be a saturated fusion system on a finite p-group P. A collection of subgroups F of P is a *conjugation family* for $\mathcal{F}$ if $\mathcal{F} = \langle\, \mathrm{Aut}_{\mathcal{F}}(U) : U \in F \,\rangle$, that is, any morphism in $\mathcal{F}$ is the composition of restrictions of automorphisms of elements of F.

Alperin's fusion theorem says that the collection of fully normalized, $\mathcal{F}$-centric, $\mathcal{F}$-radical subgroups forms a conjugation family. The following proposition may be found, for example, in [DGMP09, Proposition 2.10].

Proposition 7.25 *Let $\mathcal{F}$ be a saturated fusion system on a finite p-group P, and let F be a conjugation family. If F' is a subfamily of F such that every element of F is $\mathcal{F}$-conjugate to an element of F', then F' is a conjugation family.*

Proof Let Q be an element of $F \setminus F'$, and let R be an element of F' that is $\mathcal{F}$-conjugate to Q. Let $\mathcal{E}$ be the subsystem of $\mathcal{F}$ generated by $\mathrm{Aut}_{\mathcal{F}}(U)$, as U runs through all elements of F'. It suffices to show that $\text{Aut}_{\mathcal{E}}(Q) = \text{Aut}_{\mathcal{F}}(Q)$, since then the subsystems generated by the automorphisms of members of F' and F must be the same. It also suffices to prove that $\text{Aut}_{\mathcal{E}}(Q) \cong \text{Aut}_{\mathcal{E}}(R)$, since $\text{Aut}_{\mathcal{E}}(R) = \text{Aut}_{\mathcal{F}}(R)$.

We first claim that Q and R are $\mathcal{E}$-conjugate. By induction, on $|P : Q|$, $\text{Aut}_{\mathcal{F}}(T) = \text{Aut}_{\mathcal{E}}(T)$ if $|T| > |Q|$. Since F is a conjugation family, there is a sequence of automorphisms $\phi_1, \ldots, \phi_n$ such that the composition of various restrictions of the ϕ_i is a morphism $Q \to R$; choose such a sequence to minimize n. If the domain of any ϕ_i has order the same as Q, we may remove it without changing the fact that the composition is a morphism from Q to R, and this contradicts the choice of n. Thus the ϕ_i are all automorphisms of subgroups of order strictly greater than $|Q|$, whence they all lie inside $\mathcal{E}$, and Q and R are $\mathcal{E}$-conjugate via some morphism ϕ. Finally, if $\psi \in \text{Aut}_{\mathcal{E}}(R)$, then $\phi\psi\phi^{-1} \in \text{Aut}_{\mathcal{E}}(Q)$, and this induces a bijection, so that $\text{Aut}_{\mathcal{E}}(Q) = \text{Aut}_{\mathcal{E}}(R) = \text{Aut}_{\mathcal{F}}(R)$. $\square$

Combining Lemma 7.23 and Proposition 7.25, we get the following corollary.

Corollary 7.26 *Let $\mathcal{F}$ be a saturated fusion system on a finite p-group P. If W is a positive characteristic p-functor, then the collection of all $(\mathcal{F}, W)$-well-placed subgroups forms a conjugation family.*

The final result of the section deals with how the influence that a positive characteristic p-functor has on the local structure translates into influence on the global structure of a fusion system. In the case of finite groups, this is [Gla71, 5.5], and for blocks of finite groups, this is [KLR02, Proposition 3.2]. For fusion systems in general, this is proved in [KL08].

Proposition 7.27 (Kessar–Linckelmann [KL08, Proposition 5.3]) *Let $\mathcal{F}$ be a saturated fusion system on a finite p-group P, and let W be a positive characteristic p-functor. If, for all fully normalized subgroups Q of P, we have*

$$\text{N}_{\mathcal{F}}(Q) = \text{N}_{\mathcal{F}}(Q, W(\text{N}_P(Q))),$$

then $\mathcal{F} = \text{N}_{\mathcal{F}}(W(P))$.

Proof Write $\mathcal{N} = \text{N}_{\mathcal{F}}(W(P))$, and suppose that $\mathcal{N} < \mathcal{F}$. There is some Q such that $\text{Aut}_{\mathcal{F}}(Q) \neq \text{Aut}_{\mathcal{N}}(Q)$; by Lemma 7.23, we may assume that Q is $(\mathcal{F}, W)$-well-placed. Set $W_i = W_i(Q)$, including $W_0 = Q$. We notice that, for all $i \geq 0$,

$$\text{N}_{\mathcal{F}}(W_i) = \text{N}_{\mathcal{F}}(W_i, W_{i+1}) \leq \text{N}_{\mathcal{F}}(W_{i+1}).$$

For all sufficiently large n, $W_n = W(P)$, and so this states that $\mathrm{N}_{\mathcal{F}}(Q) \leq \mathrm{N}_{\mathcal{F}}(W(P))$; hence $\mathrm{Aut}_{\mathcal{F}}(Q) \leq \mathrm{Aut}_{\mathcal{N}}(Q)$, a contradiction. Hence $\mathcal{F} = \mathrm{N}_{\mathcal{F}}(W(P))$, as claimed. □

7.3 The ZJ-theorems

In the previous section we saw that, in any $Qd(p)$-free p-constrained group G (for p odd), and any Glauberman functor W, we have that $W(P) \trianglelefteq G$. We first define the notion of an H-free fusion system, mimicking the notion of an H-free group. By Theorem 3.70, if Q is an $\mathcal{F}$-centric subgroup of a p-group P, then there is a unique p-constrained group with no normal p'-subgroup for which $\mathrm{N}_{\mathcal{F}}(Q)$ is the fusion system: we denote this group by $L_Q^{\mathcal{F}}$.

Definition 7.28 A saturated fusion system $\mathcal{F}$ on a finite p-group P is H-*free* if, for any fully normalized, $\mathcal{F}$-centric subgroup Q of P, the group $L_Q^{\mathcal{F}}$ is H-free.

The main result that we will prove in this section is the following.

Theorem 7.29 *Any $Qd(p)$-free fusion system is soluble, and hence arises from a finite $Qd(p)$-free group.*

This theorem is the culmination of work by Kessar and Linckelmann in [KL08], Onofrei and Stancu in [OS09], and the author in [Cra09]; in particular, the method of proof comes from [Cra09], using important reductions from [KL08].

We begin with a few results on H-free fusion systems, needed in the proof of Theorem 7.29; basically, we need to show that if $\mathcal{F}$ is H-free then so are $\mathrm{N}_{\mathcal{F}}(Q)$ and $\mathcal{F}/Q$, when Q is fully normalized and strongly $\mathcal{F}$-closed respectively. We recall from the remarks at the end of Section 3.6 that if Q is a fully normalized, $\mathcal{F}$-centric subgroup of P, then the group $L_Q^{\mathcal{F}}$ is determined, up to isomorphism, by the group $\mathrm{Aut}_{\mathcal{F}}(Q)$ and the extension

$$1 \to \mathrm{Z}(Q) \to \mathrm{N}_P(Q) \to \mathrm{Aut}_P(Q) \to 1.$$

We will use this information in the next three propositions.

Proposition 7.30 (Kessar Linckelmann [KL08, Proposition 6.1]) *Let $\mathcal{F}$ be a saturated fusion system on a finite p-group P. If there is a fully normalized, centric subgroup Q of P such that $L_Q^{\mathcal{F}}$ involves H, then there is a fully normalized, centric, radical subgroup R of P such that $L_R^{\mathcal{F}}$ involves H.*

Proof Let Q be a fully normalized, $\mathcal{F}$-centric subgroup of P such that H is involved in $L_Q^{\mathcal{F}}$, and choose Q such that $|Q|$ is maximal. If Q is $\mathcal{F}$-radical, then we are done. If Q is not radical, then $O_p(\mathrm{Aut}_{\mathcal{F}}(Q)) > \mathrm{Inn}(Q)$; let R denote the preimage of $O_p(\mathrm{Aut}_{\mathcal{F}}(Q))$ in $N_P(Q)$. (As Q is fully $\mathcal{F}$-normalized, $\mathrm{Aut}_P(Q)$ is a Sylow p-subgroup of $\mathrm{Aut}_{\mathcal{F}}(Q)$, so $O_p(\mathrm{Aut}_{\mathcal{F}}(Q)) \leq \mathrm{Aut}_P(Q)$, and hence $\mathrm{Aut}_R(Q) = O_p(\mathrm{Aut}_{\mathcal{F}}(Q))$.) Since $Q < R$, we see that R is $\mathcal{F}$-centric. Also, since $\mathrm{Aut}_R(Q)$ is normal in $\mathrm{Aut}_{\mathcal{F}}(Q)$, we see that $R \lhd N_P(Q)$, so $N_P(Q) \leq N_P(R)$. By Proposition 4.8, we get a morphism $\psi : N_P(R) \to P$ such that $R\psi$ is fully normalized, and so is $Q\psi$. Hence we may assume that both Q and R are fully normalized and $\mathcal{F}$-centric. We will show that H is involved in $L_R^{\mathcal{F}}$, which contradicts the choice of Q (as $Q < R$). We will do this by proving that $L_Q^{\mathcal{F}}$ is involved in $L_R^{\mathcal{F}}$.

Let L be the preimage of $\mathrm{Aut}_{\mathcal{F}}(Q \leq R)$ in $L_R^{\mathcal{F}}$; hence L is an extension of $Z(R)$ by $\mathrm{Aut}_{\mathcal{F}}(Q \leq R)$. We claim that $L \cong L_Q^{\mathcal{F}}$, proving the result.

Consider the restriction map $f : \mathrm{Aut}_{\mathcal{F}}(Q \leq R) \to \mathrm{Aut}_{\mathcal{F}}(Q)$; we claim that this is a surjective map. Since Q is fully normalized, it has the surjectivity property, and so $\mathrm{Aut}_{\mathcal{F}}(Q \leq R) \to N_{\mathrm{Aut}_{\mathcal{F}}(Q)}(\mathrm{Aut}_R(Q))$ is surjective. Hence it suffices to show that $\mathrm{Aut}_R(Q)^\phi = \mathrm{Aut}_R(Q)$ for all $\phi \in \mathrm{Aut}_{\mathcal{F}}(Q)$, but this is clear since $\mathrm{Aut}_R(Q) = O_p(\mathrm{Aut}_{\mathcal{F}}(Q))$.

The kernel of the map f is the set of all automorphisms of R that act trivially on Q, and since Q is centric, this is $\mathrm{Aut}_{Z(Q)}(R) = Z(Q)/Z(R)$, a p-group (see Exercise 6.5). This yields the commutative diagram below.

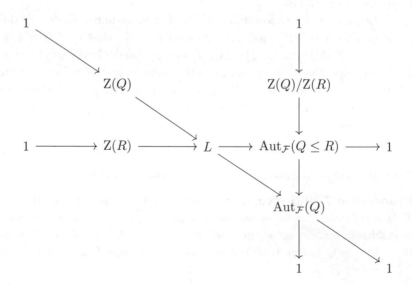

The subgroup Q is fully normalized, and so $\mathrm{Aut}_P(Q)$ is a Sylow p-subgroup of $\mathrm{Aut}_{\mathcal{F}}(Q)$; hence the inverse image of $\mathrm{Aut}_P(Q)$ under the composition $L \to \mathrm{Aut}_{\mathcal{F}}(Q \leq R) \to \mathrm{Aut}_{\mathcal{F}}(Q)$, whose kernel is a p-group, is a Sylow p-subgroup of L. This inverse image consists of all automorphisms of Q, induced by elements of P, that extend to automorphisms of R induced by elements of P. Finally, since $\mathrm{N}_P(Q) \leq \mathrm{N}_P(R)$, this inverse image is actually $\mathrm{N}_P(Q)$, and hence this is a Sylow p-subgroup of L.

We now see that $L \cong L_Q^{\mathcal{F}}$: by the remarks at the end of Section 3.6, the group $L_Q^{\mathcal{F}}$ is determined up to isomorphism by $\mathrm{Aut}_{\mathcal{F}}(Q)$ and the p-group extension of $\mathrm{Z}(Q)$ by $\mathrm{Aut}_P(Q)$. We have shown that these are the same for L, and so $L \cong L_Q^{\mathcal{F}}$, as claimed. $\qquad\square$

Having proved that we only have to show that fully normalized, centric, radical subgroups are H-free, we can attack the problem of proving that normalizers are H-free.

Proposition 7.31 (Kessar–Linckelmann [KL08, Proposition 6.3]) *Let $\mathcal{F}$ be an H-free saturated fusion system on the finite p-group P. If Q is a fully normalized subgroup of P, then $\mathrm{N}_{\mathcal{F}}(Q)$ is H-free.*

Proof Write $\mathcal{E} = \mathrm{N}_{\mathcal{F}}(Q)$, and let R be a fully $\mathcal{E}$-normalized, $\mathcal{E}$-centric, $\mathcal{E}$-radical subgroup of $\mathrm{N}_P(Q)$; since $Q \trianglelefteq \mathrm{N}_{\mathcal{F}}(Q)$, by Proposition 4.61, $Q \leq R$. By Exercise 7.6, R is $\mathcal{F}$-centric; using Proposition 4.17, we may find a fully $\mathcal{F}$-normalized subgroup S of P and an $\mathcal{F}$-isomorphism $\phi : R \to S$ such that ϕ extends to a map $\bar{\phi} : \mathrm{N}_P(R) \to \mathrm{N}_P(S)$. We will show that the model $L_R^{\mathcal{E}}$ is isomorphic with a subgroup of $L_S^{\mathcal{F}}$; since we will assume that $\mathcal{F}$ is H-free, this proves that H is not involved in $L_R^{\mathcal{E}}$, and so by Proposition 7.30 we have that $\mathcal{E}$ is H-free, as needed.

The map $\mathrm{Aut}_{\mathcal{F}}(R) \to \mathrm{Aut}_{\mathcal{F}}(S)$ obtained by conjugation by ϕ restricts to an injective map $\alpha : \mathrm{Aut}_{\mathcal{E}}(R) \to \mathrm{Aut}_{\mathcal{F}}(S)$. If A denotes the image of α in $\mathrm{Aut}_{\mathcal{F}}(S)$, then we will denote the preimage under the map $L_S^{\mathcal{F}} \to \mathrm{Aut}_{\mathcal{F}}(S)$ (which has kernel $\mathrm{Z}(S)$) of A by L. Hence L is an extension of $\mathrm{Z}(S)$ by $\mathrm{Aut}_{\mathcal{E}}(R) \cong A$. This yields the commutative diagram

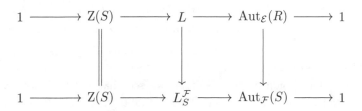

Since $L_S^{\mathcal{F}}$ is H-free, so is L (as it is a subgroup of $L_S^{\mathcal{F}}$), and therefore if $L \cong L_R^{\mathcal{E}}$, we are done.

Since S is fully $\mathcal{F}$-normalized, $\mathrm{Aut}_P(S)$ is a Sylow p-subgroup of $\mathrm{Aut}_{\mathcal{F}}(S)$, and similarly, as R is fully $\mathcal{E}$-normalized, $\mathrm{Aut}_{N_P(Q)}(R) = \mathrm{Aut}_P(Q \leq R)$ is a Sylow p-subgroup of $\mathrm{Aut}_{\mathcal{E}}(R)$. If U denotes the preimage of $\mathrm{Aut}_P(Q \leq R)$ in L, then U is a Sylow p-subgroup of L.

By the remarks at the end of Section 3.6, the group L is determined, up to isomorphism, by $\mathrm{Aut}_{\mathcal{E}}(R)$ and the short exact sequence

$$1 \to Z(S) \to U \to \mathrm{Aut}_P(Q \leq R) \to 1,$$

which is the restriction to U of the top row of the commutative diagram above. To prove that L and $L_R^{\mathcal{E}}$ are isomorphic, it therefore remains to show that this short exact sequence is equivalent to the sequence

$$1 \to Z(R) \to N_P(Q, R) \to \mathrm{Aut}_P(Q \leq R) \to 1,$$

since this is the restriction of the corresponding short exact sequence for $L_R^{\mathcal{E}}$ to Sylow p-subgroups.

To see this final step, we simply notice that the diagram

$$
\begin{array}{ccccccccc}
1 & \longrightarrow & Z(R) & \longrightarrow & N_P(Q, R) & \longrightarrow & \mathrm{Aut}_P(Q \leq R) & \longrightarrow & 1 \\
 & & \downarrow & & \downarrow & & \| & & \\
1 & \longrightarrow & Z(S) & \longrightarrow & U & \longrightarrow & \mathrm{Aut}_P(Q \leq R) & \longrightarrow & 1
\end{array}
$$

commutes. Hence $L \cong L_R^{\mathcal{E}}$, and the proposition follows. $\qquad\square$

Having proved that normalizers are H-free, we move on to quotients.

Proposition 7.32 (Kessar–Linckelmann [KL08, Proposition 6.4]) *Let $\mathcal{F}$ be a saturated fusion system on a finite p-group P, and let Q be a strongly $\mathcal{F}$-closed subgroup of P. If $\mathcal{F}$ is H-free then $\mathcal{F}/Q$ is H-free as well.*

Proof Let R be a subgroup of P containing Q, such that R/Q is a fully $\mathcal{F}/Q$-normalized, $\mathcal{F}/Q$-centric subgroup of P/Q; by Proposition 5.58, R is fully $\mathcal{F}$-normalized and $\mathcal{F}$-centric. Write $L = L_R^{\mathcal{F}}$, and $\bar{L} = L/Q$. We claim that $L_{R/Q}^{\mathcal{F}/Q}$ is a quotient of L and, since $\mathcal{F}$ is H-free, this implies that $L_{R/Q}^{\mathcal{F}/Q}$ is H-free. As R/Q was chosen arbitrarily, we see that $\mathcal{F}/Q$ is H-free, as needed.

Since L has fusion system $N_{\mathcal{F}}(R)$, we see that $\bar{L}$ has fusion system $N_{\mathcal{F}}(R)/Q$ by Theorem 5.20, and this is equal to $N_{\mathcal{F}/Q}(R/Q)$ by Exercise

5.11. Since R/Q is $\mathcal{F}/Q$-centric it is $N_{\mathcal{F}}(R)/Q$-centric, and so R/Q is p-centric in $\bar{L}$; hence

$$C_{\bar{L}}(R/Q) = Z(R/Q) \times X/Q,$$

where X/Q is some p'-group. We claim more precisely that $L_{R/Q}^{\mathcal{F}/Q} \cong L_Q^{\mathcal{F}}/X$; let $C = C_L(R/Q)$ denote the preimage of $C_{\bar{L}}(R/Q)$, and $Z = C_R(R/Q)$ be the preimage of $Z(R/Q)$ in R. We have maps from L to $\mathrm{Aut}_{\mathcal{F}}(R)$, and from $\mathrm{Aut}_{\mathcal{F}}(R)$ to $\mathrm{Aut}_{\mathcal{F}/Q}(R/Q)$; the composition of these homomorphisms has kernel C. Since C is the preimage of $C_{\bar{L}}(R/Q)$, we have that $C = ZX$, and the intersection of $Z = C_R(R/Q)$ and X is Q. Therefore the diagram

$$
\begin{array}{ccccccccc}
1 & \longrightarrow & Z(R) & \longrightarrow & L & \longrightarrow & \mathrm{Aut}_{\mathcal{F}}(R) & \longrightarrow & 1 \\
& & \downarrow & & \downarrow & & \downarrow & & \\
1 & \longrightarrow & Z(R/Q) & \longrightarrow & L/X & \longrightarrow & \mathrm{Aut}_{\mathcal{F}/Q}(R/Q) & \longrightarrow & 1
\end{array}
$$

is actually commutative.

Restricting the second row of this diagram to Sylow p-subgroups, we get the short exact sequence

$$1 \to Z(R/Q) \to N_{P/Q}(R/Q) \to \mathrm{Aut}_{P/Q}(R/Q) \to 1,$$

and hence L/K is isomorphic with $L_{R/Q}^{\mathcal{F}/Q}$, as needed.   □

We may now embark on the proof of Theorem 7.29. In fact, we shall prove something that appears slightly more general.

Theorem 7.33   *Let H be a finite group such that there exists a positive characteristic p-functor W with the property that $W(Q) \trianglelefteq G$ whenever G is H-free and Q is a Sylow p-subgroup of G. If $\mathcal{F}$ is an H-free saturated fusion system on a finite p-group P then $\mathcal{F}$ is soluble, so that $\mathcal{F} = \mathcal{F}_P(G)$ for some p-constrained, p'-reduced, H-free group G with Sylow p-subgroup P. In particular, $W(P) \trianglelefteq \mathcal{F}$.*

Proof   Let $\mathcal{F}$ be a minimal counterexample, in the sense that the number of morphisms in $\mathcal{F}$ is minimal. If $O_p(\mathcal{F}) \neq 1$, then $\mathcal{F}/O_p(\mathcal{F})$ is soluble by induction and Proposition 7.32, and hence $\mathcal{F}$ is soluble, a contradiction. Hence $O_p(\mathcal{F}) = 1$, and in particular $N_{\mathcal{F}}(Q) < \mathcal{F}$ for all fully normalized subgroups Q of P. Since each normalizer subsystem is strictly smaller than $\mathcal{F}$, by induction and Proposition 7.31, we have that (writing $\mathcal{E} = N_{\mathcal{F}}(Q)$),

$$\mathcal{E} = N_{\mathcal{E}}(W(N_P(Q)))$$

for $Q \leq P$ fully normalized; by Proposition 7.27, $\mathcal{F} = N_{\mathcal{F}}(W(P))$, a contradiction. Hence $\mathcal{F}$ is soluble, as claimed. $\qquad\square$

Specializing to the case where $H = Qd(p)$, all we need is the existence of Glauberman functors for Theorem 7.29 to hold, and by the previous section we know they do; thus Theorem 7.29 is proved.

7.4 Normal p-complement theorems

Frobenius's normal p-complement theorem, Theorem 1.12, tells us what, group theoretically, is equivalent to the statement $\mathcal{F}_P(G) = \mathcal{F}_P(P)$, namely that G possesses a normal p-complement. For us, since $O_{p'}(G)$ cannot be seen by the fusion system, we always assume that $O_{p'}(G) = 1$, and so $\mathcal{F}_P(G) = \mathcal{F}_P(P)$ implies that $G = P$.

As well as Frobenius's normal p-complement theorem, there are several other normal p-complement theorems in group theory; in Chapter 1 we saw some of them, such as the Glauberman–Thompson normal p-complement theorem.

In this section we provide analogues of the normal p-complement theorems of Thompson and Glauberman–Thompson, for all saturated fusion systems. We begin with the Thompson theorem, but before this we have a definition.

Definition 7.34 Let $\mathcal{F}$ be a saturated fusion system on a finite p-group P. We say that $\mathcal{F}$ is *trivial* if $\mathcal{F} = \mathcal{F}_P(P)$, and that $\mathcal{F}$ is *sparse* if the only saturated subsystems of $\mathcal{F}$ on P are $\mathcal{F}$ and $\mathcal{F}_P(P)$.

The idea is that fusion systems that are sparse should be minimal counterexamples to statements about whether a given fusion system is trivial, i.e., to extensions of normal p-complement theorems to all fusion systems.

For odd primes, sparse fusion systems cannot be minimal counterexamples actually, since they are constrained (in fact soluble), a fact that we shall prove in this section. Before we do so, we will extend a theorem of Glauberman to fusion systems, and use this to prove an extension of Thompson's normal p-complement theorem, Theorem 1.17.

We begin with a helpful lemma of Stancu, which can be used with sparse fusion systems in particular.

Lemma 7.35 (Stancu) *Let $\mathcal{F}$ be a saturated fusion system on a finite p-group P. If Q is a normal subgroup of $\mathcal{F}$, then*

$$\mathcal{F} = \langle P\,C_{\mathcal{F}}(Q), N_{\mathcal{F}}(Q\,C_P(Q))\rangle.$$

Proof Let R be a fully normalized, centric, radical subgroup of P, and let ϕ be an $\mathcal{F}$-automorphism of R. We see that Q is contained in R by Proposition 4.46. Since $\mathcal{F} = N_{\mathcal{F}}(Q)$, Q is strongly $\mathcal{F}$-closed, and so $\psi = \phi|_Q$ is an $\mathcal{F}$-automorphism of Q. Certainly $R \leq N_\psi$, and it is always true that $Q\,C_P(Q) \leq N_\psi$. Thus there is a homomorphism $\theta \in \mathrm{Hom}_{\mathcal{F}}(RQ\,C_P(Q), P)$ such that $\theta|_Q = \psi$; hence

$$\phi = \theta|_R \circ \left((\theta|_R)^{-1} \circ \phi\right).$$

The morphism $\theta|_R$ is a morphism in $N_{\mathcal{F}}(Q\,C_P(Q))$ (since θ acts as an automorphism on $Q\,C_P(Q)$), and $\theta|_R^{-1} \circ \phi$ lies in $P\,C_{\mathcal{F}}(Q)$. Thus $\phi \in \langle P\,C_{\mathcal{F}}(Q), N_{\mathcal{F}}(Q\,C_P(Q))\rangle$, and by Alperin's fusion theorem we get the result. $\qquad\square$

Using this lemma, we get a nice condition for when $O_p(\mathcal{F})$ is centric, which does not appear to have been noticed before, although it is hinted at in the proof of [KL08, Theorem A].

Corollary 7.36 *Let $\mathcal{F}$ be a saturated fusion system on a finite p-group, and let $Q = O_p(\mathcal{F})$. If $P\,C_{\mathcal{F}}(Q) = \mathcal{F}_P(P)$, then Q is $\mathcal{F}$-centric, and so $\mathcal{F}$ is the fusion system of a finite group.*

Proof Write $Q = O_p(\mathcal{F})$; note that $Q\,C_P(Q)$ is $\mathcal{F}$-centric by Lemma 4.42. By Lemma 7.35, and since $P\,C_{\mathcal{F}}(Q) = \mathcal{F}_P(P)$, we must have that $N_{\mathcal{F}}(Q\,C_P(Q)) = \mathcal{F}$, so that $Q\,C_P(Q) = Q$, as needed. $\qquad\square$

Now we are ready for our first major theorem of the section. In [Gla68b], Glauberman proves the following theorem.

Theorem 7.37 (Glauberman [Gla68b]) *Let G be a finite group and let P be a Sylow p-subgroup of G. Suppose that x is an element of $P \cap Z(N_G(J(P)))$, and that*

(i) *p is odd,*
(ii) *$p = 2$ and G is S_4-free, or*
(iii) *$x \in Z(P)^p$.*

Then x is weakly closed in P with respect to G.

This can be almost directly translated into fusion systems. Again, a minimal counterexample will turn out to be constrained, and so comes from a finite group. This theorem comes from [DGMP09], although the proof here was kindly supplied to us by Chris Parker and Jason Semeraro.

Theorem 7.38 (Díaz–Glesser–Mazza–Park [DGMP09]) *Let $\mathcal{F}$ be a saturated fusion system on a finite p-group P. Suppose that x is an element of $Z(N_{\mathcal{F}}(J(P)))$, and that*

(i) *p is odd,*
(ii) *$p = 2$ and $\mathcal{F}$ is S_4-free, or*
(iii) *$x \in Z(P)^p$.*

Then x lies in $Z(\mathcal{F})$.

Proof Choose a minimal counterexample to the statement such that the number of morphisms in $\mathcal{F}$ is minimal; thus x lies in $Z(N_{\mathcal{F}}(J(P)))$, but x is not central in $\mathcal{F}$. (Note that in particular x is central in P.)

Step 1: *We have that $O_p(\mathcal{F}) \neq 1$.* Write F for the conjugation family of $(\mathcal{F}, J)$-well-placed subgroups and, since x is not central in $\mathcal{F}$, we see that there is some $\phi \in \mathrm{Aut}_{\mathcal{F}}(U)$ for some U in F such that $x\phi \neq x$; choose U such that $|N_P(U)|$ is maximal subject to this condition. Since both U and $R = J(N_P(U))$ are in F, they are both fully normalized, and therefore the subsystems $N_{\mathcal{F}}(U)$, $N_{\mathcal{F}}(R)$, and $N_{\mathcal{F}}(R, U)$ are saturated (the last since $R \trianglelefteq N_P(U)$). Suppose that $N_{\mathcal{F}}(U) < \mathcal{F}$; by induction $N_{\mathcal{F}}(U)$ satisfies the theorem, and since x is not in the centre of $N_{\mathcal{F}}(U)$ by choice of U, we must have (since x is central in P) that x is not central in $N_{\mathcal{F}}(R, U)$. If x is not central in $N_{\mathcal{F}}(R, U)$, then it certainly isn't central in $N_{\mathcal{F}}(R)$, and so $N_P(U) \neq P$ since by assumption $x \in Z(N_{\mathcal{F}}(J(P)))$.

However, if $U \ntrianglelefteq P$, then $|N_P(J(N_P(U)))| > |N_P(U)|$; this contradicts the choice of U, since there is an automorphism of $J(N_P(U))$ moving x (as x is not central in its normalizer). Therefore, $\mathcal{F} = N_{\mathcal{F}}(U)$ and so $O_p(\mathcal{F}) > 1$.

Step 2: *$\mathcal{F}$ is constrained.* By Lemma 7.35, we have that $\mathcal{F}$ is generated by $P\,C_{\mathcal{F}}(U)$ and $N_{\mathcal{F}}(U\,C_P(U))$. Since x lies in the centre of $P\,C_{\mathcal{F}}(U)$ (as all morphisms in this subsystem act as P-automorphisms of U), we must have that x is not in the centre of $N_{\mathcal{F}}(U\,C_P(U))$. However, x does lie in the centre of $N_{\mathcal{F}}(U\,C_P(U), J(P))$, and so this contradicts the conclusion of the theorem. By choice of counterexample we must have that $\mathcal{F} = N_{\mathcal{F}}(U\,C_P(U))$; thus $\mathcal{F}$ has a normal, $\mathcal{F}$-centric subgroup, and so is constrained.

The result now follows from the corresponding result for groups, Theorem 7.37, together with the fact that if a constrained fusion system is S_4-free, then so is the 2-constrained, $2'$-reduced group modelling it. $\square$

In [Gla68b], Glauberman remarks that this can easily be used to prove Thompson's normal p-complement theorem, Theorem 1.17. We will do the same with the extension to fusion systems. Notice that Theorem 7.38 says that if p is odd or $\mathcal{F}$ is S_4-free, then $Z(\mathcal{F}) = Z(N_{\mathcal{F}}(J(P)))$.

Theorem 7.39 (Díaz–Glesser–Mazza–Park [DGMP09]) *Let $\mathcal{F}$ be a saturated fusion system on a finite p-group P, where p is odd or $p = 2$ and $\mathcal{F}$ is S_4-free. If both $N_{\mathcal{F}}(J(P))$ and $C_{\mathcal{F}}(Z(P))$ are trivial, so is $\mathcal{F}$.*

Proof As we have just remarked, in this case, $Z(\mathcal{F}) = Z(N_{\mathcal{F}}(J(P)))$, and since $N_{\mathcal{F}}(J(P)) = \mathcal{F}_P(P)$, we have $Z(P) = Z(\mathcal{F})$. Therefore $\mathcal{F} = C_{\mathcal{F}}(Z(P))$, and so $\mathcal{F} = \mathcal{F}_P(P)$, as claimed. $\square$

This theorem will be used to prove that, for odd primes, sparse fusion systems are soluble of p-length 2.

Corollary 7.40 (Glesser [Gle09]) *Let p be an odd prime, and let $\mathcal{F}$ be a saturated fusion system on a finite p-group P. If $\mathcal{F}$ is sparse, then $\mathcal{F}$ is soluble, of p-length at most 2.*

Proof If both $C_{\mathcal{F}}(Z(P))$ and $N_{\mathcal{F}}(J(P))$ are $\mathcal{F}_P(P)$, then by Theorem 7.39 we have that $\mathcal{F} = \mathcal{F}_P(P)$. Since $\mathcal{F}$ is sparse, this means that one of them is $\mathcal{F}$, and hence that $O_p(\mathcal{F}) > 1$. Write $Q = O_p(\mathcal{F})$; if $Q < R \trianglelefteq P$ we have that

$$N_{\mathcal{F}}(R)/Q = N_{\mathcal{F}/Q}(R/Q),$$

and since $Q = O_p(\mathcal{F})$ we see that $N_{\mathcal{F}}(R) = \mathcal{F}_P(P)$. Thus $N_{\mathcal{F}/Q}(R/Q) = \mathcal{F}_{P/Q}(P/Q)$. Therefore, by Theorem 7.39 again, $\mathcal{F}/Q = \mathcal{F}_{P/Q}(P/Q)$, and so $\mathcal{F}$ is soluble, of p-length 2, as claimed. $\square$

Using Corollary 7.40, the proof of the extension of the Glauberman–Thompson theorem is not difficult.

Theorem 7.41 (Kessar–Linckelmann [KL08]) *Let p be an odd prime, and let $\mathcal{F}$ be a saturated fusion system on a finite p-group P. Then $\mathcal{F} = \mathcal{F}_P(P)$ if and only if $N_{\mathcal{F}}(Z(J(P))) = \mathcal{F}_P(P)$.*

Proof We will prove that a minimal counterexample to the statement must be sparse, and then invoke Corollary 7.40 to prove the result. Let $\mathcal{F}$ be a minimal counterexample, in the sense that the number of morphisms in $\mathcal{F}$ is minimal, such that $N_{\mathcal{F}}(Z(J(P))) = \mathcal{F}_P(P)$ but $\mathcal{F} > \mathcal{F}_P(P)$. To see that $\mathcal{F}$ is sparse, notice that if $\mathcal{E}$ is a proper subsystem of $\mathcal{F}$ on P, then since $N_{\mathcal{E}}(Z(J(P))) = \mathcal{F}_P(P)$, we have that $\mathcal{E} = \mathcal{F}_P(P)$ by minimal choice of $\mathcal{F}$. Hence $\mathcal{F}$ is constrained, so a fusion system of a finite group, and the corresponding theorem for groups, Theorem 1.18, completes the proof. $\square$

In fact, we have a stronger result.

Theorem 7.42 *Let p be an odd prime, and let $\mathcal{F}$ be a saturated fusion system on a finite p-group P. Let $W_1, \ldots, W_n$ be a collection of positive characteristic p-functors such that, for all finite groups G with Sylow p-subgroup P, G possesses a normal p-complement if and only if $N_G(W_i(P))$ does for all $1 \le i \le n$. Then $\mathcal{F} = \mathcal{F}_P(P)$ if and only if $N_{\mathcal{F}}(W_i(P)) = \mathcal{F}_P(P)$ for all $1 \le i \le n$.*

The same proof as the previous result holds, as a minimal counterexample must be a sparse fusion system. There are several other positive characteristic p-functors (such as K^∞ and K_∞ by [Gla70, Theorem C]) that have the property needed in the hypothesis of Theorem 7.42, and so this gives a raft of normal p-complement theorems of this type.

7.5 The hyperfocal and residual subsystems

In this section we will introduce two important weakly normal subsystems, the hyperfocal subsystem – denoted by $O^p(\mathcal{F})$ – and the residual subsystem – denoted by $O^{p'}(\mathcal{F})$; both of these were introduced by Puig in [Pui06], although alternative treatments exist in the literature, namely [Asc11] and [BCGLO07]. The former subsystem has the property that the subgroup Q on which it lies is the smallest strongly $\mathcal{F}$-closed subgroup of P such that $\mathcal{F}/Q$ is the trivial fusion system, and the latter subsystem has the property that it is the smallest weakly normal subsystem of $\mathcal{F}$ on the p-group P itself.

Definition 7.43 Let $\mathcal{F}$ be a saturated fusion system on a finite p-group P. The *hyperfocal subgroup* of $\mathcal{F}$ is the subgroup

$$\mathfrak{hyp}(\mathcal{F}) = \langle [\phi, Q] : Q \le P, \phi \in O^p(\mathrm{Aut}_{\mathcal{F}}(Q)) \rangle.$$

To prove facts about the hyperfocal subgroup we need to understand the importance of the subgroup $[\phi, Q]$, and we do this with a small lemma that will help enormously.

Lemma 7.44 *Let G be a finite group and let ϕ be an automorphism of G. If H is a normal subgroup of G containing $[\phi, G]$, then H is ϕ-invariant and the induced automorphism of ϕ on G/H is trivial. Conversely, if H is a normal, ϕ-invariant subgroup of G such that ϕ acts trivially on G/H, then H contains $[\phi, G]$.*

Proof If H is a normal subgroup of G containing $[\phi, G]$, then H is ϕ-invariant since, if $x \in H$ and $h = [\phi, x^{-1}] = (x\phi)x^{-1} \in H$, then $x\phi = hx \in H$. The automorphism ϕ acts trivially on G/H if and only if $Hg = H(g\phi)$, which holds if and only if $(g\phi)g^{-1} = [\phi, g^{-1}]$ lies in H, as required. $\square$

Proposition 7.45 *Let $\mathcal{F}$ be a saturated fusion system on a finite p-group P. The subgroup $\mathfrak{hyp}(\mathcal{F})$ of P is strongly $\mathcal{F}$-closed.*

Proof Write $Q = \mathfrak{hyp}(\mathcal{F})$; we first see that Q is a normal subgroup of P, since, if R is any subgroup and $\phi \in O^p(\mathrm{Aut}_{\mathcal{F}}(R))$, and $g \in G$, we see that $[\phi, R]^g = [c_{g^{-1}}\phi c_g, R^g]$, and so $Q \trianglelefteq P$. If ϕ is a p'-automorphism of any subgroup R, then $[\phi, R] \leq Q \cap R$, so that $Q \cap R$ is a normal subgroup of R containing $[\phi, R]$; hence by Lemma 7.44, $Q \cap R$ is ϕ-invariant.

By Alperin's fusion theorem, $\mathcal{F}$ is generated by p'-automorphisms ϕ of subgroups R of P, and by the previous paragraph ϕ fixes the intersection of R with Q. Hence Q is strongly $\mathcal{F}$-closed, as required. $\square$

The hyperfocal subgroup has the useful property that $\mathcal{F}/\mathfrak{hyp}(\mathcal{F})$ is the trivial fusion system, and that if Q is any other strongly $\mathcal{F}$-closed subgroup of P such that $\mathcal{F}/Q$ is the trivial fusion system, then $\mathfrak{hyp}(\mathcal{F}) \leq Q$; this mimics the situation for groups.

Proposition 7.46 *Let $\mathcal{F}$ be a saturated fusion system on a finite p-group P, and write $Q = \mathfrak{hyp}(\mathcal{F})$. Let R be a subgroup of P.*

(i) $\mathcal{F}/Q = \mathcal{F}_{P/Q}(P/Q)$.
(ii) *If R is strongly $\mathcal{F}$-closed and $\mathcal{F}/R = \mathcal{F}_{P/R}(P/R)$, then $R \geq Q$.*
(iii) *If R is a normal in P and contains Q, then R is strongly $\mathcal{F}$-closed and $\mathcal{F}/R = \mathcal{F}_{P/R}(P/R)$.*

Proof By Alperin's fusion theorem, $\mathcal{F}/R = \mathcal{F}_{P/R}(P/R)$ if and only if there are no p'-automorphisms of subgroups S/R of P/R, and this is true if and only if, for every subgroup S of P and p'-automorphism $\phi \in$

$\mathrm{Aut}_{\mathcal{F}}(S)$, ϕ acts trivially on RS/R. By Lemma 7.44, ϕ acts trivially on RS/R if and only if $[\phi, S] \leq R$. We see therefore that $\mathcal{F}/R$ possesses no p'-automorphisms of subgroups if and only if $[\phi, S] \leq R$ for all $S \leq P$ and all p'-automorphisms ϕ of S in $\mathcal{F}$. Since $O^p(\mathrm{Aut}_{\mathcal{F}}(S))$ is the subgroup generated by all p'-elements of $\mathrm{Aut}_{\mathcal{F}}(S)$, we see that $\mathcal{F}/R$ is trivial if and only if $\mathfrak{hyp}(\mathcal{F}) \leq R$, proving the first two parts.

For the final part, any normal subgroup of P/Q is strongly $\mathcal{F}/Q$-closed, and hence Theorem 5.21 proves that if $Q \leq R \trianglelefteq P$ then R is strongly $\mathcal{F}$-closed. The third isomorphism theorem for fusion systems proves that $(\mathcal{F}/Q)/(R/Q) = \mathcal{F}/R$, and hence we get the result. $\qquad\square$

We also saw the focal subgroup in Chapter 1; this has a similar definition to the hyperfocal subgroup given above.

Definition 7.47 Let $\mathcal{F}$ be a saturated fusion system on a finite p-group P. The *focal subgroup* of $\mathcal{F}$ is the subgroup

$$\mathfrak{foc}(\mathcal{F}) = \langle [Q, \phi] \,:\, Q \leq P, \ \phi \in \mathrm{Aut}_{\mathcal{F}}(Q) \rangle.$$

Clearly we have $\mathfrak{hyp}(\mathcal{F}) \leq \mathfrak{foc}(\mathcal{F})$, so that $\mathfrak{foc}(\mathcal{F})$ is strongly $\mathcal{F}$-closed and $\mathcal{F}/\mathfrak{foc}(\mathcal{F})$ is a trivial fusion system. Also, since $\mathfrak{foc}(\mathcal{F})$ contains $[P, c_g]$ for all $g \in P$, we see that $P' \leq \mathfrak{foc}(\mathcal{F})$, so that the quotient is a trivial fusion system on an abelian p-group.

Proposition 7.48 *Let $\mathcal{F}$ be a saturated fusion system on a finite p-group P, and write $Q = \mathfrak{foc}(\mathcal{F})$. Let R be a subgroup of P such that P/R is abelian.*

(i) $\mathcal{F}/Q = \mathcal{F}_{P/Q}(P/Q)$ *and P/Q is abelian.*
(ii) *If R is strongly $\mathcal{F}$-closed and $\mathcal{F}/R = \mathcal{F}_{P/R}(P/R)$, then $R \geq Q$.*
(iii) $Q = P'\mathfrak{hyp}(\mathcal{F})$.

Proof The proof of (i) is given before the statement of the proposition. Exercise 7.9 proves (ii). To prove (iii), by (ii) we have $Q \leq P'\mathfrak{hyp}(\mathcal{F})$, and since both $\mathfrak{hyp}(\mathcal{F})$ and P' are contained in Q, we get the opposite inclusion. $\qquad\square$

Having determined some properties of the hyperfocal subgroup, we now must construct a certain weakly normal subsystem on any normal overgroup of $\mathfrak{hyp}(\mathcal{F})$. In fact, we will construct a saturated subsystem on each overgroup of $\mathfrak{hyp}(\mathcal{F})$, and show that it is the unique such saturated subsystem on that subgroup. We will follow [BCGLO07] in making the following definition.

Definition 7.49 Let $\mathcal{F}$ be a saturated fusion system on a finite p-group P. A saturated subsystem $\mathcal{E}$ of $\mathcal{F}$ is said to have *p-power index* if $\mathcal{E}$ lies on a subgroup Q of P containing $\mathfrak{hyp}(\mathcal{F})$ and, whenever R is a subgroup of Q, then

$$\mathrm{Aut}_{\mathcal{E}}(R) \geq O^p(\mathrm{Aut}_{\mathcal{F}}(R)).$$

We will first show that if $\mathfrak{hyp}(\mathcal{F}) \leq Q \leq P$ then there can be at most one saturated subsystem of $\mathcal{F}$ of p-power index on Q.

Lemma 7.50 *Let $\mathcal{F}$ be a saturated fusion system on a finite p-group P, and let Q be a subgroup of P containing $\mathfrak{hyp}(\mathcal{F})$. If $\mathcal{E}$ is a saturated subsystem of $\mathcal{F}$ on Q with p-power index, and $\mathcal{E}'$ is any other saturated subsystem of $\mathcal{F}$ on Q, then $\mathcal{E}'$ is contained in $\mathcal{E}$. In particular, there is at most one saturated subsystem of $\mathcal{F}$ of p-power index on Q.*

Proof Let R be a fully $\mathcal{E}'$-normalized subgroup of Q. Since R is fully $\mathcal{E}'$-normalized, $\mathrm{Aut}_Q(R)$ is a Sylow p-subgroup of $\mathrm{Aut}_{\mathcal{E}'}(R)$ and since, for any group G, we have $G = \langle O^p(G), S \rangle$, where $S \in \mathrm{Syl}_p(G)$, we have

$$\begin{aligned}
\mathrm{Aut}_{\mathcal{E}'}(R) &= \langle O^p(\mathrm{Aut}_{\mathcal{E}'}(R)), \mathrm{Aut}_Q(R) \rangle \\
&\leq \langle O^p(\mathrm{Aut}_{\mathcal{F}}(R)), \mathrm{Aut}_Q(R) \rangle \\
&\leq \mathrm{Aut}_{\mathcal{E}}(R).
\end{aligned}$$

Since $\mathrm{Aut}_{\mathcal{E}'}(R) \leq \mathrm{Aut}_{\mathcal{E}}(R)$ for all fully $\mathcal{E}'$-normalized subgroups $R \leq Q$, and $\mathcal{E}'$ is generated by such groups by Alperin's fusion theorem, we see that $\mathcal{E}'$ must be contained in $\mathcal{E}$, as claimed. $\square$

We now come to the existence of a subsystem of p-power index on each overgroup of $\mathfrak{hyp}(\mathcal{F})$; we will construct a weakly normal map on every normal subgroup of P containing $\mathfrak{hyp}(\mathcal{F})$ and show that the resulting subsystem has p-power index; since every subgroup is subnormal, this will prove the existence of a subsystem of p-power index on each overgroup of $\mathfrak{hyp}(\mathcal{F})$.

Theorem 7.51 (Puig [Pui06, Theorem 7.4]) *Let $\mathcal{F}$ be a saturated fusion system on a finite p-group P, and let Q be a subgroup of P containing $\mathfrak{hyp}(\mathcal{F})$.*

(i) *There is a unique saturated subsystem $\mathcal{F}_Q$ of $\mathcal{F}$ on Q of p-power index.*

(ii) *If Q is normal in P then $\mathcal{F}_Q$ is weakly normal in $\mathcal{F}$.*

(iii) *If $Q/\mathfrak{hyp}(\mathcal{F})$ is characteristic in $P/\mathfrak{hyp}(\mathcal{F})$ then $\mathcal{F}_Q$ is weakly characteristic in $\mathcal{F}$.*

Proof We proceed in a series of steps. Suppose firstly that Q is a normal subgroup of P containing $\mathfrak{hyp}(\mathcal{F})$, so that Q is strongly $\mathcal{F}$-closed by Proposition 7.45. We will define a fusion system $\mathcal{F}_Q$ on Q via an $\mathcal{F}$-invariant map $A(-)$, and show that it is a weakly normal map.

Step 1: *Construction of the $\mathcal{F}$-invariant map.* If $U \leq Q$ is fully $\mathcal{F}$-normalized, define $A(U)$ to be $O^p(\operatorname{Aut}_{\mathcal{F}}(U)) \cdot \operatorname{Aut}_Q(U)$. If U and V are fully $\mathcal{F}$-normalized subgroups of Q, $\mathcal{F}$-conjugate via an isomorphism ϕ, then $O^p(\operatorname{Aut}_{\mathcal{F}}(U))^\phi = O^p(\operatorname{Aut}_{\mathcal{F}}(V))$ and $\operatorname{Aut}_P(U)^\phi = \operatorname{Aut}_P(V)$, proving that $A(U)^\phi = A(V)$. Clearly $\operatorname{Aut}_Q(U) \leq A(U)$ as well, and hence there is an $\mathcal{F}$-invariant map $A(-)$ extending this by Lemma 6.18. Let $\mathcal{F}_Q$ be the subsystem generated by $A(-)$; since $Q \geq \mathfrak{hyp}(\mathcal{F})$ and by construction $O^p(\operatorname{Aut}_{\mathcal{F}}(U))$ is contained in $\operatorname{Aut}_{\mathcal{F}_Q}(U)$ for all $U \leq Q$, we see that $\mathcal{F}_Q$ has p-power index and is $\mathcal{F}$-invariant. It remains to show that $\mathcal{F}_Q$ is saturated, which will be done by proving that $A(-)$ is a weakly normal map.

Step 2: *If R is a fully $\mathcal{F}$-normalized subgroup of Q then $\operatorname{Aut}_Q(R)$ is a Sylow p-subgroup of $A(R)$.* We will show that $\operatorname{Aut}_Q(R)$ contains a Sylow p-subgroup of $O^p(\operatorname{Aut}_{\mathcal{F}}(R))$. If this is true then $\operatorname{Aut}_Q(R)$ is clearly a Sylow p-subgroup of $O^p(\operatorname{Aut}_{\mathcal{F}}(R)) \operatorname{Aut}_Q(R) = A(R)$. Since R is fully $\mathcal{F}$-normalized, $\operatorname{Aut}_P(R)$ is a Sylow p-subgroup of $\operatorname{Aut}_{\mathcal{F}}(R)$.

We will apply the hyperfocal subgroup theorem, Theorem 1.33, to $\operatorname{Aut}_{\mathcal{F}}(R)$; this states that $A = \operatorname{Aut}_P(R) \cap O^p(\operatorname{Aut}_{\mathcal{F}}(R))$ is generated by the sets $[\operatorname{Aut}_S(R), \phi]$, where S runs over all subgroups of $\mathrm{N}_P(R)$ (which can be chosen to contain $\mathrm{C}_P(R)$) and ϕ is any p'-element of $\mathrm{N}_{\operatorname{Aut}_{\mathcal{F}}(R)}(\operatorname{Aut}_S(R))$. By replacing S by SR, which can only increase the size of $[\operatorname{Aut}_S(R), \phi]$ (since $\operatorname{Aut}_R(R)$ is a normal subgroup of $\operatorname{Aut}_{\mathcal{F}}(R)$, if ϕ normalizes $\operatorname{Aut}_S(R)$ it normalizes $\operatorname{Aut}_{SR}(R)$), we may assume that A is generated by the sets $[\operatorname{Aut}_S(R), \phi]$, where $R\,\mathrm{C}_P(R) \leq S \leq \mathrm{N}_P(R)$ and ϕ is a p'-element of $\mathrm{N}_{\operatorname{Aut}_{\mathcal{F}}(R)}(\operatorname{Aut}_S(R))$. Let S be such a subgroup and ϕ be such an automorphism.

Since $\mathcal{F}$ is saturated and R is fully $\mathcal{F}$-normalized, by Theorem 6.9, ϕ extends to an automorphism ψ of S, which can be chosen to be a p'-automorphism by raising to a suitable power. Therefore the set $[\operatorname{Aut}_S(R), \phi]$ is simply the image of $[S, \psi]$ in $\operatorname{Aut}_{\mathcal{F}}(R)$. Finally, $\mathfrak{hyp}(\mathcal{F})$ contains $[S, \psi]$, and hence so does Q, and therefore $\operatorname{Aut}_Q(R)$ contains $[\operatorname{Aut}_S(R), \phi]$; thus $\operatorname{Aut}_Q(R)$ contains A, completing the proof.

Step 3: *If R is a fully $\mathcal{F}$-normalized subgroup of Q, then every element of $A(R)$ extends to an element of $A(R\,\mathrm{C}_Q(R))$.* If $\phi \in O^p(\operatorname{Aut}_{\mathcal{F}}(R))$, then ϕ has an extension $\bar{\phi}$ to $R\,\mathrm{C}_Q(R)$. By choosing a suitable p-power of $\bar{\phi}$, we can produce a p'-automorphism $\bar{\phi}$ extending ϕ. If $c_g \in \mathrm{Aut}_Q(R)$, then g normalizes R, whence g normalizes $R\,C_Q(R)$, and so c_g lies in $\mathrm{Aut}_Q(R\,C_Q(R))$. Hence every element of $A(R) = \mathrm{O}^p(\mathrm{Aut}_{\mathcal{F}}(R))\,\mathrm{Aut}_Q(R)$ extends to an element of $A(R\,C_Q(R))$, as required.

Step 4: *If $R \trianglelefteq S$ are subgroups of Q, then the restriction map $A(R \leq S) \to \mathrm{Aut}(R)$ has image inside $\mathrm{N}_{A(R)}(\mathrm{Aut}_S(R))$. If $\phi \in A(R)$ extends to an automorphism of S, then $\phi \in \mathrm{N}_{A(R)}(\mathrm{Aut}_S(R))$, so it suffices to show that if $\phi \in A(R \leq S)$ then $\phi|_R \in A(R)$.*

Firstly, we assume that S is fully $\mathcal{F}$-normalized. By Step 2, the subgroups $\mathrm{Aut}_P(S)$ and $\mathrm{Aut}_Q(S)$ are Sylow p-subgroups of $\mathrm{Aut}_{\mathcal{F}}(S)$ and $A(S)$ respectively. Since $\mathrm{Aut}_{\mathcal{F}}(S) = \mathrm{Aut}_P(S)A(S)$, any two Sylow p-subgroups of $\mathrm{Aut}_{\mathcal{F}}(S)$ are conjugate by an element of $A(S)$, and so there is $\chi \in A(S)$ such that $\mathrm{Aut}_P(S)^\chi$ contains a Sylow p-subgroup of $\mathrm{Aut}_{\mathcal{F}}(R \leq S)$. Writing $\hat{R} = R^{\chi^{-1}}$, we have that $\mathrm{Aut}_P(\hat{R} \leq S)^\chi$ is a Sylow p-subgroup of $\mathrm{Aut}_{\mathcal{F}}(R \leq S)$. Since $\mathrm{Aut}_P(\hat{R} \leq S)^\chi \cap \mathrm{Aut}_Q(S)^\chi = \mathrm{Aut}_Q(\hat{R} \leq S)^\chi$, we see that $\mathrm{Aut}_Q(\hat{R} \leq S)^\chi$ is a Sylow p-subgroup of $A(R \leq S)$. Hence

$$A(R \leq S) = \mathrm{Aut}_Q(\hat{R} \leq S)^\chi \, \mathrm{O}^p\left(A(R \leq S)\right).$$

Clearly the restriction of $\mathrm{Aut}_Q(\hat{R} \leq S)$ to $\mathrm{Aut}(\hat{R})$ is contained in $A(\hat{R})$, so the restriction of $\mathrm{Aut}_Q(\hat{R} \leq S)^\chi$ is contained in $A(\hat{R})^\chi = A(R)$. Also, the restriction of $\mathrm{O}^p(A(R \leq S))$ to R is contained within $\mathrm{O}^p(\mathrm{Aut}_{\mathcal{F}}(R))$, and by construction this is contained in $A(R)$ as well.

For the general case, fix R and S, and let $\phi : S \to \bar{S}$ be an $\mathcal{F}$-automorphism with $\bar{S}$ fully $\mathcal{F}$-normalized. Writing $\bar{R} = R\phi$, we have the following diagram.

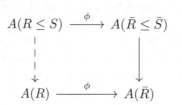

Both of the horizontal maps are isomorphisms, so the dashed arrow is an induced map, which agrees with the restriction map from $A(R \leq S)$ to $\mathrm{Aut}(R)$. This completes the proof of Step 4.

Step 5: *If R is fully $\mathcal{F}$-normalized and $C_Q(R) \leq R$, then for $R \leq S \leq \mathrm{N}_Q(R)$ the restriction map $A(R \leq S) \to \mathrm{N}_{A(R)}(\mathrm{Aut}_S(R))$ is surjective.* Since R is fully $\mathcal{F}$-normalized, the map $\mathrm{Aut}_{\mathcal{F}}(R \leq S) \to \mathrm{N}_{\mathrm{Aut}_{\mathcal{F}}(R)}(\mathrm{Aut}_S(R))$ is surjective. Hence, this must map $\mathrm{O}^p(\mathrm{Aut}_{\mathcal{F}}(R \leq S))$

onto $O^p\left(N_{\mathrm{Aut}_{\mathcal{F}}(R)}(\mathrm{Aut}_S(R))\right)$. On the one hand, $O^p(\mathrm{Aut}_{\mathcal{F}}(R \leq S)) \leq O^p(\mathrm{Aut}_{\mathcal{F}}(R)) \leq A(R)$, and so therefore $O^p(\mathrm{Aut}_{\mathcal{F}}(R \leq S)) \leq A(R \leq S)$. Since the restrictions of elements of $A(R \leq S)$ lie in $A(R)$, it must be that $O^p\left(N_{\mathrm{Aut}_{\mathcal{F}}(R)}(\mathrm{Aut}_S(R))\right)$ lies inside $N_{A(R)}(\mathrm{Aut}_S(R))$; we must have a surjective map

$$O^p\left(A(R \leq S)\right) \to O^p\left(N_{A(R)}(\mathrm{Aut}_S(R))\right).$$

As any group G is generated by $O^p(G)$ and a Sylow p-subgroup of G, it remains to show that a Sylow p-subgroup of $A(R \leq S)$ surjects onto a Sylow p-subgroup of $N_{A(R)}(\mathrm{Aut}_S(R))$.

Let K be the subgroup $N_{A(R)}(\mathrm{Aut}_S(R))$ of $\mathrm{Aut}(R)$, and notice that $\mathrm{Aut}_S(R)$ is contained in K. Since $A(R) \cap \mathrm{Aut}_P(R)$ contains $\mathrm{Aut}_Q(R)$, and $\mathrm{Aut}_Q(R)$ is a Sylow p-subgroup of $A(R)$ by Step 2, we see that $\mathrm{Aut}_Q(R) = A(R) \cap \mathrm{Aut}_P(R)$, and so $\mathrm{Aut}_P^K(R) = \mathrm{Aut}_Q^K(R)$ (as $K \leq A(R)$).

Firstly, assume that R is fully K-normalized, so that $\mathrm{Aut}_P^K(R) = \mathrm{Aut}_Q^K(R)$ is a Sylow p-subgroup of K by Lemma 4.35. Writing out the definition of $\mathrm{Aut}_Q^K(R)$, we see that

$$N_{\mathrm{Aut}_Q(R)}(\mathrm{Aut}_S(R)) \in \mathrm{Syl}_p\left(N_{A(R)}(\mathrm{Aut}_S(R))\right).$$

Let $X = N_Q(R) \cap N_Q(S)$. There are surjective maps $f : X \to \mathrm{Aut}_Q(R \leq S)$ and $f' : X \to N_{\mathrm{Aut}_Q(R)}(\mathrm{Aut}_S(R))$, given by the maps $c_S : X \to \mathrm{Aut}(S)$ and $c_R : X \to \mathrm{Aut}(R)$. By Step 4, we also have the restriction map $f'' : \mathrm{Aut}_Q(R \leq S) \to N_{\mathrm{Aut}_Q(R)}(\mathrm{Aut}_S(R))$, and clearly we have that $ff'' = f'$. Since f' is surjective, we have that ff'' is surjective, and so f'' is surjective. As $N_{\mathrm{Aut}_Q(R)}(\mathrm{Aut}_S(R))$ is a Sylow p-subgroup of $N_{A(R)}(\mathrm{Aut}_S(R))$, this completes the proof in the case where R is also fully K-normalized.

To finish the proof, suppose that R is not fully K-normalized, and let $\phi : R \to \bar{R}$ be an $\mathcal{F}$-isomorphism where $\bar{R}$ is a fully K^ϕ-normalized subgroup of Q. By Lemma 4.36, there is a map $\psi : R N_P^K(R) \to N_P(R\phi)$ and an automorphism $\chi \in K^\phi$ such that $\psi\chi$ extends ϕ; replacing ϕ by ψ (and noticing that $K^\phi = K^\psi$ and $N_P^K(R) = N_Q^K(R)$), we may assume that ϕ extends to a map $\bar{\phi} : N_Q^K(R) \to N_Q(\bar{R})$. Let $\theta : \bar{R} \to R$ be some map that extends to $\bar{\theta} : N_Q(\bar{R}) \to N_Q(R)$ (which exists as R is fully $\mathcal{F}$-normalized), and let $\hat{S} = S\bar{\phi}\bar{\theta}$. We claim that R is fully $\mathcal{F}$-normalized and fully $K^{\phi\theta}$-normalized. The former claim is true by hypothesis, and the latter claim is true because $N_Q^{K^\phi}(\bar{R})$ is embedded in $N_Q(R)$ by the map $\bar{\theta}$, so the image R of $\bar{R}$ must be fully $K^{\phi\theta}$-normalized.

By the first argument, we know that the restriction map

$\alpha : A(R \leq \hat{S}) \rightarrow \mathrm{N}_{A(R)}(\mathrm{Aut}_{\hat{S}}(R))$ is surjective. As in the previous step, we have the commutative diagram given below.

$$
\begin{array}{ccc}
A(R \leq \hat{S}) & \xrightarrow{\bar{\phi}\bar{\theta}} & A(R \leq S) \\
{\scriptstyle \alpha}\big\downarrow & & \big\downarrow \\
\mathrm{N}_{A(R)}(\mathrm{Aut}_{\hat{S}}(R)) & \xrightarrow{\phi\theta} & \mathrm{N}_{A(R)}(\mathrm{Aut}_S(R))
\end{array}
$$

Since the horizontal arrows are isomorphisms and the leftmost vertical map is surjective, the rightmost vertical map is also surjective, completing the proof of Step 5.

Step 6: *Completion of the proof.* Completing the proof of (ii), notice that by the previous steps we have verified the five properties for $A(-)$ to be a weakly normal map. To complete the proof of (i), suppose that Q is not a normal subgroup of P, and let R be a subgroup of P with $|R : Q| = p$. By induction on $|P : Q|$ there is a subsystem $\mathcal{F}_R$ of p-power index on R, and $Q \trianglelefteq R$, so we have a subsystem $\mathcal{F}_Q$ of p-power index in $\mathcal{F}_R$. It remains to see that $\mathcal{F}_Q$ has p-power index in $\mathcal{F}$, not just in $\mathcal{F}_R$. However, Q contains $\mathfrak{hyp}(\mathcal{F})$ by hypothesis, and certainly, for any $U \leq Q$, we have

$$
\mathrm{Aut}_{\mathcal{F}_Q}(U) \geq \mathrm{O}^p(\mathrm{Aut}_{\mathcal{F}_R}(U)) = \mathrm{O}^p(\mathrm{Aut}_{\mathcal{F}}(U)),
$$

since $\mathrm{Aut}_{\mathcal{F}_R}(U)$ contains $\mathrm{O}^p(\mathrm{Aut}_{\mathcal{F}}(U))$. This proves that $\mathcal{F}_Q$ has p-power index in $\mathcal{F}$, completing (i).

Finally, to see (iii), note that any element of $\mathrm{Aut}(\mathcal{F})$ must fix $\mathfrak{hyp}(\mathcal{F})$, since it is the smallest subgroup whose quotient system is trivial, and so every element of $\mathrm{Aut}(\mathcal{F})$ induces an automorphism on $\mathcal{F}/\mathfrak{hyp}(\mathcal{F})$. If $Q/\mathfrak{hyp}(\mathcal{F})$ is characteristic in $P/\mathfrak{hyp}(\mathcal{F})$, then every element of $\mathrm{Aut}(\mathcal{F})$ must fix Q, and since $\mathcal{F}_Q$ is the largest saturated subsystem of $\mathcal{F}$ on Q, it must be fixed by all automorphisms of $\mathcal{F}$, proving (iii). $\qquad\square$

Having established the existence of such a subsystem, we make the following definition.

Definition 7.52 Let $\mathcal{F}$ be a saturated fusion system on a finite p-group P. The *hyperfocal subsystem* of $\mathcal{F}$, denoted by $\mathrm{O}^p(\mathcal{F})$, is the unique saturated subsystem of p-power index on $\mathfrak{hyp}(\mathcal{F})$.

The first thing we need to do is establish that taking the hyperfocal subgroup again does not affect things. More generally, we should check that p-power index subsystems of a given fusion system really do have the same hyperfocal subsystem as the original fusion system. The first parts of this next theorem are given in [Pui06, 7.5], and the rest in [Asc11, (7.19)], although the method of proof is different.

Theorem 7.53 (Puig [Pui06], Aschbacher [Asc11]) *Let $\mathcal{F}$ be a saturated fusion system on a finite p-group P, and let $\mathcal{E}$ be a saturated subsystem of $\mathcal{F}$, on the subgroup Q of P.*

(i) $O^p(\mathcal{E}) \leq O^p(\mathcal{F})$.

(ii) $O^p(O^p(\mathcal{F})) = O^p(\mathcal{F})$.

(iii) *If $\mathcal{E}$ lies on P (i.e., $Q = P$), then $O^p(\mathcal{E}) = O^p(\mathcal{F})$ implies $\mathcal{E} = \mathcal{F}$.*

(iv) *$\mathcal{E}$ has p-power index if and only if $O^p(\mathcal{E}) = O^p(\mathcal{F})$.*

Proof For this proof, write S for the subgroup $\mathfrak{hyp}(\mathcal{F})$, and T for the subgroup $\mathfrak{hyp}(\mathcal{E})$. Clearly, we have $T \leq S$ by the definition.

Let $\mathcal{E}_S$ denote the subsystem of $\mathcal{E}$ of p-power index on the subgroup S. By Lemma 7.50, $O^p(\mathcal{F})$ is the largest saturated subsystem of $\mathcal{F}$ on S, and so $\mathcal{E}_S \leq O^p(\mathcal{F})$. Hence $O^p(\mathcal{E}) \leq \mathcal{E}_S \leq O^p(\mathcal{F})$, proving (i).

To prove that $O^p(O^p(\mathcal{F})) = O^p(\mathcal{F})$, let $\mathcal{E}$ denote $O^p(\mathcal{F})$. We will prove that T is strongly $\mathcal{F}$-closed and that $\mathcal{F}/T$ is trivial, showing that $S = T$ by Proposition 7.46. Since $O^p(\mathcal{E})$ is weakly characteristic in $\mathcal{E}$, which is in turn weakly characteristic in $\mathcal{F}$, we see that $O^p(\mathcal{E})$ is weakly characteristic in $\mathcal{F}$ (by Proposition 5.44), proving in particular that T is strongly $\mathcal{F}$-closed.

We quotient out by T, so we may assume that $O^p(\mathcal{E}) = 1$, i.e., that $\mathcal{E} = \mathcal{F}_S(S)$ by Proposition 7.46. Since $\mathcal{E}$ is weakly normal in $\mathcal{F}$, we get that $\mathcal{F} = N_{\mathcal{F}}(S)$ by Theorem 5.37. However, $\mathrm{Aut}_{\mathcal{F}}(S)$ must be a p-group, since $\mathrm{Aut}_{\mathcal{E}}(S)$ is a p-group and $O^p(\mathrm{Aut}_{\mathcal{F}}(S)) \leq \mathrm{Aut}_{\mathcal{E}}(S)$, by the definition of a p-power index subsystem. This implies that S is hypercentral; Lemma 5.61 now implies that $\mathcal{F} = \mathcal{F}_P(P)$, since $\mathcal{F}/S = \mathcal{F}_{P/S}(P/S) = \mathcal{F}_P(P)/S$ (and clearly $\mathcal{F}_P(P) \leq \mathcal{F}$). This contradicts the statement that $\mathfrak{hyp}(\mathcal{F}) = S$, unless $S = T$, proving (ii).

We turn our attention to (iii), so assume that $Q = P$. Suppose that $O^p(\mathcal{E}) = O^p(\mathcal{F})$; we aim to prove that $\mathcal{E} = \mathcal{F}$. Notice that, if R is a subgroup of S, then $O^p(\mathrm{Aut}_{\mathcal{E}}(R)) = O^p(\mathrm{Aut}_{\mathcal{F}}(R))$, and so any p'-automorphism of R lying in $\mathcal{F}$ also lies in $\mathcal{E}$. Let U be an $\mathcal{F}$-centric, $\mathcal{F}$-radical subgroup of P, and let $R = U \cap S$. Since S is normal in P, R is normal in U, and in fact, since S is strongly $\mathcal{F}$-closed, $\mathrm{Aut}_{\mathcal{F}}(U) = \mathrm{Aut}_{\mathcal{F}}(R \leq U)$. For a contradiction, we choose U to have largest order subject to $\mathrm{Aut}_{\mathcal{E}}(U) < \mathrm{Aut}_{\mathcal{F}}(U)$, so we may assume that $\mathrm{Aut}_{\mathcal{F}}(V) = \mathrm{Aut}_{\mathcal{E}}(V)$ for all subgroups of P of order greater than $|U|$. (This allows the case $U = P$.) Since both $\mathcal{F}$ and $\mathcal{E}$ are saturated, the $\mathcal{F}$-conjugacy class and $\mathcal{E}$-conjugacy class of U coincide.

If $\tilde{U}$ is any fully $\mathcal{F}$-normalized subgroup of P that is $\mathcal{F}$-conjugate (hence $\mathcal{E}$-conjugate) to U via $\phi : \tilde{U} \to U$, then $\mathrm{Aut}_P(\tilde{U})$ is a Sylow p-subgroup of both $\mathrm{Aut}_{\mathcal{E}}(\tilde{U})$ and $\mathrm{Aut}_{\mathcal{F}}(\tilde{U})$. As $\mathrm{Aut}_P(\tilde{U})^{\phi}$ is a Sylow p-subgroup of both $\mathrm{Aut}_{\mathcal{E}}(U)$ and $\mathrm{Aut}_{\mathcal{F}}(U)$, it suffices to show that any p'-automorphism of U lying in $\mathcal{F}$ also lies in $\mathcal{E}$, i.e., that $\mathrm{O}^p(\mathrm{Aut}_{\mathcal{E}}(U)) = \mathrm{O}^p(\mathrm{Aut}_{\mathcal{F}}(U))$. If R is not fully $\mathcal{F}$-normalized, then we may choose a fully $\mathcal{F}$-normalized subgroup $\bar{R}$ of S and a map $\mathrm{N}_P(R) \to \mathrm{N}_P(\bar{R})$ that maps U to $\bar{U}$, another $\mathcal{F}$-centric, $\mathcal{F}$-radical subgroup of P. Since we need to show that $\mathrm{O}^p(\mathrm{Aut}_{\mathcal{E}}(U)) = \mathrm{O}^p(\mathrm{Aut}_{\mathcal{F}}(U))$, we may pass to $\bar{U}$ without a problem, so from now on we may assume that R is fully $\mathcal{F}$-normalized.

Let K denote the subgroup of all elements of $\mathrm{Aut}_{\mathcal{F}}(U)$ that act trivially on R. This is a normal subgroup of $\mathrm{Aut}_{\mathcal{F}}(U)$; if $\phi \in K$ is a p'-automorphism, then ϕ acts trivially on R. Also, the image of ϕ in $\mathcal{F}/S$ is a p-automorphism, and hence is trivial, so that ϕ acts trivially on $U/R \cong U/(\tilde{U} \cap S)$; thus K is a normal p-subgroup of $\mathrm{Aut}_{\mathcal{F}}(U)$ by Exercise 4.5. As U is $\mathcal{F}$-radical, $\mathrm{O}_p(\mathrm{Aut}_{\mathcal{F}}(U)) = \mathrm{Inn}(U)$, and in particular $K \leq \mathrm{Aut}_{\mathcal{E}}(U)$.

Let ϕ be a p'-automorphism in $\mathrm{Aut}_{\mathcal{F}}(U)$; the restriction $\hat{\phi}$ of ϕ to R is a non-trivial p'-automorphism of R (as ϕ acts trivially on U/R), and since $\hat{\phi}$ is a p'-automorphism, $\hat{\phi} \in \mathrm{Aut}_{\mathcal{E}}(R)$. If θ is another automorphism of U such that $\theta|_R = \hat{\phi}$, then $\phi\theta^{-1} \in K$. Since $K \leq \mathrm{Aut}_{\mathcal{E}}(R)$, ϕ lies in $\mathrm{Aut}_{\mathcal{E}}(U)$ if and only if θ does, and hence it suffices to show that there is a single extension of $\hat{\phi}$ to U in $\mathcal{E}$.

Notice that, since ϕ acts trivially on U/R, $\hat{\phi}$ acts trivially on $\mathrm{Aut}_U(R)$. As $\mathcal{E}$ is saturated and R is fully $\mathcal{F}$-normalized, $\hat{\phi}$ extends to an automorphism ψ of the preimage in $N_{\hat{\phi}}$ of $\mathrm{Aut}_U(R)$, namely

$$N = \mathrm{Aut}_U(R)\,\mathrm{C}_P(R) = \mathrm{N}_P^L(R),$$

where $L = \mathrm{Aut}_U(R)$. By raising ψ to an appropriate power, we may assume that ψ is a p'-automorphism extending ϕ. The intersection $N \cap S$ is clearly $\mathrm{N}_S^L(R)$, and, as before, ψ must act trivially on $N/(N \cap S)$ (as this action is the image of ψ in $\mathcal{F}/S$).

We claim that ψ also acts trivially on $(N \cap S)/R$; if this is true then ψ acts trivially on N/R, and so in particular induces an automorphism of U; this will complete the proof, as then $\phi \in \mathrm{Aut}_{\mathcal{E}}(U)$. Since $\hat{\phi}$ acts trivially on L, we must have that ψ acts trivially on $(N \cap S)/R\,\mathrm{C}_S(R)$;

if $C_S(R) \leq R$ then we are done, but this is true by Exercise 5.14, completing the proof of (iii).

It remains to prove (iv). Let $\mathcal{E}$ be a subsystem of p-power index in $\mathcal{F}$. By (i), $\mathfrak{hyp}(\mathcal{E}) \leq \mathfrak{hyp}(\mathcal{F})$. If we show the opposite inclusion, then by the uniqueness of subsystems of p-power index, we must have that $O^p(\mathcal{F}) = O^p(\mathcal{E})$, since clearly $O^p(\mathcal{F})$ is a subsystem of p-power index in $\mathcal{E}$. We claim that this inclusion follows from (ii). To see this, suppose that $\mathfrak{hyp}(\mathcal{E}) < \mathfrak{hyp}(\mathcal{F})$; the image of $O^p(\mathcal{F})$ in $\mathcal{E}/\mathfrak{hyp}(\mathcal{E})$ is then a trivial fusion system. However, by (ii), $O^p(\mathcal{E})$ has no trivial fusion systems as quotients, so this is not possible. Hence if $\mathcal{E}$ is a subsystem of p-power index in $\mathcal{F}$, then $O^p(\mathcal{E}) = O^p(\mathcal{F})$.

Conversely, suppose that $\mathcal{E}$ is a saturated subsystem of $\mathcal{F}$ such that $O^p(\mathcal{E}) = O^p(\mathcal{F})$, and let $\mathcal{F}_Q$ denote the subsystem of $\mathcal{F}$ on Q – the subgroup on which $\mathcal{E}$ is defined – of p-power index. Since $\mathcal{E}$ is contained in $\mathcal{F}_Q$ by Lemma 7.50, we may assume without loss of generality that $\mathcal{E}$ lies on P itself, since $O^p(\mathcal{F}_Q) = O^p(\mathcal{F})$; this case follows by (iii), completing the proof. $\qquad\square$

The other weakly normal subsystem to be introduced in this section is the residual subsystem. It is the smallest weakly normal subsystem of a fusion system $\mathcal{F}$ on the group P itself. Essentially, the proof of the existence of this system is given in Theorem 6.25, and here we prove some ancillary results, including the proof of the existence of various subsystems in between $\mathcal{F}$ and the residual subsystem, which we will denote by $O^{p'}(\mathcal{F})$. This includes being able to extend a saturated fusion system to a larger one on the same p-group.

In Exercise 7.1, we see that, if $\mathcal{E}$ is an $\mathcal{F}$-invariant subsystem on P, then the set of $\mathcal{F}$-centric subgroups and $\mathcal{E}$-centric subgroups coincides, as does the set of fully $\mathcal{F}$-normalized subgroups and fully $\mathcal{E}$-normalized subgroups. We want to extend this to essential subgroups; in order to do this we need a lemma on strongly p-embedded subgroups.

Lemma 7.54 *Let G be a finite group with a strongly p-embedded subgroup, and let P be a fixed Sylow p-subgroup of G. If H is a normal subgroup of G containing P, then H possesses a strongly p-embedded subgroup if and only if G does.*

Proof Let M be a strongly p-embedded subgroup of G containing P; then $M \cap H$ contains P, and if $g \notin M$ then $(M \cap H) \cap (M \cap H)^g$ is a p'-group, so that $M \cap H$ is a strongly p-embedded subgroup of H.

Conversely, let M be a strongly p-embedded subgroup of H; then $N_G(M)$ contains P, and is a proper subgroup of G since $N_H(M) = M < H$. We claim that $N_G(M)$ is strongly p-embedded. Let g be an element of $G \setminus N_G(M)$, and write $L = N_G(M) \cap N_G(M)^g$. Since L normalizes M, we have that $LM/M \cong L/(L \cap M)$. Clearly $L \cap M = L$; hence $M = LM$, so that $L \leq M$. Similarly, $L \leq M^g$, and hence $L \leq M \cap M^g$, a p'-group, as required. □

The end result of this lemma is that it is possible to understand the essential subgroups of weakly normal subsystems that lie on the same p-group as the fusion system itself.

Proposition 7.55 *Let $\mathcal{F}$ be a saturated fusion system on a finite p-group P. If $\mathcal{E}$ is a weakly normal subsystem of $\mathcal{F}$ on P itself, then a subgroup Q of P is $\mathcal{F}$-essential if and only if it is $\mathcal{E}$-essential.*

Proof If $\mathcal{E}$ is an $\mathcal{F}$-invariant subsystem on P, we know that $\mathrm{Out}_\mathcal{E}(Q) \trianglelefteq \mathrm{Out}_\mathcal{F}(Q)$ by Lemma 5.33, and also $\mathrm{Out}_P(Q) \leq \mathrm{Out}_\mathcal{E}(Q)$, so we see that $\mathrm{Out}_\mathcal{E}(Q)$ has a strongly p-embedded subgroup by Lemma 7.54. Since every $\mathcal{F}$-centric subgroup is $\mathcal{E}$-centric, every $\mathcal{F}$-essential subgroup is $\mathcal{E}$-essential, proving one direction.

Conversely, let Q be an $\mathcal{E}$-essential subgroup of P. Since $\mathrm{Out}_\mathcal{E}(Q)$ possesses a strongly p-embedded subgroup and $\mathrm{Out}_\mathcal{E}(Q) \trianglelefteq \mathrm{Out}_\mathcal{F}(Q)$, by Lemma 7.54 we see that $\mathrm{Out}_\mathcal{F}(Q)$ has a strongly p-embedded subgroup. Since every $\mathcal{E}$-centric subgroup is $\mathcal{F}$-centric, the proof is complete. □

The existence of the subsystem $O^{p'}(\mathcal{F})$ of $\mathcal{F}$ is guaranteed by the following theorem, which was originally proved by Puig in [Pui06, Theorem 6.11], is proved again in [BCGLO07, Theorem 5.4], and as a consequence of Aschbacher's theorem on the intersections of subsystems [Asc11, Theorem 1]. In the text here, it is an obvious corollary of Theorem 6.25.

Theorem 7.56 *Let $\mathcal{F}$ be a saturated fusion system on a finite p-group P. The collection of weakly normal subsystems of $\mathcal{F}$ on P, ordered by inclusion, has a unique minimal element, called the* residual subsystem, *and denoted by $O^{p'}(\mathcal{F})$.*

Technically, if we are to be consistent with the nomenclature from group theory, $O^p(\mathcal{F})$ and $O^{p'}(\mathcal{F})$ should be called the p-residual and p'-residual subsystems respectively; however, for historical reasons we have the focal and hyperfocal subgroups and hence it is natural to extend these to subsystems. Thus we may drop the 'p'-' and simply refer to $O^{p'}(\mathcal{F})$ as the residual subsystem of $\mathcal{F}$.

As well as finding subsystems of $\mathcal{F}$ on the group P, we can also consider oversystems of $\mathcal{F}$ on P. By Proposition 5.73, if $\hat{\mathcal{F}}$ is an oversystem of $\mathcal{F}$, also on P, in which $\mathcal{F}$ is weakly normal, then $|\operatorname{Aut}_{\hat{\mathcal{F}}}(P) : \operatorname{Aut}_{\mathcal{F}}(P)|$ has index prime to p. Also, if $\mathcal{F}$ is $\hat{\mathcal{F}}$-invariant, then $\operatorname{Aut}_{\hat{\mathcal{F}}}(P) \le \operatorname{Aut}(\mathcal{F})$. Hence we are looking for p'-subgroups of $\operatorname{Aut}(\mathcal{F})/\operatorname{Aut}_{\mathcal{F}}(P)$. In the next theorem, we construct all saturated fusion systems that contain a given saturated subsystem as a weakly normal subsystem on the same p-group.

Theorem 7.57 (Puig [Pui06, Proposition 6.10]) *Let $\mathcal{F}$ be a saturated fusion system on a finite p-group P.*

(i) *Let H be a subgroup of $\operatorname{Aut}(\mathcal{F})$, containing $\operatorname{Aut}_{\mathcal{F}}(P)$, such that the quotient $H/\operatorname{Aut}_{\mathcal{F}}(P)$ is a p'-group. There is a unique saturated fusion system $\hat{\mathcal{F}}$ on P, containing $\mathcal{F}$, such that $\operatorname{Aut}_{\hat{\mathcal{F}}}(P) = H$, and $\mathcal{F}$ is $\hat{\mathcal{F}}$-invariant. In this case, we have*

$$\operatorname{Hom}_{\hat{\mathcal{F}}}(Q, P) = \operatorname{Hom}_{\mathcal{F}}(Q, P) \cdot H$$

for all subgroups Q of P.

(ii) *Let $\hat{\mathcal{F}}$ be a saturated fusion system on P, containing $\mathcal{F}$, such that $\mathcal{F}$ is $\hat{\mathcal{F}}$-invariant. The group $H = \operatorname{Aut}_{\hat{\mathcal{F}}}(P)$ is a subgroup of $\operatorname{Aut}(\mathcal{F})$ containing $\operatorname{Aut}_{\mathcal{F}}(P)$ such that $H/\operatorname{Aut}_{\mathcal{F}}(P)$ is a p'-group.*

Proof Let $\hat{\mathcal{F}}$ be the object whose morphism sets $\operatorname{Hom}_{\hat{\mathcal{F}}}(Q, P)$ are all morphisms ϕ such that ϕ lies in $\operatorname{Hom}_{\hat{\mathcal{F}}}(Q, P) = \operatorname{Hom}_{\mathcal{F}}(Q, P) \cdot H$. We will show that this is a saturated fusion system on P.

Our first claim is that $\hat{\mathcal{F}}$ is a fusion system on P. To see this, notice firstly that every morphism in $\hat{\mathcal{F}}$ may be written as $\phi\chi$, where ϕ is a morphism in $\mathcal{F}$ and χ is (the restriction of) an automorphism of $\mathcal{F}$ lying in H. If $\phi\chi_1$ and $\psi\chi_2$ are morphisms in $\hat{\mathcal{F}}$ that are composable (so that the image of $\phi\chi_1$ is contained within the domain of $\psi\chi_2$) then we have

$$\phi\chi_1\psi\chi_2 = \phi(\psi)^{\chi_1^{-1}}\chi_1^{-1}\chi_2;$$

since $\chi_1 \in \operatorname{Aut}(\mathcal{F})$, we see that $\phi(\psi)^{\chi_1^{-1}}$ is a morphism in $\mathcal{F}$, and hence the map above lies in $\hat{\mathcal{F}}$, as claimed. Hence $\hat{\mathcal{F}}$ is a category.

Clearly the first two axioms of a fusion system are satisfied; if $\phi\chi$ is a map in $\hat{\mathcal{F}}$, so are the maps ϕ and χ, and hence $(\phi\chi)^{-1} = \chi^{-1}\phi^{-1}$ lies in $\hat{\mathcal{F}}$ since both ϕ^{-1} and χ^{-1} lie in the category $\hat{\mathcal{F}}$. Thus $\hat{\mathcal{F}}$ is a fusion system on P, as claimed.

We next show that $\mathcal{F}$ is $\hat{\mathcal{F}}$-invariant, so let $\phi : Q \to R$ be a morphism in $\mathcal{F}$ with $Q \le R$ subgroups of P, and let $\psi : R \to S$ be an isomorphism in $\hat{\mathcal{F}}$, where S is a subgroup of P. Since ψ lies in $\hat{\mathcal{F}}$, we may write $\psi = \theta\chi$, where θ is a morphism in $\mathcal{F}$ and $\chi \in H$. Hence the map $\psi^{-1}\phi\psi$ may be written as $(\theta^{-1}\phi\theta)^{\chi}$; since both ϕ and θ lie in $\mathcal{F}$, so does $\theta^{-1}\phi\theta$, and since $H \leq \mathrm{Aut}(\mathcal{F})$, the conjugate of $\theta^{-1}\psi\theta$ by χ also lies in $\mathcal{F}$, proving that $\mathcal{F}$ is $\hat{\mathcal{F}}$-invariant.

We now need to show that $\hat{\mathcal{F}}$ is saturated. We will show that P is fully $\hat{\mathcal{F}}$-automized, and that every fully $\hat{\mathcal{F}}$-normalized subgroup is $\hat{\mathcal{F}}$-receptive. This will prove saturation via Definition 4.25, and leave us with only uniqueness of $\hat{\mathcal{F}}$ to prove. We see that $\mathrm{Aut}_{\hat{\mathcal{F}}}(P) = H$, and $H/\mathrm{Aut}_{\mathcal{F}}(P)$ is a p'-group, so that P is fully $\hat{\mathcal{F}}$-automized (since it is fully $\mathcal{F}$-automized).

Let R be a fully $\hat{\mathcal{F}}$-normalized subgroup of P. Since the $\hat{\mathcal{F}}$-conjugacy classes are unions of $\mathcal{F}$-conjugacy classes, this means that R is fully $\mathcal{F}$-normalized, and hence is $\mathcal{F}$-receptive by Theorem 4.21. Let Q be some subgroup of P, and let $\phi\chi : Q \to R$ be an isomorphism in $\hat{\mathcal{F}}$, with ϕ a morphism in $\mathcal{F}$ and $\chi \in \mathrm{Aut}(\mathcal{F})$. Denote by R_1 the image of ϕ, so that $R_1\chi = R$. Since χ is an automorphism of $\mathcal{F}$, R_1 is also $\mathcal{F}$-receptive by Lemma 5.50. If $N_{\phi\chi} = N_{\phi}$, then ϕ extends to a map $\bar{\phi} : N_{\phi\chi} \to N_P(R_1)$, and hence we get a map $\bar{\phi}\chi : N_{\phi\chi} \to N_P(R)$, proving that R is $\hat{\mathcal{F}}$-receptive. The equality between $N_{\phi\chi}$ and N_{ϕ} is clear, because

$$N_{\phi\chi}c_Q = \mathrm{Aut}_P(Q) \cap \mathrm{Aut}_P(Q)^{\chi^{-1}\phi^{-1}}$$
$$= \mathrm{Aut}_P(Q) \cap \mathrm{Aut}_P(R_1)^{\phi^{-1}} = N_{\phi}c_Q.$$

This proves that every fully $\hat{\mathcal{F}}$-normalized subgroup is $\hat{\mathcal{F}}$-receptive, completing the proof of saturation.

We end with a proof of the uniqueness of $\hat{\mathcal{F}}$. Notice that $\hat{\mathcal{F}}$ is the smallest possible oversystem of $\mathcal{F}$ with the specified automorphism group of P. Hence if $\mathcal{E}$ is some other saturated fusion system on P, containing $\mathcal{F}$, with $\mathrm{Aut}_{\mathcal{E}}(P) = H$, then $\hat{\mathcal{F}} \leq \mathcal{E}$. We proceed by induction, proving that if Q is a fully $\mathcal{F}$-normalized, $\mathcal{F}$-centric subgroup of P, then $\mathrm{Aut}_Q(\hat{\mathcal{F}}) = \mathrm{Aut}_Q(\mathcal{E})$. Since $\mathcal{F}$, $\hat{\mathcal{F}}$ and $\mathcal{E}$ all have the same fully normalized, centric subgroups, by Alperin's fusion theorem we have $\hat{\mathcal{F}} = \mathcal{E}$.

If $Q = P$ then the result holds, so we assume that $Q < P$, and that the result holds for any fully normalized, $\mathcal{F}$-centric subgroup of P of larger order than $|Q|$. Let K be the subgroup $\mathrm{Aut}_P(Q)^{\phi}$, a p-subgroup of $\mathrm{Aut}_{\mathcal{E}}(Q)$, which is hence contained in $\mathrm{Aut}_{\mathcal{F}}(Q)$ (since $\mathrm{Aut}_{\mathcal{F}}(Q)$ is a normal subgroup of $\mathrm{Aut}_{\mathcal{E}}(Q)$ of p'-index). Hence there is an automorphism ψ in $\mathrm{Aut}_{\mathcal{F}}(Q)$ such that $K^{\psi} = \mathrm{Aut}_P(Q)$. Hence $\mathrm{Aut}_P(Q)^{\phi\psi} = \mathrm{Aut}_P(Q)$, so that $N_{\phi\psi} = N_P(Q)$, which strictly contains Q. Since $\mathcal{E}$ is saturated,

$\phi\psi$ extends to $\overline{\phi\psi}$, an $\mathcal{E}$-automorphism of $N_P(Q)$, which by Alperin's fusion theorem and induction lies in $\hat{\mathcal{F}}$. Thus $\overline{\phi\psi}|_Q = \phi\psi$ lies in $\hat{\mathcal{F}}$, and so ϕ lies in $\hat{\mathcal{F}}$.

Hence $\mathrm{Aut}_{\hat{\mathcal{F}}}(Q) = \mathrm{Aut}_{\mathcal{E}}(Q)$ for all fully normalized, centric subgroups Q of p, and so Alperin's fusion theorem proves that $\hat{\mathcal{F}} = \mathcal{E}$.

The proof of (ii) is easy; as both $\mathcal{F}$ and $\hat{\mathcal{F}}$ are saturated, $|\mathrm{Aut}_{\hat{\mathcal{F}}}(P) : \mathrm{Aut}_{\mathcal{F}}(P)|$ must be prime to p. Since $\mathcal{F}$ is $\hat{\mathcal{F}}$-invariant, $\mathrm{Aut}_{\hat{\mathcal{F}}}(P) \leq \mathrm{Aut}(\mathcal{F})$ by Lemma 5.40, completing the proof of (ii). $\qquad\square$

We can use this theorem to complete our understanding of the $\mathcal{F}$-invariant subsystems on the same p-group, and construct all subsystems of $\mathcal{F}$ containing $O^{p'}(\mathcal{F})$.

Theorem 7.58 *Let $\mathcal{F}$ be a saturated fusion system on a finite p-group P, let $\mathcal{F}_0 = O^{p'}(\mathcal{F})$, and write H for the p'-group $\mathrm{Aut}_{\mathcal{F}}(P)/\mathrm{Aut}_{\mathcal{F}_0}(P)$. There is a one-to-one correspondence between subgroups K of H and saturated subsystems $\mathcal{E}$ of $\mathcal{F}$ containing $\mathcal{F}_0$, given in the one direction by associating $\mathcal{E}$ with $\mathrm{Aut}_{\mathcal{E}}(P)/\mathrm{Aut}_{\mathcal{F}_0}(P)$, and in the other by associating K with the subsystem $\mathcal{E}$ given by*

$$\mathrm{Hom}_{\mathcal{E}}(Q, P) = \mathrm{Hom}_{\mathcal{F}_0}(Q, P) \cdot K.$$

Proof By Theorem 7.57, there exists a unique saturated fusion system $\mathcal{E}$ on P containing $\mathcal{F}_0$ and with a given subgroup K as $\mathrm{Aut}_{\mathcal{E}}(P)/\mathrm{Aut}_{\mathcal{F}_0}(P)$, with the structure given in the statement of the theorem. One simply has to notice that each of these is contained in $\mathcal{F}$, since $\mathcal{F}$ has the property that $\mathrm{Hom}_{\mathcal{F}}(Q, P) = \mathrm{Hom}_{\mathcal{F}_0}(Q, P) \cdot H$. $\qquad\square$

In the next chapter, we will need to understand which elements of $\mathrm{Aut}_{\mathcal{E}}(T)$ lie in $O^{p'}(\mathcal{E})$, where $\mathcal{E}$ is a subsystem on T. The next result basically determines this. This result may be found in [BCGLO07], although the very explicit description, and the proof given here, are due to Oliver and may be found in [AKO11].

Theorem 7.59 *Let $\mathcal{F}$ be a saturated fusion system on a finite p-group P, and let $\mathcal{F}_0$ denote the subsystem of $\mathcal{F}$ generated by the groups $O^{p'}(\mathrm{Aut}_{\mathcal{F}}(Q))$ for Q an $\mathcal{F}$-centric subgroup of P. An isomorphism $\phi : A \to B$ in $\mathcal{F}$ between $\mathcal{F}$-centric subgroups A and B of P lies in $O^{p'}(\mathcal{F})$ if and only if there is some $\mathcal{F}$-centric subgroup C of A such that $\phi|_C$ lies in $\mathcal{F}_0$.*

Proof Let $\mathcal{F}_0$ be as above, and let $\mathcal{F}'$ be the subsystem of $\mathcal{F}$ generated by all morphisms whose restrictions to some $\mathcal{F}$-centric subgroup of P lie in $\mathcal{F}_0$. Let A be an $\mathcal{F}$-centric subgroup of P and let B be a fully $\mathcal{F}$-normalized subgroup of P that is $\mathcal{F}$-conjugate to A. Since $\mathrm{Aut}_{\mathcal{F}'}(B)$ contains all p-automorphisms in $\mathrm{Aut}_{\mathcal{F}}(B)$, $\mathrm{Aut}_P(B)$ is a subgroup of $\mathrm{Aut}_{\mathcal{F}'}(B)$, so is a Sylow p-subgroup of it; therefore B is fully $\mathcal{F}'$-automized. Let $\phi : A \to B$ be an isomorphism in $\mathcal{F}'$; notice that the subgroup N_ϕ is the same for $\mathcal{F}'$ and $\mathcal{F}$. Since $\mathcal{F}$ is saturated, ϕ extends to a morphism ψ with domain N_ϕ, and by construction ψ lies in $\mathcal{F}'$, so that B is receptive in $\mathcal{F}'$.

We claim that $\mathrm{Aut}_{\mathcal{F}'}(U) \trianglelefteq \mathrm{Aut}_{\mathcal{F}}(U)$ for U an $\mathcal{F}$-centric subgroup of P. To see this, if ϕ is an automorphism of U in $\mathcal{F}'$ and $\psi \in \mathrm{Aut}_{\mathcal{F}}(U)$, then there exists an $\mathcal{F}$-centric subgroup V of U such that $\phi|_V \in \mathrm{Aut}_{\mathcal{F}_0}(V)$; clearly therefore $(\phi^\psi)|_V \in \mathrm{Aut}_{\mathcal{F}_0}(V)$, so that $\phi^\psi \in \mathrm{Aut}_{\mathcal{F}'}(U)$. Also, if $\theta : U \to \bar{U}$ is an isomorphism in $\mathcal{F}$ for some subgroup $\bar{U}$ of P, then clearly $\mathrm{Aut}_{\mathcal{F}'}(U)^\theta = \mathrm{Aut}_{\mathcal{F}'}(\bar{U})$; therefore the assignment $A(U) = \mathrm{Aut}_{\mathcal{F}'}(U)$ can be extended to an invariant map by Lemma 6.22.

We will show that $A(-)$ is a weakly normal map. Clearly $\mathrm{Aut}_P(P)$ is a Sylow p-subgroup of $A(P)$, since it is a Sylow p-subgroup of $\mathrm{Aut}_{\mathcal{F}}(P)$. Also, any fully $\mathcal{F}$-normalized $\mathcal{F}$-centric subgroup U of P is $\mathcal{F}'$-receptive, so that U has the surjectivity property in $\mathcal{F}'$. Hence $A(-)$ satisfies the final condition to be a weakly normal map.

As $\mathcal{F}'$ is generated by morphisms between $\mathcal{F}$-centric subgroups of P, $A(-)$ generates the subsystem $\mathcal{F}'$, and so $\mathcal{F}'$ is a weakly normal subsystem of $\mathcal{F}$; hence $O^{p'}(\mathcal{F}) \leq \mathcal{F}'$. If Q is a fully $\mathcal{F}$-normalized subgroup of P, then since $\mathrm{Aut}_{O^{p'}(\mathcal{F})}(Q)$ is a normal subgroup of $\mathrm{Aut}_{\mathcal{F}}(Q)$ containing $\mathrm{Aut}_P(Q)$, a Sylow p-subgroup of $\mathrm{Aut}_{\mathcal{F}}(Q)$, we see that $\mathrm{Aut}_{O^{p'}(\mathcal{F})}(Q)$ contains $O^{p'}(\mathrm{Aut}_{\mathcal{F}}(Q))$. Hence $\mathcal{F}_0 \leq O^{p'}(\mathcal{F})$.

Let Q and R be $\mathcal{F}$-centric subgroups of P, and let $\phi : Q \to R$ be an isomorphism in $\mathcal{F}$. If ψ_1 and ψ_2 are extensions in $\mathcal{F}$ of ϕ to an overgroup $\bar{Q}$ of Q contained in $N_P(Q)$, then ψ_1 and ψ_2 have the same image, namely the preimage $\bar{R}$ of $\mathrm{Aut}_{\bar{Q}}(Q)^\phi$ in $N_P(R)$. We see that $\psi_1\psi_2^{-1}$ is an automorphism of $\bar{Q}$ that acts trivially on Q, so is a p-automorphism of $\bar{Q}$ by Exercise 6.4. Therefore if $\mathcal{E}$ is a subsystem of $\mathcal{F}$ containing all p-automorphisms of all $\mathcal{F}$-centric subgroups of P (such as the subsystems $\mathcal{F}'$ or $O^{p'}(\mathcal{F})$ of $\mathcal{F}$) and ψ_1 is a morphism in $\mathcal{E}$ then so is ψ_2.

Let A and B be $\mathcal{F}$-centric subgroups of P, and let $\phi : A \to B$ be an isomorphism in $\mathcal{F}'$. There exists an $\mathcal{F}$-centric subgroup C of A such that $\phi|_C$ lies in $\mathcal{F}_0$, hence in $O^{p'}(\mathcal{F})$. Choose C to be a maximal $\mathcal{F}$-centric subgroup of A such that $\phi|_C$ lies in $O^{p'}(\mathcal{F})$. Clearly, $N = N_{\phi|_C} \cap A > C$, and so there are extensions $\psi_1 = \phi|_N$ in $\mathcal{F}'$ and ψ_2 in $O^{p'}(\mathcal{F})$ of ϕ to N (since both $\mathcal{F}'$ and $O^{p'}(\mathcal{F})$ are saturated, and using Exercise 8.7). By the previous paragraph, $\psi_1 = \phi|_N$ also lies in $\mathrm{O}^{p'}(\mathcal{F})$; hence $C = A$, and so ϕ lies in $\mathrm{O}^{p'}(\mathcal{F})$. Thus $\mathrm{O}^{p'}(\mathcal{F}) = \mathcal{F}'$, completing the proof. □

This theorem yields the following corollary.

Corollary 7.60 *Let $\mathcal{F}$ be a saturated fusion system on a finite p-group P, and let $\mathcal{E}$ be a weakly normal subsystem of $\mathcal{F}$, on the strongly $\mathcal{F}$-closed subgroup Q of P. An automorphism $\alpha \in \mathrm{Aut}_\mathcal{E}(Q)$ lies in $\mathrm{O}^{p'}(\mathcal{E})$ if and only if there is some $\mathcal{E}$-centric subgroup R of Q such that $\alpha|_R$ is the restriction of the composition of p-automorphisms in $\mathcal{E}$ of subgroups of Q.*

7.6 Bisets

A group action (of groups on sets and other objects) is one of the fundamental notions in group theory. It is natural to consider actions of (finite) groups on both the left and the right, in which case a set X becomes a biset.

Definition 7.61 Let G and H be groups. A (G, H)-*biset* is a set X, with a group action on the left by G and on the right by H, such that for all $x \in X$, $g \in G$, and $h \in H$,

$$(gx)h = g(xh).$$

Two (G, H)-bisets X and Y are said to be *isomorphic* if there is a bijection $f : X \to Y$ that commutes with both the left G-action and the right H-action.

Before we discuss bisets more, let us give an example of a biset.

Example 7.62 Let G and H be finite groups, let K be a subgroup of G, and let $\phi : K \to H$ be a homomorphism. The product set $G \times H$ becomes a (G, H)-biset, with the operation given by (assuming $g \in G$, $h \in H$, and $(x, y) \in G \times H$)

$$g(x, y)h = (gx, yh).$$

We would like to identify $K \leq G$ with $K\phi \leq H$, producing a 'quotient' biset. To do this, we declare an equivalence relation $\sim$ on $G \times H$ by $(gk, h) \sim (g, (k\phi)h)$, and form the set $(G \times H)/\sim$. This inherits the biset structure from $G \times H$, simply because if $(x_1, y_1) \sim (x_2, y_2)$ then

$(gx_1, y_1h) \sim (gx_2, y_2h)$, as is easily seen, so that G and H act on the equivalence classes of $\sim$. We denote this (G, H)-biset by $G \times_{(K, \phi)} H$.

The left action of G on $G \times_{(K, \phi)} H$ is free (i.e., the stabilizer of a point as a left G-set is trivial) and the right action of H on it is also free if ϕ is injective. If we are allowed to act by both G and H simultaneously then we do have a stabilizer, though. To see this, assume that the point being stabilized is $(1, 1)$. Acting by $g \in G$ and $h \in H$, for $(1, 1)$ to be fixed we must have $g(1, 1)h = (g, h) \sim (1, 1)$. However, it is clear that those elements of $G \times H$ that are equivalent to $(1, 1)$ are $(k^{-1}, k\phi)$, and so for g and h to stabilize $(1, 1)$, we must have $g \in K$ and $h = (g^{-1})\phi$.

Since both G and H act on a biset, we should be able to produce a right action of $G \times H$. However, in order to do this, we have to invert the action of G, because it will act from the 'wrong' side, so we actually get an action of $G^{\mathrm{op}} \times H$. To get this, if (g, h) is a pair in $G \times H$, then it acts on a (G, H)-biset X as $g^{-1}xh$. By equating bisets with standard sets, we may use results like the orbit-stabilizer theorem from group actions to get results on 'transitive' bisets.

If X is a (G, H)-biset and $x \in X$, then the *orbit* of x is the set of all $y \in X$ such that $y = gxh$ for some $g \in G$ and $h \in H$. The orbits of X partition the set X; X is said to be *transitive* if it has only one orbit. If X and Y are (G, H)-bisets, then the *disjoint union*, $X \amalg Y$, is also a (G, H)-biset, and every biset is the disjoint union of transitive bisets, namely the bisets acting on each orbit. These transitive sub-bisets of any biset are called the *transitive factors* of a biset.

If x is a point in the biset X, then the *stabilizer* is the collection of all pairs (g^{-1}, h) in $G \times H$ such that $gxh = x$. (The use of g^{-1} in the product is so that this is a subgroup of $G \times H$.) This is also the stabilizer of the corresponding $(G \times H)$-set discussed above. By the orbit-stabilizer theorem, the $(G \times H)$-set X is isomorphic with the cosets of $(G \times H)/\operatorname{Stab}_{G \times H}(x)$.

Example 7.63 Let us return to our example of $G \times_{(K, \phi)} H$. From $(1, 1)$ (modulo $\sim$), the pair (g, h) can be reached by multiplying by g on the left and h on the right, and so this is a transitive biset. The stabilizer was calculated in the previous example, and as a subgroup of $G \times H$ it is $\{(x, x\phi) : x \in K\}$, which can be thought of as a twisted diagonal subgroup of $G \times H$. If Δ_K^ϕ denotes this twisted diagonal subgroup, then $G \times_{(K, \phi)} H$ is isomorphic with the cosets of $(G \times H)/\Delta_K^\phi$.

We have produced a supply of transitive bisets in the bisets $G \times_{(K, \phi)} H$.

The next result proves that we have indeed found them all. (The proof of this result is Exercise 7.11.)

Proposition 7.64 *Let G and H be finite groups.*

(i) *If X is a transitive (G, H)-biset with G acting freely then there exists a subgroup K of G and an injective homomorphism $\phi : K \to H$ such that $X \cong G \times_{(K,\phi)} H$.*

(ii) *Two bisets $G \times_{(K,\phi)} H$ and $G \times_{(L,\psi)} H$ are isomorphic if and only if there are $g \in G$ and $h \in H$ such that $K^g = L$, and $(x\phi)^h = (x^g)\psi$ for all $x \in K$.*

If $K \leq G$ and $L \leq H$, a (G, H)-biset X can be thought of as a (K, L)-biset: this is called a *restriction* of X, and denoted by $X|_{(K,L)}$. If we have an injective homomorphism ϕ from A to G, then we may induce a left action on a (G, H)-biset by ϕ, and get an (A, H)-biset, via

$$axh = (a\phi)xh;$$

the (A, H)-biset obtained this way will be denoted by $X|_{(\phi,H)}$. Similarly, if $\phi : A \to H$ is an injective map, we may induce a right action on a (G, H)-biset by ϕ, to get a (G, A)-biset, which we will denote by $X|_{(G,\phi)}$.

Let $\mathcal{F}$ be a saturated fusion system on a p-group P. In [BLO03b], Broto, Levi and Oliver proved an important theorem on bisets, constructing a (P, P)-biset X that has three properties related to the structure of $\mathcal{F}$; these properties were first considered by Linckelmann and Webb in unpublished work, and the importance of finding such bisets was first recognized by them.

Theorem 7.65 (Broto–Levi–Oliver [BLO03b, Proposition 5.5]) *Let $\mathcal{F}$ be a saturated fusion system on a finite p-group P. There is a (P, P)-biset Y such that*

(i) *every transitive factor of Y is of the form $P \times_{(Q,\phi)} P$ for some $Q \leq P$ and some $\phi \in \mathrm{Hom}_{\mathcal{F}}(Q, P)$,*

(ii) *for each $Q \leq P$ and $\phi \in \mathrm{Hom}_{\mathcal{F}}(Q, P)$, the bisets $Y|_{(Q,P)}$ and $Y|_{(\phi,P)}$ are isomorphic as (Q, P)-bisets, as are the bisets $Y|_{(P,Q)}$ and $Y|_{(P,\phi)}$ (as (P, Q)-bisets), and*

(iii) $|Y|/|P| \equiv 1 \bmod p$.

As in [Rag06], we will call such a biset a *characteristic biset of $\mathcal{F}$*. We will construct such a biset later in this section, but we will first discuss the 'correct' setting to view characteristic bisets, namely the double Burnside ring. This is a ring, constructed in a purely formal way, that contains the collection of all (G, G)-bisets for any finite group G. In order to describe it, we need to describe how to 'add' and 'multiply' bisets.

Let $A^+(G, H)$ denote the set of all (G, H)-bisets on which G acts freely. If X and Y are two (G, H)-bisets in $A^+(G, H)$, then $X + Y$ is simply the disjoint union $X \amalg Y$ of the two, another (G, H)-biset in $A^+(G, H)$. Multiplication is harder, and only defined in some circumstances (like multiplication of matrices). If A, B, and C are finite groups, X is a (A, B)-biset in $A^+(G, H)$ and Y is a (B, C)-biset in $A^+(B, C)$, then $X \circ Y$ is a (A, C)-biset in $A^+(A, C)$. To describe it, we will define it on the transitive bisets on which A acts freely, which as we noted in Proposition 7.64 are the bisets $A \times_{(K, \phi)} B$ for some $K \le A$ and homomorphism $\phi : K \to B$. In this case,

$$(A \times_{(K, \phi)} B) \circ (B \times_{(L, \psi)} C) = \coprod_{t \in T} (A \times_{(K\phi \cap L^t)\phi^{-1}, \phi c_t \psi} C),$$

where T is a set of $(K\phi, L)$-double coset representatives.

If A, B, and C are all a single finite group G, then this defines both an additive and a multiplicative structure on the set of all left-free (G, G)-bisets. We will turn this into a ring formally, by attaching additive inverses of all transitive bisets. More formally, let $\mathcal{X}$ denote the set of all isomorphism classes of transitive (G, G)-bisets on which G acts freely on the left, and let $A(G, G)$ denote the free abelian group with generators $\mathcal{X}$. We embed the collection of all (G, G)-bisets into $A(G, G)$ by defining the disjoint union X of a collection of transitive bisets to be their sum in $A(G, G)$. This produces a subset of $A(G, G)$ that is identical to $A^+(G, G)$ (so we will identify them), and every element of $A(G, G)$ may be written as the difference of two elements of $A^+(G, G)$.

Finally, we define a multiplication $\circ$ on $A(G, G)$ by defining it on the generators $\mathcal{X}$ as above, then extend to a multiplication on all of $A(G, G)$ by linearity. With this multiplication $A(G, G)$ forms a ring, called the *double Burnside ring* of G.

If $\mathcal{F}$ is a fusion system on the p-group P, then we will form a subring of the double Burnside ring $A(P, P)$, denoted by $A_{\mathcal{F}}(P, P)$, which consists of the subring generated by the subset $\mathcal{X}_{\mathcal{F}}$ of $\mathcal{X}$ of elements of the form $P \times_{(Q, \phi)} P$, where $\phi \in \mathrm{Hom}_{\mathcal{F}}(Q, \phi)$. (This is closed under multiplication as each of the bisets on the right-hand side of the multiplication formula above lie in $\mathcal{X}_{\mathcal{F}}$.)

Let $\varepsilon : A(G, G) \to \mathbb{Z}$ be the function obtained by counting the number of orbits under the left P-action. It is easy to see (since G acts freely on the left) that $\varepsilon(X \circ Y) = \varepsilon(X)\varepsilon(Y)$, and also that $\varepsilon(X) = |X|/|G|$. Also, recall that if X is a (G, H)-biset then we have (K, L)-bisets $X_{(K,L)}$ for $K \leq G$ and $L \leq H$, and if $\phi : K \to G$ and $\psi : L \to H$ are injective maps, then we also have the (K, L)-biset $X_{(\phi,\psi)}$ induced by ϕ and ψ. (Here we slightly extend our notation, and write $X_{(\phi,\psi)}$ for $(X|_{(\phi,H)})|_{(K,\psi)}$.) We extend these maps to all of $A(G, G)$ in the obvious way.

Definition 7.66   Let $\mathcal{F}$ be a fusion system on a finite p-group P. An element X of $A(P, P)$ is *left $\mathcal{F}$-stable* if, whenever $\phi : Q \to P$ is a morphism in $\mathcal{F}$, the (Q, P)-bisets $X|_{(Q,P)}$ and $X|_{(\phi,P)}$ are isomorphic. Similarly, X is *right $\mathcal{F}$-stable* if, whenever $\phi : Q \to P$ is a morphism in $\mathcal{F}$, the (P, Q)-bisets $X|_{(P,Q)}$ and $X|_{(P,\phi)}$ are isomorphic. The biset X is *$\mathcal{F}$-stable* if it is both left and right $\mathcal{F}$-stable, or equivalently, whenever $\phi : Q \to P$ and $\psi : R \to P$ are morphisms in $\mathcal{F}$, the bisets $X|_{(Q,R)}$ and $X|_{(\phi,\psi)}$ are isomorphic.

With this definition we may state the theorem of Broto, Levi and Oliver in the setting of the double Burnside ring.

Corollary 7.67   *Let $\mathcal{F}$ be a saturated fusion system on a finite p-group P. There exists an element X of $A(P, P)$ such that*

(i) *$X \in A_{\mathcal{F}}(P, P)$,*
(ii) *X is $\mathcal{F}$-stable, and*
(iii) *$\varepsilon(X) = 1$.*

Furthermore, X may be chosen to lie in $A^+(P, P)$. An element X satisfying the three properties above is a characteristic element *for $\mathcal{F}$.*

We will also consider the ring $\mathbb{Z}_{(p)}$, which is the subring of $\mathbb{Q}$ obtained by adjoining the inverses of all primes not equal to p in $\mathbb{Z}$. The ring $A(P, P)_{(p)}$ with coefficients in $\mathbb{Z}_{(p)}$ (which may formally be constructed as $A(P, P) \otimes_{\mathbb{Z}} \mathbb{Z}_{(p)}$) will also be important, and we will consider characteristic elements in $A(P, P)_{(p)}$. (We will extend the function $\varepsilon : A(P, P) \to \mathbb{Z}$ to $\varepsilon : A(P, P)_{(p)} \to \mathbb{Z}_{(p)}$.) In this ring, a *characteristic element* is an $\mathcal{F}$-stable element $X \in A_{\mathcal{F}}(P, P)_{(p)}$ such that $\varepsilon(X)$ is prime to p.

If X and Y are characteristic elements then so is $X \circ Y$ [Rag06, Lemma 4.5]. In particular, if X is a characteristic element with $\varepsilon(X) = 1$ then so are the powers X^p, X^{p^2}, X^{p^3}, and so on. Using this sequence, one may construct a characteristic element ω that is also an idempotent in $A(P, P)_{(p)}$. Moreover, we have the following result.

Theorem 7.68 (Ragnarsson [Rag06, Proposition 5.6]) *Let $\mathcal{F}$ be a saturated fusion system on a finite p-group P. There exists a unique characteristic idempotent $\omega_{\mathcal{F}}$ in $A(P, P)_{(p)}$.*

In a certain sense, the converse of the previous theorem is also true, and we can determine saturation in this case. This was proved by Puig first, using characteristic bisets.

Theorem 7.69 (Puig [Pui09, Proposition 21.9]) *Let $\mathcal{F}$ be a fusion system on a finite p-group P. There is a characteristic biset for $\mathcal{F}$ if and only if $\mathcal{F}$ is saturated.*

Independently, Ragnarsson and Stancu proved the following theorem, working in the ring $A(P, P)_{(p)}$.

Theorem 7.70 (Ragnarsson–Stancu [RaS09, Theorem A]) *Let $\mathcal{F}$ be a fusion system on a finite p-group P. There is a characteristic idempotent for $\mathcal{F}$ in $A(P, P)_{(p)}$ if and only if $\mathcal{F}$ is saturated.*

Although Theorem 7.70 proves that saturation may be detected in the double Burnside ring, it is not totally satisfactory because the definition of a characteristic idempotent still requires that we have the fusion system $\mathcal{F}$ beforehand. In [RaS09], Ragnarsson and Stancu produce a condition on an idempotent in $A(P, P)_{(p)}$ that guarantees that it is the characteristic idempotent of some saturated fusion system $\mathcal{F}$, without needing $\mathcal{F}$ a priori. We say that a biset is *bifree* if it is free under both the left and right actions, and we extend this by linearity to all of $A(P, P)$.

If X and Y are (P, P)-bisets, by $(X \times Y)_\Delta$ we denote the $(P \times P, P)$-biset with underlying set $X \times Y$, with $P \times P$ acting on the left by $(g, h)(x, y) = (gx, hy)$, and P acting on the right diagonally, so that $(x, y)g = (xg, yg)$. A (P, P)-biset X *satisfies Frobenius reciprocity* if the two bisets $(X \times X)_\Delta$ and $(X \times 1)_\Delta \circ X$ are isomorphic, where $\circ$ is the multiplication in the double Burnside ring and 1 is the unit in $A(P, P)_{(p)}$.

Theorem 7.71 (Ragnarsson–Stancu [RaS09, Theorems B and C]) *Let P be a finite p-group, and let X be a bifree element of $A(P, P)_{(p)}$. If $\varepsilon(X)$ is not divisible by p and X satisfies Frobenius reciprocity, then X is a characteristic element for a saturated fusion system $\mathcal{F}$ over P.*

Furthermore, there is a bijection between bifree idempotents X in $A(P, P)_{(p)}$ that satisfy Frobenius reciprocity and with $\varepsilon(X) = 1$, and saturated fusion systems over P.

This striking characterization of saturated fusion systems suggests that the double Burnside ring, and characteristic bisets, are an important tool by which one may study saturated fusion systems.

Having discussed how the characteristic biset is of importance in the theory of fusion systems, we will end this section by proving Theorem 7.65.

We begin with a lemma that enables us to find P-sets with particular properties which, when applied to $\mathcal{F} \times \mathcal{F}$, result in the biset needed by Theorem 7.65. Recall that if G acts on a set X and H is a subgroup of G, then the set of points of X fixed by all elements of H is denoted by X^H. It is the set of all $x \in X$ such that H is contained in the stabilizer $\mathrm{Stab}_H(x)$. In the proof of this lemma, the transitive G-set on the (right) cosets of H is denoted X_H; clearly $(X_H)^K$ is empty unless K is contained within a conjugate of H.

Lemma 7.72 (Broto–Levi–Oliver [BLO03b, Lemma 5.4]) *Let $\mathcal{F}$ be a saturated fusion system on a finite p-group P, and let $\mathcal{H}$ be a union of $\mathcal{F}$-conjugacy classes of subgroups of P, closed under taking subgroups. Let X be a set with a right P-action such that, if Q_1 and Q_2 are $\mathcal{F}$-conjugate and not in $\mathcal{H}$, then $|X^{Q_1}| = |X^{Q_2}|$. There exists a set Y with right P-action such that*

  (i) $X \subseteq Y$ *as P-sets,*
 (ii) $|Y^{Q_1}| = |Y^{Q_2}|$ *for all $\mathcal{F}$-conjugate subgroups Q_1 and Q_2 of P, and*
(iii) $Y^Q = X^Q$ *for all subgroups Q of P not in $\mathcal{H}$.*

For any such set Y, if $Q \leq P$ and $\phi \in \mathrm{Hom}_{\mathcal{F}}(Q, P)$, then Y with the standard action $\cdot$ of Q is isomorphic (as a Q-set) to the Q-set Y with action $y \circ g = y \cdot (g\phi)$ for $g \in Q$ and $y \in Y$.

Proof We proceed by induction on the number of subgroups in $\mathcal{H}$, noting that the case $\mathcal{H} = \emptyset$ is obvious. Let Q be a fully $\mathcal{F}$-normalized subgroup of P in $\mathcal{H}$ of smallest index, write $\mathcal{Q}$ for the $\mathcal{F}$-conjugacy class containing Q, and $\mathcal{H}' = \mathcal{H} \setminus \mathcal{Q}$. By induction the result holds for $\mathcal{H}'$. By appending a sufficient number of the transitive P-sets X_Q to X, we may assume that if R is $\mathcal{F}$-conjugate to Q, then $|X^Q| \geq |X^R|$. (By the remark before this lemma, appending copies of X_Q to X does not alter the sets X^U for U not in $\mathcal{H}$.)

Let R be $\mathcal{F}$-conjugate to Q, and let $\phi : \mathrm{N}_P(R) \to \mathrm{N}_P(Q)$ be an $\mathcal{F}$-morphism with $R\phi = Q$. If $U \leq \mathrm{N}_P(R)$ strictly contains Q, then by maximality of Q, $|X^U| = |X^{U\phi}|$. Comparing the induced action of

$N_P(R)/R$ on X^R and on X^Q (acting via the isomorphism $\phi : R \to Q$), we see that the sets of orbits that are not of order $|N_P(R)/R|$ (i.e., have trivial stabilizer in $N_P(R)/R$) have the same order; therefore $|X^R| \equiv |X^Q| \mod |N_P(R)/R|$. We may therefore set a_R to be the integer

$$a_R = \frac{|X^Q| - |X^R|}{|N_P(R)/R|},$$

which is positive since $|X^Q| \geq |X^R|$. Finally, take a_Q to be 0, and let X' be the P-set

$$X' = X \amalg \left(\coprod_{R \in Q} a_R \cdot X_R \right),$$

where X_R again refers to the transitive P-set on the cosets of R, and $a_R \cdot X_R$ is the disjoint union of a_R copies of X_R.

Let S be a subgroup of P. If S is not in $\mathcal{H}$ then $X^S = (X')^S$, since none of the P-sets added has S lying inside a point stabilizer. If S is $\mathcal{F}$-conjugate to Q, then we have added a_S copies of X_S to X; this adds $a_S|N_P(S)/S|$ points to X^S, so that $|(X')^S| = |(X')^Q|$ for all $S \in Q$. Therefore the set X' satisfies the conditions of the lemma applied to $\mathcal{H}'$, and so by induction there exists a P-set Y with this property.

To see the last remark, notice that two P-sets A and B are isomorphic if and only if $|A^Q| = |B^Q|$ for all subgroups Q of P. This latter statement holds because, if $|A^Q| = |B^Q|$ for all $Q \leq P$, then the number of orbits of A and B that are isomorphic with X_R are equal for all subgroups R of P. This completes the proof. $\qquad\square$

We now turn our attention to proving Theorem 7.65. Let R be a subgroup of P, and let ϕ be a morphism in $\mathrm{Hom}_{\mathcal{F}}(R, P)$. As above, write Δ_R^ϕ for the subgroup $\{(x, x\phi) : x \in R\}$ of $P \times P$, and let $\mathcal{H}$ denote the set of subgroups Δ_R^ϕ, as R ranges over all proper subgroups of P and $\phi \in \mathrm{Hom}_{\mathcal{F}}(R, P)$.

Since $\mathcal{F}$ is saturated, $|\mathrm{Out}_{\mathcal{F}}(P)|$ is prime to p, so we may choose $k \in \mathbb{N}$ such that $k \cdot |\mathrm{Out}_{\mathcal{F}}(P)| \equiv 1 \mod p$. Define the biset X by

$$X = k \cdot \coprod_{\phi \in \mathrm{Out}_{\mathcal{F}}(P)} (P \times_{(P,\phi)} P).$$

(If ϕ and ϕ' are $\mathcal{F}$-automorphisms whose image in $\mathrm{Out}_{\mathcal{F}}(P)$ is the same, then $P \times_{(P,\phi)} P$ and $P \times_{(P,\phi')} P$ are isomorphic as (P, P)-bisets.) Notice that $|X| = k|\mathrm{Out}_{\mathcal{F}}(P)| \cdot |P|$, and so $|X|/|P| \equiv 1 \mod p$.

We will use Lemma 7.72 to construct a biset Y containing X. The fusion system $\mathcal{F} \times \mathcal{F}$ is a saturated fusion system over $P \times P$, by Theorem

6.29; the collection $\mathcal{H}$ is closed under the conjugacy induced by $\mathcal{F} \times \mathcal{F}$, and obviously all subgroups of elements of $\mathcal{H}$ also lie in $\mathcal{H}$. As X is a (P, P)-biset, it has a right $(P \times P)$-action. Finally, if a subgroup Q of $P \times P$ is a proper subgroup of the stabilizer of any point in X, then $Q \in \mathcal{H}$, and so if Q is a subgroup not in $\mathcal{H}$ or of the form Δ_P^ϕ, then $X^Q = \emptyset$.

Since all of the conditions of Lemma 7.72 are satisfied for X and $\mathcal{H}$, there exists a $(P \times P)$-set Y satisfying the three conclusions given there. The third one implies that $Y^Q = \emptyset$ for Q not of the form Δ_R^ϕ for some subgroup R of P and $\phi \in \mathrm{Hom}_{\mathcal{F}}(R, P)$, and so Y is the disjoint union of $(P \times P)$-sets of the form $(P \times P)/\Delta_R^\phi$. Hence the first condition of Theorem 7.65 is satisfied by Y.

To show that Y satisfies the second condition of Theorem 7.65, the remark at the end of Lemma 7.72 states that if Q is a subgroup of P (thought of as $Q \times 1 \le P \times P$) and $\phi \in \mathrm{Hom}_{\mathcal{F}}(Q, P)$, then the Q-sets Y under the standard action and under the induced action of (ϕ, id) are isomorphic. This implies that (as (P, P)-bisets) $Y|_{(Q,P)}$ and $Y|_{(\phi,P)}$ are isomorphic. The other isomorphism is exactly the same.

Finally, we need to show that $|Y|/|P| \equiv 1 \bmod p$, or equivalently $|Y| \equiv |X| \bmod p|P|$. To prove this, it suffices to show that every orbit of $Y \setminus X$ has order a multiple of $p|P|$, but this is true since they are of the form $(P \times P)/\Delta_R^\phi$ for R a proper subgroup of P, and this has order $|P|^2/|R|$. This completes the proof of Theorem 7.65.

While we have proved that there is always a characteristic biset for any saturated fusion system $\mathcal{F}$, in the case where $\mathcal{F} = \mathcal{F}_P(G)$ the group G itself forms a characteristic biset, once one takes enough copies of G to satisfy (iii) (see Exercise 7.12).

7.7 The transfer

The transfer was introduced at the turn of the twentieth century by Burnside, and produces a homomorphism from a finite group G into the abelianization H/H' of any subgroup H of G. We will make this slightly more general by including a homomorphism from H to an abelian group A. The central theorem on which the transfer is based is the following.

Theorem 7.73 *Let G be a finite group and let H be a subgroup of G. Let $\phi : H \to A$ be a homomorphism, where A is an abelian group. Let $T = \{t_1, t_2, \ldots, t_n\}$ denote a right transversal to H in G, and for any*

$g \in G$, write $h_i(g)$ for the element of H such that $t_i g = h_i(g) t_j$ for the appropriate j. The mapping $G \to A$ given by

$$g\tau = \prod_{i=1}^{n} h_i(g)\phi$$

does not depend on the choice of right transversal T, and is a homomorphism from G to A.

We will not prove this theorem here; for a proof, see for example [Gor80, Theorem 7.3.2]. The transfer takes a homomorphism from H to A and produces a homomorphism from G to A. We will denote the transfer by $t_H^G : \operatorname{Hom}(H, A) \to \operatorname{Hom}(G, A)$. We will apply this in the case where G is a p-group P and H is a subgroup Q of P.

Let $\mathcal{F}$ be a saturated fusion system on the finite p-group P. If $Y = P \times_{(Q,\phi)} P$ is a transitive (P, P)-biset, we define $t_Y : \operatorname{Hom}(P, A) \to \operatorname{Hom}(P, A)$ by $t_Y : \alpha \to (\phi\alpha)t_Q^P$. If Y is the disjoint union of two bisets, we simply multiply the resulting maps. In particular, if $X = \amalg_{i \in I} P \times_{(Q_i, \phi_i)} P$ is a characteristic biset for $\mathcal{F}$, then the *transfer with respect to X* is a function $t_X : \operatorname{Hom}(P, A) \to \operatorname{Hom}(P, A)$, defined by

$$t_X : \alpha \mapsto \prod_{i \in I} (\phi_i \alpha)t_{Q_i}^P.$$

In other words, for each $P \times_{(Q,\phi)} P$ that appears in X, the map $\phi\alpha$ is from Q to A, and therefore $(\phi\alpha)t_Q^P$ is a map from P to A; hence we take the product of all of these maps in $\operatorname{Hom}(P, A)$, and this is αt_X.

Of course, we can apply the formula above for the transfer to get an explicit formula for the transfer with respect to X in terms of the transversals for the Q_i.

We start by understanding the relationship between the focal subgroup and quotients.

Lemma 7.74 *Let $\mathcal{F}$ be a saturated fusion system on a finite p-group P. If Q is a strongly $\mathcal{F}$-closed subgroup of P, then*

$$\mathfrak{foc}(\mathcal{F}/Q) = Q\mathfrak{foc}(\mathcal{F})/Q \cong \mathfrak{foc}(\mathcal{F})/\mathfrak{foc}(\mathcal{F}) \cap Q.$$

Proof We will use the third isomorphism theorem many times in this proof, and so we do not explicitly reference every time we use it. Write R for the preimage in P of $\mathfrak{foc}(\mathcal{F}/Q)$, and write $S = \mathfrak{foc}(\mathcal{F})$. We will show that $QS = R$. Firstly, $\mathcal{F}/S$ is a trivial fusion system on an abelian p-group, and so therefore is $\mathcal{F}/QS$, as it is a quotient of $\mathcal{F}/S$. Also, $\mathcal{F}/QS$ is a quotient of $\mathcal{F}/Q$, and therefore $(\mathcal{F}/Q)/(QS/Q)$ is a trivial fusion system on an abelian p-group, so that $QS/Q \geq R/Q$, by Proposition 7.48. To see the converse, since $(\mathcal{F}/Q)/(R/Q)$ is the trivial fusion system on an abelian p-group, and so therefore is $\mathcal{F}/R$; hence $R \geq S$, and so (as $Q \leq R$), $R \geq QS$. Hence $R = QS$, and so

$$\mathfrak{foc}(\mathcal{F}/Q) = Q\mathfrak{foc}(\mathcal{F})/Q,$$

with the final isomorphism being the second isomorphism theorem. $\square$

If $z \in Z(\mathcal{F})$ then we can calculate the effect of the transfer map, at least on one particular homomorphism in $\mathrm{Hom}(P, P/P')$.

Lemma 7.75 *Let $\mathcal{F}$ be a saturated fusion system on a finite p-group P, and let X be a characteristic biset for $\mathcal{F}$. Let π denote the canonical surjection $\pi : P \to P/P'$, and let $\tau = \pi t_X$. If $n = |X|/|P|$ and $z \in Z(\mathcal{F})$, then $z\tau = P'z^n$.*

Proof Let $Y = P \times_{(Q,\phi)} P$ be a transitive sub-biset of X. Let $T = \{t_1, \ldots, t_d\}$ be a right transversal to Q in P. We have, for $z \in Z(\mathcal{F})$, the equation $t_i z = h_i(z) t_j$ as in the definition of the transfer for groups. Then

$$z(\pi t_Y) = z\left((\phi\pi)t_Q^P\right) = \prod_{t_i \in T} (h_i(z))(\phi\pi).$$

The element z permutes the integers $1, \ldots, d$ via the equation $t_i z = h_i(z) t_j$ sending i to j; let σ denote this permutation. Write I for the set of orbits of σ, $r = |I|$, and s_i for a set of orbit representatives of σ. Notice that $h_i(z) = t_i z t_{i\sigma}^{-1}$, and so

$$z(\pi t_Y) = \prod_{t_i \in T} t_i z t_{i\sigma}^{-1}.$$

Consider each orbit A in I. Taking the product over all elements of I, we get the equation

$$\prod_{t \in A} t_i z t_{i\sigma}^{-1} = \prod_{j=1}^{|A|} s_{i\sigma^{j-1}} z s_{i\sigma^j}^{-1} = s_i z^{|A|} s_i^{-1} = z^{|A|},$$

with the last equality following since $z \in Z(P)$. Hence each orbit A in I contributes exactly $z^{|A|}$, and so

$$z(\pi t_Y) = \prod_{A \in I} z^{|A|} \pi = z^d \pi = z^{|P:Q|} \pi.$$

This calculates πt_Y for each $Y \subseteq X$. If $X = \amalg_i P \times_{(Q_i, \phi_i)} P$, notice that

$|X|/|P| = n = \sum_i |P : Q_i|$, and hence

$$z(\pi t_X) = \prod_i z^{|P:Q_i|} = z^n \pi = P' z^n,$$

as required. □

Using this lemma, we can prove the following interesting fact, a useful application of the transfer for fusion systems.

Proposition 7.76 *Let $\mathcal{F}$ be a saturated fusion system on a finite p-group P. For all $i \geq 1$, $\mathfrak{foc}(\mathcal{F}) \cap \mathrm{Z}_i(\mathcal{F}) = P' \cap \mathrm{Z}_i(\mathcal{F})$.*

Proof We first prove the case where $i = 1$. One containment – that $P' \cap \mathrm{Z}(\mathcal{F})$ is contained within $\mathfrak{foc}(\mathcal{F}) \cap \mathrm{Z}(\mathcal{F})$ – is clear, and so suppose that $z \in \mathfrak{foc}(\mathcal{F}) \cap \mathrm{Z}(\mathcal{F})$. As in Lemma 7.75, let X be a characteristic biset for $\mathcal{F}$, let $\pi : P \to P/P'$ be the canonical surjection, and let $\tau = \pi t_X$. For any subgroup Q of P and $\phi \in \mathrm{Hom}_{\mathcal{F}}(Q, P)$, the (Q, P)-bisets $X_{(Q,P)}$ and $X_{(\phi,P)}$ are isomorphic, and hence for $x \in Q$ the maps $x\tau$ and $(x\phi)\tau$ are equal. Hence, for $x \in P$ and $\phi \in \mathrm{Hom}(\langle x \rangle, P)$, $(x^{-1}(x\phi))\tau = 1\tau = P'$. Since $z \in \mathfrak{foc}(\mathcal{F})$, z is expressible as a product of such elements, and hence $z\tau = P'$. On the other hand, by Lemma 7.75, $z\tau = P'z^n$ for some p'-integer n. Hence $z^n \in P'$ and, since n is prime to p, $z \in P'$, as needed.

For $i > 1$, we leave the proof to Exercise 7.13. □

We now turn our attention to control of transfer. Before we define this explicitly, we will see it happening in a situation.

Proposition 7.77 *Let W be a characteristic p-functor such that, if G is any p-constrained group with $\mathrm{O}_{p'}(G) = 1$ and P is a Sylow p-subgroup of G, then $\mathrm{O}_p(G) \cap G' = \mathrm{O}_p(G) \cap \mathrm{N}_G(W(P))'$. For any p-constrained group G with $\mathrm{O}_{p'}(G) = 1$ and Sylow p-subgroup P, we have, writing $\mathcal{F} = \mathcal{F}_P(G)$, that*

$$\mathfrak{foc}(\mathcal{F}) = \mathfrak{foc}(\mathrm{N}_{\mathcal{F}}(W(P))), \quad i.e., \quad P \cap G' = P \cap \mathrm{N}_G(W(P))'.$$

Proof Write $Q = \mathrm{O}_p(G)$, which coincides with $\mathrm{O}_p(\mathcal{F})$, since G is p-constrained. Choose G to be a counterexample to the statement with the number of morphisms in $\mathcal{F}$ minimal. Since $\mathcal{F}/Q$ has fewer morphisms than $\mathcal{F}$, the theorem holds for $\mathcal{F}/Q$; writing $\bar{H}$ for the image of a given subgroup H in G/Q, we have $\mathcal{F}/Q = \mathcal{F}_{\bar{P}}(\bar{G})$, and $\bar{P} \cap \bar{G}' = \bar{P} \cap \mathrm{N}_{\bar{G}}(W(\bar{P}))'$. Let H denote the preimage of $\mathrm{N}_{\bar{G}}(W(\bar{P}))$ in G, so that $\bar{P} \cap \bar{G}' = \bar{P} \cap \bar{H}'$. This clearly shows that $P \cap G' \leq H'Q$. Hence $P \cap G'$ is contained in $P \cap H'Q = (P \cap H')Q$, by the modular law [Asc00, (1.14)]. We claim that in fact $P \cap G' = P \cap H'$.

Suppose that this claim is true; as $\bar{G}$ has no non-trivial normal p-subgroups, $\bar{H} < \bar{G}$, and since $W(\bar{P})$ char $\bar{P}$, $\bar{P} \leq \bar{H}$, so that H is a proper subgroup of G containing P. By Corollary 3.73, since $|H| < |G|$, $\mathcal{F}_P(H) < \mathcal{F}$. Clearly $\mathrm{C}_H(Q) \leq Q$ and $Q \trianglelefteq H$ so that $\mathcal{F}_P(H)$ is also constrained; hence by minimality of $\mathcal{F}$ (and the focal subgroup theorem),

$$P \cap G' = P \cap H' = \mathfrak{foc}(\mathcal{F}_P(H)) = \mathfrak{foc}(\mathrm{N}_{\mathcal{F}_P(H)}(W(P)))$$
$$\leq \mathfrak{foc}(\mathrm{N}_{\mathcal{F}}(W(P))) \leq \mathfrak{foc}(\mathcal{F}) = P \cap G'.$$

Hence all inequalities are really equalities, and so

$$\mathfrak{foc}(\mathcal{F}) = \mathfrak{foc}(\mathrm{N}_{\mathcal{F}}(W(P))).$$

It remains to prove the claim that $P \cap G' = P \cap H'$. Using the modular law again, we get

$$P \cap G' = Q(P \cap H') \cap (P \cap G') = (Q \cap (P \cap G'))(P \cap H') = (Q \cap G')(P \cap H').$$

The fact that $\mathrm{N}_G(W(P))$ is contained in H implies that $Q \cap G' = Q \cap H'$, and so

$$P \cap G' = (Q \cap H')(P \cap H') = P \cap H',$$

as needed. $\qquad\qquad\square$

If we consider the control of transfer to be when the focal subgroup $P \cap G'$ is determined by some local subgroup, then this result gives us a condition for transfer in p-constrained groups to be locally controlled. It motivates us to make the following definition.

Definition 7.78 Let W be a positive characteristic p-functor.

(i) We say that W *controls constrained transfer* if, for all p-constrained groups G, we have

$$\mathrm{O}_p(G) \cap G' = \mathrm{O}_p(G) \cap \mathrm{N}_G(W(P))'.$$

(ii) If $\mathcal{F}$ is a saturated fusion system on the finite p-group P, then W *controls transfer* in $\mathcal{F}$ if $\mathfrak{foc}(\mathcal{F}) = \mathfrak{foc}(\mathrm{N}_{\mathcal{F}}(W(P)))$.

Proposition 7.77 states that our terminology is consistent, in that any positive characteristic p-functor that controls constrained transfer controls transfer in constrained fusion systems. In fact, we will show that if W controls constrained transfer then it controls transfer in *any* saturated fusion system. The last thing that is needed is that there in fact do exist functors that control constrained transfer; the functors K^{∞} and K_{∞} that were introduced earlier in this chapter perform this task, at least for $p \geq 5$. There are no such functors for $p = 2$, as we shall see at the end of this section, but for $p = 3$ the question is still unresolved.

We begin with a lemma that mirrors Proposition 7.27, in that we can lift a result about normalizers to one about the fusion system itself.

Lemma 7.79 (Díaz–Glesser–Mazza–Park [DGMP10, Lemma 5.1]) *Let $\mathcal{F}$ be a saturated fusion system on a finite p-group P, and let W be a positive characteristic p-functor. If W controls transfer in $\mathrm{N}_{\mathcal{F}}(Q)$ for every non-trivial, $(\mathcal{F}, W)$-well-placed subgroup Q of P, then W controls transfer in $\mathcal{F}$.*

Proof Let Q be a non-trivial, $(\mathcal{F}, W)$-well-placed subgroup of P; since W controls transfer in $\mathrm{N}_{\mathcal{F}}(Q)$ and $W(\mathrm{N}_P(Q)) \trianglelefteq \mathrm{N}_P(Q)$, we see that

$$\mathfrak{foc}(\mathrm{N}_{\mathcal{F}}(Q)) = \mathfrak{foc}(\mathrm{N}_{\mathcal{F}}(Q, W(\mathrm{N}_P(Q)))) \leq \mathfrak{foc}(\mathrm{N}_{\mathcal{F}}(W(\mathrm{N}_P(Q)))).$$

In other words, we see that $\mathfrak{foc}(\mathrm{N}_{\mathcal{F}}(W_i(Q))) \leq \mathfrak{foc}(\mathrm{N}_{\mathcal{F}}(W_{i+1}(Q)))$ for all i and since, for all sufficiently large i, we have $W_i(Q) = W(P)$, we see that

$$\mathfrak{foc}(\mathrm{N}_{\mathcal{F}}(Q)) \leq \mathfrak{foc}(\mathrm{N}_{\mathcal{F}}(W(P))),$$

for any $(\mathcal{F}, W)$-well-placed subgroup Q of P. Alperin's fusion theorem implies that the left-hand side of this inequality is actually $\mathfrak{foc}(\mathcal{F})$, whence we get the result. $\square$

We may now embark on the proof of the main theorem in this section.

Theorem 7.80 (Díaz–Glesser–Mazza–Park [DGMP10]) *Let $\mathcal{F}$ be a saturated fusion system on a finite p-group P. If W is a positive characteristic p-functor that controls constrained transfer, then W controls transfer in $\mathcal{F}$.*

Proof Choose $\mathcal{F}$ to be a counterexample to the theorem, minimal subject to the number of morphisms, so that any subsystem or quotient satisfies the theorem.

Firstly, if $\mathrm{O}_p(\mathcal{F}) = 1$, then for any $(\mathcal{F}, W)$-well-placed subgroup S of P we have $\mathrm{N}_{\mathcal{F}}(S) < \mathcal{F}$, whence by induction W controls transfer in $\mathrm{N}_{\mathcal{F}}(S)$; by Lemma 7.79, we see that W controls transfer in $\mathcal{F}$. Write $Q = \mathrm{O}_p(\mathcal{F})$, which is non-trivial. By Proposition 7.77, $R = Q\,\mathrm{C}_P(Q)$ strictly contains Q, so that R is not a normal subgroup of $\mathcal{F}$. In particular, W controls transfer in $\mathrm{N}_{\mathcal{F}}(R)$. If Q is not hypercentral, i.e., $P\,\mathrm{C}_{\mathcal{F}}(Q) < \mathcal{F}$, then W controls transfer in $P\,\mathrm{C}_{\mathcal{F}}(Q)$ as well, by induction, and so W controls transfer in $\mathcal{F}$, which is generated by these two subsystems by Lemma 7.35. Hence $Q = \mathrm{Z}_{\infty}(\mathcal{F})$.

Let V be the preimage of $W(P/Q)$ in P. Since $Q = O_p(\mathcal{F})$ and $W(P/Q)$ is non-trivial, $V > Q$ and so $N_{\mathcal{F}}(V) < \mathcal{F}$. Hence, by induction,

$$\mathfrak{foc}(N_{\mathcal{F}}(V)) = \mathfrak{foc}(N_{\mathcal{F}}(V, W(P))) \leq \mathfrak{foc}(N_{\mathcal{F}}(W(P))).$$

However, also $\mathfrak{foc}(\mathcal{F}) = \mathfrak{foc}(N_{\mathcal{F}}(V))$: to see this, note that $\mathfrak{foc}(\mathcal{F}) \cap Q = \mathfrak{foc}(N_{\mathcal{F}}(V)) \cap Q$ by Proposition 7.76, and hence $\mathfrak{foc}(\mathcal{F}) = \mathfrak{foc}(N_{\mathcal{F}}(V))$ by Lemma 7.74, so that $\mathfrak{foc}(\mathcal{F}) = \mathfrak{foc}(N_{\mathcal{F}}(W(P)))$, as needed. $\qquad\square$

The remaining part of the proof is that K^∞ and K_∞ control constrained transfer for $p \geq 5$. This entirely group-theoretic property is proved in, for example, [DGMP10, Theorem 1.2], and we will not reproduce this here.

If W is a positive characteristic p-functor that controls transfer in a fusion system $\mathcal{F}$, then we can get some information about $\mathcal{F}$ from this.

Proposition 7.81 (Díaz–Glesser–Mazza–Park [DGMP10, Proposition 6.1]) *Let $\mathcal{F}$ be a saturated fusion system on a finite p-group P, and suppose that $\mathrm{Aut}_{\mathcal{F}}(P)$ is a p-group. If W is a positive characteristic p-functor that controls transfer in $\mathcal{F}$ and all quotients of $\mathcal{F}$, then $\mathfrak{foc}(\mathcal{F}) < P$.*

Proof Let $\mathcal{F}$ be a counterexample, such that the number of morphisms in $\mathcal{F}$ is minimal. Since $\mathfrak{foc}(\mathcal{F}) = P$, and $\mathfrak{foc}(\mathcal{F}) = \mathfrak{foc}(N_{\mathcal{F}}(W(P)))$, we must have that $N_{\mathcal{F}}(W(P)) = \mathcal{F}$, so that $W(P) \trianglelefteq \mathcal{F}$. If Z is a strongly $\mathcal{F}$-closed central subgroup of $W(P)$, then we may quotient out by Z: by Lemma 7.74, $P/Q = \mathfrak{foc}(\mathcal{F})/Q = \mathfrak{foc}(\mathcal{F}/Q)$, but by minimal choice of $\mathcal{F}$, $\mathfrak{foc}(\mathcal{F}/Q) < P/Q$, a contradiction. Therefore $Z = P$ and so P is abelian. However, in abelian groups $\mathcal{F} = N_{\mathcal{F}}(P)$ by Burnside's theorem on control of fusion (see Exercise 1.8), and if $\mathrm{Aut}_{\mathcal{F}}(P)$ is a p-group then $N_{\mathcal{F}}(P) = \mathcal{F}_P(P)$. We see that $\mathfrak{foc}(\mathcal{F}) = 1$, and so the proof is complete. $\qquad\square$

By using the functor K_∞, which we know controls transfer for all saturated fusion systems for odd primes, we get the following corollary.

Corollary 7.82 *Let P be a finite p-group with $p \geq 5$ and $\mathrm{Aut}(P)$ a p-group. If $\mathcal{F}$ is a saturated fusion system on P then $\mathfrak{foc}(\mathcal{F}) < P$, and so in particular $\mathcal{F}$ cannot be simple.*

We can now see why there can be no functor that controls constrained transfer for the prime 2; the fusion systems of $\mathrm{PSL}_2(q)$ are simple, and yet the automorphism group of a dihedral 2-group is a 2-group, contradicting the conclusion of the corollary.

Exercises

7.1 Let $\mathcal{F}$ be a saturated fusion system on a finite p-group P, and let $\mathcal{E}$ be an $\mathcal{F}$-invariant subsystem of $\mathcal{F}$, on the subgroup Q of P.

(i) Prove that, if R is an $\mathcal{F}$-centric subgroup of Q, then R is $\mathcal{E}$-centric, and if R is a fully $\mathcal{F}$-normalized subgroup of Q and $Q = P$, then R is fully $\mathcal{E}$-normalized.

(ii) Prove that, if $\mathcal{E}$ is weakly normal in $\mathcal{F}$, then a subgroup R of Q is $\mathcal{F}$-centric if and only if it is $\mathcal{E}$-centric, and if $Q = P$, then a subgroup R of Q is fully $\mathcal{F}$-normalized if and only if it is fully $\mathcal{E}$-normalized.

7.2 Let P be a finite p-group, and let $\mathcal{F}$ be a saturated fusion system on P with $\mathcal{F} = N_{\mathcal{F}}(P)$. Prove that $\mathcal{F} = \mathcal{F}_P(G)$ for some group G with a normal Sylow p-subgroup P, and so in particular is of the form $G = P \rtimes A$ for A a p'-group.

7.3 Let p be an odd prime, and let G be a subgroup of $\mathrm{GL}_2(p)$ acting on a 2-dimensional $\mathbb{F}_p$-vector space V.

(i) Suppose that $p \mid |G|$, and that $O_p(G) = 1$. Prove that G contains $\mathrm{SL}_2(p)$.

(ii) Prove that $\mathrm{SL}_2(p)$ acts transitively on the $p + 1$ subspaces of V of dimension 1.

Now suppose that $\mathcal{F}$ is a saturated fusion system on a finite p-group P, and that Q is a 2-generator, $\mathcal{F}$-essential subgroup of P.

(iii) Prove that $\mathrm{Inn}(Q)$ is the kernel of the natural map $\mathrm{Aut}(Q) \to \mathrm{Aut}(Q/\,\Phi(Q))$.

(iv) Deduce that if Q is an $\mathcal{F}$-essential subgroup in some saturated fusion system $\mathcal{F}$, then all maximal subgroups of Q are isomorphic.

7.4 Let p be an odd prime, and let P be a metacyclic p-group. Suppose that P is not cyclic.

(i) Prove that P has exactly $p^2 - 1$ elements of order p.

(ii) Deduce that P has at most one subgroup isomorphic to $C_{p^n} \times C_{p^n}$ for a given n, and hence this is characteristic.

7.5 Let p be an odd prime, and let P be an extraspecial group of order p^3.

(i) If P has exponent p, prove that P is not metacyclic.

(ii) If P has exponent p^2, prove that P is resistant.

7.6   Let $\mathcal{F}$ be a saturated fusion system on a finite p-group P, and let Q be a fully normalized subgroup of P. Prove that a subgroup R of $N_P(Q)$ containing Q is fully $N_{\mathcal{F}}(Q)$-centralized if and only if it is fully $\mathcal{F}$-centralized, and therefore R is $N_{\mathcal{F}}(Q)$-centric if and only if it is $\mathcal{F}$-centric.

7.7   Let P be the 2-group $C_{2^n} \wr C_2$. Prove that P is not metacyclic.

7.8   Let P be a Suzuki 2-group containing exactly three involutions, and let ϕ be an automorphism of order 3 permuting them transitively. Let A be a maximal, ϕ-invariant, abelian normal subgroup of P.

(i)   Prove that $C_P(A) = A$, and so $|A| = 16$.

(ii) Prove that P has order 64.

At this point, either [GLS05, Lemma 2.2.3], or a computer calculation using the 267 groups of order 64, proves that there is a unique Suzuki 2-group of order 64 with three involutions. (It turns out that there are three groups of order 64 with $\Omega_1(P) = Z(P) = P' = \Phi(P)$, and an automorphism of odd order. However, for two of these groups the odd-order automorphisms fix $Z(P)$, and so they are not Suzuki 2-groups. For a description of the 2-groups with three involutions that have an odd-order automorphism, see [CG10].)

7.9   Let $\mathcal{F}$ be a saturated fusion system on a finite p-group P. Let Q be a strongly $\mathcal{F}$-closed subgroup of P such that P/Q is abelian and $\mathcal{F}/Q$ is trivial. Prove that $\mathfrak{foc}(\mathcal{F}) \leq Q$. (This is Proposition 7.48(ii). As a hint, mimic the proof of Proposition 7.46(ii).)

7.10 Let $\mathcal{F}$ be a saturated fusion system on a finite p-group P, and let Q be a strongly $\mathcal{F}$-closed subgroup of P. Let $\mathcal{E}$ be a subsystem of $\mathcal{F}$ on Q. Suppose that $\mathrm{Aut}_{\mathcal{F}}(Q) \leq \mathrm{Aut}(\mathcal{E})$, and let R and S be two subgroups of Q. As in the proof of Lemma 5.42, we say that $\phi \in \mathrm{Hom}_{\mathcal{F}}(R, S)$ has an $\alpha\beta$-decomposition if there exists $\alpha \in \mathrm{Aut}_{\mathcal{F}}(Q)$ and $\beta \in \mathrm{Hom}_{\mathcal{E}}(R\alpha, S)$ such that $\phi = \alpha\beta$.

Prove that if $\phi : R \to S$ and $\psi : S \to T$ have $\alpha\beta$-decompositions, then so do $\phi|_U$, $\phi\psi$, and ϕ^{-1} (if ϕ is an isomorphism), for any subgroup U of R. Consequently, prove that if all $\mathcal{F}$-automorphisms of subgroups of Q have $\alpha\beta$-decompositions then $\mathcal{E}$ is $\mathcal{F}$-invariant.

7.11 Prove Proposition 7.64.

7.12 Let G be a finite group and let P be a Sylow p-subgroup of G. We may think of G as a (P, P)-biset, and both left and right actions of P are free. If $|G : P| = n$, let m be an integer such that $mn \equiv$

1 mod p. Prove that the biset X consisting of m copies of G is a characteristic biset for $\mathcal{F}_P(G)$.

7.13 Let $\mathcal{F}$ be a saturated fusion system on a finite p-group P. In Proposition 7.76 we stated that $\mathfrak{foc}(\mathcal{F}) \cap Z_i(\mathcal{F}) = P' \cap Z_i(\mathcal{F})$ for all $i \geq 1$ but proved only the case $i = 1$. Prove the case $i > 1$.

7.14 Let $\mathcal{F}_1, \ldots, \mathcal{F}_n$ be saturated fusion systems and let $\mathcal{F}$ be their direct product. Prove that

 (i) $O^{p'}(\mathcal{F}) = O^{p'}(\mathcal{F}_1) \times O^{p'}(\mathcal{F}_2) \times \cdots \times O^{p'}(\mathcal{F}_n)$,
 (ii) $O^p(\mathcal{F}) = O^p(\mathcal{F}_1) \times O^p(\mathcal{F}_2) \times \cdots \times O^p(\mathcal{F}_n)$,
 (iii) $O_p^{(i)}(\mathcal{F}) = O_p^{(i)}(\mathcal{F}_1) \times O_p^{(i)}(\mathcal{F}_2) \times \cdots \times O_p^{(i)}(\mathcal{F}_n)$ for all i, and
 (iv) $Z_i(\mathcal{F}) = Z_i(\mathcal{F}_1) \times Z_i(\mathcal{F}_2) \times \cdots \times Z_i(\mathcal{F}_n)$ for all i.

7.15 Let $\mathcal{F}$ be a saturated fusion system on a finite p-group P, and let α be a p'-automorphism of $\mathcal{F}$. Suppose that α acts trivially on $P/Z(\mathcal{F})$. Prove that α acts trivially on $\mathfrak{foc}(\mathcal{F})$, and hence if $\mathcal{F}$ is perfect then $\alpha = 1$.

7.16 Let $\mathcal{F}$ be a saturated fusion system on a finite p-group P, and let $\mathcal{E}$ be a weakly normal subsystem of $\mathcal{F}$ on the subgroup T of P. Suppose that $\mathfrak{hyp}(\mathcal{F}) \leq T$ and let Q be a fully $\mathcal{F}$-normalized subgroup of P such that $Q \cap T = 1$.

 (i) Prove that, if $K \leq \operatorname{Aut}(Q)$, then $N_{\mathcal{F}}^K(Q) = X_K \times N_Q^K(Q)$ for some weakly normal subsystem X_K of $N_{\mathcal{F}}^K(Q)$ on $N_T^K(Q)$.
 (ii) Prove that $X_K = X_1$, where the last term is the direct factor corresponding to $K = \{1\}$ in the previous part.
 (iii) Prove that X_K is the subsystem of p-power index of $\mathcal{F}$ on the subgroup $N_T^K(Q)$.

(Hint: use Exercise 5.14.)

8

Local theory of fusion systems

The local theory of fusion systems – analysing the normalizers and centralizers of a fusion system to get information about the system itself – is mostly the product of a series of papers by Michael Aschbacher [Asc08a], [Asc08b], [Asc10], [Asc11]. In this chapter we will give an exposition of some of this theory, although many of the more difficult theorems are beyond the scope of this work.

In the first section, we give the definition of a normal subsystem. This differs from a weakly normal subsystem by the inclusion of an axiom on extending certain automorphisms beyond the strongly $\mathcal{F}$-closed subgroup on which the normal subsystem is defined. We give an example of a weakly normal subsystem that is not normal.

We continue by proving some of the basic facts for normal subsystems in Section 8.1, before considering the relationship between weakly normal and normal subsystems in Section 8.2. In Section 8.3, we examine intersections of weakly normal and normal subsystems. We do not have the space to prove the main theorem in this direction, namely that if $\mathcal{E}_1$ and $\mathcal{E}_2$ are normal subsystems on T_1 and T_2 respectively, then there is a normal subsystem $\mathcal{E}$ on $T = T_1 \cap T_2$, with $\mathcal{E} \leq \mathcal{E}_1 \cap \mathcal{E}_2$. However, we will use this to develop a general theory of intersections of weakly normal subsystems, as well as of normal subsystems.

In Section 8.4, we examine constrained fusion systems $\mathcal{F}$, and we prove the important result that for these systems there is a one-to-one correspondence between the normal subsystems of $\mathcal{F}$ and the normal subgroups of the unique p-constrained, p'-reduced group $L^{\mathcal{F}}$ modelling $\mathcal{F}$. (The example we give of a weakly normal subsystem that is not normal gives a counterexample to this statement for weakly normal subsystems. This represents a significant deficiency when it comes to adapting techniques from local finite group theory to fusion systems.)

Section 8.5 deals with central products of normal subsystems, and this is very useful in developing the theory of the generalized Fitting subsystem in Section 8.6. Finally, we prove a version for fusion systems of L-balance in the last section of this chapter.

8.1 Normal subsystems

Weakly normal subsystems, as defined in Chapter 5, are a very natural object to consider. However, in the following example, we find that they do not behave exactly like normal subgroups, in that they do not behave well with respect to taking products. We saw direct products of fusion systems in Section 5.4; we did not prove that, if $\mathcal{E}_1$ and $\mathcal{E}_2$ are weakly normal subsystems of a fusion system that intersect trivially, then $\mathcal{E}_1 \times \mathcal{E}_2$ is a (weakly normal) subsystem of the fusion system. This is simply because it isn't true.

Example 8.1 Let G be the group generated by $x = (1,2,3)$, $y = (4,5,6)$, and $(1,2)(4,5)$. This is a group of order 18, an elementary abelian group of order 9 with a diagonal action of the cyclic group C_2 on the two factors, making each into an S_3. Let $\mathcal{F} = \mathcal{F}_P(G)$, where P is the Sylow 3-subgroup $P = \langle x, y \rangle$. Let $\mathcal{E}_1$ denote the subsystem on $X = \langle x \rangle$ consisting of id_X and the map $\psi_x : x \mapsto x^{-1}$, and let $\mathcal{E}_2$ denote the corresponding subsystem on $Y = \langle y \rangle$ (with ψ_y inverting the elements of Y).

We notice that $\mathcal{E}_1$ and $\mathcal{E}_2$ are both saturated subsystems of $\mathcal{F}$, being the fusion system of S_3 in characteristic 3. Secondly, the $\mathcal{E}_i$ are weakly normal in $\mathcal{F}$: to see this, consider what the elements of $\mathcal{F}$ are. There are identity maps, the inversion maps on X and Y, and the map ϕ that inverts all elements of P. To prove weak normality, we merely need to show that $\phi^{-1}\psi_x\phi = \psi_x$, which is clearly true, and $\phi^{-1}\psi_y\phi = \psi_y$, which is similar.

Hence we have two weakly normal subsystems whose intersection is trivial. However, we do not have the entire direct product $\mathcal{E}_1 \times \mathcal{E}_2$ inside $\mathcal{F}$, since we do not have the map that acts as ψ_x on X and the identity on Y, for example.

In the example above, we can quite easily repair this difficulty by requiring that, if ϕ is a map in a 'normal' subsystem $\mathcal{E}$ on a subgroup Q, and there is some subgroup R such that $R \cap Q = 1$ and R and Q commute, then ϕ should extend to a map on QR that acts trivially on

R. This will mean that if $\mathcal{E}_1$ is a normal subsystem on Q and $\mathcal{E}_2$ is a normal subsystem on R, then $\mathcal{E}_1 \times \mathcal{E}_2$ should exist. (We know that ϕ always extends to *some* map $\psi : QR \to P$, since $R \leq C_P(Q) \leq N_\phi$, but this map need not act trivially on R, as we saw in the example above.)

In [Asc08a], Aschbacher made the following definition.

Definition 8.2 (Aschbacher [Asc08a])   Let $\mathcal{F}$ be a saturated fusion system on a finite p-group P, and let T be a strongly $\mathcal{F}$-closed subgroup of P. A subsystem $\mathcal{E}$ on T is *normal* in $\mathcal{F}$ if $\mathcal{E}$ is weakly normal in $\mathcal{F}$, and each $\phi \in \operatorname{Aut}_\mathcal{E}(T)$ extends to $\bar{\phi} \in \operatorname{Aut}_\mathcal{F}(T \, C_P(T))$ such that $[\bar{\phi}, C_P(T)] \leq Z(T)$. Write $\mathcal{E} \trianglelefteq \mathcal{F}$ if $\mathcal{E}$ is a normal subsystem of $\mathcal{F}$.

This condition on automorphisms can be expressed in a few different ways, as Exercise 8.3 shows. In particular, this exercise shows that $[\bar{\phi}, C_P(T)] \leq Z(T)$ is equivalent to $[\bar{\phi}, C_P(T)] \leq T$, and is also equivalent to the statement that $\bar{\phi}$ acts trivially on $C_P(T)/Z(T)$.

We see therefore that the set of such automorphisms might be important, so we make the following notational definition.

Definition 8.3   Let $\mathcal{F}$ be a saturated fusion system on a finite p-group P, and let T be a strongly $\mathcal{F}$-closed subgroup of P. If $\mathcal{E}$ is a subsystem of $\mathcal{F}$ on T, then by $\Gamma_{\mathcal{F},\mathcal{E}}(T)$ we mean the set of all $\phi \in \operatorname{Aut}_\mathcal{E}(T)$ that extend to a map $\bar{\phi} \in \operatorname{Aut}_\mathcal{F}(T \, C_P(T))$ such that $[\bar{\phi}, C_P(T)] \leq Z(T)$.

We have developed the notation of $\Gamma_{\mathcal{F},\mathcal{E}}(T)$ simply to make long-winded statements about automorphisms extending to particular subgroups and acting trivially on the particular quotient much shorter, and this is all that is happening here. This subgroup $\Gamma_{\mathcal{F},\mathcal{E}}(T)$ has some useful properties, which we will often use implicitly, since they are so basic. The proof is omitted since all parts are trivial.

Lemma 8.4   *Let $\mathcal{F}$ be a saturated fusion system on a finite p-group P, and let T and $\bar{T}$ be strongly $\mathcal{F}$-closed subgroups of P such that $T \, C_P(T) \leq \bar{T}$. Let $\mathcal{E}$ and $\bar{\mathcal{E}}$ be saturated subsystems of $\mathcal{F}$ on T and $\bar{T}$ respectively.*

(i) *$\Gamma_{\mathcal{F},\mathcal{E}}(T)$ is a normal subgroup of $\operatorname{Aut}_\mathcal{E}(T)$, and in particular $\Gamma_{\mathcal{F},\mathcal{F}}(T)$ is normal in $\operatorname{Aut}_\mathcal{F}(T)$.*

(ii) *$\Gamma_{\bar{\mathcal{E}},\mathcal{E}}(T) \leq \Gamma_{\mathcal{F},\mathcal{E}}(T)$.*

(iii) *If $\mathcal{E}$ is weakly normal in $\mathcal{F}$, then $\mathcal{E}$ is normal in $\mathcal{F}$ if and only if $\Gamma_{\mathcal{F},\mathcal{E}}(T) = \operatorname{Aut}_\mathcal{E}(T)$, or equivalently $\Gamma_{\mathcal{F},\mathcal{F}}(T) \geq \operatorname{Aut}_\mathcal{E}(T)$.*

(iv) *If $\mathcal{E}$ is weakly normal in $\mathcal{F}$ and $\mathcal{E} \leq \bar{\mathcal{E}}$, then $\mathcal{E}$ is normal in $\mathcal{F}$ if and only if $\mathcal{E}$ is normal in $\bar{\mathcal{E}}$.*

It is easy to prove that if G is a finite group with Sylow p-subgroup P, and $H \trianglelefteq G$, then $\mathcal{F}_{P \cap H}(H)$ is a normal subsystem of $\mathcal{F}_P(G)$.

Lemma 8.5 *Let G be a finite group with Sylow p-subgroup P. If H is a normal subgroup of G with Sylow p-subgroup $T = P \cap H$, then $\mathcal{F}_T(H) \trianglelefteq \mathcal{F}_P(G)$.*

Proof Let $\mathcal{E} = \mathcal{F}_T(H)$. By Lemma 5.32, $\mathcal{E}$ is weakly normal in $\mathcal{F}$. Firstly, notice that $[N_H(T), C_P(T)] \leq H \cap C_G(T) = C_H(T)$, and so $N_H(T)$ normalizes $C_H(T) C_P(T)$; thus $C_H(T) C_P(T)$ is a normal subgroup of $N_H(T) C_P(T)$. Apply the Frattini argument to $N_H(T) C_P(T)$ and the subgroup $C_H(T) C_P(T)$: this states that

$$N_H(T) C_P(T) = C_H(T) C_P(T) N_{N_H(T)}(C_P(T))$$
$$= C_H(T) C_P(T) N_H(T, C_P(T)).$$

Let ϕ be an automorphism in $\mathrm{Aut}_{\mathcal{E}}(T)$; then $\phi = c_g$ for some $g \in N_H(T)$. By the above decomposition, we may write $g = xh$ for some $x \in C_H(T) C_P(T) \leq C_{HP}(T)$ and $h \in N_H(T, C_P(T))$. The action of g and h on T are the same; notice that $[h, C_P(T)] \leq H$ since $h \in H$ and $H \trianglelefteq G$, and also $[h, C_P(T)] \leq C_P(T)$ since h normalizes $C_P(T)$. Therefore $[h, C_P(T)] \in H \cap C_P(T) = C_{P \cap H}(T) = Z(T)$, as needed. $\square$

We will examine various other constructions that we have made in previous chapters. Theorem 5.37 proved that $\mathcal{F} = N_{\mathcal{F}}(Q)$ if and only if $\mathcal{F}_Q(Q)$ is weakly normal in $\mathcal{F}$. This result may be generalized to include normality.

Lemma 8.6 *Let $\mathcal{F}$ be a saturated fusion system on a finite p-group P. If T is a strongly $\mathcal{F}$-closed subgroup of P and $x \in T$, then $c_x \in \mathrm{Aut}_{\mathcal{F}_T(T)}(T)$ lies in $\Gamma_{\mathcal{F},\mathcal{F}}(T)$. Consequently, $\mathcal{F}_T(T) \prec \mathcal{F}$ if and only if $\mathcal{F}_T(T) \trianglelefteq \mathcal{F}$. In particular, if $\mathcal{F} = N_{\mathcal{F}}(T)$ then $\mathcal{F}_T(T) \trianglelefteq \mathcal{F}$, so this holds for all strongly $\mathcal{F}$-closed subgroups T of $O_p(\mathcal{F})$.*

Moreover, if $\mathcal{E}$ is a weakly normal subsystem of $\mathcal{F}$ on T, then $\mathcal{E}$ is normal in $\mathcal{F}$ if and only if all p'-automorphisms of $\mathrm{Aut}_{\mathcal{E}}(T)$ lie in $\Gamma_{\mathcal{F},\mathcal{F}}(T)$.

Proof Let T and x be as in the statement, and notice that $c_x \in \mathrm{Aut}_{\mathcal{F}}(P)$. In addition, $[c_x, P] \leq T$, and so c_x lies in $\Gamma_{\mathcal{F},\mathcal{F}}(T)$. Finally, $\mathrm{Aut}_{\mathcal{F}_T(T)}(T)$ consists solely of conjugation maps c_x for $x \in T$, so we have the result.

To see the last statement, since $\mathrm{Aut}_T(T) \leq \Gamma_{\mathcal{F},\mathcal{F}}(T)$, $\mathrm{Aut}_T(T)$ is a Sylow p-subgroup of $\mathrm{Aut}_{\mathcal{E}}(T)$, and all p'-automorphisms of $\mathrm{Aut}_{\mathcal{E}}(T)$ lie in $\Gamma_{\mathcal{F},\mathcal{F}}(T)$, we see that $\mathrm{Aut}_{\mathcal{E}}(T) \leq \Gamma_{\mathcal{F},\mathcal{F}}(T)$, as needed. The converse is obvious. $\qquad\square$

With respect to quotients, we get a similar statement to Lemma 5.59.

Proposition 8.7 *Let $\mathcal{F}$ be a saturated fusion system on a finite p-group P, and let T be a strongly $\mathcal{F}$-closed subgroup of P. If $\mathcal{E}$ is a normal subsystem of $\mathcal{F}$, then the image of $\mathcal{E}$ in $\mathcal{F}/T$ is normal in $\mathcal{F}/T$.*

Proof This is Exercise 8.4. $\qquad\square$

The next object of study is normalizers.

Proposition 8.8 *Let $\mathcal{F}$ be a saturated fusion system on a finite p-group P, and let Q be a fully normalized subgroup of P. The subsystems $\mathrm{C}_{\mathcal{F}}(Q)$ and $Q\,\mathrm{C}_{\mathcal{F}}(Q)$ are normal in $\mathrm{N}_{\mathcal{F}}(Q)$.*

Proof By Exercise 5.10, both $\mathrm{C}_{\mathcal{F}}(Q)$ and $Q\,\mathrm{C}_{\mathcal{F}}(Q)$ are weakly normal in $\mathrm{N}_{\mathcal{F}}(Q)$. In the case of $\mathcal{E} = Q\,\mathrm{C}_{\mathcal{F}}(Q)$, notice that $\mathrm{C}_P(Q\,\mathrm{C}_P(Q)) = \mathrm{Z}(Q\,\mathrm{C}_P(Q))$, so clearly $Q\,\mathrm{C}_{\mathcal{F}}(Q) \trianglelefteq \mathrm{N}_{\mathcal{F}}(Q)$.

Thus let $\mathcal{E} = \mathrm{C}_{\mathcal{F}}(Q)$ and $T = \mathrm{C}_P(Q)$, and let α be a p'-element of $\mathrm{Aut}_{\mathcal{E}}(T)$. We may assume without loss of generality that $\mathcal{F} = \mathrm{N}_{\mathcal{F}}(Q)$. Since α lies in $\mathcal{E}$, it extends to an automorphism β of $Q\,\mathrm{C}_P(Q)$ in $\mathcal{F}$ such that $\beta|_Q = \mathrm{id}$. We claim that N_β contains $\mathrm{C}_P(T)$. Let $g \in \mathrm{C}_P(T)$, and consider the automorphism $(c_g)^\beta \in \mathrm{Aut}_{\mathcal{F}}(Q\,\mathrm{C}_P(Q))$. Since β acts trivially on Q, $c_g|_Q = (c_g)^\beta|_Q$. Also, since g acts trivially on T, $c_g|_T = (c_g)^\beta|_T = \mathrm{id}_T$, so that $c_g = (c_g)^\beta$ on $Q\,\mathrm{C}_P(Q)$. Therefore β extends to an automorphism $\gamma \in \mathrm{Aut}_{\mathcal{F}}(T\,\mathrm{C}_P(T))$. Since β is a p'-automorphism, we may choose γ to be a p'-automorphism.

Finally, consider the action of γ on $\mathrm{C}_P(T)$. By the calculation above, γ acts trivially on $T\,\mathrm{C}_P(T)/Q\,\mathrm{C}_P(Q)$, and we know that γ acts trivially on Q, so that (by Exercise 4.5) we have that γ acts trivially on $T\,\mathrm{C}_P(T)/T$, i.e., $[\gamma, \mathrm{C}_P(T)] \leq T$. This implies that $[\gamma, \mathrm{C}_P(T)] \leq \mathrm{Z}(T)$ by Exercise 8.3, proving that $\alpha \in \Gamma_{\mathcal{F},\mathcal{F}}(T)$. By Lemma 8.6, $\mathcal{E} \trianglelefteq \mathcal{F}$. $\qquad\square$

We turn our attention to the subsystems $\mathrm{O}^p(\mathcal{F})$ and $\mathrm{O}^{p'}(\mathcal{F})$. Weakly normal subsystems on P itself are obviously normal, and so it is clear that $\mathrm{O}^{p'}(\mathcal{F})$ is a normal subsystem of $\mathcal{F}$. The problem of $\mathrm{O}^p(\mathcal{F})$ (and the unique subsystems of p-power index on normal overgroups of $\mathfrak{hyp}(\mathcal{F})$) is only slightly more complicated.

Proposition 8.9 *Let $\mathcal{F}$ be a saturated fusion system on a finite p-group P, and let T be a strongly $\mathcal{F}$-closed subgroup of P containing $\mathfrak{hyp}(\mathcal{F})$. If*

$\mathcal{F}_T$ *is the unique saturated subsystem of p-power index on T, then $\mathcal{F}_T$ is normal in $\mathcal{F}$.*

Proof The subsystem $\mathcal{F}_T$ is weakly normal in $\mathcal{F}$ by Theorem 7.51. Let α be an automorphism of T in $\mathcal{F}_T$, and let $\beta \in \mathrm{Aut}_{\mathcal{F}}(T\,\mathrm{C}_P(T))$ be an extension of α to $T\,\mathrm{C}_P(T)$. By Lemma 8.6, we may assume that α, and hence β, are p'-automorphisms of T. Consider the image of β in $\mathcal{F}/T = \mathcal{F}_{P/T}(P/T)$; this must be the identity map, since it is a p'-automorphism, and so β acts trivially on $T\,\mathrm{C}_P(T)/T$. Hence $[\beta, \mathrm{C}_P(T)] \leq T$, so that $[\beta, \mathrm{C}_P(T)] \leq \mathrm{Z}(T)$ by Exercise 8.3. Hence $\mathcal{F}_T \trianglelefteq \mathcal{F}$ by Lemma 8.6, as needed. □

To prove the more useful and interesting properties of normal subsystems, we need an important theorem, due to Aschbacher, Theorem 8.36, which states that, in a *constrained* fusion system $\mathcal{F}$, there is a bijection between normal subgroups of the unique model of $\mathcal{F}$ and normal subsystems of $\mathcal{F}$. Having said that, in the next section we will determine the relationship between weakly normal and normal subsystems without this theory.

We end with a definition, extending a weakly normal map. It is obvious, and does not need to be remarked upon.

Definition 8.10 Let $\mathcal{F}$ be a saturated fusion system on a finite p-group P, and let $A(-)$ be a weakly normal map on the subgroup T of P. We say that $A(-)$ is a *normal map* if $A(T) \leq \Gamma_{\mathcal{F},\mathcal{F}}(T)$.

8.2 Weakly normal and normal subsystems

In this section we will understand the relationship between weakly normal and normal subsystems, proving in effect that a weakly normal subsystem is a normal subsystem with some automorphisms attached.

Let $\mathcal{F}$ be a saturated fusion system on a finite p-group P, and let $\mathcal{E}$ be a weakly normal subsystem on a strongly $\mathcal{F}$-closed subgroup T of P. If every automorphism $\alpha \in \mathrm{Aut}_{\mathcal{E}}(T)$ lies in $\Gamma_{\mathcal{F},\mathcal{E}}(T)$ then $\mathcal{E} \trianglelefteq \mathcal{F}$, as we stated in Lemma 8.4.

Definition 8.11 Let $\mathcal{F}$ be a saturated fusion system on a finite p-group P, and let $\mathcal{E}$ be a weakly normal subsystem on a strongly $\mathcal{F}$-closed subgroup T of P. Let α be an $\mathcal{E}$-automorphism of T. An $\mathcal{E}$-centric subgroup $Q \leq T$ is a *detecting subgroup* for α if there is a morphism $\beta : Q\,\mathrm{C}_P(T) \to P$ in $\mathcal{F}$ such that $\beta|_Q = \alpha|_Q$ and $[\beta, \mathrm{C}_P(T)] \leq \mathrm{Z}(Q\alpha)$.

Not every weakly normal subsystem is normal, as we have seen. However, if $\mathcal{E} = O^{p'}(\mathcal{E})$ is a weakly normal subsystem of $\mathcal{F}$ then we will be able to prove that $\mathcal{E}$ is actually normal in $\mathcal{F}$. This is of course equivalent to showing that, if $\mathcal{E} = O^{p'}(\mathcal{E})$, then T is a detecting subgroup for all $\alpha \in \mathrm{Aut}_{\mathcal{E}}(T)$. In the previous chapter, we determined a condition on automorphisms in $\mathrm{Aut}_{\mathcal{E}}(T)$ that meant that a given automorphism α lies in $O^{p'}(\mathcal{E})$ (Corollary 7.60). This means that we have a handle on $\alpha|_Q$ for some $\mathcal{E}$-centric subgroup Q of T, namely that it is the (restriction of) a composition of p-automorphisms lying in $\mathcal{E}$. In fact, we can show that these subgroups Q are detecting subgroups.

Proposition 8.12 *Let $\mathcal{F}$ be a saturated fusion system on a finite p-group P, and let $\mathcal{E} = O^{p'}(\mathcal{E})$ be a weakly normal subsystem on a strongly $\mathcal{F}$-closed subgroup T of P. If α is an $\mathcal{E}$-automorphism of T then there is a detecting subgroup for α.*

Proof Let Q be an $\mathcal{E}$-centric subgroup of T such that $\alpha|_Q$ is the restriction of the composition of p-automorphisms $\phi_1, \ldots, \phi_d$ of subgroups $S_1, \ldots, S_d$ of T, with ϕ_i in $\mathcal{E}$. Write $Q_1 = Q$ and $Q_i = Q\phi_1 \ldots \phi_{i-1}$ for $2 \leq i \leq d+1$, and write $R = Q_{d+1} = Q\alpha$. We will show that each ϕ_i extends to an automorphism ψ_i of $S_i \, \mathrm{C}_P(T)$ such that $[\psi_i, \mathrm{C}_P(T)] \leq T$. The composition of the restrictions $\psi_i|_{Q_i \, \mathrm{C}_P(T)}$ is a map $\psi : Q \, \mathrm{C}_P(T) \to R \, \mathrm{C}_P(T)$ such that $[\psi, \mathrm{C}_P(T)] \leq T$, or equivalently $[\psi, \mathrm{C}_P(T)] \leq \mathrm{Z}(R)$ by Exercise 8.3.

Let S be a fully $\mathcal{F}$-normalized, $\mathcal{E}$-centric subgroup of T. If $g \in \mathrm{N}_T(S)$ then c_g acts as an automorphism of $S \, \mathrm{C}_P(T)$ such that $[c_g, \mathrm{C}_P(T)] \leq T$, by Exercise 8.2(iii). If χ is an element of $\mathrm{Aut}_{\mathcal{E}}(S)$, then χ extends to some automorphism $\bar{\chi} \in \mathrm{Aut}_{\mathcal{F}}(S \, \mathrm{C}_P(T))$ (since $\mathrm{C}_P(T) \leq \mathrm{C}_P(S)$, via Exercise 8.2(i)). Notice that $(c_g)^{\chi}$ acts trivially on $\mathrm{Z}(S) \, \mathrm{C}_P(T)/\mathrm{Z}(S)$, because this is true for c_g, and χ acts on $\mathrm{Z}(S) \, \mathrm{C}_P(T)/\mathrm{Z}(S)$. Therefore all p-automorphisms of S in $\mathcal{E}$ extend to automorphisms of $S \, \mathrm{C}_P(T)$ in $\mathcal{F}$ that act trivially on $\mathrm{Z}(S) \, \mathrm{C}_P(T)/\mathrm{Z}(S)$.

Let $\bar{S}$ be any subgroup of T that is $\mathcal{F}$-conjugate to S, via some isomorphism τ. Since $\mathcal{F}$ is saturated and S is fully $\mathcal{F}$-normalized, τ extends to a map $\bar{\tau} : \bar{S} \, \mathrm{C}_P(\bar{S}) \to S \, \mathrm{C}_P(S)$, which restricts to an isomorphism $\bar{S} \, \mathrm{C}_P(T) \to S \, \mathrm{C}_P(T)$. Hence any automorphism of $\bar{S} \, \mathrm{C}_P(T)$, taken through $\bar{\tau}$, becomes an automorphism of $S \, \mathrm{C}_P(T)$, and vice versa. Thus any p-automorphism of $\bar{S}$ extends to an automorphism of $\bar{S} \, \mathrm{C}_P(T)$ acting trivially on $\mathrm{Z}(\bar{S}) \, \mathrm{C}_P(T)/\mathrm{Z}(\bar{S})$. In particular, this applies to ϕ_i and S_i, completing the proof. $\qquad\square$

If $\mathcal{E} = \mathrm{O}^{p'}(\mathcal{E})$ lies on the subgroup T of P, and $\alpha \in \mathrm{Aut}_{\mathcal{E}}(T)$, then there is a detecting subgroup for α. If α is a p'-automorphism of T, then we will prove that if Q is a detecting subgroup for α then so is $\mathrm{N}_T(Q)$. In order to do this we will need to move from a map with domain Q to map with domain $\mathrm{N}_T(Q)$.

Proposition 8.13 *Let $\mathcal{F}$ be a saturated fusion system on a finite p-group P, and let $\mathcal{E} = \mathrm{O}^{p'}(\mathcal{E})$ be a weakly normal subsystem on a strongly $\mathcal{F}$-closed subgroup T of P. Let α be an $\mathcal{E}$-automorphism of T and let Q be a detecting subgroup for α; write $R = Q\alpha$. If $\phi : Q\,\mathrm{C}_P(T) \to R\,\mathrm{C}_P(T)$ is an isomorphism in $\mathcal{F}$ such that $\phi|_Q = \alpha|_Q$ and $[\phi, \mathrm{C}_P(T)] \leq \mathrm{Z}(R)$, then there is an extension $\theta : \mathrm{N}_T(Q)\,\mathrm{C}_P(T) \to \mathrm{N}_T(R)\,\mathrm{C}_P(T)$ of ϕ in $\mathcal{F}$.*

Proof Let X be a fully $\mathcal{F}$-normalized subgroup of P that is $\mathcal{F}$-conjugate to $R\,\mathrm{C}_P(T)$; then $X = S\,\mathrm{C}_P(T)$ where $S = X \cap T$, by Exercise 8.2(i). Suppose that $\rho : R \to S$ is an isomorphism such that $N_\rho \geq \mathrm{N}_T(R)$, and let σ be any extension of ρ to $\sigma : R\,\mathrm{C}_P(T) \to S\,\mathrm{C}_P(T)$. We first claim that N_σ contains $\mathrm{N}_T(R)$.

Let g be an element of $\mathrm{N}_T(R)$, and choose $h \in \mathrm{N}_T(S)$ such that $c_g^\rho = c_h$. We claim that, as an element of $\mathrm{Aut}_{\mathcal{F}}(S\,\mathrm{C}_P(T))$, $c_g^\sigma c_h^{-1}$ acts trivially on S and $Y_S = \mathrm{Z}(S)\,\mathrm{C}_P(T)/\mathrm{Z}(S)$. To see this, clearly c_g^σ and c_h act the same on S; also, c_h acts trivially on Y_S by Exercise 8.2(iii), so we need that c_g^σ acts trivially on Y_S. However, c_g acts trivially on $Y_R = \mathrm{Z}(R)\,\mathrm{C}_P(T)/\mathrm{Z}(R)$ by Exercise 8.2(iii), and σ maps Y_R to Y_S by Exercise 8.2(ii), so that this holds.

Let K denote the set of all elements of $\mathrm{Aut}_{\mathcal{F}}(S\,\mathrm{C}_P(T))$ that act trivially on S and Y_S. The set K is a p-group, since any p'-automorphism acting trivially on both $\mathrm{Z}(S)$ and Y_S acts trivially on $\mathrm{Z}(S)\,\mathrm{C}_P(T)$, so that it acts trivially on $S\,\mathrm{C}_P(T)$ (since it acts trivially on S by hypothesis). In fact, since every $\mathcal{F}$-automorphism of $S\,\mathrm{C}_P(T)$ induces an automorphism of $S = T \cap S\,\mathrm{C}_P(T)$ and Y_S, K is a normal p-subgroup of $\mathrm{Aut}_{\mathcal{F}}(S\,\mathrm{C}_P(T))$.

As $S\,\mathrm{C}_P(T)$ is fully $\mathcal{F}$-normalized, and K is a normal p-subgroup of $\mathrm{Aut}_{\mathcal{F}}(S\,\mathrm{C}_P(T))$, we see that $K \leq \mathrm{Aut}_P(S\,\mathrm{C}_P(T))$, so that $c_g^\sigma c_h^{-1}$, which acts trivially on S and Y_S, lies in $\mathrm{Aut}_P(S\,\mathrm{C}_P(T))$. Therefore, $c_g^\sigma \in \mathrm{Aut}_P(S\,\mathrm{C}_P(T))$, and so N_σ contains c_g; thus $\mathrm{N}_T(Q) \leq N_\sigma$, as we claimed.

Choose a morphism $\tau : \mathrm{N}_T(R)\,\mathrm{C}_P(T) \to \mathrm{N}_T(S)\,\mathrm{C}_P(T)$ (which exists because we may choose a morphism $R\,\mathrm{C}_P(T) \to S\,\mathrm{C}_P(T)$ that extends to

$N_T(R) \, C_P(T))$. We apply the previous claim to the morphism $(\phi|_Q)\tau$ and $\phi\tau$. Since $\phi|_Q$ extends to an isomorphism $N_T(Q) \to N_T(R)$, we see that $N_{(\phi|_Q)\tau}$ contains $N_T(Q)$, and so therefore does $N_{\phi\tau}$. Hence there is an extension $\theta_1 : N_T(Q) \, C_P(T) \to N_T(S) \, C_P(T)$ of $\phi\tau$ in $\mathcal{F}$. We claim that $N_T(Q)\theta_1 = N_T(R)\tau$. If this is true, then $\theta = \theta_1\tau^{-1} : N_T(Q) \, C_P(T) \to N_T(R) \, C_P(T)$ is an isomorphism in $\mathcal{F}$ (since the two groups have the same order) that extends ϕ (since on $Q \, C_P(T)$ it acts as $\phi\tau\tau^{-1} = \phi$).

Write $N = N_T(Q)\theta_1$ and $M = N_T(R)\tau$. The subgroups N and M are determined by the maps $\theta_1|_Q$ and $\tau|_R$ respectively, since N and M are the preimages in $\mathrm{Aut}_T(S)$ of the subgroups $\mathrm{Aut}_T(Q)^{\theta_1|_Q}$ and $\mathrm{Aut}_T(R)^{\tau|_R}$. However, since $\mathrm{Aut}_T(Q)^{\phi|_Q} = \mathrm{Aut}_T(R)$, we see that

$$\mathrm{Aut}_T(Q)^{\theta_1} = \mathrm{Aut}_T(Q)^{(\phi|_Q)\tau} = \mathrm{Aut}_T(R)^{\tau},$$

so that $N = M$.

Hence θ is an isomorphism from $N_T(Q) \, C_P(T)$ to $N_T(R) \, C_P(T)$ in $\mathcal{F}$ that extends ϕ, as claimed. $\qquad\square$

Using this proposition we can, given a detecting subgroup Q, prove that $N_T(Q)$ is also a detecting subgroup.

Proposition 8.14 *Let $\mathcal{F}$ be a saturated fusion system on a finite p-group P, and let $\mathcal{E} = O^{p'}(\mathcal{E})$ be a weakly normal subsystem on a strongly $\mathcal{F}$-closed subgroup T of P. Let α be a p'-automorphism of T in $\mathcal{E}$. If Q is a detecting subgroup for α then so is $N_T(Q)$.*

Proof As $\mathcal{F}$ is saturated, there is an extension $\beta \in \mathrm{Aut}_{\mathcal{F}}(T \, C_P(T))$ of α. The automorphism β acts on $C_P(T)$, so we may raise β to an appropriate power to get that $\beta|_T = \alpha$ and $\beta|_{C_P(T)}$ is a p'-automorphism.

Since Q is a detecting subgroup for α, there is an isomorphism $\phi : Q \, C_P(T) \to R \, C_P(T)$ in $\mathcal{F}$ (where $R = Q\alpha$), such that $\alpha|_Q = \phi|_Q$ and $[\phi, C_P(T)] \leq Z(R)$. By Proposition 8.13, there is an $\mathcal{F}$-morphism $\theta : N_T(Q) \, C_P(T) \to N_T(R) \, C_P(T)$ extending ϕ. Let γ be the restriction of β to $N_T(Q) \, C_P(T)$. Since $Q\beta = R$, $N_T(Q)\beta = N_T(R)$, so that β has image $N_T(R) \, C_P(T)$, the same as θ. We will consider the map $\chi = \theta^{-1}\gamma$, an automorphism of $N_T(R) \, C_P(T)$.

Since $\theta|_Q$ and $\gamma|_Q$ are identical, $\chi|_R$ is the identity. Therefore, since χ induces an automorphism of $N_T(R)$ and $C_T(R) \leq R$, by Exercise 6.4 we have that $\chi|_{N_T(R)}$ is a p-automorphism.

Let $Y_Q = Z(Q) \, C_P(T)/Z(Q)$ and $Y_R = Z(R) \, C_P(T)/Z(R)$. Since both θ and γ map Y_Q to Y_R, χ induces an automorphism of Y_R; we claim that this is a p'-automorphism. To see this, notice that, for $x \in C_P(T)$, we have $(Z(Q)x)\theta = Z(R)x$ and $(Z(Q)x)\gamma = Z(R)(x\beta)$. Therefore $(Z(R)x)\chi = Z(R)(x\beta)$, so that the order of the automorphism that χ induces on Y_R divides the order of $\beta|_{C_P(T)}$, which itself is a p'-automorphism.

Thus we have that χ acts like a p-automorphism on $N_T(R)$ and like a p'-automorphism on Y_R. Therefore there exists an integer n such that $\chi^n|_{N_T(R)} = \chi|_{N_T(R)}$, and χ^n acts like the identity on Y_R. Consider $\theta\chi^n$; this map acts like γ on $N_T(Q)$ (since it acts like $\theta\chi^n = \theta\chi = \theta\theta^{-1}\gamma = \gamma$), and since χ acts like the identity on Y_R, $\theta\chi^n$ acts like θ on Y_Q, i.e., $[\theta\chi^n, C_P(T)] \leq T$. Hence $\theta\chi^n$ is an extension of $\alpha|_{N_T(Q)}$ to $N_T(Q) C_P(T)$ such that $[\theta\chi^n, C_P(T)] \leq Z(N_T(Q))$ (using Exercise 8.3). Thus $N_T(Q)$ is a detecting subgroup for α, as required. $\qquad\square$

Propositions 8.12 and 8.14 prove that if α is a p'-automorphism in $\mathcal{E}$ then T is a detecting subgroup for α, so that α extends to an automorphism $\beta \in \mathrm{Aut}_{\mathcal{F}}(T C_P(T))$ such that $[\beta, C_P(T)] \leq Z(T)$. Therefore all p'-automorphisms in $\mathrm{Aut}_{\mathcal{E}}(T)$ lie in $\Gamma_{\mathcal{F},\mathcal{E}}(T)$, and so $\mathcal{E} \trianglelefteq \mathcal{F}$ by Lemma 8.6. We have proved the following theorem.

Theorem 8.15 (Craven [Cra10b]) *Let $\mathcal{F}$ be a saturated fusion system on a finite p-group P. If $\mathcal{E}$ is a weakly normal subsystem of $\mathcal{F}$ then $O^{p'}(\mathcal{E})$ is a normal subsystem of $\mathcal{F}$.*

This theorem gives us a structural understanding of the relationship between weakly normal and normal subsystems of a saturated fusion system. It has the following important corollary. Recall that a simple fusion system is one with no proper, non-trivial weakly normal subsystems.

Corollary 8.16 (Craven [Cra10b]) *A saturated fusion system is simple if and only if it has no proper, non-trivial normal subsystems.*

There are other corollaries to this theorem, some of which are useful in this chapter. We begin with an observation, whose proof is trivial.

Proposition 8.17 *Let $\mathcal{F}$ be a saturated fusion system on a finite p-group P. If $\mathcal{E}$ is a weakly normal subsystem of $\mathcal{F}$ contained in a saturated subsystem $\mathcal{F}'$ of $\mathcal{F}$, then $\mathcal{E}$ is weakly normal in $\mathcal{F}'$.*

This is *not* true for normality, as the following example demonstrates.

Example 8.18 We will expand upon the situation given in Example 8.1, so we assume that notation there. Let L be generated by x, y, and the two involutions $g = (1, 2)$ and $h = (4, 5)$. Thus $L \cong S_3 \times S_3$, and $G = \langle x, y, gh \rangle$. Let $K_1 = \langle x, g \rangle$ and $K_2 = \langle y, h \rangle$. In Example 8.1, we showed that the $\mathcal{E}_i$ are not normal in $\mathcal{F} = \mathcal{F}_P(G)$. However, since $\mathcal{E}_1 = \mathcal{F}_X(K_1)$

and $\mathcal{E}_2 = \mathcal{F}_Y(K_2)$, and $K_i \trianglelefteq L$, we *do* have that $\mathcal{E}_i \trianglelefteq \hat{\mathcal{F}} = \mathcal{F}_P(L)$. Thus even though $\mathcal{E}_1$ is a normal subsystem of $\hat{\mathcal{F}}$ contained in $\mathcal{F}$, $\mathcal{E}_1$ is not normal in $\mathcal{F}$. Indeed, in this case $\mathcal{F}$ is even normal in $\hat{\mathcal{F}}$, but this does not help.

Given this negative result, the following corollary to Theorem 8.15 shows that all is not lost.

Corollary 8.19 *Let $\mathcal{F}$ be a saturated fusion system on a finite p-group P, and let $\mathcal{E}$ be a normal subsystem of $\mathcal{F}$. If $\mathcal{E} = \mathrm{O}^{p'}(\mathcal{E})$ and $\mathcal{F}'$ is a saturated subsystem of $\mathcal{F}$ such that $\mathcal{E} \leq \mathcal{F}'$, then $\mathcal{E} \trianglelefteq \mathcal{F}'$.*

Many of the subsystems we are interested in do satisfy $\mathcal{E} = \mathrm{O}^{p'}(\mathcal{E})$, and so we will be able to apply Corollary 8.19.

8.3 Intersections of subsystems

In group theory, the intersection of two subgroups is a subgroup, and the intersection of two normal subgroups is a normal subgroup. While the intersection of two subsystems is a subsystem, and the intersection of two invariant subsystems is an invariant subsystem, it is not true that the intersection of two saturated subsystems is saturated, as we showed in Example 5.34, and in the following example.

Example 8.20 Let $p = 2$, let H be the group S_5, and let V be a faithful $\mathbb{F}_2 H$-module, say the regular module. Let $G = V \rtimes (H_1 \times H_2)$, with both H_i being isomorphic to H, and acting as H acts on V. Finally, let P denote a Sylow p-subgroup of G, and let $Q_i = H_i \cap P$.

Let G_i denote $\langle V, H_i \rangle$, and write $\mathcal{F} = \mathcal{F}_P(G)$ and $\mathcal{E}_i = \mathcal{F}_{Q_i}(G_i)$. Since $G_i \trianglelefteq G$, $\mathcal{E}_i \trianglelefteq \mathcal{F}$. Notice that $\mathcal{E} = \mathcal{E}_1 \cap \mathcal{E}_2$ is a fusion system on V, with $\mathrm{Aut}_{\mathcal{E}}(V) \cong S_5$. There are two saturated subsystems of this, the fusion systems of $V \rtimes C_3$ and $V \rtimes C_5$, which are maximal subject to being saturated subsystems of $\mathcal{E}$.

Therefore, there need not be a maximal saturated subsystem of the intersection of two saturated subsystems, even if both subsystems are normal.

Hence in general there is no hope of there being an intersection subsystem. However, sometimes when we apply the fact that the intersection of two normal subgroups is a normal subgroup, we don't need that the intersection itself is normal, just that it contains a 'large' normal subgroup, in some sense.

Since we know that the intersection of two normal subsystems need not be normal (or even saturated), we could ask whether, if $\mathcal{E}_1$ and $\mathcal{E}_2$ are normal subsystems in $\mathcal{F}$ on T_1 and T_2 respectively, there is a normal subsystem $\mathcal{E}$ in $\mathcal{F}$, on the subgroup $T = T_1 \cap T_2$, contained in $\mathcal{E}_1 \cap \mathcal{E}_2$. This would be a 'large' subsystem of $\mathcal{E}_1 \cap \mathcal{E}_2$ that is normal in $\mathcal{F}$, and might well fulfil the requirements of an intersection subsystem.

It is an important result in the theory of fusion systems that this large normal subsystem actually does exist.

Theorem 8.21 (Aschbacher [Asc11]) *Let $\mathcal{F}$ be a saturated fusion system on a finite p-group P, and let T_1 and T_2 be strongly $\mathcal{F}$-closed subgroups of P. If $\mathcal{E}_1$ and $\mathcal{E}_2$ are normal subsystems on T_1 and T_2 respectively, then there is a subsystem, $\mathcal{E}_1 \wedge \mathcal{E}_2$, normal in $\mathcal{F}$ and contained within $\mathcal{E}_1 \cap \mathcal{E}_2$, on the subgroup $T_1 \cap T_2$ of P. Furthermore, $\mathcal{E}_1 \wedge \mathcal{E}_2$ is the largest normal subsystem of $\mathcal{F}$ that is normal in both $\mathcal{E}_1$ and $\mathcal{E}_2$.*

This result requires a lot of the local theory of fusion systems contained in [Asc08a] and [Asc11]. As such, we will unfortunately not be able to prove it in this book. However, we will develop a theory of intersections of subsystems which requires only a special case of Theorem 8.21, namely that if T_1 and T_2 are strongly $\mathcal{F}$-closed subgroups of P, and both T_i support normal subsystems of $\mathcal{F}$, then there is a normal subsystem of $\mathcal{F}$ on $T_1 \cap T_2$.

In Theorem 6.25, we proved that there is a type of intersection subsystem, when the two subsystems are both defined on the same p-group as the fusion system itself. We now extend this result to a much more general case. The method of proof is similar, but more involved.

Theorem 8.22 (Craven [Cra10b]) *Let $\mathcal{F}$ be a saturated fusion system on a finite p-group P, and let $\mathcal{E}_1$ and $\mathcal{E}_2$ be saturated subsystems of $\mathcal{F}$, on the strongly $\mathcal{F}$-closed subgroups T and $\bar{T}$ of P respectively, with $T \leq \bar{T}$. Suppose that $\mathcal{E}_2$ is weakly normal in $\mathcal{F}$. There exists a saturated subsystem on T, weakly normal in $\mathcal{E}_1$, contained in $\mathcal{E}_1 \cap \mathcal{E}_2$ and denoted by $\mathcal{E}_1 \curlywedge \mathcal{E}_2$, such that for any subgroup Q of T with $\mathrm{C}_T(Q) \leq Q$, we have*

$$\mathrm{Aut}_{\mathcal{E}_1 \curlywedge \mathcal{E}_2}(Q) = \mathrm{Aut}_{\mathcal{E}_1}(Q) \cap \mathrm{Aut}_{\mathcal{E}_2}(Q).$$

Furthermore, if $\mathcal{E}_1$ is also weakly normal in $\mathcal{F}$ then $\mathcal{E}_1 \curlywedge \mathcal{E}_2$ is also weakly normal in $\mathcal{F}$, and $\mathcal{E}_1 \curlywedge \mathcal{E}_2$ is the largest weakly normal subsystem of $\mathcal{F}$ contained in $\mathcal{E}_1 \cap \mathcal{E}_2$. The subsystem $\mathcal{E}_1 \curlywedge \mathcal{E}_2$ is the weak intersection *of $\mathcal{E}_1$ and $\mathcal{E}_2$.*

Proof Let $A(-)$ be the map such that $A(Q) = \mathrm{Aut}_{\mathcal{E}_1}(Q) \cap \mathrm{Aut}_{\mathcal{E}_2}(Q)$ if Q is a fully $\mathcal{E}_1$-normalized, $\mathcal{E}_1$-centric subgroup of T; clearly $A(Q)$ is a normal subgroup of $\mathrm{Aut}_{\mathcal{E}_1}(Q)$, since $\mathrm{Aut}_{\mathcal{E}_2}(Q) \trianglelefteq \mathrm{Aut}_{\mathcal{F}}(Q)$. Notice that if R is some other fully $\mathcal{E}_1$-normalized, $\mathcal{E}_1$-centric subgroup of T and $\phi : Q \to R$ is an isomorphism, then $A(Q)^\phi = A(R)$. Also, since $\mathrm{Aut}_T(Q)$ is contained in $\mathrm{Aut}_{\mathcal{E}_i}(Q)$ for $i = 1, 2$, $\mathrm{Aut}_T(Q) \leq A(Q)$. Hence by Lemma 6.22, we may extend $A(-)$ to an invariant map with respect to $\mathcal{E}_1$, and it remains to show conditions (iii) and (v) for $A(-)$ to be a weakly normal map. (It is clear that $A(Q) = \mathrm{Aut}_{\mathcal{E}_1}(Q) \cap \mathrm{Aut}_{\mathcal{E}_2}(Q)$ holds for any subgroup Q of T with $\mathrm{C}_T(Q) \leq Q$, not just for fully $\mathcal{E}_1$-normalized subgroups.)

Since $\mathrm{Aut}_T(T)$ is a Sylow p-subgroup of $\mathrm{Aut}_{\mathcal{E}_1}(T)$ and is contained in $\mathrm{Aut}_{\mathcal{E}_2}(T)$, $\mathrm{Aut}_T(T)$ is a Sylow p-subgroup of $A(T)$. Hence it remains to show condition (v).

Let Q be a fully $\mathcal{E}_1$-normalized subgroup of T with $\mathrm{C}_T(Q) \leq Q$, and let R be an overgroup of Q contained in $\mathrm{N}_T(Q)$. Let ϕ be an automorphism in $\mathrm{N}_{A(Q)}(\mathrm{Aut}_R(Q))$. Since $\mathcal{E}_1$ is saturated, there is some $\psi_1 \in \mathrm{Aut}_{\mathcal{E}_1}(R)$ extending ϕ. We claim that there is $\psi_2 \in \mathrm{Aut}_{\mathcal{E}_2}(R)$ also extending ϕ. Let $\bar{Q}$ be a fully $\mathcal{E}_2$-normalized subgroup of T that is $\mathcal{E}_2$-conjugate to Q, and choose $\alpha : Q \to \bar{Q}$ such that α extends to an isomorphism (also denoted by α) from R to some subgroup $\bar{R}$ of T. The map $\bar{\phi} \in \mathrm{Aut}_{\mathcal{E}_1}(\bar{Q})$ extends to an automorphism $\overline{\psi_2}$ of $\bar{R}$; hence there exists $\psi_2 \in \mathrm{Aut}_{\mathcal{E}_2}(R)$ such that $\psi_2^\alpha = \overline{\psi_2}$, and ψ_2 extends ϕ.

Let S be a fully $\mathcal{F}$-normalized subgroup of T that is $\mathcal{F}$-conjugate to R via $\beta : R \to S$. Notice that $\mathrm{Aut}_T(R)^\beta$ is a Sylow p-subgroup of $\mathrm{Aut}_{\mathcal{E}_1}(S)$ and is contained in $\mathrm{Aut}_{\mathcal{E}_2}(S)$. Hence $\mathrm{Aut}_{\mathcal{E}_2}(S)$ is a normal subgroup of p'-index of $A = \mathrm{Aut}_{\mathcal{E}_1}(R) \mathrm{Aut}_{\mathcal{E}_2}(R)$, and so all p-subgroups of A are p-subgroups of $\mathrm{Aut}_{\mathcal{E}_2}(R)$.

Let K be the subgroup of A consisting of all automorphisms of R acting trivially on Q, which must therefore act trivially on $\mathrm{Aut}_R(Q) \cong R/Q$; K is a p-subgroup of A by Exercise 4.5. Since all p-subgroups of A are p-subgroups of $\mathrm{Aut}_{\mathcal{E}_2}(R)$, we see that K is contained in $\mathrm{Aut}_{\mathcal{E}_2}(R)$. Notice that $\psi_1 \psi_2^{-1}$ acts trivially on Q so lies in K, and therefore lies in $\mathrm{Aut}_{\mathcal{E}_2}(R)$. Hence $\psi_1 \in \mathrm{Aut}_{\mathcal{E}_2}(R)$, so $\psi_1 \in A(R)$, as needed. Hence condition (v) of being a weakly normal map is satisfied, proving that the subsystem generated by $A(-)$ is a weakly normal subsystem of $\mathcal{E}_1$.

Finally, suppose that $\mathcal{E}_1$ is also $\mathcal{F}$-invariant. Let Q be a fully $\mathcal{F}$-normalized subgroup of T such that $\mathrm{C}_T(Q) \leq Q$; define $A(Q)$ to be the intersection of $\mathrm{Aut}_{\mathcal{E}_1}(Q)$ and $\mathrm{Aut}_{\mathcal{E}_2}(Q)$, and extend $A(-)$ to an invariant map in $\mathcal{F}$ via Lemma 6.22. We notice that $A(-)$ is identical to the map defined above, and so the final two conditions, (iii) and (v), are also satisfied by $A(-)$. Hence $A(-)$ is a weakly normal map with respect to $\mathcal{F}$, so generates a subsystem that is weakly normal in $\mathcal{F}$, as needed. To see that $\mathcal{E}_1 \curlywedge \mathcal{E}_2$ is the largest weakly normal subsystem of $\mathcal{F}$ contained in $\mathcal{E}_1 \cap \mathcal{E}_2$, notice that

$$\mathrm{Aut}_{\mathcal{E}_1 \curlywedge \mathcal{E}_2}(T) = \mathrm{Aut}_{\mathcal{E}_1}(T) \cap \mathrm{Aut}_{\mathcal{E}_2}(T) = \mathrm{Aut}_{\mathcal{E}_1 \cap \mathcal{E}_2}(T),$$

and therefore if $\mathcal{E}$ is another weakly normal subsystem on T contained in $\mathcal{E}_1 \cap \mathcal{E}_2$ then $\mathrm{Aut}_{\mathcal{E}}(T) \leq \mathrm{Aut}_{\mathcal{E}_1 \curlywedge \mathcal{E}_2}(T)$. Let $\mathcal{E}' = \mathcal{E} \curlywedge (\mathcal{E}_1 \curlywedge \mathcal{E}_2)$; applying Theorem 7.57 to $\mathcal{E}'$, we have that $\mathcal{E}$ and $\mathcal{E}_1 \curlywedge \mathcal{E}_2$ are oversystems of $\mathcal{E}'$ with $\mathrm{Aut}_{\mathcal{E}}(T) \leq \mathrm{Aut}_{\mathcal{E}_1 \curlywedge \mathcal{E}_2}(T)$, so that $\mathcal{E} \leq \mathcal{E}_1 \curlywedge \mathcal{E}_2$, as required. □

This theorem has a variety of corollaries. The first can be obtained either from this theorem, or via Theorems 8.15 and 8.21.

Corollary 8.23 *Let $\mathcal{F}$ be a saturated fusion system on a finite p-group P, and let T be a strongly $\mathcal{F}$-closed subgroup of P. If $\mathcal{E}_1$ and $\mathcal{E}_2$ are two weakly normal subsystems of $\mathcal{F}$, both on T, then $O^{p'}(\mathcal{E}_1) = O^{p'}(\mathcal{E}_2)$. Consequently, there is a unique minimal weakly normal subsystem of $\mathcal{F}$ on T.*

We denote the smallest weakly normal subsystem (and via Theorem 8.15 the smallest normal subsystem) of $\mathcal{F}$ on the strongly $\mathcal{F}$-closed subgroup T by $\mathcal{R}_{\mathcal{F}}(T)$.

Corollary 8.24 *Let $\mathcal{F}$ be a saturated fusion system on a finite p-group P, and let T be a strongly $\mathcal{F}$-closed subgroup of P. If $\mathcal{E}_1$ and $\mathcal{E}_2$ are weakly normal subsystems on T such that $\mathrm{Aut}_{\mathcal{E}_1}(T) = \mathrm{Aut}_{\mathcal{E}_2}(T)$, then $\mathcal{E}_1 = \mathcal{E}_2$.*

To prove this, we simply apply Theorem 7.57 to the subsystem $\mathcal{R}_{\mathcal{F}}(T)$ and $\mathrm{Aut}_{\mathcal{E}_i}(T)$, or consider the subsystems $\mathcal{E}_1 \curlywedge \mathcal{E}_2$ and $\mathcal{E}_1$, using Proposition 5.73. The next corollary relates these subsystems $\mathcal{R}_{\mathcal{F}}(T)$.

Corollary 8.25 *Let $\mathcal{F}$ be a saturated fusion system on a finite p-group P, and let T and $\bar{T}$ be strongly $\mathcal{F}$-closed subgroups of P with $T \leq \bar{T}$. If both T and $\bar{T}$ support normal subsystems then $\mathcal{R}_{\mathcal{F}}(T) \leq \mathcal{R}_{\mathcal{F}}(\bar{T})$.*

Our final corollary concerns arbitrary saturated subsystems.

Corollary 8.26 *Let $\mathcal{F}$ be a saturated fusion system on a finite p-group P, let T be a strongly $\mathcal{F}$-closed subgroup of P, and let $\mathcal{E}$ be a saturated subsystem of $\mathcal{F}$, on the subgroup T, satisfying $\mathcal{E} = O^{p'}(\mathcal{E})$. If $\bar{\mathcal{E}}$ is any weakly normal subsystem of $\mathcal{F}$ on a subgroup $\bar{T}$ of P, and $T \leq \bar{T}$, then $\mathcal{E} \leq \bar{\mathcal{E}}$.*

The subsystem $\mathcal{E}_1 \curlywedge \mathcal{E}_2$ has so far only been defined in the case where the underlying subgroups of the $\mathcal{E}_i$ are nested. In the general case we need Theorem 8.21 to prove that, if $\mathcal{E}_i$ lies on T_i, then there are weakly normal subsystems lying on $T_1 \cap T_2$. If this is true then we may consider the set of all weakly normal subsystems on $T_1 \cap T_2$ that are contained in $\mathcal{E}_1 \cap \mathcal{E}_2$. If this collection has a unique maximal element (under inclusion) then we may set $\mathcal{E}_1 \curlywedge \mathcal{E}_2$ to be this maximal element.

In order to prove this, we need to define another subsystem of $\mathcal{F}$. Let $\mathcal{F}$ be a saturated fusion on the finite p-group P, and let T be a strongly $\mathcal{F}$-closed subgroup of P. It is an easy corollary of Theorem 7.57 that the weakly normal subsystems of $\mathcal{F}$ on the subgroup T are in one-to-one correspondence with the normal subgroups of $\mathrm{Aut}_{\mathcal{F}}(T)$ containing $\mathrm{Aut}_{\mathcal{R}_{\mathcal{F}}(T)}(T)$ and such that $p \nmid |\mathrm{Aut}_{\mathcal{F}}(T) : \mathrm{Inn}(T)|$. In particular, there is a largest weakly normal subsystem of $\mathcal{F}$ on T, which we denote by $\mathcal{R}^{\mathcal{F}}(T)$.

In summary, there is a unique smallest and largest weakly normal subsystem of $\mathcal{F}$ on T (provided there is a weakly normal subsystem of $\mathcal{F}$ on T), which we denote by $\mathcal{R}_{\mathcal{F}}(T)$ and $\mathcal{R}^{\mathcal{F}}(T)$ respectively. The weakly normal subsystems of $\mathcal{F}$ on T lie in between these two, and form a lattice identical to the lattice of normal subgroups of $\mathrm{Aut}_{\mathcal{F}}(T)$ lying between $\mathrm{Aut}_{\mathcal{R}_{\mathcal{F}}(T)}(T)$ and $\mathrm{Aut}_{\mathcal{R}^{\mathcal{F}}(T)}(T)$.

Using this information, we may now define the subsystem $\mathcal{E}_1 \curlywedge \mathcal{E}_2$ whenever $\mathcal{E}_1$ and $\mathcal{E}_2$ are weakly normal subsystems of a saturated fusion system $\mathcal{F}$.

Theorem 8.27 (Craven [Cra10b]) *Let $\mathcal{F}$ be a saturated fusion system on a finite p-group P, and let T_1 and T_2 be strongly $\mathcal{F}$-closed subgroups of P. Let $\mathcal{E}_i$ be a weakly normal subsystem of $\mathcal{F}$ on T_i for $i = 1, 2$, and write $T = T_1 \cap T_2$. There exists a unique largest weakly normal subsystem of $\mathcal{F}$, on the subgroup T, contained in $\mathcal{E}_1 \cap \mathcal{E}_2$. We denote this subsystem by $\mathcal{E}_1 \curlywedge \mathcal{E}_2$. (This notation extends that of Theorem 8.22.)*

Proof By Theorem 8.21, there are weakly normal subsystems defined on T, and so in particular we may construct the subsystem $\mathcal{R}^{\mathcal{F}}(T)$. Since T is contained in both T_i, we may define $\mathcal{E}_1 \curlywedge \mathcal{E}_2$ to be the subsystem $(\mathcal{R}^{\mathcal{F}}(T) \curlywedge \mathcal{E}_1) \curlywedge \mathcal{E}_2$ in the sense of Theorem 8.22. This is a weakly normal subsystem of $\mathcal{F}$, on the subgroup T, and contained in $\mathcal{E}_1 \cap \mathcal{E}_2$.

It remains to show that any other weakly normal subsystem $\mathcal{E}$ of $\mathcal{F}$ on T is contained in $\mathcal{E}_1 \curlywedge \mathcal{E}_2$. If $\mathcal{E}$ is contained in $\mathcal{E}_1$ then $\mathcal{E}$ is contained in $\mathcal{R}^{\mathcal{F}}(T) \curlywedge \mathcal{E}_1$ since it is clearly contained in $\mathcal{R}^{\mathcal{F}}(T)$. Also, if it is contained in $\mathcal{E}_2$ as well, then it must be contained in $(\mathcal{R}^{\mathcal{F}}(T) \curlywedge \mathcal{E}_1) \curlywedge \mathcal{E}_2$, since it is contained in both terms. This completes the proof. □

Having defined the 'intersection' of two weakly normal subsystems in general, we want to know how this interacts with normality. If $\mathcal{E}_1$ and $\mathcal{E}_2$ are normal subsystems of $\mathcal{F}$ then $\mathcal{E}_1 \curlywedge \mathcal{E}_2$ need not be normal in $\mathcal{F}$. In order to get a normal subsystem, we need a version for normal subsystems of $\mathcal{R}^{\mathcal{F}}(T)$, i.e., we need a largest normal subsystem of $\mathcal{F}$ on T.

Let $\mathcal{F}$ be a saturated fusion system on a finite p-group P, and let T be a strongly $\mathcal{F}$-closed subgroup of P. Suppose that there is a normal subsystem on T, and let $\Gamma = \Gamma_{\mathcal{F},\mathcal{F}}(T)$, a normal subgroup of $\mathrm{Aut}_{\mathcal{F}}(T)$ by Lemma 8.4. In the set of all normal subgroups of $\mathrm{Aut}_{\mathcal{F}}(T)$ lying between $\mathrm{Aut}_{\mathcal{R}_{\mathcal{F}}(T)}(T)$ and $\mathrm{Aut}_{\mathcal{R}^{\mathcal{F}}(T)}(T)$, we have a subset of those normal subgroups that also lie inside Γ. Since $\mathcal{R}_{\mathcal{F}}(T)$ is a normal subsystem of $\mathcal{F}$, this set is non-empty, and it is easy to see that, again, this set has a unique maximal element under inclusion. We will denote the largest normal subsystem of $\mathcal{F}$ on T by $\overline{\mathcal{R}}^{\mathcal{F}}(T)$. (We will use $\overline{\mathcal{R}}_{\mathcal{F}}(T)$ to denote the smallest normal subsystem of $\mathcal{F}$ on T, so $\overline{\mathcal{R}}_{\mathcal{F}}(T) = \mathcal{R}_{\mathcal{F}}(T)$.)

As we have mentioned in Theorem 8.21 before, in [Asc11] Aschbacher constructs a normal subsystem $\mathcal{E}_1 \wedge \mathcal{E}_2$, given two normal subsystems $\mathcal{E}_1$ and $\mathcal{E}_2$. If $\mathcal{F}$ is the fusion system of a finite group and $\mathcal{E}_1$ and $\mathcal{E}_2$ are subsystems corresponding to normal subgroups, then $\mathcal{E}_1 \wedge \mathcal{E}_2$ is the subsystem of the intersection of the two normal subgroups [Asc11, (3.9)]. This subsystem $\mathcal{E}_1 \wedge \mathcal{E}_2$ is the largest normal subsystem of $\mathcal{F}$, contained in $\mathcal{E}_1 \cap \mathcal{E}_2$, that is also normal in both $\mathcal{E}_i$. Using our notation therefore, if $\mathcal{E}_i$ lies on T_i and $T = T_1 \cap T_2$, we get the formula

$$\mathcal{E}_1 \wedge \mathcal{E}_2 = \overline{\mathcal{R}}^{\mathcal{E}_1}(T) \curlywedge \overline{\mathcal{R}}^{\mathcal{E}_2}(T) \curlywedge \overline{\mathcal{R}}^{\mathcal{F}}(T),$$

which we can take as the definition of $\mathcal{E}_1 \wedge \mathcal{E}_2$.

Example 8.28 Let G be the group $S_3 \times S_3$, let H be the subgroup of index 2 that does not contain either S_3 factor, and let K be the first S_3 factor. Let P be the Sylow 3-subgroup of G, and let $Q = P \cap K$ be the Sylow 3-subgroup of K. Let $\mathcal{F} = \mathcal{F}_P(G)$, $\mathcal{E}_1 = \mathcal{F}_P(H)$ and $\mathcal{E}_2 = \mathcal{F}_Q(K)$.

Firstly, since H and K are normal subgroups of G, $\mathcal{E}_i \trianglelefteq \mathcal{F}$. Certainly $\mathcal{E}_2 < \mathcal{E}_1$, so that $\mathcal{E}_1 \curlywedge \mathcal{E}_2 = \mathcal{E}_2$. Also, $\mathcal{E}_1 \curlywedge \mathcal{E}_2 = \mathcal{E}_2 \trianglelefteq \mathcal{F}$. However, $H \cap K = Q$, so that $\mathcal{E}_1 \wedge \mathcal{E}_2 = \mathcal{F}_Q(Q)$.

Therefore, if $\mathcal{E}_1$ and $\mathcal{E}_2$ come from normal subgroups of a group, then even if $\mathcal{E}_1 \curlywedge \mathcal{E}_2$ is a normal subsystem of $\mathcal{F}$ (and not just a weakly normal subsystem) then $\mathcal{E}_1 \curlywedge \mathcal{E}_2$ need not be $\mathcal{E}_1 \wedge \mathcal{E}_2$, the subsystem of the intersection of those normal subgroups. Also, even if the subsystem $\mathcal{E}_1 \curlywedge \mathcal{E}_2$ is normal in $\mathcal{F}$ then it need not be normal in each $\mathcal{E}_i$, since $\mathcal{E}_2$ is not normal in $\mathcal{E}_1$, as it is easy to see.

Using Theorem 8.22 we can develop a theory of subnormal subsystems. We begin by defining a subnormal subsystem, both in the case of weak normality and normality.

Definition 8.29 Let $\mathcal{F}$ be a saturated fusion system on a finite p-group P. A saturated subsystem $\mathcal{E}$ is *weakly subnormal* in $\mathcal{F}$ if there is a series

$$\mathcal{E} = \mathcal{E}_0 \leq \mathcal{E}_1 \leq \cdots \leq \mathcal{E}_r = \mathcal{F}$$

such that, for all i, $\mathcal{E}_i \prec \mathcal{E}_{i+1}$. If in addition $\mathcal{E}_i \trianglelefteq \mathcal{E}_{i+1}$ for all i, then $\mathcal{E}$ is *subnormal* in $\mathcal{F}$. The *defect* of a (weakly) subnormal subsystem is the length of a smallest series of the form above between $\mathcal{E}$ and $\mathcal{F}$ (where the series above has length r).

The obvious extension to Theorem 8.15 is the following.

Corollary 8.30 *Let $\mathcal{F}$ be a saturated fusion system on a finite p-group P, and let T be a strongly $\mathcal{F}$-closed subgroup of P. If $\mathcal{E}$ is a weakly subnormal subsystem of $\mathcal{F}$, on the subgroup of T, then $\mathrm{O}^{p'}(\mathcal{E})$ is subnormal. In particular, every weakly subnormal subsystem $\mathcal{E}$ of $\mathcal{F}$ with $\mathrm{O}^{p'}(\mathcal{E}) = \mathcal{E}$ is subnormal.*

Proof We may assume that $\mathrm{O}^{p'}(\mathcal{E}) = \mathcal{E}$. Let

$$\mathcal{E} = \mathcal{E}_0 \prec \mathcal{E}_1 \prec \cdots \prec \mathcal{E}_r = \mathcal{F}$$

be a series of subsystems of $\mathcal{F}$, with $\mathcal{E}_i$ on the subgroup T_i. By Corollary 8.25, $\mathcal{R}_{\mathcal{E}_{i+1}}(T_i) \leq \mathcal{R}_{\mathcal{E}_{i+1}}(T_{i+1})$ for all i, so that $\mathrm{O}^{p'}(\mathcal{E}_i) \leq \mathrm{O}^{p'}(\mathcal{E}_{i+1})$ for all i. In particular, this means that there is a series

$$\mathcal{E} = \mathrm{O}^{p'}(\mathcal{E}_0) \trianglelefteq \mathrm{O}^{p'}(\mathcal{E}_1) \trianglelefteq \cdots \trianglelefteq \mathrm{O}^{p'}(\mathcal{E}_r) = \mathrm{O}^{p'}(\mathcal{F}),$$

with each normal in the next by Theorem 8.15. Hence $\mathcal{E}$ is subnormal, as claimed. $\square$

We turn to defining the weak intersection for (weakly) subnormal subsystems, not just for (weakly) normal subsystems.

Proposition 8.31 *Let $\mathcal{F}$ be a saturated fusion system on a finite p-group P, and let $\mathcal{D}$ and $\mathcal{E}$ be weakly subnormal subsystems of $\mathcal{F}$ on* *subgroups T and U of P respectively. There is a saturated subsystem of $\mathcal{F}$ on $T \cap U$, contained in $\mathcal{D} \cap \mathcal{E}$, that is weakly subnormal in $\mathcal{D}$ and $\mathcal{E}$ (and hence in $\mathcal{F}$). If $\mathcal{E}$ is weakly normal in $\mathcal{F}$ then this subsystem may be chosen to be weakly normal in $\mathcal{D}$.*

Proof Let $\mathcal{D} = \mathcal{D}_r \prec \mathcal{D}_{r-1} \prec \cdots \prec \mathcal{D}_0 = \mathcal{F}$ and $\mathcal{E} = \mathcal{E}_s \prec \mathcal{E}_{s-1} \prec \cdots \prec \mathcal{E}_0 = \mathcal{F}$ be chains of saturated subsystems, each weakly normal in the next, and choose them so that r and s are minimal. Suppose firstly that $s = 1$, so that $\mathcal{E}$ is weakly normal in $\mathcal{F}$. The subsystem $\mathcal{F}_1 = \mathcal{D}_1 \curlywedge \mathcal{E}$ is weakly normal in both $\mathcal{D}_1$ and $\mathcal{E}$. The subsystem $\mathcal{F}_2 = \mathcal{D}_2 \curlywedge \mathcal{F}_1$ is weakly normal in $\mathcal{D}_2$ and $\mathcal{F}_1$. Inductively, define $\mathcal{F}_i = \mathcal{D}_i \curlywedge \mathcal{F}_{i-1}$ for $2 \le i \le r$. Notice that $\mathcal{F}_i \prec \mathcal{F}_{i-1}$, and so $\mathcal{F}_r$ is weakly subnormal in $\mathcal{E}$ and is weakly normal in $\mathcal{D}_r = \mathcal{D}$. Also, by construction it lies on $T \cap U$ and in $\mathcal{D} \cap \mathcal{E}$, so the proposition is true in this case.

Next, we assume that $s > 1$. Define $\mathcal{F}_1$ in this case to be the subsystem defined in the previous paragraph with respect to $\mathcal{D}$ and $\mathcal{E}_1$, and in general define $\mathcal{F}_i$ to be the subsystem defined in the previous paragraph with respect to $\mathcal{F}_{i-1}$ and $\mathcal{E}_i$, for $2 \le i \le s$. The subsystem $\mathcal{F}_s$ is weakly subnormal in $\mathcal{D}$ and $\mathcal{E}$, and the other properties are equally clear; this completes the proof. □

Denote by $\mathcal{E}_1 \curlywedge \mathcal{E}_2$ the weakly subnormal subsystem of $\mathcal{E}_1$, $\mathcal{E}_2$ and $\mathcal{F}$ constructed in the previous proposition, and call it the *weak intersection* of $\mathcal{E}_1$ and $\mathcal{E}_2$, as before. Of course, *a priori*, this definition depends on the choice of chains connecting the $\mathcal{E}_i$ and $\mathcal{F}$. However, the subsystem $\mathcal{E}_1 \curlywedge \mathcal{E}_2$ is independent of the choice of chains, and is set as Exercise 8.6.

8.4 Constraint and normal subsystems

If $\mathcal{F}$ is a constrained fusion system, then by Theorem 3.70 there is a unique p-constrained group G with $O_{p'}(G) = 1$ and $\mathcal{F}_P(G) = \mathcal{F}$. If N is a normal subgroup of G then $\mathcal{F}_{P \cap N}(N)$ is a normal subsystem of $\mathcal{F}$; remarkably, the converse [Asc08a, Theorem 1] is true, so if $\mathcal{E}$ is a normal subsystem of $\mathcal{F}$ then there is normal subgroup H of G such that $\mathcal{E} = \mathcal{F}_{P \cap H}(H)$ under the obvious embedding of $\mathcal{F}_{P \cap H}(H)$ into $\mathcal{F}_P(G)$. In this section we will prove this fact; the method here broadly follows that in [Asc08a], but some alterations have been made.

We begin with an important, but easy, lemma.

Lemma 8.32 *Let G be a p-constrained group with $O_{p'}(G) = 1$. If H is a normal subgroup of G then H is also p-constrained and $O_{p'}(H) = 1$.*

Proof Since $O_{p'}(H)$ char $H \trianglelefteq G$, $O_{p'}(H)$ is a normal p'-subgroup of G, and hence is trivial. Any component of H is clearly a component of G, and since G has no components by Proposition 5.92(ii), neither does H. Therefore H is p-constrained. $\qquad\square$

We will use this result to prove that, if we have two distinct normal subgroups of a p-constrained group G, then they yield distinct normal subsystems of the fusion system of G.

Lemma 8.33 *Let G be a p-constrained finite group such that $O_{p'}(G) = 1$, and let P be a Sylow p-subgroup of G. Let H_1 and H_2 be normal subgroups of G with $H_i \cap P = T_i$, and let $\mathcal{E}_i$ denote the subsystem of $\mathcal{F} = \mathcal{F}_P(G)$ obtained by embedding $\mathcal{F}_{T_i}(H_i)$ in $\mathcal{F}$. If $\mathcal{E}_1 = \mathcal{E}_2$ then $H_1 = H_2$.*

Proof Suppose that $\mathcal{E}_1 = \mathcal{E}_2$, and let $H = H_1 H_2$. Since $\mathcal{E}_1 = \mathcal{E}_2$, $T_1 = T_2$, so call this subgroup T. Since T is a Sylow p-subgroup of both H_i, it is a Sylow p-subgroup of H. Let $Q = O_p(G)$, and notice that $Q \cap T = O_p(H) = O_p(H_i)$. Write $Z = Z(Q \cap T)$; as H is p-constrained, $C_H(Q \cap T) = Z$.

Since $\mathrm{Aut}_{\mathcal{E}_i}(Q \cap T) = H_i c_{Q \cap T}$, where $c_{Q \cap T}$ is the natural map from H to $\mathrm{Aut}_H(Q \cap T)$ (with kernel $Z(Q \cap T)$), we see that $H_1/Z = H_2/Z$ as subgroups of H/Z, and hence $H_1 = H_2$, as required. $\qquad\square$

Having proved that distinct normal subgroups yield distinct normal subsystems, we now need to prove the rest of the statement; that is, given a normal subsystem $\mathcal{E}$ of a constrained fusion system $\mathcal{F}$, we need to construct a normal subgroup H of the (unique) p-constrained model G, such that $\mathcal{E}$ is the subsystem of $\mathcal{F}$ corresponding to H.

Let $\mathcal{F}$ be a constrained fusion system on a finite p-group P, with model G, and let T be a strongly $\mathcal{F}$-closed subgroup of P, on which a normal subsystem $\mathcal{E}$ of $\mathcal{F}$ is defined. Not only is it not clear that there is a normal subgroup H of G with $T = P \cap H$ and $\mathcal{F}_T(H) = \mathcal{E}$, it is not even clear that there is any normal subgroup H such that $H \cap P = T$. In particular, if this latter statement holds then the normal closure of T, $\langle T^G \rangle$, will be a normal subgroup that has Sylow p-subgroup T.

Proposition 8.34 *Let $\mathcal{F}$ be a constrained fusion system on a finite p-group P, and let $G = L^{\mathcal{F}}$ be the model of $\mathcal{F}$. Let T be a strongly $\mathcal{F}$-closed subgroup of P and $\mathcal{E}$ be a normal subsystem of $\mathcal{F}$, on T. If $H = \langle T^G \rangle$, then $H \cap P = T$, so that $T \in \mathrm{Syl}_p(H)$. Furthermore, $\mathcal{F}_T(H) = \overline{\mathcal{R}}_{\mathcal{F}}(T)$.*

Proof Write $Q = O_p(G) = O_p(\mathcal{F})$ and let $U = T \cap Q$. We first show that $T\, C_H(U)$ is a Sylow p-subgroup of H, i.e., $T\, C_H(U) = P \cap H$.

Since $U \leq Q \trianglelefteq G$, $U^g \leq Q$ for all $g \in G$, and since T is strongly $\mathcal{F}$-closed and $U \leq T$, $U^g \leq T$ if $U^g \leq P$, so that $U^g = U$ for all $g \in G$, proving that $U \trianglelefteq G$. Indeed, we also see that $[T, Q] \leq U$, since it is contained in Q as $Q \trianglelefteq G$, and since T is strongly closed in P with respect to G, $[T, Q] \leq T$ if $[T, Q] \leq P$; hence $[T, Q] \leq T \cap Q = U$. Therefore all elements of T act trivially on Q/U, so all conjugates of T act trivially on Q/U. We must have that $H = \langle T^G \rangle$ acts trivially on Q/U, i.e., $[H, Q] \leq U$.

As both Q and U are normal subgroups of G, so are $C_G(Q/U)$ and $C_G(U)$. Therefore $C_G(Q/U) \cap C_G(U)$ is a normal subgroup of G; if x is a p'-element of G that centralizes both U and Q/U then it centralizes Q by Exercise 4.5, but as G is p-constrained, $C_G(Q) \leq Q$, so that $x = 1$. Hence $C_G(Q/U) \cap C_G(U)$ is a normal p-subgroup of G, so is contained in Q. Since all elements of H centralize Q/U (as $[H, Q] \leq U$),

$$C_H(U) \leq C_G(Q/U) \cap C_G(U) \leq Q.$$

Consider the map $c_U : H \to \mathrm{Aut}(U)$. The kernel of this map is $C_H(U) \leq Q$. Since $\mathcal{E}$ is a normal subsystem of $\mathcal{F}$ on T, $\mathrm{Aut}_T(U)$ is a Sylow p-subgroup of $\mathrm{Aut}_{\mathcal{E}}(U)$, and so in particular $\mathrm{Aut}_T(U)$ is a Sylow p-subgroup of $A = \langle \mathrm{Aut}_T(U)^{\mathrm{Aut}_{\mathcal{F}}(U)} \rangle$, the smallest normal subgroup of $\mathrm{Aut}_{\mathcal{F}}(U)$ containing $\mathrm{Aut}_T(U)$. We claim that $H c_U = A$. To see this, notice that $H c_U$ contains $\mathrm{Aut}_T(U)$ and is a normal subgroup of $\mathrm{Aut}_{\mathcal{F}}(U)$, so $H c_U \geq A$. Conversely, the preimage of A under c_U is a normal subgroup of G containing T, so it contains $\langle T^G \rangle = H$. In particular, $\mathrm{Aut}_T(U)$ is a Sylow p-subgroup of $H c_U$, so since $\ker c_U$ is a p-group, the preimage of $\mathrm{Aut}_T(U)$, namely $T\, C_H(U)$, is a Sylow p-subgroup of H, proving our claim.

Next, we will show that actually T is a Sylow p-subgroup of H. Notice that $T \cap C_H(U) = Z(U)$, so write $\bar{T} = T/Z(U)$ and so on. We first note that $\overline{C_H(U)} \leq Z(\bar{H})$, or equivalently that $[H, C_H(U)] \leq Z(U)$. To see this, as $C_H(U) \leq Q$, it normalizes T, so that $[T, C_H(U)] \leq T \cap C_H(U) = C_T(U) = Z(U)$ (as $C_H(U) \trianglelefteq H$). Since $H = \langle T^G \rangle$, we see that $[H, C_H(U)] \leq Z(U)$, proving our claim.

Since T is strongly $\mathcal{F}$-closed, T is strongly closed in P with respect to G, so it is strongly closed in $T\, C_H(U)$ with respect to H. Therefore, by Exercise 8.8, there exists a normal subgroup $\bar{L}$ of $\bar{H}$ such that $\bar{H} =$

$\bar{L} \times \overline{C_H(U)}$, and $\bar{T} \leq \bar{L}$. Taking preimages in H, we see that H contains a normal subgroup L such that $L \cap C_H(U) = Z(U)$ and $T \leq L$.

Since H is a normal subgroup of G, we have by the Frattini argument $G = N_G(T\,C_H(U))H$, and so $g \in G$ may be written as $g = kh$, with $k \in N_G(T\,C_H(U))$ and $h \in H$. Therefore

$$T^g = T^{kh} \leq (T\,C_H(U))^{kh} = (T\,C_H(U))^h = T^h\,C_H(U).$$

As T is strongly closed in P with respect to G, so strongly closed in $T\,C_H(U)$ with respect to G, T^h is also strongly closed in $(T\,C_H(U))^h = T^h\,C_H(U)$ with respect to G. Hence $T^g = T^h$, so in particular $T^g \leq L$ as L is a normal subgroup of H containing T. We see therefore that $H = \langle T^G \rangle \leq L$, so that $H = L$. In particular, T is therefore a Sylow p-subgroup of H.

Finally, we prove that $\mathcal{F}_T(H) = \overline{\mathcal{R}}_\mathcal{F}(T)$. Since $\mathcal{F}_T(H)$ is a normal subsystem of $\mathcal{F}$ on T, we may assume that $\mathcal{E} = \mathcal{F}_T(H)$. By Corollary 7.60, the automorphisms ϕ of T that lie in $O^{p'}(\mathcal{E})$ are those such that there exists an $\mathcal{E}$-centric subgroup R of T with $\phi|_R$ a product of p-automorphisms in $\mathcal{E}$ of overgroups of R. Clearly U is $\mathcal{E}$-centric (since $C_H(U) \leq U$), and any $h \in H$ may be written as the product of conjugates $t_i^{g_i}$ of elements t_i of T (hence p-automorphisms); as $U \trianglelefteq G$, $t_i^{g_i}$ normalizes U, whence

$$c_h|_U = c_{t_1^{g_1}}|_U \ldots c_{t_d^{g_d}}|_U,$$

proving that $c_h \in \mathrm{Aut}_{O^{p'}(\mathcal{E})}(T)$. Therefore $\mathrm{Aut}_H(T) = \mathrm{Aut}_{O^{p'}(\mathcal{E})}(T)$, so that $\mathcal{E} = O^{p'}(\mathcal{E})$, as claimed. $\qquad\square$

Having produced a normal subgroup corresponding to the subsystem $\overline{\mathcal{R}}_\mathcal{F}(T)$ for each relevant T, we turn our attention to producing a normal subgroup corresponding to *every* normal subsystem of $\mathcal{F}$.

Proposition 8.35 *Let $\mathcal{F}$ be a constrained fusion system on a finite p-group P, and let G be the model of $\mathcal{F}$. Let T be a strongly $\mathcal{F}$-closed subgroup of P and $\mathcal{E}$ be a normal subsystem of $\mathcal{F}$, on T. There is a normal subgroup H of G with $T = P \cap H$, such that $\mathcal{F}_T(H) = \mathcal{E}$.*

Proof Let $L = \langle T^G \rangle$, $Q = O_p(G)$ and $U = Q \cap T$; by Proposition 8.34 we have that $\mathcal{F}_T(L) = O^{p'}(\mathcal{E})$, so if there exists a normal subgroup H of G containing L such that $\mathrm{Aut}_H(T) = \mathrm{Aut}_\mathcal{E}(T)$ then $\mathcal{F}_T(H) = \mathcal{E}$. We therefore look for such a normal subgroup H. Notice that such a normal subgroup H acts trivially on Q/U since $[H, Q] \leq H \cap Q = U$; i.e., we have that H is contained in $G_1 = C_G(Q/U)$.

Firstly, we want to show that if $\phi \in \operatorname{Aut}_{\mathcal{E}}(T)$ then there exists $g \in G_1$ such that $c_g = \phi$ on T and $[\mathrm{C}_P(T), g] \leq \mathrm{Z}(T)$. If $\phi \in \operatorname{Aut}_T(T)$ then this is clear, so we may assume that ϕ has p'-order. Since $\mathcal{E}$ is a normal subsystem of $\mathcal{F}$, we may choose $g \in G$ (of p'-order) such that $[\mathrm{C}_P(T), g] \leq \mathrm{Z}(T)$, so that $[T\,\mathrm{C}_P(T), g] \leq T$. Also, as $Q \trianglelefteq G$, $\operatorname{Aut}_Q(T) \trianglelefteq \operatorname{Aut}_{\mathcal{F}}(T)$, and also $\operatorname{Aut}_{\mathcal{E}}(T) \trianglelefteq \operatorname{Aut}_{\mathcal{F}}(T)$, so that $[\operatorname{Aut}_Q(T), \operatorname{Aut}_{\mathcal{E}}(T)] \leq \operatorname{Aut}_Q(T) \cap \operatorname{Aut}_{\mathcal{E}}(T) = \operatorname{Inn}(T)$. Therefore $[Q, g] \leq T\,\mathrm{C}_P(T)$. Since Q is a normal subgroup of G, we get that $[Q, g] \leq Q \cap T\,\mathrm{C}_P(T)$, and so g acts trivially on $Q/(Q \cap T\,\mathrm{C}_P(T))$. Also, since $[T\,\mathrm{C}_P(T), g] \leq T$, by taking intersections with Q we see that g acts trivially on $(Q \cap T\,\mathrm{C}_P(T))/(Q \cap T)$. As g has p'-order, we therefore get that g acts trivially on $Q/(Q \cap T) = Q/U$ by Exercise 4.5; hence $g \in G_1$.

Let A denote the preimage of $\operatorname{Aut}_{\mathcal{E}}(T)$ in $\mathrm{N}_{G_1}(T)$. (Notice that by the previous paragraph $\operatorname{Aut}_A(T) = \operatorname{Aut}_{\mathcal{E}}(T)$.) The kernel K of the natural map $A \to \operatorname{Aut}_A(T)$ is $\mathrm{C}_{G_1}(T) = \mathrm{C}_G(Q/U) \cap \mathrm{C}_G(T)$. In particular, $K \leq \mathrm{C}_G(U) \cap \mathrm{C}_G(Q/U)$, so by the proof of the previous proposition, $K \leq Q$ and is in particular a p-group. Therefore a Sylow p-subgroup of A is TK.

If we prove that $\mathrm{O}^p(A) \cap TK \leq T$, then $H = L\,\mathrm{O}^p(A)$ will have Sylow p-subgroup T and have $\operatorname{Aut}_H(T) = \operatorname{Aut}_{\mathcal{E}}(T)$, since $\operatorname{Aut}_A(T) = \operatorname{Aut}_{T\,\mathrm{O}^p(A)}(T)$. To prove that $X = \mathrm{O}^p(A) \cap TK \leq T$ we will use the hyperfocal subgroup theorem, which states that X is generated by $[h, x]$, where $x \in R \leq TK$ runs over all subgroups of TK and h is an element of p'-order in $\mathrm{N}_A(R)$ (see Theorem 1.33).

Since $A \leq \mathrm{N}_G(T)$, every element $h \in A$ normalizes T and also therefore normalizes TK as $K \leq \mathrm{C}_G(T)$ and so A has a normal Sylow p-subgroup. Hence $\mathrm{O}^p(A) \cap TK$ is generated by $[h, TK]$, where h is a p'-element of A. Let $\phi \in \operatorname{Aut}_{\mathcal{E}}(T)$ be a p'-automorphism; by the second paragraph of this proof, we may choose a p'-element $h \in A$ such that $c_h|_T = \phi$ and $[\mathrm{C}_P(T), h] \leq \mathrm{Z}(T)$. If h' is any other p'-element of A such that $c'_h|_T = \phi$, then $h^{-1}h'$ lies in $\mathrm{C}_A(T) = K$, so acts trivially on T and acts as a p-automorphism on $\mathrm{C}_P(T)/\mathrm{Z}(T)$ (since K is a p-group). Since h acts trivially on $\mathrm{C}_P(T)/\mathrm{Z}(T)$, we see that h' acts as a p-automorphism on $\mathrm{C}_P(T)/\mathrm{Z}(T)$. However, h' is a p'-element, so it must therefore also act trivially on $\mathrm{C}_P(T)/\mathrm{Z}(T)$; hence any p'-element of A acts trivially on $\mathrm{C}_P(T)/\mathrm{Z}(T)$, i.e., $[h', \mathrm{C}_P(T)] \leq \mathrm{Z}(T)$, so that $[h', T\,\mathrm{C}_P(T)] \leq T$.

We therefore see that $[h, TK] \leq T$ for any p'-element $h \in A$, so that $\mathrm{O}^p(A) \cap TK \leq T$. This proves that $H = L\,\mathrm{O}^p(A)$ has Sylow p-subgroup T and $\operatorname{Aut}_H(T) = \operatorname{Aut}_{\mathcal{E}}(T)$, completing the proof. $\qquad\square$

Combining Lemma 8.33 and Proposition 8.35, we get the next result.

Theorem 8.36 (Aschbacher [Asc08a, Theorem 1]) *Let G be a p-constrained finite group such that $O_{p'}(G) = 1$, and let $\mathcal{F} = \mathcal{F}_P(G)$, where P is a Sylow p-subgroup of G. If $\mathcal{E}$ is a normal subsystem of $\mathcal{F}$, then there is a unique normal subgroup H of G such that $\mathcal{E} = \mathcal{F}_{P \cap H}(H)$. Consequently, there is a bijection between the set of normal subgroups of G and the set of normal subsystems of $\mathcal{F}$.*

To illustrate the power of this theorem, we use it to prove two results, Propositions 8.39 and 8.40, that we will use later in the chapter. To prove these, we need some information on two subsystems, $\mathcal{D}(U)$ and $\mathcal{E}(U)$, introduced by Aschbacher in [Asc08a]. Formally, their definitions are as follows.

Definition 8.37 Let $\mathcal{F}$ be a saturated fusion system on a finite p-group P, and let T be a strongly $\mathcal{F}$-closed subgroup of P. If U is a fully $\mathcal{F}$-normalized subgroup of T, write $\bar{U}$ for $U \, C_T(U)$ and $V = \bar{U} \, C_P(\bar{U})$. Define

$$\mathcal{D}(U) = N_{\mathcal{F}}(V, U).$$

Let $\mathcal{E}$ be a subsystem of $\mathcal{F}$ on T. Define

$$\mathcal{E}(U) = N_{\mathcal{E}}(\bar{U}, U).$$

Informally, one may think of $\mathcal{D}(U)$ as being the 'largest' constrained subsystem of $\mathcal{F}$ in which U is normal and $\mathcal{E}(U)$ as the 'largest' constrained subsystem of $\mathcal{E}$ in which U is normal. We say 'largest' in the sense that this subsystem is the largest canonical such subsystem. Some fundamental properties about these subsystems are collected in Exercise 8.9. In particular, if $\mathcal{E}$ is weakly normal in $\mathcal{F}$ then $\mathcal{E}(U)$ is weakly normal in $\mathcal{D}(U)$. The obvious extension to this is next.

Lemma 8.38 *Let $\mathcal{F}$ be a saturated fusion system on a finite p-group P, and let $\mathcal{E}$ be a normal subsystem of $\mathcal{F}$, defined on the subgroup T of P. If Q is a fully $\mathcal{F}$-normalized subgroup of T then $\mathcal{E}(Q) \trianglelefteq \mathcal{D}(Q)$.*

Proof Let Q be a counterexample to the statement with $n = |T : Q|$ minimal. If $n = 1$, then $Q = T$. In this case $\mathcal{E}(T) = N_{\mathcal{E}}(T)$ and $\mathcal{D}(T) = N_{\mathcal{F}}(T \, C_P(T))$. As $\mathcal{E}$ is normal in $\mathcal{F}$, any $\mathcal{E}$-automorphism of T lies in $\Gamma_{\mathcal{F},\mathcal{F}}(T)$. Also, since all $\mathcal{F}$-automorphisms of $T \, C_P(T)$ lie in $\mathcal{D}(T)$, we clearly see that

$$\mathrm{Aut}_{\mathcal{E}(T)}(T) = \mathrm{Aut}_{\mathcal{E}}(T) = \Gamma_{\mathcal{F},\mathcal{E}}(T) = \Gamma_{\mathcal{D}(T),\mathcal{E}(T)}(T).$$

Hence $\mathcal{E}(T) \trianglelefteq \mathcal{D}(T)$, proving that $n > 1$.

Let $R = \mathrm{N}_T(Q)$ and choose $\phi \in \mathrm{Hom}_{\mathcal{F}}(R, T)$ such that $S = R\phi$ is fully $\mathcal{F}$-normalized. We may choose ϕ such that it extends to a map $\mathrm{N}_P(R) \to \mathrm{N}_P(S)$, which we will also call ϕ. Let α be an automorphism in $\mathrm{Aut}_{\mathcal{E}(Q)}(R)$; we need to show that α lies in $\Gamma_{\mathcal{D}(Q),\mathcal{E}(Q)}(R)$. The map $\beta = \alpha^\phi$ is an automorphism of S, so extends to an automorphism $\bar{\beta}$ of $S\,\mathrm{C}_P(S)$ in $\mathcal{F}$. Since T is strongly $\mathcal{F}$-closed, $\bar{\beta}$ acts as an automorphism of $S\,\mathrm{C}_P(S) \cap T = S\,\mathrm{C}_T(S)$, and therefore also acts as an automorphism of $S\,\mathrm{C}_T(S)\,\mathrm{C}_P(S\,\mathrm{C}_T(S))$. We see therefore that β lies in $\mathcal{D}(S)$. By our choice of Q, $\mathcal{E}(S)$ is a normal subsystem of $\mathcal{D}(S)$.

By Exercise 8.9(ii), $\mathcal{D}(S)$ is constrained, and hence we may consider the model $G = L^{\mathcal{D}(S)}$ of $\mathcal{D}(S)$. As $\mathcal{E}(S)$ is a normal subsystem of $\mathcal{D}(S)$, there is a unique normal subgroup $H \trianglelefteq G$ such that $\mathcal{F}_{\mathrm{N}_T(S)}(H) = \mathcal{E}(S)$. Since $\beta \in \mathrm{Aut}_{\mathcal{E}(S)}(S)$, there exists $h \in H$ such that $c_h|_S = \beta$. As Q is fully $\mathcal{F}$-normalized it is fully $\mathcal{E}$-normalized by Exercise 5.13(i), and so $R = \mathrm{N}_T(Q)$ is $\mathcal{E}$-centric; hence S is $\mathcal{E}$-centric, and so

$$[h, \mathrm{C}_P(S)] \le \mathrm{C}_H(S) = \mathrm{Z}(S).$$

(Since S is $\mathcal{E}$-centric, S is $\mathcal{F}_{\mathrm{N}_T(S)}(H)$-centric, and S is a normal subgroup of $\mathcal{E}(S)$, so that $S \le \mathrm{O}_p(H)$; hence $\mathrm{C}_H(S) = \mathrm{Z}(S) \times \mathrm{O}_{p'}(\mathrm{C}_H(S))$, and since $\mathrm{O}_{p'}(S) = 1$, $\mathrm{C}_H(S) = \mathrm{Z}(S)$.) Hence $\bar{\beta} = c_h|_{S\,\mathrm{C}_P(S)}$ is an extension of β to $S\,\mathrm{C}_P(S)$ such that $[\bar{\beta}, \mathrm{C}_P(S)] \le \mathrm{Z}(S)$. Let $\bar{\alpha} = \bar{\beta}^{\phi^{-1}}$, an extension of α to $R\,\mathrm{C}_P(R)$ such that $[\bar{\alpha}, \mathrm{C}_P(R)] \le \mathrm{Z}(R)$. We see that $\alpha \in \Gamma_{\mathcal{D}(Q),\mathcal{E}(Q)}(R)$, as needed. Hence $\mathcal{D}(Q) \trianglelefteq \mathcal{E}(Q)$, contradicting the choice of counterexample. $\qquad\square$

The next result has a very similar proof to Lemma 8.38, and it is left as Exercise 8.10.

Proposition 8.39 *Let $\mathcal{F}$ be a saturated fusion system on a finite p-group P, and let $\mathcal{E}$ be a normal subsystem of $\mathcal{F}$, defined on the subgroup T of P. If Q is an $\mathcal{E}$-centric subgroup of T and α is an $\mathcal{E}$-automorphism of Q, then α extends to an automorphism $\bar{\alpha}$ in $\mathcal{F}$ of $Q\,\mathrm{C}_P(Q)$ with $[\bar{\alpha}, \mathrm{C}_P(Q)] \le \mathrm{Z}(Q)$.*

In Proposition 5.44, we proved that if $\mathcal{E} \prec \mathcal{E}' \prec \mathcal{F}$ with $\mathcal{E}$ on T, and $\mathrm{Aut}_{\mathcal{E}'}(T) \le \mathrm{Aut}(\mathcal{E})$, then $\mathcal{E} \prec \mathcal{F}$. There is a corresponding result for normality.

Proposition 8.40 (Aschbacher [Asc11, (7.4)]) *Let $\mathcal{F}$ be a saturated fusion system on a finite p-group P, and let T and U be strongly $\mathcal{F}$-closed subgroups of P with $T \le U$. If $\mathcal{E}'$ is a normal subsystem of $\mathcal{F}$ on U and*

$\mathcal{E}$ is a normal subsystem of $\mathcal{E}'$ on T such that all elements of $\mathrm{Aut}_{\mathcal{F}}(U)$ induce automorphisms of $\mathcal{E}$, then $\mathcal{E}$ is normal in $\mathcal{F}$.

Proof By Proposition 5.44, $\mathcal{E}$ is weakly normal in $\mathcal{F}$, so it remains to show that $\mathrm{Aut}_{\mathcal{E}}(T) = \Gamma_{\mathcal{F},\mathcal{E}}(T)$. Write $Q = T\,\mathrm{C}_U(T)$, and define $R = Q\,\mathrm{C}_P(Q) = T\,\mathrm{C}_U(T)\,\mathrm{C}_P(T\,\mathrm{C}_U(T))$. Since T is strongly $\mathcal{F}$-closed, $\mathcal{D}(T) = \mathrm{N}_{\mathcal{F}}(R)$. Since $\mathcal{E}'$ is normal in $\mathcal{F}$, we have that $\mathcal{E}'(T) \trianglelefteq \mathcal{D}(T)$ by Lemma 8.38. Writing $G = L^{\mathcal{D}(T)}$, there exists a normal subgroup H of G such that $\mathcal{F}_U(H) = \mathcal{E}'(T)$. Similarly, since $\mathcal{E}$ is normal in $\mathcal{E}'$, there is a normal subgroup K of H such that $\mathcal{E}(T) = \mathcal{F}_T(K) \trianglelefteq \mathcal{F}_U(H) = \mathcal{E}'(T)$. Since $\mathcal{E}(T)$ is a saturated subsystem on T with $\mathcal{E}(T) = \mathrm{N}_{\mathcal{E}(T)}(T)$, K has a normal Sylow p-subgroup T. Let A denote a Hall p'-subgroup of K, so that $K = TA$.

Write $X = \mathrm{C}_H(T)$. Since T is a normal subgroup of the p-constrained group H, $X = \mathrm{C}_H(T) \trianglelefteq H$ is also p-constrained. Note that $[X, A]$ is contained in X because $X \trianglelefteq H$, and $[X, A]$ is contained in K because $A \leq K \trianglelefteq H$. Hence $[X, A] \leq X \cap K = \mathrm{C}_K(T) = \mathrm{Z}(T)$. Next, we see that $[\mathrm{C}_P(T), A]$ must be contained in H since $H \trianglelefteq G$, and also in $\mathrm{C}_G(T)$ since A normalizes T, and hence $[\mathrm{C}_P(T), A] \leq X$.

Finally, we use [Gor80, Theorem 5.3.6], which states that, since A has p'-order, we have $[\mathrm{C}_P(T), A] = [\mathrm{C}_P(T), A, A]$. Therefore

$$[\mathrm{C}_P(T), A] = [\mathrm{C}_P(T), A, A] \leq [X, A] \leq \mathrm{Z}(T).$$

Hence A, a Hall p'-subgroup of $\mathrm{Aut}_{\mathcal{E}}(T)$, lies in $\Gamma_{\mathcal{F},\mathcal{E}}(T)$, and so by Lemma 8.6 we have $\mathrm{Aut}_{\mathcal{E}}(T) = \Gamma_{\mathcal{F},\mathcal{E}}(T)$. Thus $\mathcal{E}$ is normal in $\mathcal{F}$, as claimed. $\square$

Corollary 5.52, that if $\mathcal{E}$ is weakly characteristic in $\mathcal{E}'$, which is in turn weakly normal in $\mathcal{F}$, then $\mathcal{E}$ is weakly normal in $\mathcal{F}$, also holds for normal subsystems. We of course need a definition first.

Definition 8.41 Let $\mathcal{F}$ be a saturated fusion system on a finite p-group P. A subsystem $\mathcal{E}$ of $\mathcal{F}$ is *characteristic* if it is normal and, for any automorphism α of $\mathcal{F}$, we have that $\mathcal{E}\alpha = \mathcal{E}$. If $\mathcal{E}$ is characteristic in $\mathcal{F}$ then we write $\mathcal{E}\operatorname{char}\mathcal{F}$.

Corollary 8.42 *Let $\mathcal{F}$ be a saturated fusion system on a finite p-group P, and let $\mathcal{E}$ and $\mathcal{E}'$ be subsystems of $\mathcal{F}$. If $\mathcal{E}'\operatorname{char}\mathcal{E} \trianglelefteq \mathcal{F}$, then $\mathcal{E}' \trianglelefteq \mathcal{F}$, and if $\mathcal{E}'\operatorname{char}\mathcal{E}\operatorname{char}\mathcal{F}$ then $\mathcal{E}'\operatorname{char}\mathcal{F}$.*

We end with a theorem on quotienting by central subgroups, namely that it induces a bijection of normal subsystems.

Theorem 8.43 (Aschbacher [Asc08a, Lemma 8.10]) *Let $\mathcal{F}$ be a satu-rated fusion system on a finite p-group P, and let Z be a normal subgroup of P contained in $\mathrm{Z}(\mathcal{F})$. There is a one-to-one correspondence between normal subsystems of $\mathcal{F}$ on subgroups of P containing Z and normal subsystems of $\mathcal{F}/Z$, given by $\mathcal{E} \mapsto \mathcal{E}/Z$.*

We are unable to prove this theorem here because it relies on various local characterizations for a subsystem to be normal, also proved by Aschbacher in [Asc08a], and we refer to that article for a proof.

8.5 Central products

In Example 8.1 we proved that if $\mathcal{E}_1$ and $\mathcal{E}_2$ are weakly normal subsystems of a saturated fusion system $\mathcal{F}$ on strongly $\mathcal{F}$-closed subgroups T_1 and T_2 respectively, with $T_1 \cap T_2 = 1$, then we need not have that the direct product $\mathcal{E}_1 \times \mathcal{E}_2$ is a subsystem of $\mathcal{F}$, never mind a weakly normal subsystem of $\mathcal{F}$.

However, if the $\mathcal{E}_i$ are normal in $\mathcal{F}$ then we do indeed get that $\mathcal{E}_1 \times \mathcal{E}_2$ is not only a subsystem of $\mathcal{F}$, but a normal subsystem as well. We will actually prove something more general, a result that we will need in the remainder of the chapter, namely that *central* products exist, in the same sense as they do for groups.

We begin with a lemma about direct products of fusion systems, which were introduced in Chapter 6.

Lemma 8.44 *Let $\mathcal{F}_1$ and $\mathcal{F}_2$ be saturated fusion systems on the finite p-groups P_1 and P_2 respectively, and let $\mathcal{F} = \mathcal{F}_1 \times \mathcal{F}_2$, on the p-group $P = P_1 \times P_2$. Write $\hat{\mathcal{F}}_1$ for the subcategory of $\mathcal{F}$ with objects all subgroups of the form $Q \times P_2$ for $Q \le P_1$, with*

$$\mathrm{Hom}_{\hat{\mathcal{F}}_1}(Q \times P_2, R \times P_2) = \{(\phi, \mathrm{id}) \mid \phi \in \mathrm{Hom}_{\mathcal{F}_1}(Q, R)\},$$

where (ϕ, id) denotes the morphism in $\mathcal{F}$ acting as ϕ on Q and as the identity on P_2. We define $\hat{\mathcal{F}}_2$ similarly. The fusion system $\mathcal{F}$ is gener-ated by $\hat{\mathcal{F}}_1$ and $\hat{\mathcal{F}}_2$. Consequently, any quotient $\mathcal{F}/T$ is generated by the images of $\hat{\mathcal{F}}_1$ and $\hat{\mathcal{F}}_2$ in $\mathcal{F}/T$, where T is any strongly $\mathcal{F}$-closed subgroup of P.

This follows immediately from the definition of the direct product of two fusion systems. We will apply this lemma to find a direct product of two trivially intersecting subsystems.

Proposition 8.45 *Let $\mathcal{F}$ be a saturated fusion system on a finite p-group P, and let $\mathcal{E}_1$ and $\mathcal{E}_2$ be normal subsystems of $\mathcal{F}$, on the subgroups T_1 and T_2 of P respectively. If $T_1 \cap T_2 = 1$ then the fusion system $\mathcal{E} = \mathcal{E}_1 \times \mathcal{E}_2$ is a normal subsystem of $\mathcal{F}$.*

Proof Let ϕ be an automorphism of an $\mathcal{E}_i$-centric subgroup Q_i of T_i for $i = 1$ or $i = 2$. By Proposition 8.39, ϕ extends to an automorphism of $Q_i \, \mathrm{C}_P(Q_i)$ in $\mathcal{F}$ that acts trivially on $\mathrm{C}_P(Q_i)/\mathrm{Z}(Q_i)$, so acts trivially on T_{3-i}. Therefore, since $\mathcal{E}_i$ is saturated, any isomorphism $\phi : Q_i \to R_i$ (with $R_i \leq T_i$) in $\mathcal{E}_i$ extends to a morphism $\hat{\phi}$ in $\mathcal{F}$ that acts trivially on T_{3-i} by Alperin's fusion theorem. Let $\hat{\mathcal{E}}_i$ be the subcategory of $\mathcal{F}$ consisting of those maps $\hat{\phi}$, for ϕ a morphism in $\mathcal{E}_i$. By Lemma 8.44, the subsystem $\mathcal{E}$ generated by $\hat{\mathcal{E}}_1$ and $\hat{\mathcal{E}}_2$ is isomorphic with $\mathcal{E}_1 \times \mathcal{E}_2$.

It remains to show that $\mathcal{E}$ is normal in $\mathcal{F}$. Let $A_i(-)$ denote the weakly normal map corresponding to $\mathcal{E}_i$. Define a map $A(-)$ on T by the following: for $Q \leq T$, write Q_i for the projection of Q along T_i. The set $A(Q)$ is defined to be all automorphisms of Q induced by elements of $A_1(Q_1) \times A_2(Q_2)$. Clearly $A(Q) \leq \mathrm{Aut}_{\mathcal{E}}(Q)$ for all $Q \leq T$, and since both $\mathcal{E}_i$ are generated by automorphisms of subgroups of T_i, and $\mathcal{E}$ is generated by maps of the form (ϕ, ψ) for ϕ a map in $\mathcal{E}_1$ and ψ a map in $\mathcal{E}_2$, the subsystem generated by $A(-)$ must be $\mathcal{E}$.

To see that $A(-)$ is an $\mathcal{F}$-invariant map, let $\phi : Q \to R$ be an isomorphism in $\mathcal{F}$, and write ϕ_i for the projection of ϕ along T_i. We first claim that $\phi_i : Q_i \to R_i$ (where R_i is the projection of R along T_i) lies inside $\mathcal{F}$. To see this, let $\bar{\phi}_i$ denote the image of ϕ_i in the quotient system $\mathcal{F}/T_{3-i}$. Since $\mathcal{F} \to \mathcal{F}/T_{3-i}$ is a morphism of fusion systems, there is a morphism $\psi_i : Q_i T_{3-i} \to R_i T_{3-i}$ in $\mathcal{F}$ whose image in $\mathcal{F}/T_{3-i}$ is $\bar{\phi}_i$. However, since $T_i \cap T_{3-i} = 1$ and T_i is strongly $\mathcal{F}$-closed, so that $Q_i \psi_i = R_i$, we see that $\psi_i|_{Q_i} = \phi_i$; hence ϕ_i is a morphism in $\mathcal{F}$, as claimed.

Since ϕ_i is a morphism in $\mathcal{F}$, $A_i(Q_i)^{\phi_i} = A_i(R_i)$, and so $A(Q)^{\phi} = A(R)$, proving that $A(-)$ satisfies the first property of being an $\mathcal{F}$-invariant map. Let Q be a subgroup of T, and let Q_i be the projection of Q along T_i. Notice that $\mathrm{N}_T(Q) \leq \mathrm{N}_{T_1}(Q_1) \times \mathrm{N}_{T_2}(Q_2)$. As each $\mathrm{Aut}_{T_i}(Q_i)$ is contained in $A_i(Q_i)$, we see that $\mathrm{Aut}_T(Q)$ is contained in $A(Q)$, proving the second condition. Therefore $\mathcal{E}$ is $\mathcal{F}$-invariant, and hence weakly normal in $\mathcal{F}$ since it is saturated.

Finally, let $\alpha = (\alpha_1, \alpha_2)$ be an $\mathcal{E}$-automorphism of T, with $\alpha_i \in \mathrm{Aut}_{\mathcal{E}_i}(T_i)$. Since $\mathcal{E}_i$ is normal in $\mathcal{F}$, there is an extension $\bar{\alpha}_i$ of α_i to an $\mathcal{F}$-automorphism of $T_i \, \mathrm{C}_P(T_i)$ acting trivially on $\mathrm{C}_P(T_i)/\mathrm{Z}(T_i)$. Let

$\bar{\alpha} = \bar{\alpha}_1 \bar{\alpha}_2$; since $\bar{\alpha}_i$ acts trivially on $\mathrm{C}_P(T_i)/\mathrm{Z}(T_i)$, certainly it acts trivially on $\mathrm{C}_P(T)/\mathrm{Z}(T)$, so that $\bar{\alpha}$ acts trivially on $\mathrm{C}_P(T)/\mathrm{Z}(T)$. To see that $\bar{\alpha}$ extends α, notice that $\bar{\alpha}_i$ acts trivially on T_{3-i}. Hence $\mathcal{E} \trianglelefteq \mathcal{F}$, as required. □

The direct product is good, but in order to produce an analogue of the generalized Fitting subgroup we need to prove that central products of subsystems are subsystems. We should first define a central product of two fusion systems.

Definition 8.46 Let $\mathcal{F}_1$ and $\mathcal{F}_2$ be saturated fusion systems on the finite p-groups P_1 and P_2 respectively. If Z is a subgroup of $\mathrm{Z}(\mathcal{F}_1) \times \mathrm{Z}(\mathcal{F}_2)$ such that $Z \cap P_i = 1$, then $\mathcal{F}_1 \times \mathcal{F}_2/Z$ is the *central product* with respect to Z, and is denoted by $\mathcal{F}_1 \times_Z \mathcal{F}_2$.

Central products for fusion systems are like central products for groups and, indeed, the central product of two groups yields a central product of their corresponding fusion systems (see Exercise 8.11).

Example 8.47 Let Q and R be two generalized quaternion groups, and let P be the central product of Q and R, amalgamating the two centres. If $\mathcal{E}$ is a saturated fusion system on Q, and $\mathcal{F}$ is a saturated fusion system on R, then $\mathcal{E} \times_Z \mathcal{F}$ is a saturated fusion system on P, where Z is the central subgroup of order 2 of $Q \times R$ that intersects both Q and R trivially.

We now give the analogue of Lemma 8.44 for central products, which follows immediately from the last statement of Lemma 8.44.

Lemma 8.48 Let $\mathcal{F}_1$ and $\mathcal{F}_2$ be saturated fusion systems on the finite p-groups P_1 and P_2 respectively, and let $\mathcal{F} = \mathcal{F}_1 \times \mathcal{F}_2$, on the p-group $P = P_1 \times P_2$. Write $\hat{\mathcal{F}}_1$ for the subcategory of $\mathcal{F}$ with objects all subgroups of the form $Q \times P_2$ for $Q \leq P_1$, with

$$\mathrm{Hom}_{\hat{\mathcal{F}}_1}(Q \times P_2, R \times P_2) = \{(\phi, \mathrm{id}) \mid \phi \in \mathrm{Hom}_{\mathcal{F}_1}(Q, R)\},$$

where (ϕ, id) denotes the morphism in $\mathcal{F}$ acting as ϕ on Q and as the identity on P_2. We define $\hat{\mathcal{F}}_2$ similarly. Let Z be a subgroup of $\mathrm{Z}(\mathcal{F}_1) \times \mathrm{Z}(\mathcal{F}_2)$ such that $Z \cap \mathrm{Z}(\mathcal{F}_i) = 1$ for $i = 1, 2$. The fusion system $\mathcal{F}_1 \times_Z \mathcal{F}_2 = \mathcal{F}/Z$ is generated by $\hat{\mathcal{F}}_1/Z$ and $\hat{\mathcal{F}}_2/Z$.

Using this result, we may now embark on a proof that central products of normal subsystems always exist inside saturated fusion systems, using Proposition 8.45 as a preliminary result, and a guide as to the steps needed.

Theorem 8.49 (Aschbacher [Asc11, Theorem 3]) *Let $\mathcal{F}$ be a saturated fusion system on a finite p-group P, and let $\mathcal{E}_1$ and $\mathcal{E}_2$ be normal subsystems of $\mathcal{F}$, on the subgroups T_1 and T_2 of P respectively. If $T_1 \cap T_2 \leq Z(\mathcal{E}_1) \cap Z(\mathcal{E}_2)$ then the subsystem $\mathcal{E} = \mathcal{E}_1 \times_Z \mathcal{E}_2$ is a normal subsystem of $\mathcal{F}$, where Z is the subgroup of $Z(T_1) \times Z(T_2)$ such that $(T_1 \times T_2)/Z = T_1 T_2$.*

Proof Let ϕ_i be an automorphism of an $\mathcal{E}_i$-centric subgroup Q_i of T_i for $i = 1$ or $i = 2$. We claim that ϕ_i extends to an automorphism of $Q_i T_{3-i}$ that acts trivially on T_{3-i}. Firstly, $Q_i \cap T_{3-i} \leq T_1 \cap T_2 \leq Z(T_i)$, so that ϕ_i acts trivially on $Q_i \cap T_{3-i}$. By Proposition 8.39, ϕ extends to an automorphism of $Q_i C_P(Q_i)$ in $\mathcal{F}$ that acts trivially on $C_P(Q_i)/Z(Q_i)$, and hence ϕ_i extends to an automorphism $\bar{\phi}_i$ of $Q_i T_{3-i}$ in $\mathcal{F}$ that acts trivially on $T_{3-i}/Q_i \cap T_{3-i}$. If ϕ_i is a p'-automorphism then $\bar{\phi}_i$ acts trivially on T_{3-i} by Exercise 4.5. If $\phi_i \in \mathrm{Aut}_{T_i}(Q_i)$ then clearly also in this case $\phi_i = c_g$ (for some $g \in T_i$) extends to a morphism, also c_g, acting trivially on T_{3-i} (as T_1 and T_2 commute). Hence any automorphism in the subgroup of $\mathrm{Aut}_{\mathcal{E}_i}(Q_i)$ generated by all p'-automorphisms and $\mathrm{Aut}_{T_i}(Q_i)$ extends to an automorphism in $\mathcal{F}$ acting trivially on T_{3-i}. If Q_i is fully $\mathcal{E}_i$-normalized then this is all of $\mathrm{Aut}_{\mathcal{E}_i}(Q_i)$, and if Q_i is not fully $\mathcal{E}_i$-normalized, we choose $\bar{Q}_i$ to be fully $\mathcal{F}$-normalized (and hence fully $\mathcal{E}_i$-normalized by Exercise 5.13) and $\psi : Q_i \to \bar{Q}_i$ an isomorphism. Since $\mathcal{F}$ is saturated, ψ extends to an isomorphism $\bar{\psi} : Q_i T_{3-i} \to \bar{Q}_i T_{3-i}$, whence we see (via conjugation by $\bar{\psi}$) that every automorphism of Q_i extends to an automorphism of $Q_i T_{3-i}$ acting trivially on T_{3-i} (since the result is true for $\bar{Q}_i T_{3-i}$).

Therefore, since $\mathcal{E}_i$ is saturated, any morphism $\phi_i : Q_i \to R_i$ (with $R_i \leq T_i$) in $\mathcal{E}_i$ extends to a morphism $\hat{\phi}_i$ in $\mathcal{F}$ that acts trivially on T_{3-i} by Alperin's fusion theorem. Let $\hat{\mathcal{E}}_i$ be the subcategory of $\mathcal{F}$ consisting of those maps $\hat{\phi}_i$, for ϕ_i a morphism in $\mathcal{E}_i$. By Lemma 8.48, the subsystem $\mathcal{E}$ generated by $\hat{\mathcal{E}}_1$ and $\hat{\mathcal{E}}_2$ is isomorphic with $\mathcal{E}_1 \times_Z \mathcal{E}_2$.

We must now show that $\mathcal{E}$ is normal in $\mathcal{F}$. Notice that $\bar{T} \leq Z(\mathcal{E})$, and so $\mathcal{E} \leq C_{\mathcal{F}}(\bar{T})$. We first prove that $\mathcal{E}$ is normal in $C_{\mathcal{F}}(\bar{T})$. As $\bar{T} \leq T_i$, the images $\bar{\mathcal{E}}_i$ of $\mathcal{E}_i$ in $\bar{\mathcal{F}} = \mathcal{F}/\bar{T}$ are two normal subsystems of $\bar{\mathcal{F}}$ that intersect trivially, and so by Proposition 8.45 their direct product $\bar{\mathcal{E}}_1 \times \bar{\mathcal{E}}_2$ is a normal subsystem of $\bar{\mathcal{F}}$. Also, the image $\bar{\mathcal{E}}$ of $\mathcal{E}$ in $\bar{\mathcal{F}}$ is this direct product, and so by Theorem 8.43 we see that $\mathcal{E} \trianglelefteq C_{\mathcal{F}}(\bar{T})$.

To conclude the proof that $\mathcal{E}$ is a normal subsystem of $\mathcal{F}$, we note firstly that $\mathcal{E} \trianglelefteq C_{\mathcal{F}}(\bar{T}) \trianglelefteq \mathcal{F}$ (as $\mathcal{F} = N_{\mathcal{F}}(\bar{T})$ and using Proposition 8.8). Let α be an $\mathcal{F}$-automorphism of T. As both $\mathcal{E}_i$ are normal in $\mathcal{F}$,

α induces automorphisms of each $\mathcal{E}_i$, and so induces an automorphism of the subcategories $\hat{\mathcal{E}}_i$. As $\mathcal{E}$ is generated by the $\hat{\mathcal{E}}_i$, this implies that α induces an automorphism of $\mathcal{E}$; hence $\mathcal{E} \trianglelefteq \mathcal{F}$ by Proposition 8.40. $\qquad\square$

Corollary 8.50 *Let $\mathcal{F}$ be a saturated fusion system on a finite p-group P, and let $\mathcal{E}_1, \ldots, \mathcal{E}_r$ be normal subsystems of $\mathcal{F}$, with $\mathcal{E}_i$ a subsystem on T_i. Write $\bar{T}_i$ for the product of all T_j except for T_i. If, for all i, $T_i \cap \bar{T}_i \leq Z(\mathcal{E}_i)$, then there is a normal subsystem $\mathcal{E}$ of $\mathcal{F}$, a central product of the $\mathcal{E}_i$.*

We end this section with an example of direct products appearing as subsystems.

Example 8.51 Let H be the group $\mathrm{SL}_2(q)$, with q an odd prime power. A Sylow 2-subgroup Q of H is generalized quaternion, say Q_{2^n}. Let G be the group $\mathrm{SL}_2(q) \wr S_3$. This has a Sylow 2-subgroup P isomorphic with $Q_{2^n} \wr C_2 \times Q_{2^n}$. Let K denote the base group of the wreath product, and write $R = P \cap K$.

The fusion system $\mathcal{F} = \mathcal{F}_P(G)$ has a surjective morphism onto the trivial fusion system $\mathcal{F}_{C_2}(C_2)$, and $\mathfrak{hyp}(\mathcal{F}) = R$. The hyperfocal subsystem $O^p(\mathcal{F})$ is the fusion system of $H \wr C_3$, and this contains $\mathcal{E} = \mathcal{F}_R(K)$ as a normal subsystem; indeed, $O^{p'}(O^p(\mathcal{F})) = \mathcal{E}$. Notice that $\mathcal{E}$ is isomorphic with the direct product of three copies of the fusion system of H, and that these subsystems are subnormal in $\mathcal{F}$.

8.6 The generalized Fitting subsystem

The results in this section are from [Asc11]; we will explicitly reference the important results.

The generalized Fitting subgroup was introduced in Section 5.7. In this section we develop analogues of quasisimple subsystems, components, the layer, and the generalized Fitting subsystem. The analogues of quasisimple subsystems and components are obvious.

Definition 8.52 Let $\mathcal{F}$ be a saturated fusion system on a finite p-group P.

(i) $\mathcal{F}$ is *quasisimple* if $\mathcal{F}$ is perfect (i.e., $\mathfrak{foc}(\mathcal{F}) = P$) and $\mathcal{F}/Z(\mathcal{F})$ is simple.

(ii) A saturated subsystem $\mathcal{E}$ of $\mathcal{F}$ is a *component* if $\mathcal{E}$ is quasisimple and subnormal in $\mathcal{F}$. Denote by $\mathrm{Comp}(\mathcal{F})$ the set of all components of $\mathcal{F}$.

We begin by giving an example of a quasisimple fusion system.

Example 8.53 Let P be the quaternion group Q_{2^n} for $n \geq 4$, and let $\mathcal{F}$ be the fusion system on P with non-trivial automorphisms on both classes of Q_8 subgroups (see Theorem 4.54). Since P has a single involution x, $\mathrm{Z}(\mathcal{F}) = \langle x \rangle$, and $\mathcal{F}/\mathrm{Z}(\mathcal{F})$ is the simple fusion system on $D_{2^{n-1}}$. This fusion system $\mathcal{F}$ is a quasisimple fusion system.

In fact, if G is any quasisimple group, $\mathrm{Z}(G)$ is a p-group, and $G/\mathrm{Z}(G)$ has a simple fusion system at the prime p, then the fusion system of G is quasisimple. This yields many quasisimple fusion systems for the prime 2, for example.

Quasisimple fusion systems are good analogues of quasisimple groups, because of Theorem 8.43. Using this result and Theorem 8.49, we can easily prove the next result.

Proposition 8.54 *Let $\mathcal{F}$ be a saturated fusion system on a finite p-group P.*

(i) *If $\mathcal{F}/\mathrm{Z}(\mathcal{F})$ is simple and $\mathcal{E}$ is a weakly normal subsystem of $\mathcal{F}$, then either $\mathcal{F}$ is the central product of $\mathrm{O}^p(\mathcal{E})$ and $\mathrm{Z}(\mathcal{F})$ or $\mathcal{E} \leq \mathrm{Z}(\mathcal{F})$ (so that $\mathcal{F}$ is quasisimple).*

(ii) *If $\mathcal{F}$ is quasisimple and $\mathcal{E}$ is a weakly subnormal subsystem of $\mathcal{F}$, then either $\mathcal{E} = \mathcal{F}$ or $\mathcal{E} \leq \mathrm{Z}(\mathcal{F})$.*

The next lemma is an immediate consequence of Theorem 8.43, Corollary 8.30 and the previous proposition.

Lemma 8.55 *Let $\mathcal{F}$ be a saturated fusion system on a finite p-group P. Every weakly subnormal, quasisimple subsystem of $\mathcal{F}$ is a component.*

Using the intersections of weakly subnormal subsystems we can prove that, if a component of a fusion system is not a component of a weakly subnormal subsystem, then the underlying subgroups of the component and the subsystem have a small intersection. This lemma is easy, and the proof is omitted.

Lemma 8.56 *Let $\mathcal{F}$ be a saturated fusion system on a finite p-group P. If $\mathcal{C}$ is a component of $\mathcal{F}$ on T and $\mathcal{E}$ is any weakly subnormal subsystem of $\mathcal{F}$ on U then either $\mathcal{C}$ is a component of $\mathcal{E}$ or $T \cap U \leq \mathrm{Z}(T)$. In particular, if $\mathcal{E}$ is another component of $\mathcal{F}$ then $T \cap U \leq \mathrm{Z}(T) \cap \mathrm{Z}(U)$, and the only component of $\mathcal{F}$ lying on T is $\mathcal{C}$ itself.*

This result is good, but we need more. We actually need that T and U commute. In order to prove this, we need a result on when automorphisms act trivially, which will be needed to show that the action of U on T is trivial.

Lemma 8.57 *Let $\mathcal{F}$ be a saturated fusion system on a finite p-group P, and let $\mathcal{E}$ be a saturated subsystem of $\mathcal{F}$, on the subgroup Q of P. Let X be a strongly $\mathcal{F}$-closed subgroup of P such that X centralizes $Q/Z(\mathcal{F})$. We have that $\mathcal{E} \leq \mathrm{N}_{\mathcal{F}}(X)$, and if $\mathcal{E} = \mathrm{O}^{p'}(\mathcal{E}) = \mathrm{O}^p(\mathcal{E})$ is (weakly) subnormal in $\mathcal{F}$ (using Corollary 8.30) then $\mathcal{E} \leq \mathrm{C}_{\mathcal{F}}(X)$.*

Proof Write $Z = Z(\mathcal{E})$, and let U be an $\mathcal{E}$-centric subgroup of Q, so that $Z \leq U$. Notice that $X \leq \mathrm{C}_P(U/Z) \leq \mathrm{N}_P(U)$, and so we may choose an $\mathcal{F}$-isomorphism $\alpha : U \to \bar{U}$ with $\bar{U}$ fully $\mathcal{F}$-normalized that extends to an isomorphism $UX \to \bar{U}X$ (as X is strongly $\mathcal{F}$-closed). Furthermore, there is an isomorphism $\beta : \bar{U} \to \bar{U}$ in $\mathrm{N}_{\mathcal{F}}(\bar{U})$ such that $\bar{Z} = Z\alpha\beta$ is fully $\mathrm{N}_{\mathcal{F}}(\bar{U})$-normalized, and β may be chosen to extend to $\mathrm{N}_P(U, Z)$ (which contains X), so there is a map $\bar{\alpha}\bar{\beta} : UX \to \bar{U}X$ such that $\bar{U} = U\bar{\alpha}\bar{\beta}$ is fully $\mathcal{F}$-normalized and $\bar{Z} = Z\bar{\alpha}\bar{\beta}$ is fully $\mathrm{N}_{\mathcal{F}}(\bar{U})$-normalized. Obviously all $\phi \in \mathrm{Aut}_{\mathcal{E}}(U)$ extend to automorphisms in $\mathrm{Aut}_{\mathcal{F}}(UX)$ if and only if the corresponding automorphisms $\phi^{\bar{\alpha}\bar{\beta}}$ of $\mathrm{Aut}_{\mathcal{F}}(\bar{U})$ extend to automorphisms in $\mathrm{Aut}_{\mathcal{F}}(\bar{U}X)$. Hence we may assume that U is fully $\mathcal{F}$-normalized and Z is fully $\mathrm{N}_{\mathcal{F}}(U)$-normalized, in which case $\mathrm{Aut}_P(Z \leq U) = \mathrm{N}_{\mathrm{Aut}_P(U)}(Z)$ is a Sylow p-subgroup of $\mathrm{Aut}_{\mathcal{F}}(Z \leq U)$.

Let $\phi \in \mathrm{Aut}_{\mathcal{E}}(U)$ and $x \in X$. Notice that x centralizes U/Z and ϕ centralizes $Z = Z(\mathcal{F})$, and so $[x, \phi]$ centralizes both U/Z and Z. The set of all $\mathcal{F}$-automorphisms that act trivially on both U/Z and Z is a normal p-subgroup of $\mathrm{Aut}_{\mathcal{F}}(Z \leq U)$ by Exercise 4.5, and so is contained in $\mathrm{Aut}_P(Z \leq U)$. Hence $\mathrm{Aut}_X(U)^{\phi} \leq \mathrm{Aut}_P(U)$, and so ϕ extends to a map in $\mathrm{Aut}_{\mathcal{F}}(UX)$ (as X is strongly $\mathcal{F}$-closed). Since this holds for all $\mathcal{E}$-centric subgroups U of Q, we see that $\mathcal{E} \leq \mathrm{N}_{\mathcal{F}}(X)$ by Alperin's fusion theorem.

From now on, we may assume that $\mathcal{F} = \mathrm{N}_{\mathcal{F}}(X)$ and $\mathcal{E} = \mathrm{O}^{p'}(\mathcal{E}) = \mathrm{O}^p(\mathcal{E})$, and so $\mathrm{C}_P(X)$ is strongly $\mathcal{F}$-closed. Since $\mathcal{E}$ centralizes Z, we see that $\mathcal{E} \leq \mathrm{N}_{\mathcal{F}}(Z)$, we may even assume that $\mathcal{F} = \mathrm{N}_{\mathcal{F}}(Z) = \mathrm{N}_{\mathcal{F}}(X)$. Notice that in this case $X \cap Z$ is also a strongly $\mathcal{F}$-closed subgroup of $\mathrm{O}_p(\mathcal{F})$, so that $\mathcal{F} = \mathrm{N}_{\mathcal{F}}(X \cap Z)$. Let K denote the subgroup of $\mathrm{Aut}_{\mathcal{F}}(X)$ of all elements that act trivially on $X/(X \cap Z)$; this is clearly a normal subgroup of $\mathrm{Aut}_{\mathcal{F}}(X)$, and so $\mathrm{N}_{\mathcal{F}}^K(X)$ is a weakly normal subsystem of $\mathcal{F}$, by Exercise 5.10. Since X acts trivially on Q/Z, $[Q, X] \leq Z$, and $[Q, X] \leq X$ as X is strongly $\mathcal{F}$-closed, so that Q acts trivially on

$X/(X \cap Z)$, and hence $Q \leq N_P^K(X)$. Therefore we may form the weak intersection $\mathcal{E} \wedge N_{\mathcal{F}}^K(X)$, a weakly normal subsystem of $\mathcal{E}$ on Q. Since $\mathcal{E} = O^{p'}(\mathcal{E})$, we see that this intersection is simply $N_{\mathcal{F}}^K(X)$, so $\mathcal{E} \leq N_{\mathcal{F}}^K(X)$. As $X \cap Z$ is a central subgroup of $\mathcal{E}$, all elements of $\mathcal{E}$ centralize $X \cap Z$ as well as $X/(X \cap Z)$; so in particular any p'-automorphism of Q in $\mathcal{E}$ extends to a map $\phi \in \text{Aut}_{\mathcal{F}}(QX)$ acting trivially on X (since it acts trivially on $X/X \cap Z$ and Z); in particular, in the group $Y = QX \rtimes \langle \bar{\phi} \rangle$, $\bar{\phi}$ and X commute. Let x be an element of X.

For $g \in Q$, notice that since $\bar{\phi}$ and x commute we see that $[[\bar{\phi}, x], g] = 1$, and since x centralizes Q/Z we have $[g, x] \in Z$, so that $[[g, x], \bar{\phi}] = 1$. Therefore by the three subgroup lemma [Asc00, (8.7)] we see that $[[g, \bar{\phi}], x] = 1$; in other words, $\bar{\phi}$ centralizes $Q/\text{C}_Q(X)$, so ϕ does.

The image $\bar{\mathcal{E}}$ of $\mathcal{E}$ in $\mathcal{F}/\text{C}_P(X)$ is isomorphic with $\mathcal{E}/(\text{C}_P(X) \cap Q) = \mathcal{E}/\text{C}_Q(X)$. We will show that $\bar{\mathcal{E}} = \mathcal{F}_A(A)$ for some abelian p-group A, so then $Q = \mathfrak{foc}(\mathcal{E})$ must lie inside $\text{C}_Q(X)$, as needed. To see that $Q' \leq \text{C}_P(X)$, for $x \in X$ let $\lambda_x : Q \to Z$ be defined by $g\lambda_x = [g, x]$. This is a homomorphism by the same proof as in Proposition 5.68, and so $Q' \leq \text{C}_Q(x)$; thus $Q' \leq \text{C}_Q(X)$, as claimed. Let $A = Q/\text{C}_Q(X)$, an abelian group.

Since $\bar{\mathcal{E}}$ is a fusion system on A, it remains to show that $\bar{\mathcal{E}} = \mathcal{F}_A(A)$. If $\bar{\mathcal{E}}$ is non-trivial, there is a p'-element α of $\text{Aut}_{\bar{\mathcal{E}}}(Q)$ such that α acts non-trivially on $Q/\text{C}_Q(X)$; however, we showed above that all p'-automorphisms $\alpha \in \text{Aut}_{\mathcal{E}}(Q)$ act trivially on $Q/\text{C}_Q(X)$. Therefore $\mathcal{E}/\text{C}_Q(X)$ is a trivial fusion system, so that $Q = \text{C}_Q(X)$, as required. $\square$

Theorem 8.58 (Aschbacher [Asc11, (9.6)]) *Let $\mathcal{F}$ be a saturated fusion system on a finite p-group P. If $\mathcal{C}$ is a component of $\mathcal{F}$ on T and $\mathcal{E}$ is a weakly subnormal subsystem of $\mathcal{F}$ on the subgroup U of P then either $\mathcal{C}$ is a component of $\mathcal{E}$ or $[T, U] = 1$.*

Proof Notice that $T < P$ by Corollary 8.56, and so $\mathcal{C}$ is a component of some proper normal subsystem $\mathcal{F}_1$ of $\mathcal{F}$ (perhaps with $\mathcal{C} = \mathcal{F}_1$). Choose $\mathcal{F}$ to be a minimal counterexample to the theorem. Let $\mathcal{E}_1$ be a proper weakly normal subsystem of $\mathcal{F}$ containing $\mathcal{E}$ (it could be that $\mathcal{E} = \mathcal{E}_1$). If $\mathcal{C}$ is a component of $\mathcal{E}_1$ then by choice of minimal counterexample, the theorem holds (with $\mathcal{E}_1$ in place of $\mathcal{F}$), and so $\mathcal{C}$ is not a component of $\mathcal{E}$. Writing U_1 for the subgroup on which $\mathcal{E}_1$ is defined, if the theorem holds for $\mathcal{F}$ and $\mathcal{E}_1$, then $[T, U_1] = 1$, so that $[T, U] = 1$, and the theorem is true. Hence we may assume that $\mathcal{E} = \mathcal{E}_1$ is a weakly normal subsystem of $\mathcal{F}$.

Suppose that u is an element of $U \setminus C_P(T)$. We have that $\mathcal{C}c_u$ is a component of $\mathcal{F}_1$; if $\mathcal{C} \neq \mathcal{C}c_u$ then, since the theorem holds for $\mathcal{F}_1$, we must have that T and T^u commute, so $T^u \leq C_P(T)$. We aim to show that T/Z is abelian, where $Z = Z(\mathcal{C})$. Since $\mathcal{C}/Z$ is a simple fusion system, this is a contradiction. The statement that T/Z is abelian is equivalent to $[T, T] \leq Z$.

Notice that $[T, u]$ is contained in $\bar{U} = TT^u \cap U$ (since U is normal in P), and since $T^u \leq C_P(T)$ we see that this intersection $\bar{U}$ is contained in $T C_P(T)$, of which T is a normal subgroup. Therefore $[T, \bar{U}] \leq T \cap U \leq Z$ by Corollary 8.56. However, for each $t \in T$, $[t, u] = t^{-1}t^u$, so that $[T, [t, u]] = [T, t]$. Hence

$$[T, T] = [T, [T, u]] \leq Z,$$

as claimed. Thus $\mathcal{C}c_u = \mathcal{C}$, and hence U normalizes T. In addition, $[T, U] \leq T \cap U \leq Z$ again by Corollary 8.56, and so U is a strongly $\mathcal{F}$-closed subgroup that centralizes $U/Z(\mathcal{C})$, with $\mathcal{C} = O^{p'}(\mathcal{C}) = O^p(\mathcal{C})$. Applying Lemma 8.57, we see that U centralizes T. $\qquad\square$

We can use this theorem to produce a couple of corollaries that we will use later. The first corollary to this theorem is that we can understand the components of constrained fusion systems; they don't have any. (This can also be deduced from Proposition 8.31.)

Corollary 8.59 *Let $\mathcal{F}$ be a saturated fusion system on a finite p-group P. If $\mathcal{F}$ is constrained then $\mathrm{Comp}(\mathcal{F})$ is empty.*

Proof Let $Q = O_p(\mathcal{F})$. If $\mathcal{C}$ is a component of $\mathcal{F}$ on T, then either $\mathcal{C}$ is a component of $\mathcal{E} = \mathcal{O}_p(\mathcal{F})$ or $[Q, T] = 1$. First assume that $[Q, T] = 1$. Since $C_P(Q) \leq Q$, this means that $T \leq Q$. Therefore $\mathcal{E} \curlywedge \mathcal{C}$ is a weakly subnormal subsystem of $\mathcal{C}$ on $T \cap Q = T$. By Proposition 8.54, $\mathcal{E} \curlywedge \mathcal{C} = \mathcal{C}$, so that $\mathcal{C} \leq \mathcal{E}$. This is also the conclusion if $\mathcal{C}$ is a component of $\mathcal{E}$.

As $\mathcal{E} = \mathcal{F}_Q(Q)$ is soluble, it cannot have any quasisimple subsystems since quasisimple subsystems are clearly not soluble. Hence $\mathrm{Comp}(\mathcal{F})$ is empty, as claimed. $\qquad\square$

The second is about components and quasisimple subsystems, and identifying quasisimple subsystems that are components.

Corollary 8.60 *Let $\mathcal{F}$ be a saturated fusion system on a finite p-group P, and let $\mathcal{C}$ be a component of $\mathcal{F}$, on the subgroup T of P. If $\mathcal{E}$ is a weakly normal subsystem of $\mathcal{F}$, on the subgroup Q of P, and $T \leq Q$, then $\mathcal{C}$ is a component of $\mathcal{E}$.*

Proof Assume that $\mathcal{C}$ is not a component of $\mathcal{E}$. By Theorem 8.58, $[T, Q] = 1$, so since $T \leq Q$, we have that T is abelian, contradicting the fact that $\mathcal{C}$ is quasisimple. □

Having produced some results on quasisimple subsystems, we now turn our attention to constructing the layer of $\mathcal{F}$; this is the central product of all components of $\mathcal{F}$. However, while we have proved that the central product of normal subsystems exists, we have not proved that the central product of *subnormal* subsystems exists inside a fusion system.

Definition 8.61 Let $\mathcal{F}$ be a saturated fusion system on a finite p-group P. Let X denote the set of weakly normal subsystems of $\mathcal{F}$ containing all components of $\mathcal{F}$. The *layer* of $\mathcal{F}$, denoted by $E(\mathcal{F})$, is the subsystem $\curlywedge X$, the weak intersection of all elements of X.

A few things aren't particularly clear about the subsystem $E(\mathcal{F})$. Let $\mathscr{T}$ denote the set of subgroups on which the components of $\mathcal{F}$ are defined. If T is the subgroup of P generated by the elements of $\mathscr{T}$, then T is the central product of the elements of $\mathscr{T}$ by Theorem 8.58. However, why is $E(\mathcal{F})$ on the subgroup T in the first place? Also, we want $E(\mathcal{F})$ to be the central product of the components of $\mathcal{F}$. We will prove all this now.

Theorem 8.62 (Aschbacher [Asc11, Theorem 9.8]) *Let $\mathcal{F}$ be a saturated fusion system on a finite p-group P. Let $\mathscr{T}$ denote the set of subgroups on which the components of $\mathcal{F}$ are defined and let T denote the central product of the elements of $\mathscr{T}$.*

(i) *$E(\mathcal{F})$ is a characteristic subsystem of $\mathcal{F}$ defined on the subgroup T of P.*

(ii) *$E(\mathcal{F})$ is the central product of the elements of $\mathrm{Comp}(\mathcal{F})$.*

(iii) *If $\mathcal{E}$ is a normal subsystem of $\mathcal{F}$, then $E(\mathcal{E})$ is the central product of the elements of $\mathrm{Comp}(\mathcal{F})$ that lie in $\mathrm{Comp}(\mathcal{E})$.*

Proof Let $\mathcal{F}$ be a minimal counterexample, in the sense that all proper subsystems of $\mathcal{F}$ satisfy the theorem. Write $\mathrm{Comp}(\mathcal{F}) = \{\mathcal{C}_i \mid i \in I\}$ for some indexing set I. Let $\mathcal{E}$ be any proper, normal subsystem of $\mathcal{F}$; by choice of minimal counterexample $E(\mathcal{E})$ is the central product of the components of $\mathcal{E}$, and clearly any component of $\mathcal{E}$ is a component of $\mathcal{F}$, so that $E(\mathcal{E})$ is a central product of the $\mathcal{C}_j$ for $j \in J$, where $J \subseteq I$. Choose $\mathcal{E}$ so that $|J|$ is maximal. If $J = I$ then $E(\mathcal{E})$ contains all components of $\mathcal{F}$, $E(\mathcal{E})$ char $\mathcal{E} \lhd \mathcal{F}$ so that $E(\mathcal{E}) \lhd \mathcal{F}$, and hence the theorem is true.

Hence $J \neq I$; choose $a \in I \setminus J$. If $\mathcal{F}$ itself is quasisimple then $\mathrm{Comp}(\mathcal{F}) = \{\mathcal{F}\}$ by Exercise 8.12, and so $\mathcal{C}_a$ is a proper subsystem of $\mathcal{F}$. Let $\mathcal{D}$ be a normal subsystem of $\mathcal{F}$ such that $\mathcal{C}_a$ is a component of $\mathcal{D}$ (including the possibility $\mathcal{D} = \mathcal{C}_a$).

Let K denote the subset of I such that $\mathcal{C}_k$ is a component of $\mathcal{D}$ for $k \in K$. Again, by induction $E(\mathcal{D})$ is the central product of the $\mathcal{C}_k$ for $k \in K$. Let D and E be the subgroups on which $\mathcal{D}$ and $\mathcal{E}$ are defined. Since $\mathcal{E}$ is normal in $\mathcal{F}$, $\mathrm{Aut}_{\mathcal{F}}(E) \leq \mathrm{Aut}(\mathcal{E})$, and hence $\mathrm{Aut}_{\mathcal{F}}(E)$ permutes the elements of $\mathrm{Comp}(\mathcal{E})$. We may also think of this as $\mathrm{Aut}_{\mathcal{F}}(E)$ permuting the elements of J.

If k lies in $J \cap K$ then $\mathcal{C}_k$ lies inside both $\mathcal{D}$ and $\mathcal{E}$; since D and E are strongly $\mathcal{F}$-closed, if $\phi \in \mathrm{Aut}_{\mathcal{F}}(E)$ then $\mathcal{C}_k \phi$ is another component of $\mathcal{F}$ and the subgroup on which $\mathcal{C}_k \phi$ is defined lies inside both D and E; hence, by Corollary 8.60, $\mathcal{C}_k \phi$ is a component of both $\mathcal{D}$ and $\mathcal{E}$. In particular, $\mathrm{Aut}_{\mathcal{F}}(E)$ permutes the elements of $J \cap K$, and hence of $J \setminus K$. Let $\mathcal{A}$ denote the central product of the $\mathcal{C}_j$ for $j \in J \setminus K$, as a subsystem of $E(\mathcal{E})$. Notice that $\mathcal{A}$ is a normal subsystem of $E(\mathcal{E})$.

Since $\mathcal{D}$ and $\mathcal{E}$ are normal subsystems of $\mathcal{F}$, and $E(\mathcal{D}) \operatorname{char} \mathcal{D}$ and $E(\mathcal{E}) \operatorname{char} \mathcal{E}$, we have that $E(\mathcal{D})$ and $E(\mathcal{E})$ are normal subsystems of $\mathcal{F}$. By Exercise 8.13, $\mathcal{A}$ is a normal subsystem of $\mathcal{F}$. By construction, the subgroups $\bar{E}$ and A, on which $E(\mathcal{E})$ and $\mathcal{A}$ are defined respectively, commute, by Theorem 8.58. The subsystem $E(\mathcal{E}) \curlywedge \mathcal{A}$ is a weakly normal subsystem of $\mathcal{F}$ on $\bar{E} \cap A \leq \mathrm{Z}(\bar{E}) \cap \mathrm{Z}(A)$, and so $\bar{E} \cap A \leq \mathrm{Z}(E(\mathcal{E})) \cap \mathrm{Z}(\mathcal{A})$ by Exercise 8.14. Therefore a central product of $E(\mathcal{E})$ and $\mathcal{A}$ is a normal subsystem of $\mathcal{F}$ with more components than that of $\mathcal{E}$ (as this subsystem has components $\{\mathcal{C}_j \mid j \in J \cup K\}$). Therefore $\mathcal{F}$ is the central product of $E(\mathcal{E})$ and $\mathcal{A}$, and so in particular $\mathcal{F}$ is the central product of its components. This proves the first two parts; the third part follows by induction. □

Having proved that the layer of $\mathcal{F}$ is the central product of the components of $\mathcal{F}$, we turn to defining the generalized Fitting subsystem.

Proposition 8.63 *Let $\mathcal{F}$ be a saturated fusion system on a finite p-group P. If T is the subgroup of P on which $E(\mathcal{F})$ is defined then $[T, O_p(\mathcal{F})] = 1$, and $T \cap O_p(\mathcal{F}) \leq \mathrm{Z}(\mathcal{E}(\mathcal{F})) \cap \mathrm{Z}(O_p(\mathcal{F}))$, so that the central product of $E(\mathcal{F})$ and $O_p(\mathcal{F})$ is a characteristic subsystem of $\mathcal{F}$. In particular, $E(\mathcal{F}) \leq \mathrm{C}_{\mathcal{F}}(O_p(\mathcal{F}))$.*

Proof By Theorem 8.58, either some component in $\mathrm{Comp}(\mathcal{F})$ is a component of $O_p(\mathcal{F})$ or $[T, O_p(\mathcal{F})] = 1$, and since $O_p(\mathcal{F})$ is a soluble fusion system it has no components. Therefore $O_p(\mathcal{F}) \curlywedge E(\mathcal{F})$ is a weakly normal subsystem of $E(\mathcal{F})$ on a subgroup of $\mathrm{Z}(T)$, so that $O_p(\mathcal{F}) \cap T \leq$

$Z(E(\mathcal{F}))$ by Exercise 8.14. Also, $O_p(\mathcal{F}) \cap T \leq Z(O_p(\mathcal{F}))$, so that this intersection is also a central subgroup of $O_p(\mathcal{F})$. Hence the central product $\mathcal{E}$ of $O_p(\mathcal{F})$ and $E(\mathcal{F})$ is a normal subsystem of $\mathcal{F}$, and it is characteristic in $\mathcal{F}$ since both $O_p(\mathcal{F})$ and $E(\mathcal{F})$ are characteristic in $\mathcal{F}$. The last fact trivially holds since $\mathcal{E}$ is the central product of $E(\mathcal{F})$ and $O_p(\mathcal{F})$. □

This enables us to give the following definition, first considered by Aschbacher in [Asc11].

Definition 8.64 Let $\mathcal{F}$ be a saturated fusion system on a finite p-group P. The *generalized Fitting subsystem* of $\mathcal{F}$, denoted by $F^*(\mathcal{F})$, is the central product of $O_p(\mathcal{F})$ and $E(\mathcal{F})$.

Of course, one fundamental property of $F^*(G)$, the generalized Fitting subgroup of a finite group G, is that $C_G(F^*(G)) \leq F^*(G)$. For fusion systems the same result should hold; however, we do not have a definition of the centralizer of a subsystem. Let $\mathcal{F}$ be a saturated fusion system on the finite p-group P, and let $\mathcal{E}$ be a normal subsystem of $\mathcal{F}$, on the subgroup T of P. For some subgroups Q of $C_P(T)$ we will have $\mathcal{E} \leq C_{\mathcal{F}}(Q)$; for example, in Proposition 8.63, we showed that $E(\mathcal{F}) \leq C_{\mathcal{F}}(O_p(\mathcal{F}))$. It turns out (see [Asc11, Section 6]) that the set of all such subgroups of $C_P(T)$ has a unique largest member X, i.e., $\mathcal{E} \leq C_{\mathcal{F}}(Q)$ for some $Q \leq P$ if and only if $Q \leq X$. We make the following definition.

Definition 8.65 Let $\mathcal{F}$ be a saturated fusion system on a finite p-group P, and let $\mathcal{E}$ be a normal subsystem of $\mathcal{F}$, on the subgroup T of P. If $\mathcal{E} \leq C_{\mathcal{F}}(Q)$ then we say that Q *centralizes* $\mathcal{E}$. The unique largest subgroup X of $C_P(T)$ such that $\mathcal{E} \leq C_{\mathcal{F}}(X)$ is the *centralizer* of $\mathcal{E}$ and denoted by $C_P(\mathcal{E})$. Furthermore, $C_P(\mathcal{E})$ is strongly $\mathcal{F}$-closed.

If $S = C_P(\mathcal{E})$ and U is any fully $\mathcal{F}$-normalized subgroup of S with $C_P(U) \leq U$, then write

$$A(U) = O^p(\mathrm{Aut}_{C_{\mathcal{F}}(T)}(U))\,\mathrm{Aut}_S(U).$$

Extending A to an $\mathcal{F}$-invariant map, we have that $A(-)$ is a normal map. The normal subsystem of $\mathcal{F}$ generated by $A(-)$, denoted $C_{\mathcal{F}}(\mathcal{E})$, is the *centralizer* of $\mathcal{E}$ in $\mathcal{F}$.

It is this subsystem, $C_{\mathcal{F}}(\mathcal{E})$, that we will use when giving the analogue of $C_G(F^*(G)) \leq F^*(G)$.

Proposition 8.66 (Aschbacher [Asc11, (9.11)]) *Let $\mathcal{F}$ be a saturated fusion system on a finite p-group P. We have that $C_{\mathcal{F}}(F^*(\mathcal{F})) =$*

$Z(F^*(\mathcal{F}))$. *Consequently, if Q is a subgroup of P such that $F^*(\mathcal{F}) \leq C_{\mathcal{F}}(Q)$ (i.e., $Q \leq C_P(F^*(\mathcal{F})))$ then $Q \leq Z(F^*(\mathcal{F}))$.*

Proof Since $F^*(\mathcal{F}) \trianglelefteq \mathcal{F}$, the subsystem $\mathcal{E} = C_{\mathcal{F}}(F^*(\mathcal{F}))$ is a normal subsystem of $\mathcal{F}$. Consequently, every component of $\mathcal{E}$ is a component of $\mathcal{F}$, so that $E(\mathcal{E}) \leq E(\mathcal{F})$. Let T be the subgroup on which $E(\mathcal{F})$ is defined and $U = C_P(F^*(\mathcal{F}))$ be the subgroup on which $\mathcal{E}$ is defined. Certainly $U \leq C_P(T)$, so that $U \cap T$ is abelian. Since there are no quasisimple fusion systems on abelian p-groups, and $E(\mathcal{E}) \leq E(\mathcal{F})$, we see that $E(\mathcal{E})$ is a subsystem on a subgroup of $T \cap U$; hence $E(\mathcal{E}) = 1$. Also, since $\mathcal{E} \trianglelefteq \mathcal{F}$, $O_p(\mathcal{E}) = O_p(\mathcal{F}) \cap U$ (using Proposition 5.47) is central in $O_p(\mathcal{F})$ as $\mathcal{E}$ centralizes $O_p(\mathcal{F})$. Hence $O_p(\mathcal{E}) = Z(O_p(\mathcal{F})) = Z(F^*(\mathcal{F}))$. Also, since $\mathcal{E}$ centralizes $F^*(\mathcal{F})$, in fact $O_p(\mathcal{E}) = Z(\mathcal{E})$.

It remains to show that if $F^*(\mathcal{E}) = O_p(\mathcal{E}) = Z(\mathcal{E})$ then $\mathcal{E} = \mathcal{F}_Z(Z)$, where $Z = Z(\mathcal{E})$. To see this, let $\mathcal{D}$ be a minimal subnormal subsystem of $\mathcal{F}$ not contained in Z. Either $\mathcal{D}$ is perfect or it is not. If it is perfect then it is quasisimple, contradicting the statement that $\mathrm{Comp}(\mathcal{E})$ is empty. If $\mathcal{D}$ is not perfect then $O^p(\mathcal{D}) \leq Z(\mathcal{E}) = \mathcal{F}_Z(Z)$, so that $\mathcal{D}$ is the fusion system of a finite p-group (as $O^p(\mathcal{F}_Z(Z)) = 1$ and $O^p(O^p(\mathcal{E})) = O^p(\mathcal{E})$). Hence $O_p(\mathcal{E}) > Z$, a contradiction as $O_p(\mathcal{E}) = Z$; so $\mathcal{E} = \mathcal{F}_Z(Z)$, as claimed. $\qquad\square$

This allows us to provide the converse to Corollary 8.59, namely that if $\mathrm{Comp}(\mathcal{F})$ is empty then $\mathcal{F}$ is constrained.

Corollary 8.67 *A saturated fusion system $\mathcal{F}$ is constrained if and only if $\mathrm{Comp}(\mathcal{F})$ is empty.*

Proof Since $\mathcal{F}$ has no components, $F^*(\mathcal{F}) = O_p(\mathcal{F})$, and thus $O_p(\mathcal{F}) = F^*(\mathcal{F}) \leq C_{\mathcal{F}}(Q)$ if and only if $O_p(\mathcal{F}) \leq C_P(Q)$ for any subgroup Q of P; by Proposition 8.66 this is true if and only if $Q \leq Z(F^*(\mathcal{F})) = Z(O_p(\mathcal{F}))$. Hence $C_P(O_p(\mathcal{F})) \leq O_p(\mathcal{F})$, as claimed. $\qquad\square$

We end this section with a proposition on normal subsystems of $\mathcal{F}$ contained in $E(\mathcal{F})$ and central products with other normal subsystems. We start with a lemma that follows immediately from the definitions of the objects involved, so its proof is omitted.

Lemma 8.68 *Let $\mathcal{F}$ be a saturated fusion system on a finite p-group P. If $\mathcal{E}_1$ and $\mathcal{E}_2$ are normal subsystems of $\mathcal{F}$ such that there is a central product $\mathcal{E}_1\mathcal{E}_2$ of $\mathcal{E}_1$ and $\mathcal{E}_2$ in $\mathcal{F}$, then $\mathcal{E}_i \leq C_{\mathcal{F}}(\mathcal{E}_{3-i})$.*

Proposition 8.69 (Aschbacher [Asc11, (9.13)]) *Let $\mathcal{F}$ be a saturated fusion system on a finite p-group P, and let $\mathcal{E}$ be a normal subsystem of $\mathcal{F}$, on the subgroup E of P. Write $X = \mathrm{Comp}(\mathcal{F})\backslash\mathrm{Comp}(\mathcal{E})$. The central product of the elements of X is a normal subsystem $\mathcal{D}$ of $\mathcal{F}$, and we have a central product $\mathcal{D}\mathcal{E}$ of $\mathcal{D}$ and $\mathcal{E}$ in $\mathcal{F}$. In particular, if $\mathcal{C} \in \mathrm{Comp}(\mathcal{F})$ lies on $S \leq P$, then either $\mathcal{C} \in \mathrm{Comp}(\mathcal{E})$, or $\mathcal{C} \leq \mathrm{C}_{\mathcal{F}}(\mathcal{E})$ and $\mathcal{E} \leq \mathrm{C}_{\mathcal{F}}(\mathcal{C})$.*

Proof Let T be the subgroup on which $E(\mathcal{F})$ is defined. Notice that both E and T are strongly $\mathcal{F}$-closed, and so therefore is $E \cap T$. If ϕ is an $\mathcal{F}$-automorphism of T and $\mathcal{C} \in \mathrm{Comp}(\mathcal{F})$ lies on the subgroup C of P, then $C \leq E$ if and only if $C\phi \leq E$. In particular, ϕ permutes $\mathrm{Comp}(\mathcal{E})$ and hence also $\mathrm{Comp}(\mathcal{F}) \setminus \mathrm{Comp}(\mathcal{E})$. By Exercise 8.13, $\mathcal{D}$, the central product of the elements of $\mathrm{Comp}(\mathcal{F}) \setminus \mathrm{Comp}(\mathcal{E})$, is a normal subsystem of $\mathcal{F}$. Write D for the subgroup on which $\mathcal{D}$ is defined.

We must now show that $[D, E] = 1$ and $D \cap E \leq \mathrm{Z}(\mathcal{D}) \cap \mathrm{Z}(\mathcal{E})$, for then there is a central product $\mathcal{D}\mathcal{E}$ of $\mathcal{D}$ and $\mathcal{E}$. Certainly for $\mathcal{C} \in \mathrm{Comp}(\mathcal{F}) \setminus \mathrm{Comp}(\mathcal{E})$ on C, we have $[C, E] = 1$ by Theorem 8.58, and the fact that D is generated by the subgroups of P on which the components in $\mathrm{Comp}(\mathcal{F}) \setminus \mathrm{Comp}(\mathcal{E})$ are defined means that $[D, E] = 1$.

Let $\mathcal{A} = \mathcal{D} \lambda \mathcal{E}$; it is clear that $\mathrm{Comp}(\mathcal{A}) = \mathrm{Comp}(\mathcal{D}) \cap \mathrm{Comp}(\mathcal{E}) = \emptyset$. Hence $F^*(\mathcal{A}) = \mathcal{O}_p(\mathcal{A}) \leq \mathcal{O}_p(\mathcal{D})$, and since $\mathcal{D}$ is a central product of quasisimple subsystems, $\mathrm{O}_p(\mathcal{D}) = \mathrm{Z}(\mathcal{D})$. Therefore $F^*(\mathcal{A}) = \mathcal{Z}(\mathcal{A})$ which by Proposition 8.66 means that $\mathcal{A} = \mathcal{Z}(\mathcal{A}) = \mathcal{F}_{D \cap E}(D \cap E)$. Hence $\mathcal{A} \leq \mathcal{Z}(\mathcal{D})$, so that $D \cap E \leq \mathrm{Z}(\mathcal{D})$.

It remains to show that $D \cap E \leq \mathrm{Z}(\mathcal{E})$. Note that $A = D \cap E \leq \mathrm{O}_p(\mathcal{E})$ and $A \leq \mathrm{Z}(E)$ (since $[D, E] = 1$), so that $\mathrm{Aut}_E(A) = 1$. If A is not a central subgroup of $\mathcal{E}$ then $\mathrm{Aut}_{\mathcal{E}}(A) \neq 1$, so there is a non-trivial p'-element ϕ of $\mathrm{Aut}_{\mathcal{E}}(A)$. As $\mathcal{E} = \mathrm{N}_{\mathcal{E}}(A)$, ϕ extends to an automorphism of E, and since $\mathcal{E} \trianglelefteq \mathcal{F}$ and $D \leq \mathrm{C}_P(E)$, we see that ϕ extends to a p'-automorphism $\bar{\phi}$ of DE such that $[\bar{\phi}, D] \leq \mathrm{Z}(E)$. As D is strongly $\mathcal{F}$-closed, $[\bar{\phi}, D] \leq D$ and so $[\bar{\phi}, D] \leq \mathrm{Z}(E) \cap D = A$. By Exercise 7.15 a p'-automorphism that centralizes $\mathcal{D}/\mathrm{Z}(\mathcal{D})$ also centralizes $\mathfrak{foc}(\mathcal{D}) = D$, and so in particular centralizes $A \leq D$. The final statement follows from Lemma 8.68. $\qquad\square$

8.7 *L*-balance

The purpose of this section is to prove the following theorem of Aschbacher from [Asc11].

Theorem 8.70 (Aschbacher [Asc11, Theorem 7]) *Let $\mathcal{F}$ be a saturated fusion system on the finite p-group P, and let Q be a fully normalized subgroup of P. If $\mathcal{E}$ is a component of $\mathrm{N}_{\mathcal{F}}(Q)$ then $\mathcal{E}$ is contained in $E(\mathcal{F})$, i.e.,*

$$E(\mathrm{N}_{\mathcal{F}}(Q)) \leq E(\mathcal{F}).$$

The corresponding theorem for finite groups holds, but with considerable difficulties caused by the presence of *cores* – normal p'-subgroups living inside normalizers of p-subgroups. Since all normal p'-subsystems are trivial, all of the difficulties posed by cores vanish for fusion systems, and so Theorem 8.70 is quite easy to prove for fusion systems, whereas it is difficult for finite groups. However, the theorem for fusion systems does not imply the theorem for groups.

We begin with a few preliminary lemmas concerning components of local subsystems.

Lemma 8.71 *Let $\mathcal{F}$ be a saturated fusion system on a finite p-group P. If Q is a fully normalized subgroup of P, then*

$$E(\mathrm{N}_{\mathcal{F}}(Q)) = E(\mathrm{C}_{\mathcal{F}}(Q)) = E(Q\,\mathrm{C}_{\mathcal{F}}(Q)).$$

Proof As both $\mathrm{C}_{\mathcal{F}}(Q)$ and $Q\,\mathrm{C}_{\mathcal{F}}(Q)$ are normal subsystems of $\mathrm{N}_{\mathcal{F}}(Q)$ by Proposition 8.8, all components of them are components of $\mathrm{N}_{\mathcal{F}}(Q)$. Let $\mathcal{C}$ be a component of $\mathrm{N}_{\mathcal{F}}(Q)$ on the subgroup T of $\mathrm{N}_P(Q)$. Since $\mathcal{C}$ centralizes $O_p(\mathrm{N}_{\mathcal{F}}(Q))$, we see that $[T, Q] = 1$, so that $T \leq \mathrm{C}_P(Q)$. By Corollary 8.60, $\mathcal{C}$ is a component of $\mathrm{C}_{\mathcal{F}}(Q)$ and $Q\,\mathrm{C}_{\mathcal{F}}(Q)$, as claimed. □

Lemma 8.72 *Let $\mathcal{F}$ be a saturated fusion system on a finite p-group P, and let Q be a fully $\mathcal{F}$-normalized subgroup of P. If $\mathcal{C}$ is a component of $\mathrm{N}_{\mathcal{F}}(Q)$, then $\mathcal{C} \leq \mathrm{C}_{\mathcal{F}}(O_p(\mathcal{F}))$.*

Proof Write $S = O_p(\mathcal{F})$ and $\mathcal{E} = \mathrm{N}_{\mathcal{F}}(Q)$. Since $\mathcal{C}$ is a component of $\mathcal{E}$, $\mathcal{C} \leq \mathrm{C}_{\mathcal{F}}(O_p(\mathcal{E}))$ by Proposition 8.63. By Lemma 5.46, $O_p(\mathcal{E}) \geq S \cap \mathrm{N}_P(Q) = \mathrm{N}_S(Q)$.

Let U be a subgroup of T, the subgroup of P on which $\mathcal{C}$ is defined, and let ϕ be a p'-automorphism in $\mathrm{Aut}_{\mathcal{C}}(U)$. Since $\mathcal{C}$ centralizes $O_p(\mathcal{E}) = R \geq Q$, we may extend ϕ to an automorphism of UR acting trivially on R, and since $S = O_p(\mathcal{F})$, we may extend this extension to an automorphism ϕ' of URS acting trivially on R and acting as an automorphism of both U and S. By raising ϕ' to an appropriate power, we see that ϕ' is a p'-element that extends ϕ.

Note that $\mathrm{C}_S(R) \leq \mathrm{C}_S(Q)$ since $Q \leq R$, and that $\mathrm{C}_S(Q) \leq R$ since

$R \geq N_S(Q)$ above, so that ϕ' acts trivially on $C_S(R)$ since it acts trivially on R.

Consider R and $\langle \phi' \rangle$ as generating a group of automorphisms of S. Notice that $[R, \langle \phi' \rangle] = 1$, and $[\langle \phi' \rangle, C_S(R)] = 1$ since ϕ' centralizes $C_S(R)$. As R is a p-group and $\langle \phi' \rangle$ is a p'-group, we may apply Thompson's $A \times B$ lemma (see [Asc00, (24.2)]); this yields that ϕ' acts trivially on S.

Since $[U, \phi] = [U, \phi']$ and ϕ' acts trivially on S, we see that $[[U, \phi], S] = [[U, \phi'], S] = 1$ by the three subgroup lemma [Asc00, (8.7)] and the fact that $[[\phi', S], U] = [[S, U], \phi'] = 1$ (as $[S, U] \leq S$); hence for any subgroup $U \leq T$ and $\phi \in \mathrm{Aut}_{\mathcal{C}}(U)$ of p'-order, we have that $[U, \phi]$ centralizes S.

As $\mathcal{C}$ is perfect, $\mathfrak{hyp}(\mathcal{C}) = T$; therefore T is generated by subgroups of the form $[U, \phi]$ for $U \leq T$ and $\phi \in O^p(\mathrm{Aut}_{\mathcal{C}}(T))$. Each of these centralizes S, and so T centralizes S.

The above two paragraphs prove that $\mathrm{Aut}_T(U)$ and $O^p(\mathrm{Aut}_{\mathcal{C}}(U))$ lie in $C_{\mathcal{F}}(S)$ for all $U \leq T$. Hence, for any $U \leq T$,

$$\mathrm{Aut}_{\mathcal{C}}(U) = O^p(\mathrm{Aut}_{\mathcal{C}}(U)) \mathrm{Aut}_T(U) \leq C_{\mathcal{F}}(S).$$

By Alperin's fusion theorem, $\mathcal{C} \leq C_{\mathcal{F}}(S)$, as claimed. $\square$

Lemma 8.73 *Let $\mathcal{F}$ be a saturated fusion system on a finite p-group P. If Q is a fully normalized subgroup of P and $E(\mathcal{F}) \leq N_{\mathcal{F}}(Q)$, then $E(\mathcal{F}) = E(N_{\mathcal{F}}(Q))$.*

Proof By Exercise 7.14 and Theorem 8.43, $O^{p'}(E(\mathcal{F})) = E(\mathcal{F})$, and so $E(\mathcal{F})$ is normal in $N_{\mathcal{F}}(Q)$ by Corollary 8.19. Hence $E(\mathcal{F}) \leq E(N_{\mathcal{F}}(Q))$.

Conversely, suppose that $\mathcal{C}$ is a component of $N_{\mathcal{F}}(Q)$. By Lemma 8.72, we see that $\mathcal{C} \leq C_{\mathcal{F}}(O_p(\mathcal{F}))$. If $\mathcal{C}$ is not contained in $E(\mathcal{F}) \leq N_{\mathcal{F}}(Q)$ then by Theorem 8.58 the subgroup U on which $\mathcal{C}$ is defined centralizes $E(\mathcal{F})$, in the sense that $E(\mathcal{F}) \leq C_{N_{\mathcal{F}}(Q)}(U)$. Therefore $E(\mathcal{F})\mathcal{O}_p(\mathcal{F}) = F^*(\mathcal{F}) \leq C_{\mathcal{F}}(U)$, so that $U \leq Z(F^*(\mathcal{F}))$, the last step by Proposition 8.66. However, this means that U is abelian, and so U cannot support a quasisimple fusion system such as $\mathcal{C}$, a contradiction. This implies that $\mathcal{C} \leq E(\mathcal{F})$, and so $E(\mathcal{F}) = E(N_{\mathcal{F}}(Q))$, as needed. $\square$

Proposition 8.74 *Let $\mathcal{F}$ be a saturated fusion system on a finite p-group P, and let Q be a fully $\mathcal{F}$-normalized subgroup of P. Suppose that $\mathcal{F}$ is a minimal counterexample to Theorem 8.70. Let $\mathcal{C}$ be a component of $N_{\mathcal{F}}(Q)$ such that $\mathcal{C} \not\leq E(\mathcal{F})$. Suppose that $R \trianglelefteq N_P(Q)$ is a subgroup such that R centralizes $\mathcal{C}$ (i.e., $R \leq C_P(\mathcal{C})$) but $N_{\mathcal{F}}(R) < \mathcal{F}$, and that one of the following two conditions holds:*

(i) $R \le O_p(N_{\mathcal{F}}(Q))$;

(ii) *there exists a normal subsystem $\mathcal{D}$ of $N_{\mathcal{F}}(Q)$ defined on R.*

For any morphism $\phi \in \mathrm{Hom}_{\mathcal{F}}(N_P(R), P)$ with $R\phi$ fully $\mathcal{F}$-normalized, we have that $\mathcal{C}^{\phi} \le E(N_{\mathcal{F}}(R\phi))$. Thus there is a component $\mathcal{C}'$ of $N_{\mathcal{F}}(R\phi)$ such that $\mathcal{C}' \not\le E(\mathcal{F})$.

Proof Write $\mathcal{N} = N_{\mathcal{F}}(Q)$. We first claim that $\mathcal{C}$ is a component of $N_{\mathcal{F}}(Q, R)$. If $R \le O_p(N_{\mathcal{F}}(Q))$ then R centralizes $E(N_{\mathcal{F}}(Q))$, so that $E(N_{\mathcal{F}}(Q)) = E(N_{\mathcal{F}}(Q, R))$ by Lemma 8.73. Hence $\mathcal{C}$ is a component of $N_{\mathcal{F}}(Q, R)$.

In the other case, suppose that there is a normal subsystem $\mathcal{D}$ of $N_{\mathcal{F}}(Q)$ defined on R. Since the subgroup C on which $\mathcal{C}$ is defined cannot be abelian, and $\mathcal{C}$ centralizes R, we must have that $C \not\le R$, and hence $\mathcal{C} \not\le \mathcal{D}$.

Let $\mathcal{M}$ denote the product of all components of $\mathcal{N}$ not contained in $\mathcal{D}$; by Proposition 8.69, $\mathcal{M} \trianglelefteq \mathcal{N}$ and $\mathcal{M} \le C_{\mathcal{N}}(R)$, so that $\mathcal{M}$ is weakly normal in $C_{\mathcal{N}}(R)$. Since $\mathcal{M} = O^{p'}(\mathcal{M})$, $\mathcal{M}$ is normal in $C_{\mathcal{N}}(R)$ by Theorem 8.15, and so any component of $\mathcal{M}$, such as $\mathcal{C}$, is a component of $C_{\mathcal{N}}(R)$, and so of $N_{\mathcal{F}}(Q, R)$ by Lemma 8.71.

Let ϕ be any morphism in $\mathrm{Hom}_{\mathcal{F}}(N_P(R), P)$ such that $R\phi$ is fully $\mathcal{F}$-normalized. Since $R \le N_P(Q)$, we must have that $N_P(Q, R) = N_P(Q)$. It is an easy consequence of the fact that $\mathcal{E}(\mathcal{F})$ is a (weakly) normal subsystem of $\mathcal{F}$ that $\mathcal{C} \le E(\mathcal{F})$ if and only if $\mathcal{C}^{\phi} \le E(\mathcal{F})$, and so $\mathcal{C}^{\phi} \not\le E(\mathcal{F})$.

We apply Exercise 4.3 twice, to $\mathcal{F}$ and $N_{\mathcal{F}}(Q)$, to get that $\mathcal{C}^{\phi}$ is a component of $N_{\mathcal{F}}(Q\phi)$ and $N_{\mathcal{F}}(Q\phi, R\phi)$. Since R is not normal in $\mathcal{F}$, neither is $R\phi$, and hence $N_{\mathcal{F}}(R\phi) < \mathcal{F}$. By choice of minimal counterexample, Theorem 8.70 holds for $N_{\mathcal{F}}(R\phi)$, and so $\mathcal{C}^{\phi}$, a component of $N_{\mathcal{F}}(Q\phi, R\phi) = N_{N_{\mathcal{F}}(R\phi)}(Q\phi)$, must lie in $E(N_{\mathcal{F}}(R\phi))$. Since $\mathcal{C}^{\phi} \not\le E(\mathcal{F})$, we must have that $E(N_{\mathcal{F}}(R\phi)) \not\le E(\mathcal{F})$; hence some component of $N_{\mathcal{F}}(R\phi)$ does not lie in $E(\mathcal{F})$, as claimed. $\square$

We will now prove Theorem 8.70 via a series of reductions. Let $\mathcal{F}$ be a saturated fusion system on a finite p-group P, and let Q be a fully $\mathcal{F}$-normalized subgroup of P. Suppose that $\mathcal{C}$ is a component of $N_{\mathcal{F}}(Q)$ that is not contained in $E(\mathcal{F})$. We may choose $\mathcal{F}$ to be a minimal counterexample to the statement of the theorem, so in particular we may use Proposition 8.74. Notice that $\mathcal{C}$ is a component of $C_{\mathcal{F}}(Q)$ by Lemma 8.71, so in particular $\mathcal{C}$ always centralizes Q and hence any subgroup of

Q; from now on we will not justify our statements that $\mathcal{C}$ centralizes a subgroup of Q.

Step 1: *We may assume that* $Z(\mathcal{F}) = 1$. Write $Z = Z(\mathcal{F})$, write $\mathcal{E} = N_{\mathcal{F}}(Q)$, and let $\bar{\mathcal{C}}$ and $\bar{\mathcal{E}}$ denote the images of $\mathcal{C}$ and $\mathcal{E}$ under the morphism $\Phi : \mathcal{F} \to \mathcal{F}/Z$. Firstly, assume that $Z \leq Q$. We notice that $\bar{\mathcal{C}}$ is a component of $\bar{\mathcal{E}}$, since it is quasisimple and subnormal. Exercise 5.11 states that $N_{\mathcal{F}}(Q)/Z = N_{\mathcal{F}/Z}(Q/Z)$, and so by minimal choice of $\mathcal{F}$, $\bar{\mathcal{C}} \leq E(\mathcal{F}/Z)$.

Conversely, if $\bar{\mathcal{D}}$ is a component of $\mathcal{F}/Z$, then by Theorem 8.43 (applied to subnormal subsystems in the obvious way) there exists a subnormal quasisimple subsystem $\mathcal{D}$ of $\mathcal{F}$ with image $\bar{\mathcal{D}}$ in $\mathcal{F}/Z$. Hence the image of the central product $Z(\mathcal{F})E(\mathcal{F})$ in $\mathcal{F}/Z$ is $E(\mathcal{F}/Z)$. This contradicts the statement that $\mathcal{C}$ does not lie in $E(\mathcal{F})$.

Finally, assume that Z is not contained in Q. By Lemma 8.73, Q cannot normalize $E(\mathcal{F})$, so that $Q \not\leq O_p(\mathcal{F})$. In particular, $R = QZ$ is not contained in $O_p(\mathcal{F})$. We claim that $\mathcal{C} \leq C_{\mathcal{F}}(R)$. We know that $\mathcal{C}$ centralizes Q, and since every morphism in $\mathcal{F}$ extends to one acting trivially on Z, it is easy to see that $\mathcal{C} \leq C_{\mathcal{F}}(QZ)$. We then apply Proposition 8.74 (noticing that $R \leq O_p(N_{\mathcal{F}}(Q))$ and using the first condition), which says that we may choose $\mathcal{C}$ to be a component of an $\mathcal{F}$-conjugate of R, which clearly contains Z.

Step 2: *We may assume that* $O_p(\mathcal{F}) = 1$. Let $S = O_p(\mathcal{F})$ and suppose that $S > 1$. Choose $R \leq S$ of order p and contained in $Z(P)$. Certainly R is fully $\mathcal{F}$-normalized and $Q \leq N_P(R) = P$. Also, $\mathcal{C} \leq C_{\mathcal{F}}(R)$ by Lemma 8.72, so by Proposition 8.74 (using the first condition), either $\mathcal{F} = N_{\mathcal{F}}(R)$ or there exists some $\mathcal{C}' \in \mathrm{Comp}(N_{\mathcal{F}}(R))$ such that $\mathcal{C}' \not\leq E(\mathcal{F})$. This second possibility cannot occur; by Lemma 8.72, $\mathcal{C}' \leq C_{\mathcal{F}}(S) \trianglelefteq \mathcal{F}$, and so $\mathcal{C}'$ is a component of $C_{\mathcal{F}}(S)$. However, $E(C_{\mathcal{F}}(S)) = E(N_{\mathcal{F}}(S)) = E(\mathcal{F})$, a contradiction.

Hence we have that R is a normal subgroup of $\mathcal{F}$. Since $Z(\mathcal{F})$ is trivial, we know that $\mathcal{E} = C_{\mathcal{F}}(R)$ is a proper subsystem of $\mathcal{F}$. Also, $\mathcal{E} \trianglelefteq \mathcal{F}$ by Proposition 8.8, so $N_{\mathcal{E}}(Q) \trianglelefteq N_{\mathcal{F}}(Q)$ by Exercise 8.15. Since $N_{\mathcal{E}}(Q)$ is a normal subsystem of $N_{\mathcal{F}}(Q)$, both on $N_P(Q)$, by Corollary 8.60 we have that $\mathcal{C} \in \mathrm{Comp}(N_{\mathcal{E}}(Q))$. Since $\mathcal{E}$ is a proper subsystem of $\mathcal{F}$, by choice of minimal counterexample $E(N_{\mathcal{E}}(Q)) \leq E(\mathcal{E})$, and $E(\mathcal{E}) = E(\mathcal{F})$ by Corollary 8.60 again (or Lemma 8.71).

From now on, write $\mathcal{E} = E(\mathcal{F})$, and write E for the subgroup on which $\mathcal{E}$ is defined.

Step 3: *We may assume that* $Q = E$. Let $U = Q \cap E$, and suppose that $U > 1$. Since $\mathcal{C}$ centralizes U, and $\mathcal{F}$ has no normal subgroups, we apply Proposition 8.74 (using the first condition as $U \leq Q \leq O_p(N_{\mathcal{F}}(Q))$) to find a component $\mathcal{C}'$ of $N_{\mathcal{F}}(U\phi)$ (for some ϕ in $\mathcal{F}$); notice that $U\phi \leq E$, and so by choosing Q correctly, we have that either $Q \cap E = 1$ or $Q \leq E$.

Since $\mathcal{E} \lhd \mathcal{F}$, if $Q \leq E$ then by Exercise 8.15 we see that $N_{\mathcal{E}}(Q) \lhd N_{\mathcal{F}}(Q)$, and if $Q \cap E = 1$ then, as we saw in Exercise 8.16, again $N_{\mathcal{E}}(Q) \lhd N_{\mathcal{F}}(Q)$. Since $\mathcal{C}$ is a component of $N_{\mathcal{F}}(Q)$ that does not lie in $\mathcal{E}$, we see that $\mathcal{C} \leq C_{\mathcal{F}}(N_E(Q))$ by Proposition 8.69. We may now apply Proposition 8.74 (with $\mathcal{D} = N_{\mathcal{E}}(Q)$ in the second condition) again and find $\psi \in \mathrm{Hom}_{\mathcal{F}}(N_E(Q), P)$ such that $N_E(Q)\psi$ is fully $\mathcal{F}$-normalized and $\mathcal{C}'$ is a component of $N_{\mathcal{F}}(N_E(Q)\psi)$ such that $\mathcal{C}' \not\leq \mathcal{E}$.

Notice that $1 \neq N_E(Q) \leq E$, so therefore even in the case where $Q \cap E = 1$ we may choose Q to be contained in E. Repeating this procedure, we may continue replacing Q by $N_E(Q)$, and so we may assume that $Q = E$.

Write T for the subgroup on which $\mathcal{C}$ is defined.

Step 4: $T \leq C_{\mathcal{F}}(\mathcal{E}) = 1$, *the final contradiction*. Let $Z = Z(E)$, so that Z is fully $\mathcal{F}$-normalized and if $\alpha \in \mathrm{Aut}_{\mathcal{F}}(E)$ then $Z\alpha = Z$. Since $Z \leq E = Q$, $\mathcal{C}$ centralizes Z and $E \leq N_P(Z) = P$, we use Proposition 8.74 (using the first condition as $Z \leq Q \leq O_p(N_{\mathcal{F}}(Q))$); so we get a morphism $\phi \in \mathrm{Hom}_{\mathcal{F}}(N_P(Z), P)$ and a component of $N_{\mathcal{F}}(Z\phi)$ not contained in $\mathcal{E}$. Since $Z\phi = Z$, this means that we may choose $\mathcal{C}$ such that $\mathcal{C}$ is a component of $N_{\mathcal{F}}(Z)$ (and $\mathcal{C} \not\leq \mathcal{E}$), and hence a component of $C_{\mathcal{F}}(Z)$. Write $\mathcal{F}_Z$ for $C_{\mathcal{F}}(Z)$ and $\mathcal{E}_Z$ for $C_{\mathcal{E}}(Z)$, and notice that $\mathcal{E}_Z \lhd \mathcal{F}_Z$ by Exercise 8.15. Let $\mathcal{D}$ denote the central product of all components in $\mathrm{Comp}(\mathcal{F}_Z) \setminus \mathrm{Comp}(\mathcal{E}_Z)$ (which includes $\mathcal{C}$), a normal subsystem of $\mathcal{F}_Z$. If D denotes the subgroup on which $\mathcal{D}$ is defined, notice that D is strongly $N_{\mathcal{F}}(Z)$-closed (since any component of $C_{\mathcal{F}}(Z)$ is a component of $N_{\mathcal{F}}(Z)$), and that $\mathcal{D} \leq C_{\mathcal{F}}(E)$ by Proposition 8.69.

Let X be a fully $\mathcal{F}$-normalized, $\mathcal{E}$-centric subgroup of E. Certainly $C_{\mathcal{F}}(X) = C_{\mathcal{F}_Z}(X)$, and $\mathcal{D} \leq C_{\mathcal{F}}(X)$ as $\mathcal{D} \leq C_{\mathcal{F}}(E)$. Since $\mathcal{D}$ is normal in $C_{\mathcal{F}}(Z)$ it is weakly normal in $C_{\mathcal{F}}(X)$, and since $\mathcal{D} = O^{p'}(\mathcal{D})$ we see that $\mathcal{D} \lhd C_{\mathcal{F}}(X)$ by Theorem 8.15. Hence all components of $\mathcal{D}$, including $\mathcal{C}$, are components of $N_{\mathcal{F}}(X)$.

As $N_{\mathcal{E}}(X) \lhd N_{\mathcal{F}}(X)$, each component of $N_{\mathcal{F}}(X)$ either centralizes $N_{\mathcal{E}}(X)$ or lies in $N_{\mathcal{E}}(X)$; however, since X is $\mathcal{E}$-centric, $N_{\mathcal{E}}(X)$ is constrained, so has no components. Therefore $N_{\mathcal{E}}(X)$ centralizes $\mathcal{D}$, so in particular $N_{\mathcal{E}}(X) \leq C_{\mathcal{F}}(D)$.

If $\alpha \in \mathrm{Aut}_{\mathcal{F}}(E\,\mathrm{C}_P(E))$ then $D\phi = D$ and $\mathcal{E}^\phi = \mathcal{E}$, and so $\mathrm{N}_{\mathcal{E}}(X\alpha) = \mathrm{N}_{\mathcal{E}}(X)^\alpha \leq \mathrm{C}_{\mathcal{F}}(D)$. clearly every $\mathcal{F}$-automorphism of E extends to an $\mathcal{F}$-automorphism of $E\,\mathrm{C}_P(E)$, and hence $\mathrm{N}_{\mathcal{E}}(Y) \leq \mathrm{C}_{\mathcal{F}}(D)$ for Y any $\mathrm{Aut}_{\mathcal{F}}(E)$-conjugate of a fully $\mathcal{F}$-normalized, $\mathcal{E}$-centric subgroup of E. By Exercise 6.9, we have that $\mathcal{E} \leq \mathrm{C}_{\mathcal{F}}(D)$, so that $D \leq \mathrm{C}_{\mathcal{F}}(\mathcal{E})$; in particular, $T \leq \mathrm{C}_{\mathcal{F}}(\mathcal{E}) = 1$, a contradiction.

Exercises

8.1 Let $\mathcal{F}$ be a saturated fusion system on a finite p-group P, and let $\mathcal{E}$ be a weakly normal subsystem of $\mathcal{F}$, on the subgroup T of P. Prove that, if Q is an $\mathcal{E}$-centric subgroup of T, then $Q\,\mathrm{C}_P(Q) \cap T = Q$ and $Q\,\mathrm{C}_P(Q) \cap T\,\mathrm{C}_P(T) = Q\,\mathrm{C}_P(T)$.

8.2 Let $\mathcal{F}$ be a saturated fusion system on a finite p-group P, and let $\mathcal{E}$ be a weakly normal subsystem of $\mathcal{F}$, on the subgroup T of P. Let Q and R be $\mathcal{E}$-centric subgroups of T, and let $\phi : Q \to R$ be an isomorphism in $\mathcal{F}$. Let ψ be an extension of ϕ to $Q\,\mathrm{C}_P(T)$.

(i) Prove that $\mathrm{im}\,\psi = R\,\mathrm{C}_P(T)$.

(ii) Show that $Q\psi = R$, $\mathrm{Z}(Q)\psi = \mathrm{Z}(R)$, and $(\mathrm{Z}(Q)\,\mathrm{C}_P(T))\psi = \mathrm{Z}(R)\,\mathrm{C}_P(T)$.

(iii) if $g \in \mathrm{N}_T(Q)$, prove that c_g acts trivially on $\mathrm{C}_P(T)$, and so on $\mathrm{Z}(Q)\,\mathrm{C}_P(T)/\mathrm{Z}(Q)$.

8.3 Let $\mathcal{F}$ be a saturated fusion system on a finite p-group P, and let T be a strongly $\mathcal{F}$-closed subgroup of P. Let Q and R be subgroups of T, with $\mathrm{C}_T(Q) = \mathrm{Z}(Q)$ and $\mathrm{C}_T(R) = \mathrm{Z}(R)$. Let $\phi : Q\,\mathrm{C}_P(T) \to R\,\mathrm{C}_P(T)$ be a morphism in $\mathcal{F}$. Prove that the following are equivalent:

(i) $[\phi, \mathrm{C}_P(T)] \leq T$;

(ii) $[\phi, \mathrm{C}_P(T)] \leq \mathrm{Z}(R)$;

(iii) for $x \in \mathrm{C}_P(T)$, we have $(\mathrm{Z}(Q)x)\phi = \mathrm{Z}(R)x$.

8.4 Prove Proposition 8.7.

8.5 Prove that a minimal normal subsystem of a fusion system $\mathcal{F}$ is either a direct product of isomorphic simple subsystems or isomorphic with $\mathcal{F}_Q(Q)$ for some elementary abelian p-group Q.

8.6 Let $\mathcal{F}$ be a saturated fusion system on a finite p-group P. In this exercise we will prove the independence of the intersection of weakly subnormal subsystems of $\mathcal{F}$ on the choice of the weakly normal series connecting the $\mathcal{E}_i$ and $\mathcal{F}$.

   (i) Suppose that $\mathcal{E}_1$ is a weakly normal subsystem of $\mathcal{F}$ and that $\mathcal{E}_2$ is weakly subnormal in $\mathcal{F}$. Prove that the intersection $\mathcal{E}_1 \curlywedge \mathcal{E}_2$ is independent of the choice of chain connecting $\mathcal{E}_2$ and $\mathcal{F}$.

   (ii) Prove that if $\mathcal{E}_1$ is weakly subnormal in $\mathcal{F}$, then $\mathcal{E}_1 \curlywedge \mathcal{E}_2$ is independent of the choice of chains connecting the $\mathcal{E}_i$ and $\mathcal{F}$.

8.7   Let $\mathcal{F}$ be a saturated fusion system on a finite p-group P, and let $\phi : A \to B$ be an isomorphism in $\mathcal{F}$ between $\mathcal{F}$-centric subgroups A and B of P. Prove that ϕ extends to a morphism $\psi : N_\phi \to P$ in $\mathcal{F}$.

8.8   Let G be a finite group and let P be a Sylow p-subgroup of G. Suppose that $P = Q \times Z$, where $Z \leq Z(G)$ and Q is strongly closed in P with respect to G. Prove that $G = H \times Z$, where $Q \leq H$.

8.9   Let $\mathcal{F}$ be a saturated fusion system on a finite p-group P, and let T be a strongly $\mathcal{F}$-closed subgroup of P. Let U be a fully $\mathcal{F}$-normalized subgroup of T. Let $\mathcal{E}$ be a subsystem of $\mathcal{F}$ on T.

   (i) Prove that $\mathcal{D}(U)$ is a fusion system on $N_P(U)$ and $\mathcal{E}(U)$ is a subsystem on $N_T(U)$.

   (ii) Prove that $\mathcal{D}(U)$ is saturated and constrained, and that if $\mathcal{E}$ is saturated then $\mathcal{E}(U)$ is saturated and constrained.

   (iii) Prove that $\mathcal{E}(U)$ is contained in $\mathcal{D}(U)$.

   (iv) Assume that $\mathcal{E}$ is weakly normal in $\mathcal{F}$. Prove that $\mathcal{E}(U)$ is $\mathcal{D}(U)$-invariant, and so $\mathcal{E}(U) \prec \mathcal{D}(U)$.

8.10 Prove Proposition 8.39.

8.11 Let G_1 and G_2 be two finite groups, with Sylow p-subgroups P_1 and P_2 respectively. Let Z be a central p-subgroup of $G_1 \times G_2$ such that $Z \cap G_i = 1$ for $i = 1, 2$, and write $\mathcal{F}_i = \mathcal{F}_{P_i}(G_i)$. Prove that $\mathcal{F}_1 \times_Z \mathcal{F}_2$ is isomorphic to the fusion system of the central product $G_1 * G_2$ via Z.

8.12 Let $\mathcal{F}$ be a quasisimple fusion system on a finite p-group P. Prove that $\mathrm{Comp}(\mathcal{F}) = \{\mathcal{F}\}$.

8.13 Let $\mathcal{F}$ be a saturated fusion system on a finite p-group P, and let $\mathcal{E}$ be a normal subsystem of $\mathcal{F}$ on the subgroup T of P. Let X be a subset of $\mathrm{Comp}(\mathcal{E})$ such that $\mathrm{Aut}_\mathcal{F}(T)$ permutes the elements of X under the map $\mathcal{C} \mapsto \mathcal{C}\phi$ for $\mathcal{C} \in \mathrm{Comp}(\mathcal{E})$ and $\phi \in \mathrm{Aut}_\mathcal{F}(T)$. Assume that the subsystem $E(\mathcal{E})$, the central product of the components of $\mathcal{E}$, is a normal subsystem of $\mathcal{E}$, and write $\mathcal{E}'$ for the central product of the elements of X, a normal subsystem of $E(\mathcal{E})$. Prove that all elements of $\mathrm{Aut}_{\mathcal{F}}(T)$ induce automorphisms of $\mathcal{E}'$, and hence $\mathcal{E}' \trianglelefteq \mathcal{F}$.

8.14 Let $\mathcal{F}_1, \ldots, \mathcal{F}_n$ be quasisimple fusion systems, and let $\mathcal{F}$ be a central product of the $\mathcal{F}_i$, on the p-group P. If $\mathcal{E}$ is a weakly subnormal subsystem of $\mathcal{F}$ on a subgroup Z of $\mathrm{Z}(P)$, prove that $Z \leq \mathrm{Z}(\mathcal{F})$.

8.15 Let $\mathcal{F}$ be a saturated fusion system on a finite p-group P, and let $\mathcal{E}$ be a normal subsystem of $\mathcal{F}$, on the subgroup T of P. In Exercise 6.7 we proved that if Q is a fully $\mathcal{F}$-normalized subgroup of T then $\mathrm{N}_{\mathcal{E}}(Q) \prec \mathrm{N}_{\mathcal{F}}(Q)$. Prove that $\mathrm{N}_{\mathcal{E}}(Q) \trianglelefteq \mathrm{N}_{\mathcal{F}}(Q)$.

8.16 Let $\mathcal{F}$ be a saturated fusion system on a finite p-group P, and let $\mathcal{E}$ be a weakly normal subsystem of $\mathcal{F}$ on the subgroup T of P. In [Asc11, Theorem 5], Aschbacher proves that if $\mathcal{E}$ is normal in $\mathcal{F}$ then there is a saturated subsystem $\mathcal{F}'$ of $\mathcal{F}$ on P such that $\mathcal{E} \leq \mathcal{F}'$ and $\mathcal{F}'/T = \mathcal{F}_{P/T}(P/T)$. Prove the converse, that is, that if there is a saturated subsystem $\mathcal{F}'$ of $\mathcal{F}$ on P such that $\mathcal{F}'/T = \mathcal{F}_{P/T}(P/T)$, then $\mathcal{E}$ is a normal subsystem of $\mathcal{F}$.

(Remark: Suppose that $\mathcal{E}$ is actually normal in $\mathcal{F}$. If Q is a subgroup of P then there is a saturated subsystem $Q\mathcal{E}$ of $\mathcal{F}$ on QT such that $\mathcal{E} \leq Q\mathcal{E}$ and $Q\mathcal{E}/T$ is the trivial fusion system on QT/T, by [Asc11, Theorem 5]. In Exercise 7.16, we saw that if $Q \cap T = 1$, then $\mathrm{N}_{Q\mathcal{E}}(Q) = \mathrm{N}_{\mathcal{E}}(Q) \times \mathcal{F}_Q(Q)$, where we define $\mathrm{N}_{\mathcal{E}}(Q)$ to be the direct factor in this decomposition given in Exercise 7.16. In [Asc11, (8.24)], Aschbacher proves that $\mathrm{N}_{\mathcal{E}}(Q)$ is in fact a normal subsystem of $\mathrm{N}_{\mathcal{F}}(T)$.)

9

Exotic fusion systems

If all saturated fusion systems were group fusion systems, then perhaps the field of fusion systems would merely be a useful way of describing fusion in finite groups, and maybe extending such theorems to blocks of finite groups. However, exotic fusion systems offer the possibility of new structures that are yet to be discovered.

One of the more intriguing facets in the theory at the moment is that, while there are many families of exotic fusion systems for odd primes, there is only one known family of simple exotic fusion systems for the prime 2. By Proposition 5.12 and the fact that soluble fusion systems are constrained, an exotic fusion system must contain a simple subquotient, and the exotic fusion systems that we know are 'close to' simple.

Related to this is the question of whether a fusion system of a block is a fusion system of a group. Conjecture 2.37 is the conjecture commonly believed to hold, that is, that every fusion system of a block of a finite group is the fusion system of some (possibly other) finite group. The obvious way to tackle this conjecture is a two-step process, as with many conjectures in group theory: firstly we reduce to simple groups (or groups close to simple, such as quasisimple); and secondly we prove the conjecture using the classification of the finite simple groups. Neither of these steps has been completed in this case, although we have some results in this direction. In this chapter we will examine one theorem in the literature in this direction, Theorem 9.26, which shows that in certain cases such a reduction is possible.

The other problem is the existence and uniqueness of centric linking systems. For group fusion systems both of these problems are solved by Oliver in the course of the proof of the Martino–Priddy conjecture, but in general they remain open.

This chapter begins with two sections on exotic fusion systems, giving examples on the extraspecial group of order 7^3 and exponent 7, and also the Solomon fusion systems. The next two sections deal with fusion systems of blocks, proving in Section 9.3 that the fusion systems blocks of p-soluble, symmetric, and sporadic groups are fusion systems of groups, and discussing in Section 9.4 a criterion for proving that an exotic fusion system is not the fusion system of a block of a finite group. The final two sections describe centric linking systems for groups, and relate the existence and uniqueness of centric linking systems for saturated fusion systems to other conjectures in p-groups and their representation theory.

9.1 Extraspecial p-groups

In [RV04], Ruiz and Viruel classified all saturated fusion systems on the extraspecial p-group of order p^3 and exponent p, commonly denoted by p_+^{1+2}. In the process they discovered three exotic saturated fusion systems on the group 7_+^{1+2}. In this section we will perform the majority of their classification, and describe the rest.

We begin by analysing the groups p_+^{1+2}. Let P be the group generated by x and y, with both x and y of order p, and $[x, y] = z$ of order p and central in P. This is the group p_+^{1+2}, and Exercise 9.1 collects some information about it.

Our first result concerns how much information one needs about a saturated fusion system $\mathcal{F}$ on P in order to determine it uniquely. From now on, let $D_{\mathcal{F}}$ denote the set of all $\mathcal{F}$-radical subgroups of P of order p^2; notice that this is exactly the set of $\mathcal{F}$-essential subgroups of P. By Exercise 9.1(i) there are at most $p + 1$ elements of $D_{\mathcal{F}}$. It will turn out that $Y = \mathrm{Out}_{\mathcal{F}}(P)$ and $D_{\mathcal{F}}$ will determine the saturated fusion system $\mathcal{F}$ uniquely.

Proposition 9.1 *Let $\mathcal{F}$ and $\mathcal{E}$ be saturated fusion systems on the group $P \cong p_+^{1+2}$. If $\mathrm{Out}_{\mathcal{F}}(P) = \mathrm{Out}_{\mathcal{E}}(P)$ and $D_{\mathcal{F}} = D_{\mathcal{E}}$ then $\mathcal{F} = \mathcal{E}$.*

Proof Since $\mathrm{Out}_{\mathcal{F}}(P) = \mathrm{Out}_{\mathcal{E}}(P)$, we must have that $\mathrm{Aut}_{\mathcal{F}}(P) = \mathrm{Aut}_{\mathcal{E}}(P)$. By Alperin's fusion theorem, $\mathcal{F}$ is generated by $\mathrm{Aut}_{\mathcal{F}}(P)$ and $\mathrm{Aut}_{\mathcal{F}}(Q)$ for $Q \in D_{\mathcal{F}}$. We will show that if Q is in $D_{\mathcal{F}}$ then $\mathrm{Aut}_{\mathcal{F}}(Q)$ is determined by $\mathrm{Aut}_{\mathcal{F}}(P)$, proving that $\mathcal{F} = \mathcal{E}$.

If $Q = P$ then, as we have said, $\mathrm{Aut}_{\mathcal{F}}(Q) = \mathrm{Aut}_{\mathcal{E}}(Q)$. Suppose that Q is an $\mathcal{F}$-essential subgroup of P; hence $Q \cong C_p \times C_p$ by Exercise 9.1(i). Since $\mathrm{Aut}_{\mathcal{F}}(Q)$ is a subgroup of $\mathrm{GL}_2(p)$, by Exercise 7.3 we have that

$\mathrm{SL}_2(p) \leq \mathrm{Aut}_{\mathcal{F}}(Q) \leq \mathrm{GL}_2(p)$ (as Q is $\mathcal{F}$-radical). Let $G = \mathrm{GL}_2(p)$ and $H = \mathrm{SL}_2(p)$; write S for a Sylow p-subgroup of G (and hence of H). Notice that $\mathrm{N}_G(S)$ has order $p(p-1)^2$ and $\mathrm{N}_H(S)$ has order $p(p-1)$. As G/H is cyclic, there is exactly one subgroup of G of a given order containing H; therefore, if $H \leq K \leq G$, K is determined by $\mathrm{N}_K(S)$.

Now consider $\mathrm{Aut}_{\mathcal{F}}(Q)$, which lies between H and G. It is determined by $\mathrm{N}_{\mathrm{Aut}_{\mathcal{F}}(Q)}(\mathrm{Aut}_P(Q))$ which, since $\mathcal{F}$ is saturated, is the image of the natural map from $\mathrm{Aut}_{\mathcal{F}}(Q \leq P)$ to $\mathrm{Aut}(Q)$. Hence $\mathrm{Aut}_{\mathcal{F}}(Q)$ is determined by $\mathrm{Aut}_{\mathcal{F}}(P)$, as claimed. $\qquad\square$

Given this result, it makes sense to proceed by the number of subgroups in $D_{\mathcal{F}}$. The case where $D_{\mathcal{F}} = \emptyset$ is clear, since this means that $\mathcal{F}$ possesses no $\mathcal{F}$-essential subgroups. In this case, $\mathcal{F}$ is the fusion system of $P \rtimes A$ by Lemma 7.3 and Exercise 7.2, where A is a p'-subgroup of $\mathrm{Out}(P)$. In Proposition 9.4 we will (essentially) classify such subgroups A, and this will complete the determination of such saturated fusion systems.

The case where $|D_{\mathcal{F}}| = 1$ can also be settled easily using Exercise 9.3.

Lemma 9.2   *Let $\mathcal{F}$ be a saturated fusion system on a finite p-group $P = p_+^{1+2}$. If $D_{\mathcal{F}} = \{Q\}$ has exactly one element, then $\mathcal{F}$ is the fusion system of $Q \rtimes H$, where H is a subgroup of $\mathrm{GL}_2(p)$ containing $\mathrm{SL}_2(p)$.*

Proof   By Exercise 9.3, $\mathcal{F}$ is the fusion system of a finite group G such that Q is normal in G and G/Q is isomorphic to $\mathrm{Aut}_{\mathcal{F}}(Q)$. Notice that the group $Q \rtimes H$, where $H = \mathrm{Aut}_{\mathcal{F}}(Q)$ (with the obvious action of H on Q), is a finite group with such a property, and hence $\mathcal{F}$ is the fusion system of $Q \rtimes H$. $\qquad\square$

For the cases where there are at least two subgroups in $D_{\mathcal{F}}$, we have to do more work. Firstly, we will prove that, unless p is one of 3, 5, 7, and 13, there are *exactly* two subgroups in $D_{\mathcal{F}}$. We then have to consider each of these cases one at a time.

The main reason for these restrictions is the surjectivity property: if Q is an element of $D_{\mathcal{F}}$, then $\mathrm{N}_{\mathrm{Aut}_{\mathcal{F}}(Q)}(\mathrm{Aut}_P(Q))$ has order $p(p-1)r$, where $|\mathrm{Aut}_{\mathcal{F}}(Q) : \mathrm{SL}_2(p)| = r$. Since Q has the surjectivity property, there must be elements of $\mathrm{Aut}_{\mathcal{F}}(P)$ that restrict to these automorphisms of Q. We need to understand how elements of $\mathrm{Aut}_{\mathcal{F}}(P)$ act on the set X of subgroups of order p^2. By Exercise 9.1, the action of $\mathrm{Out}_{\mathcal{F}}(P)$ on X is equivalent to the corresponding subgroup of $\mathrm{GL}_2(p)$ on the set of 1-dimensional subspaces. From now on we will identify $\mathrm{Out}(P)$ and

$GL_2(p)$ via Exercise 9.1(ii), and write elements of $\mathrm{Out}_{\mathcal{F}}(P)$ as matrices acting on a 2-dimensional $\mathbb{F}_p$-vector space V.

Consider the subgroup V_Q of $\mathrm{Out}_{\mathcal{F}}(Q)$ constructed in Exercise 9.2. It fixes two maximal subgroups and permutes the others, either transitively or in three orbits. Therefore, if $\mathcal{F}$ possesses exactly two $\mathcal{F}$-essential subgroups (i.e., $|D_{\mathcal{F}}| = 2$), they must be the two subgroups Q and R fixed by V_Q (and hence by V_R also). Henceforth we assume that $|D_{\mathcal{F}}| = 2$.

As in Exercise 9.2, we can choose a generating set $\{x, y\}$ and a primitive $(p-1)$th root α of 1, such that there are elements in $Y = \mathrm{Out}_{\mathcal{F}}(P)$ of the form

$$v_Q = \begin{pmatrix} \alpha & 0 \\ 0 & \alpha^{-2} \end{pmatrix} \quad \text{and} \quad v_R = \begin{pmatrix} \alpha^{-2} & 0 \\ 0 & \alpha \end{pmatrix}.$$

As any other element of Y must either fix Q and R or swap them, the subgroup Y must be contained in the subgroup H generated by diagonal matrices and the matrix $\left(\begin{smallmatrix} 0 & 1 \\ 1 & 0 \end{smallmatrix}\right)$.

Suppose that Y consists of diagonal matrices: the element $v_Q^{-1} v_R^{-2}$ is the diagonal matrix with entries α^3 and 1, so if 3 is invertible in $\mathbb{Z}_{p-1}$ then all diagonal matrices lie in $\langle v_Q, v_R \rangle = V_Q V_R$. If 3 is not invertible in $\mathbb{Z}_{p-1}$ then we get a subgroup of index 3 in all diagonal matrices, leading to two possibilities for Y. Because $\mathcal{F}$ is determined by Y and the set $D_{\mathcal{F}} = \{\langle x, z \rangle, \langle y, z \rangle\}$, this leads to two possibilities for $\mathcal{F}$ if $3 \mid (p-1)$, both of which occur in finite groups, and a single possibility otherwise, which occurs in finite groups.

If Y contains non-diagonal matrices, then the same analysis occurs. There is a single possibility if $3 \nmid (p-1)$, and two possibilities if $3 \mid (p-1)$. Again, these appear in finite groups. We have determined all possibilities for saturated fusion systems with $|D_{\mathcal{F}}| = 2$. The determination of the finite groups that realize these fusion systems is given in [RV04, Lemma 4.9], and will not be repeated here.

Proposition 9.3 *Let $\mathcal{F}$ be a saturated fusion system on a finite p-group $P = p_+^{1+2}$. If $|D_{\mathcal{F}}| \leq 2$, then the structure of $\mathcal{F}$ is one of the possibilities in the Table 9.1. (Here, the second column lists the sizes of the $\mathcal{F}$-conjugacy classes of subgroups in $D_{\mathcal{F}}$, and Q is an example element of $D_{\mathcal{F}}$. Note that all such saturated fusion systems are realized by finite groups.)*

If the number of elements in $D_{\mathcal{F}}$ is more than 2, as we have said before, we only have four primes to consider. The reason for this is that the number of possibilities for $\mathrm{Out}_{\mathcal{F}}(P)$ when we have lots of the

Table 9.1 *Saturated fusion systems on* p_+^{1+2} *with at most two* $\mathcal{F}$-*radical maximal subgroups*

$\mathrm{Out}_{\mathcal{F}}(P)$	$D_{\mathcal{F}}$	$\mathrm{Aut}_{\mathcal{F}}(Q)$	Group	Notes
$H \leq \mathrm{GL}_2(p)$	0	N/A	$P \rtimes H$	$p \nmid |H|$
$C_{p-1} \times C_r$	1	$\mathrm{SL}_2(p) \rtimes C_r$	$Q \rtimes \mathrm{Aut}_{\mathcal{F}}(Q)$	$r \mid (p-1)$
$C_{p-1} \times C_{p-1}$	1,1	$\mathrm{GL}_2(p)$	$\mathrm{PSL}_3(p)$	$3 \nmid (p-1)$
$(C_{p-1} \times C_{p-1}) \rtimes C_2$	2	$\mathrm{GL}_2(p)$	$\mathrm{PSL}_3(p) \rtimes C_2$	
$C_{p-1} \times C_r$	1,1	$\mathrm{SL}_2(p) \rtimes C_r$	$\mathrm{PSL}_3(p)$	
$(C_{p-1} \times C_r) \rtimes C_2$	2	$\mathrm{SL}_2(p) \rtimes C_r$	$\mathrm{PSL}_3(p) \rtimes C_2$	$3 \mid (p-1)$
$C_{p-1} \times C_{p-1}$	1,1	$\mathrm{GL}_2(p)$	$\mathrm{PSL}_3(p).C_3$	$r = (p-1)/3$
$(C_{p-1} \times C_{p-1}) \rtimes C_2$	2	$\mathrm{GL}_2(p)$	$\mathrm{PSL}_3(p).S_3$	

subgroups V_Q inside it are small. We begin with a description of the p'-subgroups of $\mathrm{PGL}_2(p)$.

Proposition 9.4 (See [RV04, Lemma 4.10]) *Let G be the group* $\mathrm{PGL}_2(p)$. *The subgroups of p'-order in G are contained in one of the following:*

(i) *the dihedral groups D_{p^2+1} and D_{p^2-1};*
(ii) *the alternating and symmetric groups A_4 and S_4;*
(iii) *the alternating group A_5 if $5 \mid (p^2 - 1)$.*

Using this proposition we can prove that the only primes that we need to consider in the case where $|D_{\mathcal{F}}| > 2$ are 3, 5, 7, and 13.

Lemma 9.5 *Let $\mathcal{F}$ be a saturated fusion system on a finite p-group* $P = p_+^{1+2}$. *Suppose that Q, R, and S are distinct elements of $D_{\mathcal{F}}$. Let A denote the image of $Y = \mathrm{Out}_{\mathcal{F}}(P)$ in $\mathrm{PGL}_2(p)$.*

(i) *If $A \leq D_{p^2+1}$ or $A \leq D_{p^2-1}$ then $p = 3$ or $p = 7$.*
(ii) *If $A \leq A_4$, $A \leq S_4$, or $A \leq A_5$ then $p = 3$, $p = 5$, $p = 7$ or $p = 13$.*

Proof The subgroups V_Q, V_R, and V_S, all distinct, are inside $\mathrm{Out}_{\mathcal{F}}(P)$; hence A must contain at least two distinct subgroups of order $p - 1$ if $3 \nmid (p-1)$ and of order $(p-1)/3$ if $3 \mid (p-1)$.

If $A \leq D_{p^2+1}$ or $A \leq D_{p^2-1}$ then, since these have a unique cyclic subgroup of any order other than 2, either $p - 1 = 2$ if $3 \nmid (p-1)$ or $(p-1)/3 = 2$ if $3 \mid (p-1)$, leading to $p = 3$ or $p = 7$.

Suppose that A is contained in A_4, S_4 or A_5. In any of these three groups, the order of a cyclic subgroup is at most 5. If $3 \nmid (p-1)$, then

$p - 1 \leq 5$, yielding $p = 3$ or $p = 5$. If $3 \mid (p - 1)$, then $(p - 1)/3 \leq 5$, yielding $p = 7$ or $p = 13$. This completes the proof. □

It remains to consider each possibility case by case. This is performed in Lemmas 4.13 ($p = 3$), 4.15 ($p = 5$), 4.16 ($p = 7$) and 4.18 ($p = 13$) of [RV04], and is an analysis of which possible p'-subgroups of $GL_2(p)$ exist, using Proposition 9.4. We will perform the easiest of these analyses, $p = 3$, here, and simply list the conclusions of the other cases.

Proposition 9.6 *Let $\mathcal{F}$ be a saturated fusion system on a finite 3-group $P = 3^{1+2}_+$, and suppose that $|D_{\mathcal{F}}| > 2$. There are two possibilities for $\mathcal{F}$, both with $|D_{\mathcal{F}}| = 4$, given in the following table.*

$\mathrm{Out}_{\mathcal{F}}(P)$	$D_{\mathcal{F}}$	$\mathrm{Aut}_{\mathcal{F}}(Q)$	*Group*
D_8	$2, 2$	$GL_2(3)$	$^2F_4(2)'$
SD_{16}	4	$GL_2(3)$	J_4

Proof Before we begin, note that there are four maximal subgroups of P. The group $\mathrm{Out}(P) = GL_2(3)$ has order 48, and so the only $3'$-subgroups of $\mathrm{Out}(P)$ are 2-groups. Since a Sylow 2-subgroup of $\mathrm{Out}(P)$ is semidihedral of order 16, $Y = \mathrm{Out}_{\mathcal{F}}(P)$ is a subgroup of this. Since $\mathrm{Out}_{\mathcal{F}}(P)$ contains at least three different subgroups of order 2 (coming from V_Q for $Q \in D_{\mathcal{F}}$), and the non-trivial elements of these subgroups have determinant -1 (when thought of as elements of $GL_2(3)$), $\mathrm{Out}_{\mathcal{F}}(P)$ must contain more than three involutions. Therefore $\mathrm{Out}_{\mathcal{F}}(P)$ cannot be cyclic, Klein four or quaternion, and so the only possibilities are D_8 and SD_{16}.

If $Y \cong D_8$ then it is easy to see that Y can be chosen (by conjugating inside $GL_2(3)$) to contain all four of the matrices

$$v_Q = \begin{pmatrix} -1 & 0 \\ 0 & 1 \end{pmatrix}, \quad v_R = \begin{pmatrix} 1 & 0 \\ 0 & -1 \end{pmatrix}, \quad v_S = \begin{pmatrix} 0 & 1 \\ 1 & 0 \end{pmatrix}, \quad v_T = \begin{pmatrix} 0 & -1 \\ -1 & 0 \end{pmatrix}.$$

In this case, these are generating matrices for the four subgroups V_Q, V_R, V_S, and V_T, where Q, R, S, and T are the four maximal subgroups of P. Therefore $D_{\mathcal{F}}$ contains all maximal subgroups of P. The subgroups V_Q and V_R make $\{S, T\}$ into an $\mathcal{F}$-conjugacy class, and V_S and V_T make $\{Q, R\}$ into an $\mathcal{F}$-conjugacy class. As Y is generated by V_Q and V_R, these are the $\mathcal{F}$-conjugacy classes of maximal subgroups. Finally, we need to understand $\mathrm{Aut}_{\mathcal{F}}(Q)$; the subgroup $\mathrm{Out}_{\mathcal{F}}(Q \leq P)$ contains V_Q and V_R, so has order 4, and hence $N_{\mathrm{Aut}_{\mathcal{F}}(Q)}(\mathrm{Aut}_P(Q))$ has order 12;

Table 9.2 *Saturated fusion systems on p_+^{1+2} with at least three $\mathcal{F}$-radical maximal subgroups, for $p = 3$, $p = 5$, $p = 7$ and $p = 13$*

$\mathrm{Out}_{\mathcal{F}}(P)$	$D_{\mathcal{F}}$	$\mathrm{Aut}_{\mathcal{F}}(Q)$	Group	p
D_8	$2, 2$	$\mathrm{GL}_2(3)$	${}^2F_4(2)'$	3
SD_{16}	4	$\mathrm{GL}_2(3)$	J_4	
$4S_4$	6	$\mathrm{GL}_2(5)$	Th	5
$S_3 \times C_3$	3	$\mathrm{SL}_2(7)$	He	
$S_3 \times C_6$	3	$\mathrm{SL}_2(7) \rtimes C_2$	$He.2$	
$D_8 \times C_3$	$2, 2$	$\mathrm{SL}_2(7) \rtimes C_2$	ON	
$D_{16} \times C_3$	4	$\mathrm{SL}_2(7) \rtimes C_2$	$ON.2$	
$S_3 \times C_6$	$3, 3$	$\mathrm{SL}_2(7) \rtimes C_2$	Fi'_{24}	7
$(C_6 \times C_6) \rtimes C_2$	6	$\mathrm{SL}_2(7) \rtimes C_2$	Fi_{24}	
$(C_6 \times C_6) \rtimes C_2$	$6, 2$	$\mathrm{SL}_2(7) \rtimes C_2, \mathrm{GL}_2(7)$	None	
$D_{16} \times C_3$	$4, 4$	$\mathrm{SL}_2(7) \rtimes C_2$	None	
$SD_{32} \times C_3$	8	$\mathrm{SL}_2(7) \rtimes C_2$	None	
$4S_4 \times C_3$	6	$\mathrm{SL}_2(13) \rtimes C_4$	M	13

thus $\mathrm{Aut}_{\mathcal{F}}(Q) \cong \mathrm{GL}_2(3)$ as the normalizer of a Sylow 3-subgroup of $\mathrm{SL}_2(3)$ has order 6. Similarly, $\mathrm{Aut}_{\mathcal{F}}(S) \cong \mathrm{GL}_2(3)$, and this completes the description of the fusion system.

If $Y \cong SD_{16}$, then Y can be chosen to contain the four matrices v_Q, v_R, v_S, and v_T above, and so $\mathcal{F}$ contains the previous fusion system. In particular, all of Q, R, S, and T are $\mathcal{F}$-radical, and $\mathrm{Aut}_{\mathcal{F}}(Q) \cong \mathrm{GL}_2(3)$. Since the stabilizer of a 1-dimensional subspace in Y has order 4 and Y has order 16, Y must act transitively on the four Sylow 3-subgroups of $\mathrm{GL}_2(3)$ (which correspond to the 1-dimensional subspaces of the 2-dimensional vector space on which Y acts), and hence $\{Q, R, S, T\}$ is the sole $\mathcal{F}$-conjugacy class of maximal subgroups of P. This completes the description of the fusion system in this case.

We omit the proof that the groups mentioned do indeed possess the fusion systems specified, and refer the reader to [RV04, Lemma 4.13]. □

We now state the result for all primes. We include the case $p = 3$ to make the table complete.

Proposition 9.7 (Ruiz–Viruel [RV04]) *Let $\mathcal{F}$ be a saturated fusion system on a finite p-group $P = p_+^{1+2}$, and suppose that $|D_{\mathcal{F}}| > 2$. The possibilities for $\mathcal{F}$ are given in Table 9.2.*

Notice that for $p = 7$ there are three saturated fusion systems for which there is no finite group listed. We will not prove here that these fusion systems are exotic (as this requires the classification of the finite simple groups) but we will prove that they are saturated. To start with, we record a condition for saturation in a fusion system over p_+^{1+2}.

Proposition 9.8 *Let p be an odd prime, and let P be the group p_+^{1+2}. Let G be a subgroup of $\operatorname{Out}(P) \cong \operatorname{GL}_2(p)$, and let X be a subset of the maximal subgroups of P. Let α be a primitive $(p-1)$th root of 1. Suppose that the following hold:*

(i) *G is a p'-subgroup;*
(ii) *whenever $Q \in X$ and $g \in G$ then $Q^g \in X$;*
(iii) *for all $Q \in X$ there is a cyclic subgroup $V_Q \leq G$ generated by an automorphism $\psi \in G$, such that $x\psi = x^\alpha$ and $z\psi = z^{-\alpha}$, for some generators x and z for Q, i.e., there is an element ψ in G that normalizes Q, has order $p - 1$, and $\psi|_Q$ has determinant 1.*

There exists a unique saturated fusion system $\mathcal{F}$ on P such that $G = \operatorname{Out}_{\mathcal{F}}(P)$ and $X = D_{\mathcal{F}}$.

Proof Let $\mathcal{F}$ be the fusion system generated by G and the subgroup $\operatorname{SL}_2(p) \leq \operatorname{Aut}(Q)$ for each $Q \in X$. We claim that this fusion system is saturated; uniqueness will then follow from Proposition 9.1. Note that $\operatorname{Out}_{\mathcal{F}}(P) = G$, and so P is fully automized. Let $\mathcal{H}$ denote the set of all $\mathcal{F}$-centric subgroups of P, i.e., $\mathcal{H}$ contains all subgroups of order at least p^2. We will use Theorem 6.16, since clearly $\mathcal{F}$ is generated by $\operatorname{Aut}_{\mathcal{F}}(U)$ for $U \in \mathcal{H}$. Also, since every element of $\mathcal{H}$ is fully normalized, we will simply show that all elements of $\mathcal{H}$ have the surjectivity property.

If $U \in \mathcal{H}$ does not lie in X then $\operatorname{Aut}_{\mathcal{F}}(U)$ consists of restrictions of elements of $\operatorname{Aut}_{\mathcal{F}}(U \leq P)$, and so U has the surjectivity property. If U lies in X then $\operatorname{Aut}_{\mathcal{F}}(U)$ is the product of $A = \operatorname{SL}_2(p)$ and the restriction of all elements of $\operatorname{Aut}_{\mathcal{F}}(U \leq P)$ to U; the third condition above proves that every element of $\operatorname{N}_A(\operatorname{Aut}_P(U))$ has an extension to P, since the element $\psi|_U$ generates $\operatorname{N}_A(\operatorname{Aut}_P(U))$ (see Exercise 9.2). It is clear that since $\operatorname{Aut}_{\mathcal{F}}(U)$ is generated by $A \trianglelefteq \operatorname{Aut}_{\mathcal{F}}(U)$ and restrictions of $\mathcal{F}$-automorphisms of P, any element of $\operatorname{N}_{\operatorname{Aut}_{\mathcal{F}}(U)}(\operatorname{Aut}_P(U))$ extends to an $\mathcal{F}$-automorphism of P, proving that U has the surjectivity property in this case as well. Hence $\mathcal{F}$ is saturated, as required. $\square$

Proposition 9.9 *Let G be one of the subgroups $(C_6 \times C_6) \rtimes C_2$, $D_{16} \times C_3$ and $SD_{32} \times C_3$ of $\operatorname{GL}_2(7)$. There is a unique saturated fusion system*

$\mathcal{F}$ on 7^{1+2}_+ such that $\mathrm{Out}_{\mathcal{F}}(P) = G$ and $D_{\mathcal{F}}$ consists of all subgroups of order p^2. All of these saturated fusion systems are exotic.

Proof We will prove that, with G any of the groups mentioned and X the set of all subgroups of P of order p^2, the pair (G, X) satisfy the three conditions in the statement of Proposition 9.8, thus proving that there is such a saturated fusion system $\mathcal{F}$. Notice that in all three cases G is a p'-subgroup and X is clearly closed under G, so that the first two properties hold. We choose $\alpha = 3$, a primitive 6th root of 1 in $\mathbb{F}_7$.

Let x and y be non-central elements of P such that $\{x, y\}$ is the basis for a 2-dimensional $\mathbb{F}_7$-vector space on which $\mathrm{Out}_{\mathcal{F}}(P) \cong \mathrm{GL}_2(7)$ acts, and write $z = [x, y]$. Write $Q_i = \langle xy^i, z\rangle$ for $0 \le i \le 6$ and $Q_7 = \langle y, z\rangle$, the eight elements of X. If we construct a subgroup V_{Q_i} of Y as in Proposition 9.8, and $Q_j = Q_i^g$ for some $g \in Y$, then $V_{Q_j} = V_{Q_i}^g$, so that it suffices to find one such subgroup from each Y-conjugacy class of elements of X.

We begin with the case where $Y = \mathrm{Out}_{\mathcal{F}}(P) = (C_6 \times C_6) \rtimes C_2$, which can be thought of as generated by all diagonal matrices and the element $\left(\begin{smallmatrix} 0 & 1 \\ 1 & 0 \end{smallmatrix}\right)$. For Q_0, the element $\alpha = \left(\begin{smallmatrix} 3 & 0 \\ 0 & 4 \end{smallmatrix}\right)$ lies in Y, and its restriction to Q_0 acts as $x^\alpha = x^3$ and $z^\alpha = z^5$, so that α generates a subgroup V_{Q_0}. The other subgroups Q_i for $1 \le i \le 6$ are all easily seen to be Y-conjugate, so it remains to find a subgroup V_{Q_1}: we claim that $\left(\begin{smallmatrix} 0 & 3 \\ 3 & 0 \end{smallmatrix}\right)$ has order 6, lies in Y, and normalizes Q_1, so that this matrix generates V_{Q_1}.

Now suppose that $Y = \mathrm{Out}_{\mathcal{F}}(P) \cong D_{16} \times C_3$. We first give generators for this subgroup of $\mathrm{GL}_2(7)$: these are

$$\alpha = \begin{pmatrix} 3 & 0 \\ 0 & 4 \end{pmatrix} \quad \text{and} \quad \beta = \begin{pmatrix} 3 & 2 \\ 1 & -3 \end{pmatrix}.$$

Notice that both α and β have order 6, that $\alpha^2 = \beta^2$ is a scalar matrix, hence central, and that α^3 and β^3 are involutions that generate a dihedral group of order 16, so that $Y = \langle \alpha, \beta \rangle \cong D_{16} \times C_3$. Since α is as above, it generates V_{Q_0}. We claim that β generates V_{Q_1}; to see this, an easy calculation proves that β normalizes Q_1, and since $z^\beta = z^3$ and $(xy)^\beta = (xy)^{-2}$ modulo $\langle z\rangle$, the automorphism $\beta|_Q$ has determinant 1.

By direct calculation, α fixes Q_0 and Q_7, and interchanges Q_i with Q_{7-i} for $1 \le i \le 6$. The matrix β fixes Q_1 and Q_3, and interchanges Q_0 with Q_5, Q_2 with Q_7, and Q_4 with Q_6. Consequently the $\mathcal{F}$-conjugacy classes of the Q_i are $\{Q_0, Q_2, Q_5, Q_7\}$ and $\{Q_1, Q_3, Q_4, Q_6\}$. Since we have determined V_{Q_i} for $i = 0, 1$ we have proved that $\mathcal{F}$ is saturated.

Finally, assume that $Y \cong SD_{32} \times C_3$. This can clearly be chosen to contain $D_{16} \times C_3$, so in particular it contains the subgroups V_{Q_0} and V_{Q_1}. Hence $\mathcal{F}$ is saturated with this choice of $\mathrm{Out}_\mathcal{F}(P)$ as well. (For concreteness, the matrix $\gamma = \left(\begin{smallmatrix} 1 & 1 \\ 2 & 6 \end{smallmatrix}\right)$, together with α and β, generate $SD_{32} \times C_3$.)

We omit the proof that these fusion systems are exotic, and refer to [RV04]. □

In conclusion, we have the following theorem, collating the results of this entire section.

Theorem 9.10 (Ruiz–Viruel [RV04])   *Let $\mathcal{F}$ be a saturated fusion system on a finite p-group $P = p_+^{1+2}$. If $|D_\mathcal{F}| \leq 2$ then $\mathcal{F}$ is given by one of the possibilities in Table 9.1, and if $|D_\mathcal{F}| > 2$ then $\mathcal{F}$ is given by one of the possibilities in Table 9.2.*

9.2 The Solomon fusion system

The Solomon fusion systems are exotic saturated fusion systems on the Sylow 2-subgroups of $\mathrm{Spin}_7(r)$ for r an odd prime power. The smallest of these 2-groups is also the Sylow 2-subgroup of the sporadic simple group Co_3. In [Sol74], Solomon determines all centre-free, perfect finite groups G such that $O_{2'}(G) = 1$. In fact, [Sol74, Theorem 1.1] shows that there are only two such finite groups, Co_3 itself and $E_{16} \cdot \mathrm{GL}_4(2)$, where E_{16} is the elementary abelian group of order 16 and this group is the unique non-split extension of E_{16} by $\mathrm{GL}_4(2)$.

However, there are *three* centre-free and perfect saturated fusion systems on the Sylow 2-subgroup P of $\mathrm{Spin}_7(3)$. Two of them are $\mathcal{F}_P(Co_3)$ and $\mathcal{F}_P(E_{16} \cdot \mathrm{GL}_4(2))$, and the third is a fusion system in which all involutions are $\mathcal{F}$-conjugate, and if z is the central involution in P then $C_\mathcal{F}(z) = \mathcal{F}_P(\mathrm{Spin}_7(3))$.

In [Sol74], a presentation for this 2-group P is given. It is generated by ten elements z, s_i, and t_i for $i = 1, 2, 3$, u, v, and w. The relations are that $z^2 = 1$, that all other generators square to z (so that immediately we see that $z \in Z(P)$), and that various commutator relations are satisfied:

$$[t_i, t_{i+1}] = [s_i, t_i] = [s_1, u] = [s_2, u] = [s_1, v]$$
$$= [s_3, v] = [u, v] = [u, w] = z,$$

$$[s_1, w] = [s_2, w] = s_1 s_2 z, \quad [u, t_1] = [v, t_1] = s_1, \quad [u, t_2] = s_2,$$

$$[v, t_3] = s_3, \quad [w, t_1] = [w, t_2]^{-1} = t_1 t_2, \quad [v, w] = u.$$

All other commutators are assumed to be trivial. This generates the group P.

Some facts about this group are given now. They are easy to see, and are left as Exercise 9.4.

Lemma 9.11 *Let P be the group constructed above. Then $|P| = 1024$, $Z(P) = \langle z \rangle$, $P' = \langle s_1, s_2, s_3, t_1 t_2, u \rangle$, and $\mho^1(P') = \langle z, s_1 s_2 \rangle$.*

Another similar computational exercise is the following lemma, which is useful for finding particular maps.

Lemma 9.12 *Let P be the group constructed above. The central involution z lies in $C_P(g)'$ for every involution $g \in P$, and $z \in \mho^1(C_P(g)')$ for every involution in $P' \cup \{vs_2, s_1 t'_2, t'_1 t'_2 t'_3\}$.*

Solomon carries out an analysis of the possible fusion patterns in P, assuming that they come from perfect, centre-less finite groups with P as a Sylow p-subgroup. We will sketch his arguments here. Let $\mathcal{F}$ be a saturated fusion system on P. We will assume that $\mathcal{F}$ is centre-free; hence z is $\mathcal{F}$-conjugate to some other involution in P. We will nail down a particular involution to which z is $\mathcal{F}$-conjugate using Lemma 9.12.

Proposition 9.13 *The element z is $N_{\mathcal{F}}(\langle z, s_1 s_2 \rangle)$-conjugate to $s_1 s_2$.*

Proof Suppose that g is an involution in P that is $\mathcal{F}$-conjugate to z, via ϕ. Since z is central, we may extend ϕ to a map $\psi : C_P(g) \to P$. Since z lies in $C_P(g)'$ by Lemma 9.12, $z\psi$ lies in $P' \setminus \{z\}$, so we may assume that g lies in P'. Notice that z lies in $\mho^1(C_P(g)')$ by Lemma 9.12, so $z\psi$ lies in $\mho^1(P')$. Therefore z is $\mathcal{F}$-conjugate to some member of $\mho^1(P')$. However, this group possesses only three involutions: z, $s_1 s_2$, and $s_1 s_2 z$. Hence z is $\mathcal{F}$-conjugate to either $s_1 s_2$ or $s_1 s_2 z$. However, these two are conjugate in P, since $s_1 s_2^v = s_1^v s_2 = s_1 s_2 z$. Hence z and $s_1 s_2$ are $\mathcal{F}$-conjugate.

Finally, any map ϕ that sends $s_1 s_2$ to z extends to a map $C_P(s_1 s_2) \to P$, and this map must restrict to an automorphism of $\langle z, s_1 s_2 \rangle$; hence the two involutions are conjugate in $N_{\mathcal{F}}(\langle z, s_1 s_2 \rangle)$, as claimed. $\square$

The next stage of the analysis is to consider the saturated subsystem $C_{\mathcal{F}}(z)$. This is another saturated fusion system on P, this time with a centre. It is natural to consider the fusion system $\bar{\mathcal{F}} = C_{\mathcal{F}}(z)/\langle z \rangle$, which is a fusion system on the Sylow 2-subgroup of the alternating group A_{12} (see Exercise 9.5). In a previous paper [Sol73], Solomon determines much information about the structure of a finite group with Sylow 2-subgroup that of A_{12}, and these results are used extensively.

Let A be the subgroup generated by the s_i and the t_i, and write $\bar{P} = P/\langle z \rangle$ and $\bar{A} = A/\langle z \rangle$. Notice that $\bar{A}$ is elementary abelian of order 2^6, and so $\bar{A}$ is an $\mathbb{F}_2$-module for $H = \mathrm{Aut}_{\bar{\mathcal{F}}}(\bar{A})$. It turns out that $\bar{A}$ is an indecomposable $\mathbb{F}_2 H$-module. In [Sol74, Corollary 2.5], it is deduced from the possible finite groups with $\bar{P}$ as Sylow 2-subgroup that H is one of the groups S_4, S_5, A_6, A_7 or $\mathrm{GL}_3(2)$. This allows some more fusion of involutions to be deduced: for example, s_2 and s_3 are $\mathcal{F}$-conjugate. By proving that the centre of $\bar{\mathcal{F}}$ is trivial, Solomon deduces the following result in [Sol74, Corollary 2.8].

Lemma 9.14 *The fusion system* $\mathrm{C}_{\mathcal{F}}(z)$ *is isomorphic to* $\mathcal{F}_P(G)$, *where G is one of* $A \cdot \mathrm{GL}_3(2)$, $A \cdot A_7$, $2 \cdot \mathrm{Sp}_6(2)$ *and* $\mathrm{Spin}_7(3)$.

The centralizer of an involution in the centre of a Sylow 2-subgroup of Co_3 is $2 \cdot \mathrm{Sp}_6(2)$, and so this case certainly can occur. In [Sol74, Lemma 2.10], the case where $G = A \cdot A_7$ is discounted by proving that the fusion is not possible. The case of $G = A \cdot \mathrm{GL}_3(2)$ occurs in the fusion system of the group $E_{16} \cdot \mathrm{GL}_4(2)$ mentioned earlier.

The last case, where $\mathrm{C}_{\mathcal{F}}(z)$ is $\mathcal{F}_P(\mathrm{Spin}_7(3))$ results in $\mathcal{F}$ having a single conjugacy class of involutions, but does not yield a finite group that realizes $\mathcal{F}$. We get the following result, amalgamating results of Solomon [Sol74], Levi–Oliver [LO02], and Oliver–Ventura [OV09].

Theorem 9.15 *There are exactly three saturated fusion systems $\mathcal{F}$ on the 2-group P that are both centre-free and perfect. Two of these are the fusion systems of the simple group Co_3 and the unique non-split extension of the elementary abelian group of order 16 by $\mathrm{GL}_4(2)$. The remaining saturated fusion system is exotic.*

In the general case, where P is a Sylow 2-subgroup of $\mathrm{Spin}_7(q)$, not just $\mathrm{Spin}_7(3)$, there is a saturated fusion system on P in which all involutions are $\mathcal{F}$-conjugate and $\mathrm{C}_{\mathcal{F}}(z)$ is the fusion system of $\mathrm{Spin}_7(q)$. The existence of such a fusion system is given in [LO02] [LO05], and the fact that such groups do not exist can either be proved using the classification of the finite simple groups, or a classification-free argument given in [LO02, Proposition 3.4], due to Solomon.

Theorem 9.16 *Let P be the Sylow 2-subgroup of $\mathrm{Spin}_7(q)$ and let z be the central involution of P.*

(i) *There exists a saturated fusion system $\mathcal{F}_{\mathrm{Sol}}(r)$ on P such that all involutions are $\mathcal{F}$-conjugate and $C_{\mathcal{F}}(z)$ is the fusion system of $\mathrm{Spin}_7(q)$.*

(ii) *There exists no finite group G with Sylow 2-subgroup P such that all involutions are G-conjugate and $C_G(z)$ contains the fusion system of $\mathrm{Spin}_7(q)$.*

In Section 9.4 we will prove that the fusion systems $\mathcal{F}_{\mathrm{Sol}}(r)$ are not the fusion systems of any 2-block of any finite group, not only that they are not the fusion systems of groups. In order to prove this, we will need a few facts about $\mathcal{F}_{\mathrm{Sol}}(r)$ and the 2-group on which it is defined. We begin with some facts about the 2-groups.

Proposition 9.17 *Let P be a Sylow 2-subgroup of the finite group $\mathrm{Spin}_7(q)$, where q is odd. Write $q = a2^i \pm 1$ where $a > 1$.*

(i) *$|P| = 2^{3i+4}$, $|Z(P)| = 2$, and P has 2-rank 4.*

(ii) *Any elementary abelian subgroup Q of P of rank 4 satisfies $C_P(Q) = Q$.*

(iii) *For some elementary abelian subgroup Q of P of rank 4 we have that $\mathrm{Aut}_P(Q)$ is a Sylow 2-subgroup of $\mathrm{GL}_4(2)$, so has order 2^6.*

(iv) *P has no faithful representation of dimension less than 8 over any field of characteristic not 2.*

Proof The order of $\mathrm{Spin}_7(q)$ is $q^9(q^2-1)(q^4-1)(q^6-1)$, and the power of 2 dividing (q^2-1) and (q^6-1) is easily seen to be 2^{i+1}, and dividing (q^4-1) it is 2^{i+2}. For the rest of the facts in (i) and (ii) we refer to [LO02, Proposition A.8]. Together with [LO02, Lemma 3.1], we get (iii).

It remains to prove (iv), our proof of which will follow [Kes06, Proposition 4.1(ii)], together with a suggestion of Bob Oliver. Let $\mathbb{F}$ be a field of characteristic different from 2, which we assume is algebraically closed, and let n be the smallest dimension of a faithful representation of P. Since P is a 2-group, all representations of P over $\mathbb{F}$ are completely reducible; since $|Z(P)| = 2$, if any sum of irreducible representations acts faithfully then one of the constituents must (as $Z(P)$ lies inside all kernels of all non-faithful representations). Therefore there is a faithful irreducible representation of degree n. In particular, n is a power of 2, since P is a 2-group. Since P contains an elementary abelian subgroup of order 16, $n \geq 4$, so to prove that $n \geq 8$ it suffices to show that $n \neq 4$. Hence suppose that P has a faithful representation of degree 4, and we identify P with this subgroup of $\mathrm{GL}_4(\mathbb{F})$.

Let Q be an elementary abelian subgroup of P of rank 4 such that $\mathrm{Aut}_P(Q)$ has order 2^6. The subgroup Q can by chosen (by conjugating P inside $\mathrm{GL}_4(\mathbb{F})$) to consist of diagonal matrices, which are therefore all matrices with diagonal entries ± 1. Hence Q has four distinct 1-dimensional eigenspaces, and so $\mathrm{Aut}_{\mathrm{GL}_4(\mathbb{F})}(Q) = S_4$, as any $\mathrm{GL}_4(\mathbb{F})$-automorphism of Q must permute the eigenspaces of Q. However, this contradicts the fact that $|\mathrm{Aut}_P(Q)| = 2^6$. Thus P has no representation of degree 4, as claimed. $\qquad\square$

The last thing we need is a single fact about $\mathcal{F}_{\mathrm{Sol}}(r)$, namely that it has a particular $\mathcal{F}$-centric subgroup.

Proposition 9.18 *Let q be an odd prime power. The fusion system $\mathcal{F}_{\mathrm{Sol}}(r)$ possesses an $\mathcal{F}$-centric elementary abelian subgroup Q of order 16 such that $\mathrm{Aut}_{\mathcal{F}_{\mathrm{Sol}}(r)}(Q) = \mathrm{Aut}(Q)$.*

Proof This follows from [LO02, Lemma 3.1] and Proposition 9.17(ii). $\qquad\square$

9.3 Blocks of finite groups

In this section we will examine the fusion systems of some blocks of finite groups. One of the main things we want to understand is whether they can be exotic. The first result, an immediate corollary of the fact that constrained fusion systems come from finite groups, is the following.

Proposition 9.19 *Let G be a finite group and let P be a Sylow p-subgroup of G. Let B be a block of kG, and let $\mathcal{F}$ be the fusion system of B. If $\mathcal{F}_P(G)$ is soluble then $\mathcal{F}$ is the fusion system of a finite group. In particular, all fusion systems of p-blocks of p-soluble groups are non-exotic.*

This proposition follows from the facts that soluble fusion systems are constrained, saturated subsystems of soluble fusion systems are soluble, and that all constrained fusion systems come from finite groups.

Given this proposition, it is reasonable to ask whether for other classes of groups we can show that all blocks yield non-exotic fusion systems, and of course the groups we consider should not be p-soluble. We will show that blocks of symmetric groups have fusion systems of other finite groups. We begin with a lemma that will reduce the argument to something tractable.

Lemma 9.20 *Let Q be a p-subgroup of the symmetric group $G = S_n$ such that Q acts fixed point freely. The group algebra $k\,C_G(Q)$ has a unique block, with block idempotent 1.*

Proof Assume firstly that Q is a transitive subgroup of S_n. We will show that $C_G(Q)$ is a p-group: by Exercise 2.6, finite p-groups have only one p-block, and so we will be done in this case. However, if $g \in C_G(Q)$ then we claim that all orbits of g are of the same size; to see this, if $h \in Q$ then $g^h = g$, so that for all $x \in \{1, \ldots, n\}$, $xg^i = x$ if and only if $x(g^i)^h = x$ for any $i \in \mathbb{N}$, i.e., $xg^i = x$ if and only if $(xh^{-1})g^i = xh^{-1}$. Since Q is transitive, this means that if the orbit of x has length i then the orbit of xh^{-1}, which can be any element of $\{1, \ldots, n\}$, has length at most i. As x was chosen arbitrarily, all orbits of g have the same length. Notice that this length is a divisor of n: as Q is a transitive p-subgroup of S_n, n is a power of p, so that g has order a power of p. Since g was an arbitrary element of $C_G(Q)$, we see that $C_G(Q)$ is a p-group, as claimed.

Now suppose that Q is not transitive, but still fixed point free, and let $X_1, \ldots, X_d$ be the orbits of Q on $\{1, \ldots, n\}$. Let Q_i be the subgroup of $\mathrm{Sym}(X_i)$ given by the induced action of Q on X_i, and let $\bar{Q}$ denote the direct product of the Q_i.

Assume that $Z(\bar{Q}) \leq Q$. We claim that in this case $C_G(Q)$ is a direct product of $C_{\mathrm{Sym}(X_i)}(Q_i)$ for $1 \leq i \leq d$. Since each Q_i is transitive on X_i, $k\,C_{\mathrm{Sym}(X_i)}(Q_i)$ has a unique block, so that $k\,C_G(Q)$ has a unique block by Exercise 2.5. (Alternatively, since each factor is a p-group, the product is a p-group, so has a unique block.) To see that $C_G(Q)$ is a direct product of the $C_{\mathrm{Sym}(X_i)}(Q_i)$, we note two things. Firstly, each of the $C_{\mathrm{Sym}(X_i)}(Q_i)$ is a subgroup of $C_G(Q)$. Secondly, if $g \in C_G(Q)$, then $Z(Q_i)^g = Z(Q_i)$ since $Z(Q_i) \leq Q$; as $Z(Q)_i$ fixes all of the X_j for $j \neq i$, and does not fix X_i, we must have that $X_i^g = X_i$, since $Z(Q_i)^g$ fixes X_i^g. Thus g lies in the product of the $\mathrm{Sym}(X_i)$, hence in the product of the $C_{\mathrm{Sym}(X_i)}(Q_i)$.

Finally, we must remove the assumption that $Z(\bar{Q}) \leq Q$. If b is a block idempotent of $k\,C_G(Q)$, then $Z(\bar{Q})$ is contained in $C_G(Q)$, and is in fact a normal p-subgroup. Therefore, by Proposition 2.28, b lies in $k\,C_{C_G(Q)}(Z(\bar{Q})) = k\,C_G(QZ(\bar{Q}))$. By the previous arguments, the only block idempotent of this latter group is 1, and so $b = 1$, as claimed. $\square$

Using this lemma, we can determine the structure of the fusion systems of blocks of symmetric groups, a result which can be traced back to Puig.

Theorem 9.21 (Puig [Pui86]) *Let G be the symmetric group S_n, and let b be a block idempotent of kG, where Q is a defect group of b. Write $\mathcal{F}$ for the fusion system of the block b. There exists a non-negative integer w with $pw \leq n$, such that Q is a Sylow p-subgroup of S_{pw} and $\mathcal{F} = \mathcal{F}_Q(S_{pw})$.*

Proof Let R be a subgroup of Q and let (R, e_R) be a b-Brauer pair (including the case $R = Q$). Let X denote the set $\{1, \ldots, n\}$. Writing X^R for the set of R-fixed points of X, we have

$$C_G(R) = \mathrm{Sym}(X^R) \times C_{\mathrm{Sym}(X \setminus X^R)}(R).$$

This direct product decomposition means that $e_R = e \otimes 1$, where e is a block idempotent of $\mathrm{Sym}(X^R)$, since by the previous lemma the second factor has a unique block idempotent. Also, the same direct product decomposition occurs for the normalizer, so that

$$N_G(R) = \mathrm{Sym}(X^R) \times N_{\mathrm{Sym}(X \setminus X^R)}(R).$$

Notice that e is obviously $\mathrm{Sym}(X^R)$-stable, so e_R is $N_G(R)$-stable. Hence $N_G(R, e_R) = N_G(R)$ and therefore

$$\mathrm{Aut}_{\mathcal{F}}(R) = \mathrm{Aut}_G(R) = \mathrm{Aut}_{\mathrm{Sym}(X \setminus X^R)}(R).$$

We will next prove that Q is a Sylow p-subgroup of $\mathrm{Sym}(X \setminus X^Q)$. If this is true, then for $R \leq Q$, $X \setminus X^R$ is contained in $X \setminus X^Q$, so that $\mathrm{Sym}(X \setminus X^R) \leq \mathrm{Sym}(X \setminus X^Q)$; in other words,

$$\mathrm{Aut}_{\mathcal{F}}(R) = \mathrm{Aut}_{\mathrm{Sym}(X \setminus X^Q)}(R)$$

for all $R \leq Q$, so that $\mathcal{F} = \mathcal{F}_{\mathrm{Sym}(X \setminus X^Q)}(Q)$. As Q acts fixed point freely on $X \setminus X^Q$, this must have size a multiple of p, so that $\mathcal{F} = \mathcal{F}_{S_{pw}}(Q)$ for some $w \geq 0$, as needed.

It remains to show that Q is a Sylow p-subgroup of $\mathrm{Sym}(Y)$, where $Y = X \setminus X^Q$. Let S be a p-subgroup of $\mathrm{Sym}(Y)$ containing Q. Since Q is fixed point free on Y, so is S, and so 1 is the unique block idempotent of both $k\,C_{\mathrm{Sym}(Y)}(Q)$ and $k\,C_{\mathrm{Sym}(Y)}(S)$. Therefore there is an inclusion of Brauer pairs $(Q, 1) \leq (S, 1)$ for the group $\mathrm{Sym}(Y)$ (as these are the only Brauer pairs for these groups). Let $f = e \otimes 1$ be a block idempotent of $k\,C_G(S)$, noting that $C_G(S)$ is the direct product of $\mathrm{Sym}(X^Q)$ and $C_{\mathrm{Sym}(Y)}(S)$. Again, we have an inclusion of Brauer pairs $(Q, e_Q) \leq (S, f)$ for G (as $e_Q = e \otimes 1$ as well, so is included in f), contradicting the fact that Q is a defect group for b. Thus Q is a Sylow p-subgroup of $\mathrm{Sym}(Y)$, completing the proof. $\qquad\square$

Having understood the fusion systems of blocks of symmetric groups, we move on to considering the sporadic simple groups. In this case things are much more complicated – obviously they are, since sporadic groups are more complicated – and in order to prove that all blocks of all sporadic groups have non-exotic fusion systems we will have to use facts about the representation theory of sporadic groups. One of the most important things we will use is that it is possible (in fact, very easy for a computer) to deduce from the character table of a finite group the number of blocks and the order of the defect group of each block (see [PD77, Theorem 4.2B(iii)] for more information). Since the character tables of all sporadic groups are known [CCNPW85], we know some information about the blocks of all sporadic groups.

The theorem that we will prove is the following.

Theorem 9.22 *Let G be a sporadic simple group. If k is a field of characteristic p and B is a block of kG then the fusion system of kG is non-exotic.*

Our first reduction will be to note that we only need to consider groups G and primes $p \mid |G|$ such that $p^4 \mid |G|$. To see this, notice that all saturated fusion systems on abelian p-groups come from finite groups, so we might as well assume that the Sylow p-subgroup of G is non-abelian; in particular, $p^3 \mid |G|$. Secondly, there are five groups of order p^3: three abelian and the groups p_+^{1+2} and p_-^{1+2}. The former of these was shown in Section 9.1 to yield no exotic fusion systems unless $p = 7$, and in this case the exotic fusion systems are not contained in a fusion system of a finite group, so cannot be the fusion system of a 7-block of a finite group G where $7^4 \nmid |G|$. Finally, we showed in Exercise 7.5 that p_-^{1+2}, which has exponent p^2, is metacyclic, and therefore can support no exotic fusion systems by Theorem 7.5.

Hence we may assume that the power of the prime p dividing $|G|$ is at least p^4. The possibilities for G and p are given in the table below.

Primes	Groups
2	$M_{11}, M_{12}, M_{22}, M_{23}, M_{24}, HS, J_2, He, Ru, J_4$
2, 3	$Co_2, Co_3, McL, Suz, Th, Fi_{22}, Fi_{23}, Fi'_{24}, ON, J_3$
2, 3, 5	Co_1, HN, B, Ly
2, 3, 5, 7	M

The computer algebra package GAP contains information on blocks of sporadic groups. Using this information, we are able to deduce the following lemma immediately.

Lemma 9.23 *If G is one of the following groups, then there are no non-principal blocks with defect group of order at least p^4 for any prime p: M_{11}, M_{12}, M_{22}, M_{23}, M_{24}, HS, J_2, He, Ru, Co_2, Co_3, McL, Suz, Fi_{22}, ON, and J_3.*

The remaining groups are Th, Fi_{23}, Fi'_{24}, Co_1, HN, B, Ly, and M. For these we will have to determine whether they possess non-principal blocks of defect group of order at least p^4. We get the following result, using GAP and the ordinary character tables for the sporadic simple groups stored therein.

Lemma 9.24 *Let G be a sporadic simple group and let p be a prime dividing $|G|$. If G possesses a non-principal p-block with defect group of order at least p^4, then $p = 2$ and either*

(i) *G is HN or M and there is a block with SD_{16} defect group, or*
(ii) *G is Ly, and there is a block with defect group the Sylow 2-subgroup of $2 \cdot A_8$.*

Proof Using GAP, it is easy to confirm that the only groups with non-principal p-blocks with defect group of order at least p^4 are the ones above. In particular, $p = 2$. It remains to confirm the defect group. In [Lan78] Landrock determines the non-principal 2-blocks of the sporadic groups, and proves that the defect groups of the three blocks given in the statement of this lemma are as suggested, completing the proof. $\square$

Since we constructed all saturated fusion systems on semidihedral groups in Theorem 4.54 and proved that none is exotic, the only case we need to worry about is $G = Ly$, and b is the block idempotent of the non-principal block of kG, where k is a field of characteristic 2. Let Q be the defect group of b and let $\mathcal{F}$ be the fusion system of b. This subgroup Q is isomorphic with the Sylow 2-subgroup of three sporadic simple groups, namely M_{22}, M_{23}, and McL. From the information stored in [CCNPW85], we know that all involutions of G are conjugate, and that the centralizer of an involution is $H = 2 \cdot A_{11}$. There are exactly two 2-blocks of kH, the principal block with defect group of order 2^8, and the non-principal block, with defect group of order 2^7. If x and y are two involutions in G, and e_x and e_y are the non-principal blocks of $k \, C_G(x)$ and $k \, C_G(y)$ respectively, then any element $g \in G$ such that $x^g = y$ must also conjugate e_x to e_y. Therefore we have shown that all involutions in Q are $\mathcal{F}$-conjugate; using GAP, it is easy to prove that Q is generated by its involutions and $\mathrm{Aut}(Q)$ is a 2-group, so that $\mathcal{F}$ is simple by Corollary 5.75.

Finally, we need to quote a result from [OV09], which classifies all non-constrained, centre-free saturated fusion systems on the 2-group Q. By $\mathrm{P\Gamma L}_n(q)$ we mean the group $\mathrm{PGL}_n(q) \rtimes \langle \alpha \rangle$ and by $\mathrm{P\Sigma L}_n(q)$ we mean the group $\mathrm{PSL}_n(q) \rtimes \langle \alpha \rangle$, where α is a generator for the Galois automorphisms $\mathrm{Aut}(\mathbb{F}_q)$ of $\mathrm{PGL}_n(q)$ and $\mathrm{PSL}_n(q)$ respectively.

Theorem 9.25 (Oliver–Ventura [OV09, Theorem 5.11]) *Let P be the Sylow 2-subgroup of the Mathieu group M_{22}. If $\mathcal{F}$ is a non-constrained, centre-free saturated fusion system on P, then $\mathcal{F}$ is the fusion system of M_{22}, M_{23}, McL, $\mathrm{P\Sigma L}_3(4)$, $\mathrm{P\Gamma L}_3(4)$ or $\mathrm{PSL}_4(5)$. In particular, $\mathcal{F}$ is non-exotic.*

Hence the fusion system $\mathcal{F}$ cannot be an exotic fusion system, completing the proof of Theorem 9.22.

9.4 Block exotic fusion systems

How does one prove that a fusion system is not a fusion system of a block? To prove that a fusion system is exotic, it simply requires checking the Sylow p-subgroups of finite groups, and normally of finite simple groups. For fusion systems of blocks, firstly as we said there is no reduction to simple groups in general, and secondly checking all p-blocks of all simple groups is not easy. In the previous section we made some in-roads into the latter question; the former question does have a partial positive answer, by a theorem of Kessar and Stancu.

Theorem 9.26 (Kessar–Stancu [KS08]) *Let $\mathcal{F}_1$ and $\mathcal{F}_2$ be saturated fusion systems on a finite p-group P, and suppose that $\mathcal{F}_1$ contains $\mathcal{F}_2$. Suppose that the following three conditions are satisfied:*

(i) *if $\mathcal{F}$ is a saturated fusion system on P containing $\mathcal{F}_2$, then either $\mathcal{F} = \mathcal{F}_1$ or $\mathcal{F} = \mathcal{F}_2$;*
(ii) *there are no non-trivial strongly $\mathcal{F}_2$-closed subgroups of P;*
(iii) *the only weakly normal subsystems of $\mathcal{F}_1$ and $\mathcal{F}_2$ on P are $\mathcal{F}_1$ and $\mathcal{F}_2$ themselves.*

If either $\mathcal{F}_1$ or $\mathcal{F}_2$ appears as the fusion system of a p-block of a finite group, then either $\mathcal{F}_1$ or $\mathcal{F}_2$ appears as the fusion system of a p-block of a quasisimple group L with $Z(L)$ of p'-order.

Proving this theorem requires too much modular representation theory, and is beyond the scope of this book. However, we will sketch some of the ideas used in the proof.

Let G be a finite group, and let N be a normal subgroup of G. Clifford theory says that the simple kG-modules, when restricted to kN-modules, while not staying simple are still a direct sum of simple modules. For blocks of kG and kN, there is a similar situation. Let e be a block idempotent of kN, and notice that, for any $g \in G$, e^g is also a block idempotent of kN. The *inertia group* of e is the set $T(e)$ of all group elements g such that $e^g = e$. The element

$$f_e = \sum_{x \in X} e^x$$

is therefore a central idempotent of kG, where X is a transversal to $T(e)$ in G. Just as $1 \in kN$ is a sum of the block idempotents e of kN, we see that $1 \in kG$ is a sum of the elements f_e, as e ranges over all block idempotents of kN up to G-conjugacy. Therefore if b is a block idempotent of kG, b appears as a summand of *exactly one* of the f_e; the statement that b appears as a summand of f_e is equivalent to the statement that $bf_e = b$, which since b is primitive is equivalent to $bf_e \neq 0$. We say that the block kGb *covers* the block kNe if this is true.

The Clifford theory of blocks starts with the following theorem (see for example, [Ben98b, Theorem 6.4.1]).

Theorem 9.27 *Let G be a finite group and let k be a field of characteristic p. Let N be a normal subgroup of G. Suppose that b is a block idempotent of kG.*

(i) *The block idempotents of kN covered by b form a single G-conjugacy class of block idempotents.*

(ii) *If e is a block idempotent of kN covered by b, then some defect group of b is contained in $T(e)$.*

(iii) *There is some defect group D of b such that $D \cap N$ is a defect group of e.*

Thus if N is a normal subgroup of G, and b is a block idempotent of kG, with defect group D, then $D \cap N$ is a defect group of one of the block idempotents of kN covered by b. This allows us to reduce many questions about blocks of finite groups to blocks of simple groups.

However, this doesn't work for fusion systems, because if e is covered by b, while the defect group of e is contained in the defect group of b, the same is *not true* of the fusion system. Since the fusion system of e

need not be contained in the fusion system of b, this scraps any hope of easily going from G down to N.

However, something can still be salvaged, via the notion of a generalized Brauer pair. Let k be an algebraically closed field of characteristic p, let G be a finite group and let N be a normal subgroup of G. Let e be a G-stable block idempotent of kN, i.e., a block idempotent of kN such that $T(e) = G$. The Brauer morphism introduced in Chapter 2 can be extended to this domain: if Q is a p-subgroup of G, the restriction map $\mathrm{Br}_Q^N : kN \to k\,\mathrm{C}_N(Q)$ is an algebra morphism from $(kN)^Q$ onto $k\,\mathrm{C}_N(Q)$.

A Brauer pair for G is a pair (Q, e), where Q is a p-subgroup of G and e is a block idempotent of $k\,\mathrm{C}_G(Q)$. A *generalized (e, G)-Brauer pair* is a pair (Q, e_Q), where Q is a p-subgroup of G such that $\mathrm{Br}_Q^N(e) \neq 0$ and e_Q is a block idempotent of $k\,\mathrm{C}_N(Q)$ such that $\mathrm{Br}_Q^N(e)e_Q = e_Q$. If $G = N$, this is simply the usual definition of an e-Brauer pair.

The same ideas about inclusion of Brauer pairs and maximal Brauer pairs carry over to the generalized setting. We introduce an order relation $\leq$ on the set of generalized (e, G)-Brauer pairs, and if (Q, e_Q) and (R, e_R) are two generalized (e, G)-Brauer pairs write $(Q, e_Q) \leq (R, e_R)$ if $Q \leq R$ and for any primitive idempotent $i \in (kN)^R$ such that $\mathrm{Br}_R^N(i)e_R = e_R$, we have that $\mathrm{Br}_Q^N(i)e_Q = e_Q$. Just as with the case of Brauer pairs (Theorem 2.24), if (R, e_R) is a generalized (e, G)-Brauer pair and $Q \leq R$, then there is a unique generalized (e, G)-Brauer pair (Q, e_Q) such that $(Q, e_Q) \leq (R, e_R)$ ([BP80, Theorem 1.8(1)]). Just as with Brauer pairs, where all maximal b-Brauer pairs are conjugate (Corollary 2.31), all maximal generalized (e, G)-Brauer pairs are conjugate ([BP80, Theorem 1.14(2)]).

Using all of this machinery, we make the following definition.

Definition 9.28 Let G be a finite group and let N be a normal subgroup of G. Let k be a field of characteristic p, and let b be a block idempotent of kG covering a G-stable block idempotent e of kN. Let (P, e_P) be a maximal generalized (e, G)-Brauer pair, and for any $Q \leq P$ write (Q, e_Q) for the generalized (e, G)-Brauer pair such that $(Q, e_Q) \leq (P, e_P)$. The *generalized fusion system* of (e, G), denoted by $\mathcal{F}_{(P, e_P)}(G, N, e)$, is the category with objects all subgroups of P, and with morphisms the sets

$$\mathrm{Hom}_{\mathcal{F}_{(P, e_P)}(G, N, e)}(Q, R) = \{c_g \mid (Q, e_Q)^g \leq (R, e_R)\}.$$

Of course, we have defined the generalized fusion system to look like a fusion system, and it clearly is a fusion system, like the standard fusion system of a block. However, it is less clear that it is saturated.

Theorem 9.29 (Kessar–Stancu [KS08, Theorem 3.4]) *Let G be a finite group and let N be a normal subgroup of G. Let k be a field of characteristic p, and let b be a block idempotent of kG covering a G-stable block idempotent e of kN. Let (P, e_P) be a maximal generalized (e, G)-Brauer pair. The generalized fusion system $\mathcal{F}_{(P, e_P)}(G, N, e)$ is a saturated fusion system, and (up to isomorphism) does not depend on the choice of maximal generalized (e, G)-Brauer pair.*

Let b be a block idempotent of kG, covering a G-stable block idempotent e of kN. We now have three saturated fusion systems: the fusion systems of the block idempotents b and e, and the generalized fusion system of (e, G). In general, the fusion system of e need not be contained in the fusion system of b, because extra fusion between e-Brauer pairs is not picked up in b-Brauer pairs. However, the generalized fusion system still sees this.

Theorem 9.30 (Kessar–Stancu [KS08, Theorem 3.5]) *Let G be a finite group and let N be a normal subgroup of G. Let k be a field of characteristic p, and let b be a block idempotent of kG covering a G-stable block idempotent e of kN. If (D, e_D) denotes a maximal b-Brauer pair, then there exists a maximal generalized (e, G)-Brauer pair (P, e_P) such that $D \leq P$ and*

$$\mathcal{F}_{(D, e_D)}(G, b) \leq \mathcal{F}_{(P, e_P)}(G, N, e),$$

i.e., such that the fusion system of b is contained in the generalized fusion system of (e, G).

Furthermore, $D \cap N = P \cap N$, and if $(D \cap N, e_{D \cap N})$ is the unique generalized (e, G)-Brauer pair contained in (P, e_P), then $(D \cap N, e_{D \cap N})$ is a maximal generalized (e, N)-Brauer pair (or simply a maximal e-Brauer pair) and

$$\mathcal{F}_{(D \cap N, e_{D \cap N})}(N, N, e) = \mathcal{F}_{(D \cap N, e_{D \cap N})}(N, e) \prec \mathcal{F}_{(P, e_P)}(G, N, e),$$

i.e., the fusion system of e is a weakly normal subsystem of the generalized fusion system of (e, G).

Theorem 9.30 shows that the generalized fusion system compensates for the fact that blocks of normal subgroups need not have fusion systems that are subsystems of their covered block's fusion system. It also hints at why Theorem 9.26 holds; in the case there, there are almost no choices for the weakly normal subsystem $\mathcal{F}_{(D \cap N, e_{D \cap N})}(N, e)$ of $\mathcal{F}_{(P, e_P)}(G, N, e) = \mathcal{F}_1$, since there are only two weakly normal subsystems of $\mathcal{F}_1$. The aim of the proof of Theorem 9.26 is to prove that if $\mathcal{F}_{(D, e_D)}(G, b)$ is one of the fusion systems $\mathcal{F}_1$ and $\mathcal{F}_2$ in the statement of the theorem then so is $\mathcal{F}_{(D \cap N, e_{D \cap N})}(N, e)$, as the generalized fusion system of (e, G) cannot be larger than $\mathcal{F}_1$ by assumption. For the complete proof, we refer to [KS08].

Applying Theorem 9.26 to the three exotic fusion systems in Section 9.1 and to the Solomon fusion systems from Section 9.2, if one wants to prove that these fusion systems are not the fusion systems of blocks of any finite group, it suffices to check the blocks of the quasisimple groups.

We start with the exotic fusion systems on 7_+^{1+2}. In [KS08, Proposition 6.3] it is proved that if G is a quasisimple group such that $G/Z(G)$ is a simple group of Lie type in characteristic different from p, then kG has no p-blocks with defect group extraspecial of order p^3, for $p \geq 7$. If $G/Z(G)$ is a simple group of Lie type in defining characteristic then it is well known (see, for example, [Kes06, Lemma 5.1]) that all blocks have defect group either a Sylow p-subgroup of $G/Z(G)$ or the trivial group; if $G/Z(G)$ has extraspecial defect group 7_+^{1+2} then any block of full defect must have fusion system contained in that of G, so it cannot be one of the three exotic fusion systems.

Considering alternating groups, if p is an odd prime then the defect groups of blocks of alternating groups or double covers of alternating groups are Sylow p-subgroups of some symmetric group. It is easy to see that if a Sylow p-subgroup of S_n is of order p^3 and p is odd then $p \geq 5$ and the Sylow p-subgroup is abelian, so that extraspecial p-groups of order p^3 cannot be defect groups of these groups. It remains to check the sporadic quasisimple groups: however, the table from Section 9.3 of those sporadic groups that can support a non-principal block with exotic fusion system of order at least p^3 shows that only the Monster is a possibility; the Schur multiplier of M is trivial so the only such quasisimple group is simple, and we proved that M has no exotic fusion systems as the fusion systems of blocks. This completes the proof that the three exotic fusion systems on 7_+^{1+2} do not occur as the fusion system of a 7-block of a finite group.

Theorem 9.31 (Kessar–Stancu [KS08]) *Let $\mathcal{F}$ be one of the exotic saturated fusion systems on the 7-group 7_+^{1+2}. There is no finite group G such that $\mathcal{F}$ is the fusion system of a 7-block of G.*

We now turn our attention to the Solomon fusion systems $\mathcal{F}_{\text{Sol}}(r)$. Let Q be the finite 2-group on which $\mathcal{F}_{\text{Sol}}(r)$ is defined; then Q has 2-rank 4 by Proposition 9.17(i), and has an elementary abelian subgroup U of order 16 such that $\text{Aut}_{\mathcal{F}}(U) = \text{Aut}(U)$ by Proposition 9.18. Just these two facts together will almost prove that $\mathcal{F}_{\text{Sol}}(r)$ cannot be the fusion system of a 2-block of any quasisimple group. We will need some more information about symplectic groups as well, which we will deal with later. Our strategy follows that of Kessar [Kes06] closely (although in [Kes06] only the case $\mathcal{F}_{\text{Sol}}(3)$ is done, and we will need to modify the argument in one place), and so we omit some details and in particular do not completely reference arguments that are directly taken from there.

Lemma 9.32 *If G is a finite quasisimple group with $|Z(G)|$ odd, such that $G/Z(G)$ is an alternating group, a sporadic group, or of Lie type in characteristic 2, then there is no 2-block of G with fusion system $\mathcal{F}_{\text{Sol}}(r)$.*

Proof Let P be a Sylow 2-subgroup of G, and let Q be the 2-group on which $\mathcal{F}_{\text{Sol}}(r)$ is defined. Notice that $|Q| \geq 2^{10}$, and that Q has 2-rank 4. If $G/Z(G)$ is alternating then, since the 2-blocks of alternating groups have as defect groups Sylow 2-subgroups of (smaller) alternating groups, we need to find a Sylow 2-subgroup of an alternating group that has 2-rank 4 and order at least 2^{10}. However, the 2-rank of A_{2n+2} is easily seen to be n (as $(1,2)(3,4)$, $(3,4)(5,6)$ and so on commute), and the Sylow 2-subgroup of A_{10} has order 2^7, so this is impossible.

If $G/Z(G)$ is sporadic, Theorem 9.22 implies that $|Z(G)|$ is odd. By the information in the Atlas [CCNPW85] (or [GLS98, Table 6.1.3]), the sporadic simple groups with a non-trivial odd central extension are M_{22}, J_3, Suz, McL, ON, Fi_{22}, and Fi'_{24}, all of which have a 3-fold central extension. As the powers of 2 dividing the orders of M_{22}, J_3, McL, and ON are less than 2^{10}, these cannot support a 2-block with defect group Q, and so we are left with $3 \cdot Suz$, $3 \cdot Fi_{22}$, and $3 \cdot Fi'_{24}$. In all of these cases, GAP confirms that the only 2-blocks either have the Sylow 2-subgroup as defect group or have defect group of order less than 2^4. Hence $Q = P$. However, then $\mathcal{F}_{\text{Sol}}(r) \leq \mathcal{F}_P(G)$, contradicting Theorem 9.16. Hence $G/Z(G)$ is not a sporadic group.

Finally, we have the case where $G/Z(G)$ is a Lie type group in characteristic 2. As in [Kes06], unless $G/Z(G)$ is one of $^2F_2(2)'$ (the Tits group), $^2B_2(2)'$ or $^2G_2(2)'$, every 2-block of G has defect group either P or 1. In the former case, we again get a contradiction from Theorem 9.16, and the latter case is obviously impossible. For the remaining cases, $^2B_2(2)'$ and $^2G_2(2)'$ have Sylow 2-subgroups that are too small to possess Q as a subgroup, and $^2F_4(2)'$, with Sylow 2-subgroup of order 2^{11}, has no non-principal blocks with non-trivial defect group. Therefore $G/Z(G)$ is not of Lie type in characteristic 2. $\qquad\square$

We are left with the groups of Lie type in odd characteristic. In [Kes06] these are split into the exceptional and classical groups, and we will do the same here. Lemma 5.2 of [Kes06] proves that, unless $G/Z(G)$ is of type E_8, there is definitely no elementary abelian subgroup R of G order 2^4 such that $\mathrm{Aut}_G(R) = \mathrm{Aut}(R)$. In the case of $E_8(q)$, an analysis of maximal tori and the Weyl group is needed to prove that the subgroup R cannot be centric, completing this case. We will not repeat the proof here as it would require too many preliminaries. Although in [Kes06] only the case of $\mathcal{F}_{\mathrm{Sol}}(3)$ is considered, nothing in this argument requires this.

The remaining case is for $G/Z(G)$ to be a classical group in odd characteristic. Let G be a finite quasisimple group with $G/Z(G)$ a classical group in odd characteristic, and let b be a block idempotent of kG with fusion system $\mathcal{F}_{\mathrm{Sol}}(r)$, and defect group Q. Here an argument based on the 2-ranks of classical groups can be used to remove almost all cases. Write $\bar{G} = G/Z(G)$. The following choices are made dependent on which group $\bar{G}$ is.

- If $\bar{G} = \mathrm{PSL}_n(q)$, let $\tilde{G} = \mathrm{SL}_n(q)$ and $L = \mathrm{GL}_n(q)$.
- If $\bar{G} = \mathrm{PSU}_n(q)$, let $\tilde{G} = \mathrm{SU}_n(q)$ and $L = \mathrm{GU}_n(q)$.
- If $\bar{G} = \mathrm{P\Omega}_{2n+1}(q)$, let $\tilde{G} = \bar{G}$ and $L = \mathrm{O}_{2n+1}(q)$, the full orthogonal group.
- If $\bar{G} = \mathrm{PSp}_{2n}(q)$, let $\tilde{G} = L = \mathrm{Sp}_{2n}(q)$.
- If $\bar{G} = \mathrm{P\Omega}_{2n}^+(q)$, let $\tilde{G} = \Omega_{2n}^+(q)$ and $L = \mathrm{O}_{2n}^+(q)$.
- If $\bar{G} = \mathrm{P\Omega}_{2n}^-(q)$, let $\tilde{G} = \Omega_{2n}^-(q)$ and $L = \mathrm{O}_{2n}^-(q)$.

We see that, unless $\bar{G}$ is one of $\mathrm{PSL}_2(9)$, $\mathrm{PSU}_3(3)$, $\mathrm{P\Omega}_7(3)$, and $G_2(3)$ (where there are exceptional central extensions of $\bar{G}$ [GLS98, Table 6.1.3]), $\tilde{G}$ is a central extension of G and $\tilde{G}$ is a normal subgroup of L.

At this point we need a lemma.

Lemma 9.33 *Let G be a finite group and let b be a block idempotent of kG such that $\mathcal{F}_{\mathrm{Sol}}(r)$ is the fusion system of b, and write Q for a defect group of b. Let $\hat{G}$ be a central extension of G with kernel Z, and S be the Sylow 2-subgroup of Z. If $\hat{b}$ denotes the block idempotent whose image under the canonical map $k\hat{G} \to kG$ is b, then $\hat{b}$ has defect group $S \times Q$ and has fusion system $\mathcal{F}_{\mathrm{Sol}}(r) \times \mathcal{F}_S(S)$.*

Proof We will use Proposition 5.70. If Z is a p'-group then this is clear, so we assume that $Z = S$. Let $\mathcal{F}$ denote the fusion system of b and $\hat{\mathcal{F}}$ denote the fusion system of $\hat{b}$. Notice that $S \leq Z(\hat{\mathcal{F}})$ and $\hat{\mathcal{F}}/S = \mathcal{F}$. Since $\mathcal{F}$ contains the fusion system of $\mathrm{Spin}_7(r)$, a group with trivial Schur multiplier in the prime 2, $M(\mathcal{F})$ is trivial. Hence $\hat{\mathcal{F}} = \mathcal{F} \times \mathcal{F}_S(S)$. $\qquad \square$

Let S denote a Sylow 2-subgroup of the kernel of the map from $\tilde{G}$ to G, and let $\tilde{Q}$ be a Sylow 2-subgroup of the preimage of Q under the map $\tilde{G} \to G$. If $\tilde{b}$ denotes the block idempotent whose image under the map $k\tilde{G} \to kG$ is b, then $\tilde{b}$ has defect group $\tilde{Q}$. By the above lemma, $\tilde{Q} = Q \times S$.

There is some block idempotent f of kL with defect group R that covers $\tilde{b}$, so that $\tilde{Q} = R \cap \tilde{G}$. However, by our choice of L, we may apply the theory of the classical groups, which states that there is some semisimple element s of odd order in L such that R is a Sylow 2-subgroup of $C_L(s)$. (See [Bro86, Proposition 4.18] and [An93, Section 5A].) The structure of centralizers of semisimple elements of classical groups is well understood, and we have that $C_L(s)$ is a product of groups L_i, with L_i either a cyclic group or isomorphic to one of $\mathrm{GL}_m(q')$, $\mathrm{GU}_m(q')$, $O_{2m+1}(q')$, $\mathrm{Sp}_{2m}(q')$ or $O_{2m}^{\pm}(q')$ for q' a power of q.

Since Q is a subgroup of $C_L(s)$, and $Z(Q)$ is cyclic, there must be a subgroup isomorphic to Q in one of the L_i. Since there is no faithful representation of Q of dimension less than 8 by Proposition 9.17(iv), L_i must be a classical group of dimension at least 8. Also, since $\tilde{Q}/Q$ is cyclic and Q has 2-rank 4 by Proposition 9.17(i), $\tilde{Q}$ has 2-rank at most 5 (since it might be true that $\tilde{Q} = Q$); similarly, since $\tilde{Q} = R \cap \tilde{G}$ and $L/\tilde{G}$ is either cyclic or Klein four, the 2-rank of R, and hence L_i, is at most 7.

In [CF64], the Sylow 2-subgroups of the finite classical groups are studied, and it is shown that each of $\mathrm{GL}_m(q')$, $\mathrm{GU}_m(q')$, $O_m(q')$ (m odd) and $O_m^{\pm}(q')$ (m even) for q' a power of q has 2-rank greater than 7 when $m \geq 8$. Therefore none of these is the subgroup L_i, and hence $L_i \cong \mathrm{Sp}_{2m}(q')$ for some $m \geq 4$. The structure of the Sylow 2-subgroups of symplectic groups in [CF64] states that R_i is a direct product of groups $W_{a,b}$, where $W_{a,b}$ is a wreath product of a generalized quaternion 2-group Q_{2^a} by the iterated wreath product $C_2 \wr C_2 \wr \cdots \wr C_2$ of b copies of C_2. As before, since the centre of Q is cyclic, the fact that Q is a subgroup of R_i means that Q is isomorphic to a subgroup of one of the factors $W_{a,b}$. Since R_i has 2-rank at most 7 and Q has 2-rank 4, we must have that $Q \leq W_{a,2} = Q_{2^a} \wr D_8$.

Let X be the base group of $Q_{2^a} \wr D_8$, four copies of Q_{2^a}, and let $Z = \mathrm{Z}(X)$, an elementary abelian normal subgroup of R_i of order 16, whose automizer has order 8. If $Z \leq Q$ then Q contains a normal elementary abelian subgroup such that $Q/\mathrm{C}_Q(Z)$ has order at most 8, which contradicts Proposition 9.17, since all such subgroups of Q are self-centralizing and $|Q| \geq 2^{10}$. Therefore $Z \nleq Q$; since $|X : X \cap Q| \leq 8$, this can only happen if $a = 3$ and $|X : X \cap Q| = 8$, so that $X \cap Q$ is the direct product of three copies of Q_8. However, $|Q_8 \wr D_8| = 2^{15}$, so that $|Q| = 2^{12}$; this cannot happen by Proposition 9.17(i).

Thus $\mathcal{F}_{\mathrm{Sol}}(r)$ cannot be the fusion system of any 2-block of any odd central extension of any simple group, as needed.

Theorem 9.34 *Let r be an odd prime power, and let $\mathcal{F}$ be $\mathcal{F}_{\mathrm{Sol}}(r)$. There is no finite group G such that $\mathcal{F}$ is the fusion system of a 2-block of G.*

9.5 Abstract centric linking systems

In Chapter 3, we introduced the notion of a centric linking system for the fusion system of a finite group. At the time we only defined them in terms of the group, but in this section we will produce an abstract definition of a centric linking system that applies to *any* saturated fusion system, not just those associated to finite groups.

We begin with a piece of notation: if $\mathcal{F}$ is a fusion system on a finite p-group P, then by $\mathcal{F}^c$ we mean the full subcategory of $\mathcal{F}$ on the set of all $\mathcal{F}$-centric subgroups of P. Hence the objects of $\mathcal{F}^c$ are all $\mathcal{F}$-centric subgroups of P, and if Q and R are $\mathcal{F}$-centric subgroups then $\mathrm{Hom}_{\mathcal{F}}(Q, R) = \mathrm{Hom}_{\mathcal{F}^c}(Q, R)$.

A centric linking system is a category $\mathcal{L}$ with a surjective morphism $\pi : \mathcal{L} \to \mathcal{F}^c$. If Q is an $\mathcal{F}$-centric subgroup of P, then the standard homomorphism $Q \to \mathrm{Aut}_{\mathcal{F}}(Q)$ has as kernel $\mathrm{Z}(Q)$. The centric linking system comes with maps $\delta_Q : Q \to \mathrm{Aut}_{\mathcal{L}}(Q)$ with *no* kernel, and also $\mathrm{Aut}_{\mathcal{L}}(Q)\pi = \mathrm{Aut}_{\mathcal{F}}(Q)$, with kernel exactly $\mathrm{Z}(Q)\delta_Q$, so that we have inserted the centre of Q underneath $\mathrm{Aut}_{\mathcal{F}}(Q)$ to form $\mathrm{Aut}_{\mathcal{L}}(Q)$.

Definition 9.35 Let $\mathcal{F}$ be a fusion system on a finite p-group P. A *centric linking system* associated to $\mathcal{F}$ is a category $\mathcal{L}$, whose objects are all $\mathcal{F}$-centric subgroups of P, together with a functor $\pi : \mathcal{L} \to \mathcal{F}^c$, and monomorphisms $\delta_Q : Q \to \mathrm{Aut}_{\mathcal{L}}(Q)$ for each $\mathcal{F}$-centric subgroup $Q \leq P$, which satisfies the following conditions:

(i) the functor π is the identity on objects, and for $Q, R \in \mathcal{F}^c$, we have that $Z(Q)$ acts freely on $\operatorname{Hom}_{\mathcal{L}}(Q, R)$ by composition on the left (identifying $Z(Q)$ with $(Z(Q))\delta_Q$), and π induces a bijection

$$\operatorname{Hom}_{\mathcal{L}}(Q, R)/Z(Q) \cong \operatorname{Hom}_{\mathcal{F}}(Q, R);$$

(ii) for each $\mathcal{F}$-centric subgroup $Q \leq P$ and each $x \in P$, the image of $x\delta_Q$ under π is $c_x \in \operatorname{Aut}_{\mathcal{F}}(Q)$; and

(iii) for every $\phi \in \operatorname{Hom}_{\mathcal{L}}(Q, R)$ and $x \in Q$, writing $\psi \in \operatorname{Hom}_{\mathcal{F}}(Q, R)$ for the image of ϕ under π, we have that

$$x\delta_Q \circ \phi = \phi \circ (x\psi)\delta_R$$

as maps from Q to R.

The three conditions on the centric linking system $\mathcal{L}$ relate to making sure that the functor π and the map δ_Q are compatible with the morphism sets $\operatorname{Hom}_{\mathcal{L}}(Q, R)$ and $\operatorname{Hom}_{\mathcal{F}}(Q, R)$. However, given a saturated fusion system $\mathcal{F}$, it is not clear whether there is an associated centric linking system $\mathcal{L}$ at all, and even if there is one, it is not clear whether it is unique. Conjecturally it is though, and in the next section we will arrive at a more precise conjecture concerning the existence and uniqueness of the centric linking system.

The first thing we should do is prove that the centric linking system we constructed in Section 3.5 is actually a centric linking system in the sense of the definition above.

Proposition 9.36 *Let G be a finite group and let P be a Sylow p-subgroup of G. The centric linking system $\mathcal{L}_P^c(G)$ of G is a centric linking system for $\mathcal{F}_P(G)$ in the sense of Definition 9.35.*

Proof Let $\mathcal{F} = \mathcal{F}_P(G)$ and $\mathcal{L} = \mathcal{L}_P^c(G)$, and if Q is an $\mathcal{F}$-centric subgroup of G, so that $C_G(Q) = Z(Q) \times O_{p'}(C_G(Q))$, write $C_G'(Q) = O_{p'}(C_G(Q))$. Certainly $\mathcal{L}$ is defined on the right objects, and the functor $\pi : \mathcal{L} \to \mathcal{F}^c$ is just the map sending objects to the same objects, and with maps on morphism sets given by

$$\pi : N_G(Q, R)/\, C_G'(Q) \to N_G(Q, R)/\, C_G(Q).$$

The distinguished morphisms $\delta_Q : Q \to \operatorname{Aut}_{\mathcal{L}}(Q)$ are given by sending $g \in Q$ to the element $C_G'(Q)g$ in $N_G(Q)/\, C_G'(Q)$ (this is a genuine quotient). We need to check the conditions for $\mathcal{L}$ to be a centric linking system.

The subgroup $(Z(Q))\delta_Q$ of $\mathrm{Aut}_{\mathcal{L}}(Q)$ does indeed act freely by composition, and the map

$$\pi : \mathrm{Hom}_{\mathcal{L}}(Q, R)/Z(Q) \to \mathrm{Hom}_{\mathcal{F}}(Q, R)$$

is definitely a bijection; thus the first condition is satisfied. Also, by construction of δ_Q, it sends $g\delta_Q$ to $c_g \in \mathrm{Aut}_{\mathcal{F}}(Q)$, and so the second condition is satisfied.

The third condition is slightly more complicated. Suppose that $\phi \in \mathrm{Hom}_{\mathcal{L}}(Q, R)$ arises from $g \in \mathrm{N}_G(Q, R)$, and let $x \in Q$. Notice that $\phi\pi = c_g \in \mathrm{Hom}_{\mathcal{F}}(Q, R)$. We must prove that, as elements of $\mathrm{Hom}_{\mathcal{L}}(Q, R)$, we have $x\delta_Q \circ \phi = \phi \circ (x\psi)\delta_R$.

We have that

$$x\delta_Q \circ \phi = \mathrm{C}'_G(Q)x\,\mathrm{C}'_G(Q)g = \mathrm{C}'_G(Q)xg,$$

and

$$\phi \circ (x\psi)\delta_R = \phi \circ (xc_g)\delta_R = \phi \circ (x^g)\delta_R.$$

Now, $(x^g)\delta_R = \mathrm{C}'_G(R)(g^{-1}xg) = g^{-1}\,\mathrm{C}'_G(Q)xg$, so that this expression becomes $\mathrm{C}'_G(Q)g \circ g^{-1}\,\mathrm{C}'_G(Q)xg = \mathrm{C}'_G(Q)xg$. Therefore (iii) holds, and so $\mathcal{L}$ is a centric linking system, as required. $\square$

Proposition 3.65 shows that $|\mathcal{L}^c_P(G)|^{\wedge}_p \simeq BG^{\wedge}_p$, so one might expect that if $\mathcal{L}$ is a centric linking system associated to a saturated fusion system $\mathcal{F}$ then $|\mathcal{L}|^{\wedge}_p$ should behave like a p-completed classifying space for $\mathcal{F}$. Indeed, this is the role that $\mathcal{L}$ plays, allowing us to construct the model of a constrained fusion system, for example. We also saw in Section 3.5 (and Theorem 3.66 in particular) that the proof of the Martino–Priddy conjecture requires that the centric linking system $\mathcal{L}^c_P(G)$ is the only centric linking system that is associated to $\mathcal{F}_P(G)$. Hence the uniqueness of centric linking systems is an important topic.

The existence and uniqueness of centric linking systems are controlled by certain higher limits, as is usual. The particular functor that we need to take higher limits of is the centre functor, and in order to define it we need some more machinery.

Definition 9.37 Let $\mathcal{F}$ be a saturated fusion system on a finite p-group P. The *orbit category* $\mathcal{O}(\mathcal{F})$ of $\mathcal{F}$ is the category whose objects are the same as those of $\mathcal{F}$, and whose morphism sets are given by

$$\mathrm{Hom}_{\mathcal{O}(\mathcal{F})}(Q, R) = \mathrm{Hom}_{\mathcal{F}}(Q, R)/\mathrm{Aut}_R(R),$$

with composition of morphisms induced from that of $\mathcal{F}$. As with $\mathcal{F}$, by

$\mathcal{O}(\mathcal{F}^c)$ we denote the full subcategory of $\mathcal{O}(\mathcal{F})$ on the $\mathcal{F}$-centric subgroups of P.

This category is well defined, since if ϕ and ϕ' are morphisms in $\mathrm{Hom}_{\mathcal{F}}(Q, R)$ whose image in $\mathcal{O}(\mathcal{F})$ is the same, and ψ and ψ' are morphisms in $\mathrm{Hom}_{\mathcal{F}}(R, S)$ whose image in $\mathcal{O}(\mathcal{F})$ is the same, then $\phi\psi$ and $\phi'\psi'$ have the same image in $\mathcal{O}(\mathcal{F})$, as needed for this category to work.

With the category $\mathcal{O}(\mathcal{F}^c)$, we can define the functor that we need. This will be a contravariant functor to abelian groups.

Definition 9.38 Let $\mathcal{F}$ be a saturated fusion system on a finite p-group P. The *centre functor* is a functor from $\mathcal{O}(\mathcal{F}^c)^{\mathrm{op}}$ to the the category Ab of abelian groups,

$$\mathcal{Z}_{\mathcal{F}} : \mathcal{O}(\mathcal{F}^c)^{\mathrm{op}} \longrightarrow \mathsf{Ab},$$

by setting $\mathcal{Z}_{\mathcal{F}}(Q) = \mathrm{Z}(Q) = \mathrm{C}_P(Q)$, for each $\mathcal{F}$-centric subgroup $Q \leq P$. If $\phi : Q \to R$ is a morphism in $\mathrm{Hom}_{\mathcal{O}(\mathcal{F}^c)}(Q, R)$, since $\mathrm{Z}(R) \leq \mathrm{Z}(Q)\phi$, there is an induced injective map $\mathcal{Z}_{\mathcal{F}}(\phi) : \mathrm{Z}(R) \to \mathrm{Z}(Q)$.

The existence of a centric linking system associated to a saturated fusion system is governed by a particular 3-cocycle, an element lying in $\varprojlim_{\mathcal{O}(\mathcal{F}^c)}{}^3 \mathcal{Z}_{\mathcal{F}}$.

Theorem 9.39 (Broto–Levi–Oliver [BLO03b, Proposition 3.1]) *Let $\mathcal{F}$ be a saturated fusion system on a finite p-group P. There exists an element $\eta(\mathcal{F})$ in $\varprojlim_{\mathcal{O}(\mathcal{F}^c)}{}^3 \mathcal{Z}_{\mathcal{F}}$ such that $\eta(\mathcal{F}) = 0$ if and only if $\mathcal{F}$ has an associated centric linking system $\mathcal{L}$.*

The uniqueness of centric linking systems is governed by the set of 2-cocycles, as in the statement of the next theorem.

Theorem 9.40 (Broto–Levi–Oliver [BLO03b, Proposition 3.1]) *Let $\mathcal{F}$ be a saturated fusion system on a finite p-group P, and suppose that there is a centric linking system associated with $\mathcal{F}$. The set $\varprojlim_{\mathcal{O}(\mathcal{F}^c)}{}^2 \mathcal{Z}_{\mathcal{F}}$ acts regularly (i.e., freely and transitively) on the isomorphism classes of triples $(\mathcal{L}, \pi, \delta)$, where $\mathcal{L}$ is a centric linking system for $\mathcal{F}$ with associated maps $\pi : \mathcal{L} \to \mathcal{F}$ and $\delta_Q : Q \to \mathrm{Aut}_{\mathcal{L}}(Q)$ for Q and $\mathcal{F}$-centric subgroup of P.*

We will not prove either of these results here, but we will construct the element $\eta(\mathcal{F})$. Before we do so, we note that, using the fact that

these higher limits vanish for small p-ranks, the following theorem, also
due to Broto, Levi and Oliver, is true.

Theorem 9.41 (Broto–Levi–Oliver [BLO03b, Theorem E]) *Let $\mathcal{F}$ be
a saturated fusion system on a finite p-group P. If P has p-rank at most
$p^3 - 1$, then there is an associated centric linking system $\mathcal{L}$ to $\mathcal{F}$, and
if P has p-rank at most $p^2 - 1$, this associated centric linking system is
unique.*

This theorem is true because, if the p-rank of a group is at most $p^i - 1$,
then the ith derived functor of $\mathcal{Z}_{\mathcal{F}}$ can be proved to vanish.

We turn our attention to constructing the 3-cocycle in Theorem 9.39.
Our notation will match that of [BLO03b]: fix any section $\sigma : \mathcal{O}(\mathcal{F}^c) \to
\mathcal{F}^c$ (i.e., if $\tau : \mathcal{F}^c \to \mathcal{O}(\mathcal{F}^c)$ is the natural map then $\sigma\tau = \text{id}$), so that σ
acts like the identity on objects, and choose σ so that it sends identity
maps to identity maps. For any map ϕ in $\mathcal{O}(\mathcal{F}^c)$ we denote its image
under σ by $\tilde{\phi}$. Let Q, R, S, and T be $\mathcal{F}$-centric subgroups of P. Since
in the orbit category we quotient out by $\text{Inn}(Q)$, and so leave $\text{Out}_{\mathcal{F}}(Q)$,
and the centric linking system embeds all of Q underneath $\text{Out}_{\mathcal{F}}(Q)$, we
define

$$X(Q, R) = \text{Hom}_{\mathcal{O}(\mathcal{F}^c)}(Q, R) \times R.$$

This set $X(Q, R)$ will play the role of $\text{Hom}_{\mathcal{L}}(Q, R)$, so we define a map
$\pi_{\sigma}^{Q,R} : X(Q, R) \to \text{Hom}_{\mathcal{F}}(Q, R)$ by setting $(\phi, g)\pi_{\sigma}^{Q,R} = \tilde{\phi}c_g$. If $\phi : Q \to
R$ and $\psi : R \to S$ are composable morphisms in $\mathcal{O}(\mathcal{F}^c)$, choose $t_{\phi,\psi} \in S$
such that

$$\tilde{\phi}\tilde{\psi} = \widetilde{\phi\psi}c_{t_{\phi,\psi}}$$

and such that $t_{\phi,\psi} = 1$ if either ϕ or ψ is the identity map. Hence the $t_{\phi,\psi}$
measure the difference between the map σ being a morphism. Since the
$X(Q, R)$ are supposed to act as the morphism sets of a centric linking
system, we need a map $*$ from $X(Q, R) \times X(R, S)$ to $X(Q, S)$, and this
is given by

$$(\phi, g) * (\psi, h) = (\phi\psi, t_{\phi,\psi}(g\tilde{\psi})h).$$

We now construct the element $u_{\phi,\psi,\chi}$, which will identify how much the
composition operation $*$ deviates from being associative. Suppose that
$(\phi, g_1)\pi_{\sigma}^{Q,R} = (\phi, g_2)\pi_{\sigma}^{Q,R}$ for some $g_i \in R$ and $\phi \in \text{Hom}_{\mathcal{O}(\mathcal{F}^c)}(Q, R)$; the
element $g_1^{-1}g_2$ centralizes $Q\tilde{\phi}$, so that $g_1^{-1}g_2 \in Z(Q\tilde{\phi})$. Therefore there
is some element $u \in Z(Q)$ such that $(\phi, g_1) = (\text{id}, u) * (\phi, g_2)$. Also, we

have that the diagram

$$
\begin{array}{ccc}
X(Q,R) \times X(R,S) & \xrightarrow{\quad * \quad} & X(Q,S) \\
{\scriptstyle (\pi_\sigma^{Q,R}, \pi_\sigma^{R,S})} \Big\downarrow & & \Big\downarrow {\scriptstyle \pi_\sigma^{Q,S}} \\
\mathrm{Hom}_{\mathcal{F}}(Q,R) \times \mathrm{Hom}_{\mathcal{F}}(R,S) & \xrightarrow{\quad \circ \quad} & \mathrm{Hom}_{\mathcal{F}}(Q,S)
\end{array}
$$

commutes.

For every triple of composable maps $\phi : Q \to R$, $\psi : R \to S$, and $\chi : S \to T$ in $\mathcal{O}(\mathcal{F}^c)$, we consider the two elements $a_1 = \big((\phi,1) * (\psi,1)\big) * (\chi,1)$ and $a_2 = (\phi,1) * \big((\psi,1) * (\chi,1)\big)$, elements of $X(Q,T)$. Since composition in $\mathcal{F}$ is associative, $a_1 \pi_\sigma^{Q,T} = a_2 \pi_\sigma^{Q,T}$, and the first co-ordinate of each a_i is $\phi\psi\chi$, so that there is some $u_{\phi,\psi,\chi} \in Z(Q)$ such that

$$
a_1 = (\mathrm{id}, u_{\phi,\psi,\chi}) * a_2.
$$

These elements u form a 3-cocycle $\eta(\mathcal{F})$ with respect to the centre functor, although we will not prove this fact here. A little more manipulation shows that we actually get

$$
\big((\phi,g) * (\psi,h)\big) * (\chi,k) = (\mathrm{id}, u_{\phi,\psi,\chi}) * \big[(\phi,g) * \big((\psi,h) * (\chi,k)\big)\big]
$$

for all $g \in R$, $h \in S$, and $k \in T$.

We now explain the rest of the proof from [BLO03b]. The element $u_{\phi,\psi,\chi}$ is a 3-cocycle, and it turns out that (modulo coboundaries) it is independent of the choice of both the $t_{\phi,\psi}$ and of the section σ.

If $u = u_{\phi,\psi,\chi}$ is a coboundary, then we may choose the elements $t_{\phi,\psi}$ such that $u = 1$ for all ϕ, ψ, and χ. As $u = 1$ with respect to the choice of t, we set $\mathrm{Hom}_{\mathcal{L}}(Q,R) = X(Q,R)$, where Q and R are $\mathcal{F}$-centric subgroups of P. Because the cocycle u is 1, composition of maps between the various $\mathrm{Hom}_{\mathcal{L}}(Q,R)$ is associative, so that $\mathcal{L}$ forms a category. To prove that $\mathcal{L}$ is a centric linking system, the map $\pi : \mathcal{L} \to \mathcal{F}^c$ is given by $(\phi,g)\pi = \tilde{\phi}c_g$, and the maps $\delta_Q : Q \to \mathrm{Aut}_{\mathcal{L}}(Q)$ are simply given by $g\delta_Q = (\mathrm{id}, g)$. It is easily checked that these yield a centric linking system.

Conversely, suppose that $\mathcal{L}$ is a centric linking system for $\mathcal{F}$. Fix some section σ, which can be lifted to a section from $\mathcal{O}(\mathcal{F}^c)$ to $\mathcal{L}$ via the functor π, and so we get bijections $X(Q,R) \cong \mathrm{Hom}_{\mathcal{L}}(Q,R)$. The axioms for a centric linking system (in particular, the third) yield maps $X(Q,R) \times X(R,S) \to X(Q,S)$ for some choice of the elements t; since u measures how far this is from being associative, as this composition comes from $\mathcal{L}$, which is a category, we must have that u is a coboundary.

Hence $\mathcal{F}$ has an associated centric linking system if and only if the element u is a coboundary.

We end this section by defining a p-local finite group. These have become important objects of study in the topological approach to fusion systems, although we will say nothing more about them here.

Definition 9.42 A *p-local finite group* on P is a triple $(P, \mathcal{F}, \mathcal{L})$, where P is a finite p-group, $\mathcal{F}$ is a saturated fusion system over P, and $\mathcal{L}$ is an associated centric linking system of $\mathcal{F}$. If $(P, \mathcal{F}, \mathcal{L})$ is a p-local finite group, its *classifying space* is the space $|\mathcal{L}|_p^\wedge$.

In the case where $(P, \mathcal{F}, \mathcal{L})$ is a p-local finite group arising from a finite group G, then, as we have seen in Proposition 3.65, we have that the classifying space of $(P, \mathcal{F}, \mathcal{L})$ is homotopy equivalent to $BG_p^\wedge$. Theorem 3.64 stated that $BG_p^\wedge$ and $|\mathcal{L}_P^c(G)|_p^\wedge$ are homotopy equivalent. Related to this, we have the following very general theorem.

Theorem 9.43 (Broto–Levi–Oliver [BLO03b, Theorem A]) *Suppose that $(P, \mathcal{F}, \mathcal{L})$ and $(\bar{P}, \bar{\mathcal{F}}, \bar{\mathcal{L}})$ are p-local finite groups. If $|\mathcal{L}|_p^\wedge$ and $|\bar{\mathcal{L}}|_p^\wedge$ are homotopy equivalent, then $(P, \mathcal{F}, \mathcal{L})$ and $(\bar{P}, \bar{\mathcal{F}}, \bar{\mathcal{L}})$ are isomorphic, i.e., there is an isomorphism $\phi : P \to \bar{P}$ that induces isomorphisms of fusion systems and of centric linking systems.*

9.6 Higher limits and centric linking systems

In the last section we considered centric linking systems, and determined that the obstruction to the existence of a centric linking system is a particular 3-cocycle and that the 2-cocycles act transitively on the isomorphism classes of centric linking systems.

In the case of a group fusion system, obviously a centric linking system exists, and Oliver proved that for these saturated fusion systems all higher limits vanish if p is odd, and all higher limits *apart from the first one* vanish for $p = 2$.

Of course, one may ask these questions for an arbitrary saturated fusion system.

Conjecture 9.44 *Let $\mathcal{F}$ be a saturated fusion system on a finite p-group P. There is a unique centric linking system $\mathcal{L}$ associated to $\mathcal{F}$, and hence a unique p-local finite group associated to $\mathcal{F}$.*

This is a very reasonable conjecture, but written like this it is difficult to attack. In this section we will reformulate Conjecture 9.44 until we get a simple conjecture on failure of factorization modules for p-groups, at least for odd primes.

We now turn to a conjecture on higher limits of the centre functor, motivated by the previous section.

Conjecture 9.45 (Oliver [Oli04, Conjecture 2.2]) *Let $\mathcal{F}$ be a saturated fusion system on a finite p-group P. The higher limits*

$$\varprojlim_{\mathcal{O}(\mathcal{F}^c)}{}^i \mathcal{Z}_{\mathcal{F}}$$

of the centre functor vanish for all $i \geq 2$. If p is odd, then this higher limit also vanishes for $i = 1$.

Because of the location of the obstructions to existence and uniqueness of centric linking systems, Conjecture 9.45 implies Conjecture 9.44; this conjecture on vanishing of higher limits was proved by Oliver for all fusion systems of finite groups in [Oli04] and [Oli06].

In fact, Conjecture 9.45 follows from a purely group-theoretic conjecture given in [Oli04]. We begin with a definition. (Recall that if H and K are subgroups of a group G, we define $[H, K; 1]$ to be $[H, K]$, and inductively

$$[H, K; i] = \big[[H, K; i - 1], K\big],$$

so that $[H, K; i]$ is the commutator of H and i copies of K.)

Definition 9.46 A normal subgroup K of a finite p-group P has a *Q-series* if there is a series

$$1 = Q_0 \leq Q_1 \leq \cdots \leq Q_n = K$$

of normal subgroups of P such that $[\Omega_1(C_P(Q_{i-1})), Q_i; p-1] = 1$.

We will now prove that there is a largest such subgroup.

Lemma 9.47 *Let P be a finite p-group and let R and S be normal subgroups of P. If R and S have Q-series, so does RS.*

Proof Let $1 = Q_0 \leq \cdots \leq Q_n = R$ and $1 = Q_0' \leq \cdots \leq Q_m' = S$ be Q-series for R and S; we claim that the series

$$1 = Q_0 \leq Q_1 \leq \cdots \leq Q_n = Q_n Q_0' \leq Q_n Q_1' \leq \cdots \leq Q_n Q_m' = RS$$

is a Q-series for RS. It is clear that each of the terms in this series is a normal subgroup of P the first $n + 1$ terms (up to Q_n) inherit the right property from the Q-series for R, and so it suffices to check the property for $Q_n Q_i'$ with $i > 0$. In this case,

$$[\Omega_1(C_P(Q_n Q_{i-1}')), Q_n Q_i'; p-1] = [\Omega_1(C_P(Q_n Q_{i-1}')), Q_i'; p-1]$$
$$\leq [\Omega_1(C_P(Q_{i-1}')), Q_i'; p-1] = 1.$$

(The first equality occurs because the first term in the commutator centralizes Q_n, and the third equality holds by assumption.) Thus this is a Q-series, as claimed. □

Thus it makes sense to make the following definition.

Definition 9.48 Let P be a finite p-group. The *Oliver subgroup*, $\mathfrak{X}(P)$, is the largest normal subgroup of P possessing a Q-series.

The Oliver subgroup is obviously a characteristic subgroup, and is non-trivial since it contains the centre. (For odd primes it contains any abelian normal subgroup of P, since if A is such a subgroup, then $[[P, A], A] \leq [A, A] = 1$.) The Oliver conjecture is that, not only does it contain every *normal* abelian subgroup, it also contains all elementary abelian subgroups of maximal order, i.e., it contains the Thompson subgroup.

Conjecture 9.49 (Oliver conjecture [Oli04, Conjecture 3.9]) *If p is an odd prime and P is a finite p-group, then*

$$J(P) \leq \mathfrak{X}(P).$$

The restriction to odd primes here is because in the case where $p = 2$ the required condition becomes $[\Omega_1(C_P(Q_{i-1})), Q_i] = 1$; notice that with $i = 1$, $[\Omega_1(P), Q_1] = 1$ and so Q_1 may be taken to be $C_P(\Omega_1(P))$. If $R \geq Q_1$ is any other normal subgroup such that $1 \leq Q_1 \leq R$ is a Q-series for R, then

$$[\Omega_1(C_P(Q_1)), R] = 1.$$

However, clearly $\Omega_1(P) \leq C_P(Q_1)$, and so the left-hand term of the above commutator is $\Omega_1(P)$. Therefore R centralizes $\Omega_1(P)$ and so $Q = R$.

Therefore, in a 2-group, $\mathfrak{X}(P) = C_P(\Omega_1(P))$; if P is generated by its involutions, this simply means that $\mathfrak{X}(P) = Z(P)$, and so the subgroup contains little information about the group's structure in this case.

An interesting fact about the Oliver subgroup is that (for odd primes) it contains its centralizer, i.e., $C_P(\mathfrak{X}(P)) \leq \mathfrak{X}(P)$. To see this, we first note that, in any nilpotent group, the abelian subgroups maximal with respect to being normal are self-centralizing (Exercise 9.6); since any overgroup of a subgroup that contains its centralizer also contains its centralizer, and $\mathfrak{X}(P)$ contains all abelian normal subgroups, we must have that $\mathfrak{X}(P)$ is centric.

We will also prove that low-rank p-groups definitely satisfy the Oliver conjecture; in Theorem 9.41, a similar result about existence and uniqueness of centric linking systems was obtained by examining the higher limits directly.

Lemma 9.50 *Let P be a finite p-group and let $X = \mathfrak{X}(P)$.*

(i) *We have that $C_P(X) = Z(X)$.*

(ii) *If Q is a normal subgroup of P such that $[\Omega_1(Z(X)), Q; p-1] = 1$, then $Q \leq X$.*

(iii) *If $Z(X)$ has p-rank less than p then $X = P$, and so the Oliver conjecture is true for P.*

Proof Clearly X contains any maximal abelian normal subgroup A of P, and for such subgroups A we have $C_P(A) \leq A$ by Exercise 9.6; hence $C_P(X) \leq X$, proving (i).

The subgroup X has a Q-series $1 = Q_0 \leq \cdots \leq Q_n = X$; writing $Q_{n+1} = Q_n Q$, and since $C_P(X) = Z(X)$, we have

$$1 = [\Omega_1(C_P(X)), Q; p-1] = [\Omega_1(C_P(Q_n)), Q_{n+1}; p-1],$$

and so $1 \leq \cdots \leq Q_n \leq Q_{n+1}$ is a Q-series for QX, proving (ii).

We turn to the proof of (iii). Write $E = \Omega_1(Z(X))$, and note that $[E, P; i] < [E, P; i-1]$ unless the latter subgroup is already trivial. Since E has order at most p^{p-1}, we must have that $[E, P; p-1] = 1$, and so $P \leq X$ by (ii). $\qquad\square$

We will now show that the Oliver conjecture implies Conjecture 9.45. The main part of the argument is the following proposition, which we will state without proof; it is a special case of [Oli04, Proposition 3.5].

Proposition 9.51 *Let $\mathcal{F}$ be a saturated fusion system on a finite p-group P, where p is an odd prime. If $\mathfrak{X}(P)$ contains a weakly $\mathcal{F}$-closed, $\mathcal{F}$-centric subgroup, then all higher limits of the functor $\mathcal{Z}_{\mathcal{F}}$ over $\mathcal{O}(\mathcal{F}^c)$ vanish, so Conjecture 9.45 holds.*

With this, we can prove the implication.

Proposition 9.52 *Let $\mathcal{F}$ be a saturated fusion system on a finite p-group P, and write $J = J(P)$ and $X = \mathfrak{X}(P)$.*

(i) *The subgroup* $J \, \mathrm{C}_P(J)$ *is* $\mathcal{F}$-*centric and weakly* $\mathcal{F}$-*closed.*

(ii) *If* $J \le X$ *then Conjecture 9.45 holds for* $\mathcal{F}$.

Proof Let $K = J \, \mathrm{C}_P(J)$. Since $J(P)$ is weakly $\mathcal{F}$-closed (as any elementary abelian subgroup of maximal rank can only be $\mathcal{F}$-conjugate to other elementary abelian subgroup of maximal rank) it is fully centralized, and so, by Lemma 4.42, K is $\mathcal{F}$-centric. Suppose that $\phi : K \to R$ is an isomorphism in $\mathcal{F}$: since J is weakly $\mathcal{F}$-closed, $J\phi = J$, and since $\mathrm{C}_P(J)$ centralizes J, $\mathrm{C}_P(J)\phi$ centralizes $J\phi = J$. Hence $\mathrm{C}_P(J)\phi = \mathrm{C}_P(J)$, and K is weakly $\mathcal{F}$-closed, proving (i).

By (i) and Proposition 9.51, Conjecture 9.45 holds if we can show that $K \le X$ whenever $J \le X$. Every elementary abelian subgroup E of maximal rank lies in $J \le X$, and if E did not contain $\Omega_1(\mathrm{Z}(X))$ then $E\Omega_1(\mathrm{Z}(X))$ would be a larger-rank elementary abelian subgroup, a contradiction. Thus $\Omega_1(\mathrm{Z}(X)) \le \mathrm{Z}(J)$, and therefore

$$[\Omega_1(\mathrm{Z}(X)), J \, \mathrm{C}_P(J)] \le [\mathrm{Z}(J), J \, \mathrm{C}_P(J)] = 1;$$

thus by Lemma 9.50(ii), $\mathrm{C}_P(J) \le X$, as claimed. $\qquad\qquad\square$

The Oliver conjecture is a purely group-theoretic statement, but appears difficult to attack directly. Yet another reformulation has been proposed by Green, Héthelyi, and Lilienthal [GHL08], using the concept of a failure of factorization module. These originated in work on factorization problems in finite groups, i.e., in finding certain subgroups H and K of a finite group G such that $G = HK$.

Definition 9.53 Let G be a finite group and let V be a faithful $\mathbb{F}_p G$-module. A non-trivial elementary abelian p-subgroup A of G is said to be an *offender* if $|A| \cdot |\mathrm{C}_V(A)| \ge |V|$. If V has an offender, then V is said to be a *failure of factorization module*, or *F-module* for short.

F-modules have seen a lot of attention in finite group theory, although mainly for simple groups. Here we are interested in p-groups for p odd. In this case, we have the following conjecture, given in [GHL08].

Conjecture 9.54 *Let p be an odd prime and P be a finite p-group. If M is an F-module for P, then there is some element x of order p in $\mathrm{Z}(P)$ such that $M \downarrow_{\langle \tau \rangle}$ has no projective summand.*

There is an equivalent form of this conjecture that is sometimes easier to consider; note that if x is an element of order p in a group G and M is an $\mathbb{F}_p G$-module, then the action of x on M satisfies the polynomial

$X^p - 1$, and the minimal polynomial for this action is $X^p - 1$ if and only if $M \downarrow_{\langle x \rangle}$ has a projective summand.

Conjecture 9.55 *Let p be an odd prime and P be a finite p-group. There is no F-module M for P such that every central element of order p in P acts on M with minimal polynomial $X^p - 1$.*

It remains to show that Conjectures 9.49 and 9.54 are equivalent, as proved in [GHL08, Corollary 1.4]. Neither direction of the proof is difficult.

Lemma 9.56 (Green–Héthelyi–Lilienthal [GHL08, Lemma 2.3]) *Let p be an odd prime, and let P be a finite p-group. Suppose that V is a faithful $\mathbb{F}_p P$-module such that every central element of order p acts on V with minimal polynomial $X^p - 1$.*

(i) *If $G = V \rtimes P$, then $\mathfrak{X}(G) = V$.*
(ii) *If V is an F-module, then $J(G) \not\leq \mathfrak{X}(G)$, and so Conjecture 9.49 implies Conjecture 9.54.*

Proof Let G be as in (i); we first show that there are no abelian normal subgroups of G strictly containing V. Suppose that A is a normal subgroup strictly containing V; in $P = G/V$, we must have that $\Omega_1(Z(P)) \cap A/V \neq 1$, and so let x be an element of $A \setminus V$ such that Vx is central in P and of order p. By assumption, x acts on V with minimal polynomial $X^p - 1$; if A is abelian, then this yields a contradiction. Hence V is a subgroup maximal in G with respect to being normal and abelian. By Exercise 9.6, $V = C_G(V)$.

Since V is a normal abelian subgroup we have $V \leq \mathfrak{X}(G)$, and $1 \leq V$ is a Q-series for V. By Lemma 9.47 (or rather its proof), a Q-series for $X = \mathfrak{X}(G)$ may be chosen to start with $1 = Q_0 \leq Q_1 = V$. In this case, if Q_2 is the next term in the Q-series for X, we have

$$[\Omega_1(C_P(V)), Q_2; p - 1] = [V, Q_2; p - 1] = 1.$$

However, as we have noted earlier, as Q_2 is a normal subgroup of G and strictly contains V, there is a central element $x \in Q_2$ of order p such that x acts on V with minimal polynomial $X^p - 1$; hence $[V, x; p - 1] \neq 1$, a contradiction, proving (i).

Now suppose that in addition V is an F-module, and let A be an offender; then $|A| \cdot |C_V(A)| \geq |V|$, and the subgroup $A \times C_V(A)$ is an elementary abelian subgroup of G of at least the same order as V, not contained within $V = \mathfrak{X}(G)$. Therefore $J(G) \not\leq V$, as needed. $\qquad\square$

Having proved one implication of the equivalence, we turn to the other.

Lemma 9.57 (Green–Héthelyi–Lilienthal [GHL08, Lemma 2.2]) *Let P be a finite p-group. Suppose that $\mathfrak{X}(P) < P$, and write $G = P/\mathfrak{X}(P)$. Let $V = \Omega_1(Z(\mathfrak{X}(P)))$.*

(i) *If x is a central element of G of order p, then the action of x on V has minimal polynomial $X^p - 1$.*

(ii) *If $J(P) \nleq \mathfrak{X}(P)$ then V is an F-module, and so Conjecture 9.54 implies Conjecture 9.49.*

Proof Let $H = \langle \mathfrak{X}(P), x \rangle$, and notice that $H \trianglelefteq P$; therefore $[V, x; p - 1] \neq 1$ as $H \nleq \mathfrak{X}(P)$ (using Lemma 9.50(ii)). Since x has order p, its minimal polynomial for the action on V divides $X^p - 1$, and therefore it is $X^p - 1$, as claimed in (i).

Now suppose that A is an elementary abelian subgroup of P of maximal order not contained in $\mathfrak{X}(P)$; write $A = A_1 \times A_2 \times A_3$, where $A_1 = A \cap V$ and $A_1 \times A_2 = A \cap \mathfrak{X}(P)$, and hence $A_3 \cap \mathfrak{X}(P) = 1$. Since $V \times A_2$ is an elementary abelian subgroup of $\mathfrak{X}(P)$, we must have that $|V| \cdot |A_2| \leq |A_1| \cdot |A_2| \cdot |A_3|$. Notice that $A_1 \leq C_V(A_3)$, and hence

$$|V| \leq |A_1| \cdot |A_3| \leq |A_3| \cdot |C_V(A_3)|;$$

therefore A_3 is an offender for V, and so V is an F-module, as claimed. $\square$

The implications that we have between the conjectures are as follows: Conjectures 9.49, 9.54, and 9.55 are all equivalent, and imply Conjecture 9.45, which in turn implies the existence and uniqueness of centric linking systems.

The main problem with moving between Conjecture 9.49 and the other two equivalent formulations is that the p-group changes. If one wants to prove Conjecture 9.49 for a particular p-group P (and hence prove that any saturated fusion system on P has a unique centric linking system associated to it) one proves Conjecture 9.54 for the quotient $P/\mathfrak{X}(P)$.

In [GHL08], Conjecture 9.54 was proved for all p-groups of class 2, and in [GHM10], this was extended by proving the following result.

Theorem 9.58 (Green–Héthelyi–Mazza [GHM10]) *Let P be a finite p-group, where p is odd. If one of the following conditions holds, then Conjecture 9.54 is true for P:*

(i) *P has nilpotence class at most 4;*

(ii) *P is metabelian;*

(iii) *P has maximal class; or*

(iv) *P has rank at most p.*

Some of these are difficult to translate back into results about $\mathfrak{X}(P)$, because this theorem only deals with the quotient $P/\mathfrak{X}(P)$; if any of the above conditions hold for $P/\mathfrak{X}(P)$, then any saturated fusion system on P has a unique centric linking system associated to it.

Exercises

9.1 Let p be an odd prime and let P be the p-group p_+^{1+2}, the extraspecial group of order p^3 and exponent p considered in Section 9.1. Hence P is generated by x and y of order p with $z = [x,y]$ of order p and commuting with both x and y.

(i) Prove that the only subgroups Q of P with $C_P(Q) \leq Q$ are P and the subgroups of index p. Hence prove that these are exactly the $\mathcal{F}$-centric subgroups of P for any fusion system $\mathcal{F}$ on P. Prove that there are $p + 1$ proper $\mathcal{F}$-centric subgroups of P, each elementary abelian of order p^2.

(ii) Prove that the map $x \mapsto x^a y^b z^c$ and $y \mapsto x^i y^j z^k$ is a automorphism of P if and only if $aj - ib \neq 0$, and prove that the inner automorphisms of P are simply given by $x \mapsto xz^c$ and $y \mapsto yz^k$. Hence deduce that $\mathrm{Out}(P)$ is isomorphic with $\mathrm{GL}_2(p)$ in such a way that $\mathrm{Out}(P)$ acts naturally on $P/Z(P) \cong C_p \times C_p$ as $\mathrm{GL}_2(p)$.

(Let X denote the set of maximal subgroups of P and let Y denote the set of subspaces of dimension 1 in a 2-dimensional $\mathbb{F}_p$-vector space V. Identifying $\mathrm{Out}(P)$ with $\mathrm{GL}_2(p)$ via (ii), notice that the action of $\mathrm{Out}(P)$ on X is equivalent to the action of $\mathrm{GL}_2(p)$ on Y.)

9.2 Let P be the p-group p_+^{1+2}, as in the previous exercise, and let $\mathcal{F}$ be a saturated fusion system on P. Let Q be a subgroup of order p^2, generated by $x \notin Z(P)$ and $z \in Z(P)$. Suppose that Q is $\mathcal{F}$-radical, and let α be a primitive $(p-1)$th root of 1.

(i) Prove that there exists $\phi \in \mathrm{Aut}_{\mathcal{F}}(Q)$ such that $x\phi = x^\alpha$ and $z\phi = z^{\alpha^{-1}}$.

(ii) Prove that ϕ extends to $\psi \in \mathrm{Aut}_{\mathcal{F}}(P)$, and that ψ may be chosen so that, for some $y \in P \backslash Q$, $y\psi = y^{\alpha^{-2}}$ (and $z = [x,y]$). Moreover, if $\psi' \in \mathrm{Aut}_{\mathcal{F}}(P)$ is another extension of ϕ, prove that ψ and ψ' have the same image in $\mathrm{Out}_{\mathcal{F}}(P)$.

Let V_Q be the subgroup of $\mathrm{Out}_{\mathcal{F}}(Q)$ of order $p-1$ generated by the ψ chosen in (ii). (The subgroup $\langle \psi \rangle$ intersects $\mathrm{Inn}(P)$ trivially, so projects onto a cyclic subgroup V_Q of order $p-1$ in $\mathrm{Out}_{\mathcal{F}}(P)$.)

(iii) Let R be another $\mathcal{F}$-radical subgroup of P of order p^2, and let V_R be the corresponding subgroup of $\mathrm{Out}_{\mathcal{F}}(P)$. Show that $V_Q = V_R$ if and only if $Q = R$.

(iv) Show that V_Q acts on the set of subgroups of order p^2 in the following way: it fixes two subgroups, namely $\langle x, z \rangle$ and $\langle y, z \rangle$, and permutes the others transitively if $3 \nmid (p-1)$ and in three orbits of length $(p-1)/3$ if $3 \mid (p-1)$.

(Although there are in general many possibilities for the subgroup V_Q defined above, if A is a given p'-subgroup of $\mathrm{Out}(P) = \mathrm{GL}_2(p)$, then there is at most one possibility for V_Q lying inside A.)

9.3 Let $\mathcal{F}$ be a saturated fusion system on a finite p-group P, and suppose that $\mathcal{F}$ possesses a single $\mathcal{F}$-essential subgroup Q of P. Prove that Q is normal in $\mathcal{F}$, and hence that $\mathcal{F}$ is the fusion system of a finite group G with Sylow p-subgroup P, $Q \trianglelefteq G$, and $G/Q \cong \mathrm{Out}_{\mathcal{F}}(Q)$. Furthermore, for any group G with these properties, $\mathcal{F}_G(P) = \mathcal{F}$. (If $C_G(Q) \le Q$ and $O_{p'}(G) = 1$, then G is unique.)

9.4 Prove Lemma 9.11.

9.5 Let P be the 2-group constructed in Section 9.2, the Sylow 2-subgroup of Co_3. Find a surjective homomorphism onto the Sylow 2-subgroup of A_{12} whose kernel is $\langle z \rangle$.

9.6 Let A be a normal abelian subgroup of the nilpotent group G, and suppose that A is maximal in G subject to being a normal abelian subgroup of G. Prove that $A = C_G(A)$.

9.7 In [Oli04], Oliver defines a subgroup R of a p-group Q to be *universally weakly closed* in P (although this definition has not become standard) if, for every saturated fusion system $\mathcal{F}$ on any p-group P in which Q is a strongly $\mathcal{F}$-closed subgroup, the subgroup R is weakly $\mathcal{F}$-closed. Prove that a subgroup R is universally weakly closed in Q if, for all subgroups S of Q containing R, $R \operatorname{char} S$.

References

[AB79] Jonathan Alperin and Michel Broué. Local methods in block theory. *Ann. Math.*, 110:143–157, 1979.

[ABG70] Jonathan Alperin, Richard Brauer, and Daniel Gorenstein. Finite groups with quasi-dihedral and wreathed Sylow 2-subgroups. *Trans. Amer. Math. Soc.*, 151:1–261, 1970.

[ABG73] Jonathan Alperin, Richard Brauer, and Daniel Gorenstein. Finite simple groups of 2-rank two. *Scripta Math.*, 29:191–214, 1973.

[AKO11] Michael Aschbacher, Radha Kessar, and Bob Oliver. *Fusion Systems in Algebra and Topology*. Cambridge University Press, Cambridge, in press.

[Alp67] Jonathan Alperin. Sylow intersections and fusion. *J. Algebra*, 6:222–241, 1967.

[An93] Jian Bei An. 2-weights for classical groups. *J. Reine Angew. Math.*, 439:159–204, 1993.

[Asc93] Michael Aschbacher. Simple connectivity of p-group complexes. *Israel J. Math.*, 82:1–43, 1993.

[Asc00] Michael Aschbacher. *Finite Group Theory*, 2nd edition, volume 10 of Cambridge Studies in Advanced Mathematics. Cambridge University Press, Cambridge, 2000.

[Asc08a] Michael Aschbacher. Normal subsystems of fusion systems. *Proc. Lond. Math. Soc.*, 97:239–271, 2008.

[Asc08b] Michael Aschbacher. S_3-free 2-fusion systems. Submitted, 2008.

[Asc10] Michael Aschbacher. Generation of fusion systems of characteristic 2-type. *Invent. Math.*, 180:225–299, 2010.

[Asc11] Michael Aschbacher. The generalized Fitting subsystem of a fusion system. *Mem. Amer. Math. Soc.*, 209(986):v+110, 2011.

[BCGLO05] Carles Broto, Natàlia Castellana, Jesper Grodal, Ran Levi, and Bob Oliver. Subgroup families controlling p-local finite groups. *Proc. Lond. Math. Soc.*, 91:325–354, 2005.

[BCGLO07] Carles Broto, Natàlia Castellana, Jesper Grodal, Ran Levi, and Bob Oliver. Extensions of p-local finite groups. *Trans. Amer. Math. Soc.*, 359:3791–3858, 2007.

[Ben71] Helmut Bender. Transitive Gruppen gerader Ordnung, in denen jede Involution genau einen Punkt festläßt. *J. Algebra*, 17:527–554, 1971.

[Ben98a] David Benson. Cohomology of sporadic groups, finite loop spaces, and the Dickson invariants. In *Geometry and Cohomology in Group Theory (Durham, 1994)*, volume 252 of London Mathematical Society Lecture Note Series, pp. 10–23. Cambridge University Press, Cambridge, 1998.

[Ben98b] David Benson. *Representations and Cohomology, I: Basic Representation Theory of Finite Groups and Associative Algebras*, volume 30 of Cambridge Studies in Advanced Mathematics. Cambridge University Press, Cambridge, 1998.

[Ben98c] David Benson. *Representations and Cohomology, II: Cohomology of Groups and Modules*, volume 31 of Cambridge Studies in Advanced Mathematics. Cambridge University Press, Cambridge, 1998.

[BK72] Aldridge Bousfield and Daniel M. Kan. *Homotopy Limits, Completions and Localizations*, volume 304 of Lecture Notes in Mathematics. Springer–Verlag, Berlin, 1972.

[BLO03a] Carles Broto, Ran Levi, and Bob Oliver. Homotopy equivalences of p-completed classifying spaces of finite groups. *Invent. Math.*, 151:611–664, 2003.

[BLO03b] Carles Broto, Ran Levi, and Bob Oliver. The homotopy theory of fusion systems. *J. Amer. Math. Soc.*, 16:779–856, 2003.

[BLO04] Carles Broto, Ran Levi, and Bob Oliver. The theory of p-local groups: a survey. In *Homotopy Theory: Relations with Algebraic Geometry, Group Cohomology, and Algebraic K-Theory*, volume 346 of Contemporary Mathematics, pp. 51–84. American Mathematical Society, Providence, RI, 2004.

[BP80] Michel Broué and Lluis Puig. Characters and local structure in G-algebras. *J. Algebra*, 63:306–317, 1980.

[Bra53] Richard Brauer. A characterization of the characters of groups of finite order. *Ann. of Math.*, 57:357–377, 1953.

[Bra71] Richard Brauer. Character theory of finite groups with wreathed Sylow 2-subgroups. *J. Algebra*, 19:547–592, 1971.

[Bro86] Michel Broué. Les l-blocs des groups $GL(n, q)$ et $U(n, q^2)$ et leurs structures locales. *Astérisque*, 133–134:159–188, 1986.

[BS08] David J. Benson and Stephen D. Smith. *Classifying Spaces of Sporadic Groups*, volume 147 of Mathematical Surveys and Monographs. American Mathematical Society, Providence, RI, 2008.

[BW71] Richard Brauer and Warren Wong. Some properties of finite groups with wreathed Sylow 2-subgroup. *J. Algebra*, 19:263–273, 1971.

[CCNPW85] John Conway, Robert Curtis, Simon Norton, Richard Parker, and Robert Wilson. *The ATLAS of Finite Groups*. Oxford University Press, Oxford, 1985.

[CF64]   Roger Carter and Paul Fong. The Sylow 2-subgroups of the finite classical groups. *J. Algebra*, 1:139–151, 1964.

[CG10]   David A. Craven and Adam Glesser. Fusion systems on small p-groups. Submitted, 2010.

[Col71] Michael Collins. The characterisation of the Suzuki groups by their Sylow 2-subgroups. *Math. Z.*, 123:32–48, 1971.

[Col72] Michael Collins. The characterisation of the unitary groups $U_3(2^n)$ by their Sylow 2-subgroups. *Bull. London Math. Soc.*, 4:49–53, 1972.

[Cra09] David A. Craven. On ZJ-theorems in fusion systems. Preprint, 2009.

[Cra10a] David A. Craven. Control of fusion and solubility in fusion systems. *J. Algebra*, 323:2429–2448, 2010.

[Cra10b] David A. Craven. Normal subsystems of fusion systems. *J. Lond. Math. Soc.*, in press.

[Dad78] Everett C. Dade. Endo-permutation modules over p-groups. I. *Ann. of Math.*, 107:459–494, 1978.

[DGMP09] Antonio Díaz, Adam Glesser, Nadia Mazza, and Sejong Park. Glauberman's and Thompson's theorems for fusion systems. *Proc. Amer. Math. Soc.*, 137:495–503, 2009.

[DGMP10] Antonio Díaz, Adam Glesser, Nadia Mazza, and Sejong Park. Control of transfer and weak closure in fusion systems. *J. Algebra*, 323:382–392, 2010.

[Dor72] Larry Dornhoff. *Group Representation Theory, Part B: Modular Representation Theory.* Marcel Dekker, New York, 1972.

[DRV07] Antonio Díaz, Albert Ruiz, and Antonio Viruel. All p-local finite groups of rank two for odd prime p. *Trans. Amer. Math. Soc.*, 359:1725–1764, 2007.

[Fei82] Walter Feit. *The Representation Theory of Finite Groups.* North-Holland, Amsterdam–New York, 1982.

[GHL08] David Green, László Héthelyi, and Markus Lilienthal. On Oliver's p-group conjecture. *Algebra Number Theory*, 2:969–977, 2008.

[GHM10] David Green, László Héthelyi, and Nadia Mazza. On Oliver's p-group conjecture: II. *Math. Ann.*, 347:111–122, 2010.

[GJ99]   Paul Goerss and John Jardine. *Simplicial Homotopy Theory*, volume 174 of Progress in Mathematics. Birkhäuser, Basel, 1999.

[GL83]   Daniel Gorenstein and Richard Lyons. The local structure of finite groups of characteristic 2 type. *Mem. Amer. Math. Soc.*, 42(276):vii+731, 1983.

[Gla66] George Glauberman. Central elements in core-free groups. *J. Algebra*, 4:403–420, 1966.

[Gla68a] George Glauberman. A characteristic subgroup of a p-stable group. *Canad. J. Math.*, 20:1101–1135, 1968.

[Gla68b] George Glauberman. Weakly closed elements of Sylow subgroups. *Math. Z.*, 107:1–20, 1968.

[Gla70] George Glauberman. Prime-power factor groups of finite groups. II. *Math. Z.*, 117:46–56, 1970.

[Gla71] George Glauberman. Global and local properties of finite groups. In *Finite Simple Groups (Proceedings of the Instructional Conference, Oxford, 1969)*, pp. 1–64. Academic Press, London, 1971.

[Gla72] George Glauberman. A sufficient condition for *p*-stability. *Proc. Lond. Math. Soc.*, 25:253–287, 1972.

[Gle09] Adam Glesser. Sparse fusion systems. *Proc. Edinburgh Math. Soc.*, in press.

[GLS98] Daniel Gorenstein, Richard Lyons, and Ronald Solomon. *The Classification of the Finite Simple Groups, Number 3, Part I, Chapter A*. American Mathematical Society, Providence, RI, 1998.

[GLS05] Daniel Gorenstein, Richard Lyons, and Ronald Solomon. *The Classification of the Finite Simple Groups, Number 6, Part IV*. American Mathematical Society, Providence, RI, 2005.

[GMRS04] David Gluck, Kay Magaard, Udo Riese, and Peter Schmid. The solution of the *k*(*GV*)-problem. *J. Algebra*, 279:694–719, 2004.

[Gol70] David M. Goldschmidt. A conjugation family for finite groups. *J. Algebra*, 16:138–142, 1970.

[Gol74] David M. Goldschmidt. 2-fusion in finite groups. *Ann. of Math.*, 99:70–117, 1974.

[Gor80] Daniel Gorenstein. *Finite Groups*. Chelsea Publishing Co., New York, 2nd edition, 1980.

[GW65] Daniel Gorenstein and John Walter. The characterization of finite groups with dihedral Sylow 2-subgroups I, II, III. *J. Algebra*, 2:85–151, 218–270, 334–393, 1965.

[HH56] Philip Hall and Graham Higman. On the *p*-length of *p*-soluble groups and reduction theorems for Burnside's problem. *Proc. Lond. Math. Soc.*, 6:1–42, 1956.

[Hig53] Donald Higman. Focal series in finite groups. *Canadian J. Math.*, 5:477–497, 1953.

[Hig63] Graham Higman. Suzuki 2-groups. *Illinois J. Math.*, 7:79–96, 1963.

[Hov99] Mark Hovey. *Model Categories*, volume 63 of Mathematical Surveys and Monographs. American Mathematical Society, Providence, RI, 1999.

[Kes06] Radha Kessar. The Solomon system $\mathcal{F}_{\mathrm{sol}}(3)$ does not occur as fusion system of a 2-block. *J. Algebra*, 296:409–425, 2006.

[Kes07] Radha Kessar. Introduction to block theory. In *Group Representation Theory*, pp. 47–77. EPFL Press, Lausanne, 2007.

[KL08] Radha Kessar and Markus Linckelmann. *ZJ*-theorems for fusion systems. *Trans. Amer. Math. Soc.*, 360:3093–3106, 2008.

[KLR02] Radha Kessar, Markus Linckelmann, and Geoffrey Robinson. Local control in fusion systems of *p*-blocks of finite groups. *J. Algebra*, 257:393–413, 2002.

[KS08] Radha Kessar and Radu Stancu. A reduction theorem for fusion systems of blocks. *J. Algebra*, 319:806–823, 2008.

[Lan78] Peter Landrock. The non-principal 2-blocks of sporadic simple groups. *Comm. Algebra*, 6:1865–1891, 1978.

[Lin06] Markus Linckelmann. A note on the Schur multiplier of a fusion system. *J. Algebra*, 296:402–408, 2006.

[Lin07] Markus Linckelmann. Introduction to fusion systems. In *Group Representation Theory*, pp. 79–113. EPFL Press, Lausanne, 2007.

[LO02] Ran Levi and Bob Oliver. Construction of 2-local finite groups of a type studied by Solomon and Benson. *Geom. Topol.*, 6:917–990 (electronic), 2002.

[LO05] Ran Levi and Bob Oliver. Correction to: 'Construction of 2-local finite groups of a type studied by Solomon and Benson'. *Geom. Topol.*, 9:2395–2415 (electronic), 2005.

[Lyo72] Richard Lyons. A characterization of the group $U_3(4)$. *Trans. Amer. Math. Soc.*, 164:371–387, 1972.

[Mar86] Ursula Martin. Almost all p-groups have automorphism group a p-group. *Bull. Amer. Math. Soc.*, 15:78–82, 1986.

[May92] J. Peter May. *Simplicial Objects in Algebraic Topology*. Chicago Lectures in Mathematics. University of Chicago Press, Chicago, 1992. Reprint of the 1967 original.

[MP96] John Martino and Stewart Priddy. Unstable homotopy classification of $BG_p^\wedge$. *Math. Proc. Cambridge Phil. Soc.*, 119:119–137, 1996.

[Oli04] Bob Oliver. Equivalences of classifying spaces completed at odd primes. *Math. Proc. Cambridge Phil. Soc.*, 137:321–347, 2004.

[Oli06] Bob Oliver. Equivalences of classifying spaces completed at the prime two. *Mem. Amer. Math. Soc.*, 180(848):vi+102, 2006.

[OS09] Silvia Onofrei and Radu Stancu. A characteristic subgroup for fusion systems. *J. Algebra*, 322:1705–1718, 2009.

[OV09] Bob Oliver and Joana Ventura. Saturated fusion systems over 2-groups. *Trans. Amer. Math. Soc.*, 361:6661–6728, 2009.

[PD77] B. M. Puttaswamaiah and John Dixon. *Modular Representations of Finite Groups*. Academic Press, New York, 1977.

[Pui86] Lluís Puig. The Nakayama conjecture and the Brauer pairs. In *Séminaire sur les groupes finis, Tome III*, volume 25 of Publications Mathématiques de Université Paris VII, pp. ii, 171–189. Université Paris VII, Paris, 1986.

[Pui88] Lluís Puig. Nilpotent blocks and their source algebras. *Invent. Math.*, 93:77–116, 1988.

[Pui00] Lluís Puig. The hyperfocal subalgebra of a block. *Invent. Math.*, 141:365–397, 2000.

[Pui06] Lluís Puig. Frobenius categories. *J. Algebra*, 303:309–357, 2006.

[Pui09] Lluís Puig. *Frobenius Categories versus Brauer Blocks*, volume 274 of Progress in Mathematics. Birkhäuser Verlag, Basel, 2009.

[Rag06] Kári Ragnarsson. Classifying spectra of saturated fusion systems. *Algebr. Geom. Topol.*, 6:195–252, 2006.

[RaS09] Kári Ragnarsson and Radu Stancu. Saturated fusion systems as idempotents in the double Burnside ring. Submitted, 2009.

[Rob96] Derek J. S. Robinson. *A Course in the Theory of Groups*, Springer-Verlag, New York, 2nd edition, 1996.

[Ros78] John S. Rose. *A Course on Group Theory*. Cambridge University Press, Cambridge, 1978.

[RoS09] Kieran Roberts and Sergey Shpectorov. On the definition of saturated fusion systems. *J. Group Theory*, 12:679–687, 2009.

[RV04] Albert Ruiz and Antonio Viruel. The classification of p-local finite groups over the extraspecial group of order p^3 and exponent p. *Math. Z.*, 248:45–65, 2004.

[Ser79] Jean-Pierre Serre. *Local Fields*. Springer-Verlag, New York–Berlin, 1979.

[Sol73] Ronald Solomon. Finite groups with Sylow 2-subgroups of type $\mathfrak{A}_{12}$. *J. Algebra*, 24:346–378, 1973.

[Sol74] Ronald Solomon. Finite groups with Sylow 2-subgroups of type .3. *J. Algebra*, 28:182–198, 1974.

[Sta03] Radu Stancu. Quotients of fusion systems. Preprint, 2003.

[Sta06] Radu Stancu. Control of fusion in fusion systems. *J. Algebra Appl.*, 5:817–837, 2006.

[Ste90] Bernd Stellmacher. An analogue to Glauberman's ZJ-theorem. *Proc. Amer. Math. Soc.*, 109:925–929, 1990.

[Ste92] Bernd Stellmacher. Errata to: 'An analogue to Glauberman's ZJ-theorem'. *Proc. Amer. Math. Soc.*, 114:588, 1992.

[Ste96] Bernd Stellmacher. A characteristic subgroup of Σ_4-free groups. *Israel J. Math.*, 94:367–379, 1996.

[Suz62] Michio Suzuki. On a class of doubly transitive groups. *Ann. of Math.*, 75:105–145, 1962.

[Swi75] Robert Switzer. *Algebraic Topology – Homotopy and Homology*, volume 212 of Die Grundlehren der mathematischen Wissenschaften. Springer-Verlag, New York, 1975.

[Thé95] Jacques Thévenaz. *G-Algebras and Modular Representation Theory*. Oxford Mathematical Monographs. Oxford University Press, Oxford, 1995.

[Tho60] John Thompson. Normal p-complements for finite groups. *Math. Z.*, 72:332–354, 1960.

[Tho64] John Thompson. Normal p-complements for finite groups. *J. Algebra*, 1:43–46, 1964.

Index of notation

(Q, e) (Brauer pair), 38
$(Q, f) \leq (R, e)$, 39
$(Q, f) \leq (R, e)$ (generalized Brauer pairs), 337
$(Q, f) \trianglelefteq (R, e)$, 39
$(X \times Y)_\Delta$, 257
$(kG)^P$, 34
$(kX)_H^G$, 35
$Q \trianglelefteq \mathcal{F}$, 148
$X \amalg Y$, 253
$X \times Y$ (simplicial sets), 61
$X^\cdot \times \Delta^n$, 77
X^H (stable elements), 34
$X_p^\wedge$, 80
$X|_{(K,L)}$, 254
$X|_{(\phi,H)}$, $X|_{(G,\phi)}$, 254
$[H, K; i]$, 350
$[n]$, 57
$\langle \mathcal{C} \rangle$, 95
$\mathcal{E} \prec \mathcal{F}$, 151
$\mathcal{E} \trianglelefteq \mathcal{F}$, 272
$\mathcal{E}_1 \cap \mathcal{E}_2$, 95
$\mathcal{E}_1 \curlywedge \mathcal{E}_2$, 281, 284
$\mathcal{E}_1 \curlywedge \mathcal{E}_2$ (subnormal), 287
$\mathcal{E}_1 \wedge \mathcal{E}_2$, 281, 285
$\mathcal{F}_1 \times \mathcal{F}_2$, 211
$\mathcal{F}_1 \times_Z \mathcal{F}_2$, 297
ψ^ϕ, 135
$|X|$ (simplicial set), 64

$A_p(G)$, 120
$\mathrm{Aut}(\mathcal{F})$, 147
$\mathrm{Aut}_\mathcal{F}(Q)$, 20
$\mathrm{Aut}_H(K)$, 3
$\mathrm{Aut}_\mathcal{F}^K(Q)$, $\mathrm{Aut}_P^K(Q)$, 110
$\mathrm{Aut}_\mathcal{F}(Q \leq R)$, $\mathrm{Aut}_P(Q \leq R)$, 190

BG, 67
$\mathscr{B}(G)$, 67
$\mathcal{B}(Q)$, 38
Br, 34

$\mathbf{Cat}$, 60
$\mathrm{c}\mathscr{C}$, 75
$C_G'(Q)$, 84
$C_P(\mathcal{E})$, 306
$C_\mathcal{F}(Q)$, 108
$C_\mathcal{F}(\mathcal{E})$, 306
c_g, 3
c_H, 3
H char G, 130
$\mathcal{E}$ char $\mathcal{F}$, 294
$\mathrm{Comp}(\mathcal{F})$, 299
$\mathbf{csSet}$, 75

$\mathbf{\Delta}$, 57
δ_Q (centric linking systems), 343
Δ^n, 61
$\mathcal{D}(U)$, 292

$\mathcal{E}(U)$, 292
$E(\mathcal{F})$, 304
$E(G)$, 183
EG, 67
$\mathscr{E}(G)$, 67
$\eta(\mathcal{F})$, 348
$\exp(G)$, 4

$\mathcal{F}^c$, 343
$\mathcal{F}_{(D, e_D)}(G, N, e)$, 337
$\mathcal{F}_{(D, e_D)}(G, b)$, 47
$F(G)$, 183
$F^*(G)$, 183

$F^*(\mathcal{F})$, 306
$\mathcal{F}/Q$, 138
$\mathfrak{foc}(\mathcal{F})$, 238
$\mathcal{F}_P(G)$, 7
$\bar{\mathcal{F}}_Q$, 137
$\mathcal{F}_{\mathrm{Sol}}(r)$, 329

$\overline{\Gamma}_{\mathcal{F},\mathcal{E}}(T)$, 272

$\mathfrak{hyp}(\mathcal{F})$, 236

$\mathrm{Inn}(G)$, 3

$J(P)$ (Thompson subgroup),
   10
$J(R)$ (Jacobson radical), 29

$k(B)$, 50
K_∞, K^∞, 223

Λ^n_k, 69
$L^{\mathcal{F}}$, 88
$\mathcal{L}^c_P(G)$, 84

$\mathrm{Map}(X,Y)$ (space), 63
$\mathrm{Map}(X^\cdot, Y^\cdot)$ (cosimplicial space),
   77
$M(G)$, $M(\mathcal{F})$, 170

$\mathcal{N}\mathscr{C}$, 60
$\mathrm{N}_{\mathcal{F}}(Q)$, 108
$\mathrm{N}_G(Q,e)$, 38
$\mathrm{N}^K_{\mathcal{F}}(Q)$, 110
N_ϕ, 21
$\mathrm{N}_P(Q)\,\mathrm{C}_{\mathcal{F}}(Q)$, 115
$\mathrm{N}_{\mathcal{F}}(Q,R)$, $\mathrm{N}_P(Q,R)$, 116

$\mathcal{O}(\mathcal{F})$, 345
$\mathcal{O}(\mathcal{F}^c)$, 346
$\mathrm{O}^{(i)}_p(\mathcal{F})$, 181
$\mathrm{O}^p(\mathcal{F})$, 243
$\mathrm{O}^{p'}(\mathcal{F})$, 247
$\mathrm{O}_p(\mathcal{F})$, $\mathcal{O}_p(\mathcal{F})$, 149
$\mathrm{O}_p(G)$, $\mathrm{O}_\pi(G)$, $\mathrm{O}^p(G)$, $\mathrm{O}^\pi(G)$, 4
$\mathrm{Out}(G)$, $\mathrm{Out}_G(H)$, 3

$\mathrm{P\Gamma L}_n(q)$, 335
π (centric linking systems),
   343
$\pi(X,*)$ (simplicial sets), 71
$\mathrm{P\Sigma L}_n(q)$, 335

$Q\,\mathrm{C}_{\mathcal{F}}(Q)$, 115
$Qd(p)$, 11

$R^\cdot X$, **80**
$\mathcal{R}^{\mathcal{F}}(T)$, 284
$\mathcal{R}_{\mathcal{F}}(T)$, 283
$\overline{\mathcal{R}}^{\mathcal{F}}(T)$, 285
$\overline{\mathcal{R}}_{\mathcal{F}}(T)$, 285
$R_\infty X$, 80
$R \otimes X$, 78
RX, 78

$\mathrm{s}\mathscr{C}$, 75
$\mathbf{Set}$, 58
$\mathrm{Sing}(X)$, 66
$\mathbf{sSet}$, 59
$\mathrm{Syl}_p(G)$, 4

$T(e)$ (inertia group), 336
$\mathrm{T}_G(Q,R)$, 84
$\mathrm{Tot}(X^\cdot)$, 77
$\mathcal{T}_P(G)$, 84
Tr, 35

$\mathcal{U}(P)$, 95

$\mathfrak{X}(P)$, 351

$\mathcal{Z}_{\mathcal{F}}$, 346
$Z(\mathcal{F})$, $\mathcal{Z}(\mathcal{F})$, 161
$Z_i(\mathcal{F})$, $Z_\infty(\mathcal{F})$, 165

Index

Alperin's fusion theorem, 9, 15
  for finite groups, **13**, 14
  for fusion systems, **121**, 123
augmentation
  of a cosimplicial space, **80**
  in a k-algebra, **51**
augmentation ideal, **51**
automizer, **3**
automorphism of a fusion system, **147**

bad space, 81
balance, 309
belong (module to a block), **31**
biset, **252**
  bifree, 257
  characteristic, 254
  stable, 256
block, **27**
  bijection between $\mathcal{O}G$ and kG, 33
  Clifford theory for, 336
  cover another, 336
  defect group of, 41
  fusion system is saturated, 49
  and idempotents, 28, 29
  module belongs to it, 31
  nilpotent, 53
  with trivial defect group, 50
block fusion system, **47**
  is saturated, 49
  of sporadic groups, 333
block idempotent, **29**
Brauer morphism, **34**
Brauer pair, **38**
  b-Brauer pair, **41**
  generalized, 337
  inclusion of, 39

  maximal b-Brauer pair, **41**
  uniqueness of inclusion, 40
Brauer's $k(B)$ conjecture, **50**
Brauer's first main theorem, 45
Burnside's normal p-complement
  theorem, **9**

central product, **297**
  and finite groups, 315
  and normal subsystems, 298
centralizer subgroup of a subsystem,
  306
centralizer subsystem, **108**
  for finite groups, 108
  is saturated, 109
centre, **161**
  fusion systems without centre on
    2-groups of 2-rank 2, 178
  and normal subsystems, 295
centre functor, **346**
centric, **117**, 129
  for finite groups, **83**, 117
centric linking system, **343**
  and classifying spaces, 84
  for finite groups, **84**, 344
  and higher limits, 346
  uniqueness for finite groups, 86
characteristic biset, **254**
characteristic element of the double
  Burnside ring, **256**
characteristic subsystem, **294**
classifying space, **66**
  and centric linking systems, 84
  is fibrant, 70
  homotopy groups of, 72
  of a p-local finite group, **349**

Clifford theory of blocks, 336
codegeneracy map, **75**
coface map, **75**
R-complete space, **81**
R-completion, **80**
component
  for finite groups, **183**
  for fusion systems, **299**
H-conjugate, **3**
conjugate in a fusion system, 21, **94**
conjugation family, **14**
  fusion systems, **225**
p-constrained finite group, **86**
constrained fusion system, **87**
  bijection for normal subsystems, 291
  has a model, 87
control constrained transfer, **264**
control fusion, **215**
  control G-fusion in H, **5**
  control weak fusion in H with respect
    to G, **5**
  in groups with abelian Sylows, 6
control transfer, **264**
core of a finite group, 309
cosimplicial object, **75**
cosimplicial simplex, **76**
cosimplicial space, **75**
  $R^{\boldsymbol{\cdot}}X$, 80
cover for blocks, **336**

defect group, **41**
defect of a subnormal subsystem,
  286
defined (a subsystem on a subgroup),
  95
degeneracy map, **58**
degenerate simplex, **58**
detecting subgroup, **275**
direct product, **211**
  of normal subsystems, 296
disjoint union of two bisets, 253
double Burnside ring, **255**
  characteristic element, 256
double normalizer, **116**

essential, **119**, 129
exotic fusion system, **24**
  over p_+^{1+2}, 324
exponent of a finite group, **4**
extensions of morphisms, 98, 103
extremal, **15**

F-module, **353**
face map, **58**
factor system, **138**
failure of factorization module, **353**
family, **13**
  conjugation, 14
  weak conjugation, 14
fibrant, **68**
  simplicial groups are, 75
fibrant replacement, **73**
filling, **69**
first isomorphism theorem, 143
Fitting subgroup, **183**
focal subgroup
  for finite groups, **16**
  for fusion systems, 187, **238**
focal subgroup theorem, **16**
$\mathcal{F}$-Frattini subsystem, **154**
  and weakly normal subsystems, 156
H-free for fusion systems, **227**
Frobenius reciprocity, 257
Frobenius's normal p-complement
  theorem, **9**, 18, 20, 133
fully automized, **97**
fully centralized, 21, **101**
  and receptive, 103–105
  for finite groups, 22
fully K-normalized, **110**
fully normalized, 21, **101**
  for finite groups, 22
  proper pairs, **194**
  and receptive/fully automized, 103,
    105
function space, **63**
  adjunction with product, 63, 77
  cosimplicial spaces, **77**
fused
  elements of a group, **4**
fusion system, 20, **94**
  of a block, **47**
  centralizer, 108
  constrained have models, 87
  on dihedral 2-groups, 125
  exotic, 24
  generated by morphisms, **95**
  of $GL_3(2)$, 7
  for groups, 7
  of a group is saturated, 100
  group fusion systems are saturated,
    22
  H-free, 227

$\mathcal{H}$-generated, 193
normalizer, 108
on p_+^{1+2}, 326
of a p-soluble group, 185
on quaternion 2-groups, 125
saturated, 22, 100
on semidihedral 2-groups, 125
simple, 160
soluble, 181
on Sylow 2-subgroup of Co_3, 328
universal, 95
on wreathed 2-groups, 180

generalized Brauer pair, **337**
generalized Fitting subgroup, **183**
generalized Fitting subsystem, **306**
generalized fusion system (of a block),
    337
$\mathcal{H}$-generated, **193**
generated (as a fusion system), **95**
generated by an invariant map, **202**
geometric realization, **64**
    and singular simplices functor, 66
Glauberman functor, **222**
    K^∞ and K_∞ are, 224
    for the prime 2, 222
    ZJ is, 222
Glauberman's Z^*-theorem, 167
Glauberman's ZJ-theorem, 12, 222
Glauberman–Thompson normal
        p-complement theorem
    for finite groups, **11**
    for fusion systems, **235**
Goldschmidt
    simple groups with strongly closed
        abelian subgroups, 185
Goldschmidt group, **185**
good space, 81
group fusion system
    centric, 117
    fully centralized, 22
    fully normalized, 22
    is saturated, 22
Grün
    first theorem, 9, **24**
    second theorem, **132**

homocyclic, **175**
homology groups of a simplicial set,
    79
homotopy (simplicial sets), **70**

homotopy category, 74
homotopy equivalence for simplicial
        sets, **74**
homotopy groups for simplicial sets,
    71
horn, **69**
    filling of, 69
    inner, outer, 89
hypercentral subgroup theorem, 165
hypercentre, **165**
hyperfocal subgroup
    for finite groups, **19**
    for fusion systems, **236**
hyperfocal subgroup theorem, **19**
hyperfocal subsystem, **243**

idempotent, **27**
    and blocks, 28, 29
    of a block, 29
    lifting of, 33
inclusion of Brauer pairs, **39**
    uniqueness, 40
indecomposable ideal of k-algebra, **27**
induces a morphism of fusion systems,
    135
inductively saturated, **191**
inertia group of a block, **336**
injective morphism of fusion systems,
    136
inner horn, **89**
intersection
    of normal subsystems, 281
    of weakly normal subsystems, 281
intersection subsystem, **95**
invariant map, **202**
invariant subsystem, **151**
involution, **5**
isomorphic
    bisets, 252
    in a fusion system, 20, **94**
isomorphism theorems
    first, 143
    second, 143
    third, 144

Jacobson radical, **29**

Kan complex, *see* fibrant
Kelley product, **61**
kernel of a morphism of fusion systems,
    136

layer, **304**
  is a central product, 304
  for finite groups, **183**
  and normalizers, 309
p-local subgroup, 4

map
  codegeneracy, 75
  coface, 75
Martino–Priddy conjecture, 55, 317
Maschke's theorem, 30
metacyclic, **217**
model, **87**
model category, 74
p-modular system, 32
morphism (extensions of), 98, 103
morphism of fusion systems, **135**
  $\bar{\mathcal{F}}_Q$, 137
  injective, 136
  kernel, 136
  kernels are strongly closed, 136
  surjective, 136
  when $\mathcal{F} = \mathrm{N}_{\mathcal{F}}(Q)$, 138

nerve, **60**
  products of, 61
p-nilpotent, **9**
nilpotent block, **53**
non-degenerate simplex, **58**
normal p-complement, **9**
normal map, **275**
normal subgroup, 87, **148**
  equivalent conditions, 129, 130, 150,
    153, **153**
normal subsystem, **272**
  and central products, 298
  and constrained fusion systems, 291
  and direct products, 296
  intersection of, 281
  and normal subgroups of fusion
    systems, 273
  and normal subgroups of groups, 273,
    291
  and normalizers, 274
  and quotients, 274
  quotient by central subgroups, 295
  transitivity, 293
  and weakly normal subsystems, 279
K-normalizer subgroup, **110**
normalizer subsystem, **108**
  for finite groups, 108

  is saturated, 109
K-normalizer subsystem, **111**
  is saturated, 111

offender, **353**
Oliver conjecture, **351**
Oliver subgroup, **351**
orbit category, **345**
orbit of a biset, **253**
orthogonal idempotent, **28**
outer horn, **89**

p-complement theorems for fusion
  systems, 235
p-local finite group, **349**
p-modular system, 32
p-nilpotent, **9**
p-perfect finite group, **168**
p-power index, **239**
path connected simplicial set, 89
perfect fusion system, **168**
positive characteristic p-functor, **222**
  K^{∞}, K_{∞}, **223**
primitive idempotent, **28**
principal block, **33**
  fusion system of, 53
product
  adjunction with function space, 63, 77
  of simplicial sets, 61
proper Q-pair, **194**

Q-series, **350**
$Qd(p)$-free fusion system, 227
  is soluble, 227
quasisimple fusion system, **299**
Quillen equivalence, 74
quotient
  $\bar{\mathcal{F}}_Q$, 137
  $\mathcal{F}/Q$, 138
  $\mathcal{F}/Q$ is saturated, 139
  $\mathcal{F}/Q = \bar{\mathcal{F}}_Q$, 141
  for finite groups, 145

R-bad, **81**
R-completion, **80**
R-good, **81**
radical, **118**, 129
receptive, **99**
  for finite groups, 99
  and fully centralized, 103–105
R-receptive, **189**

π-reduced, 88
reduction modulo p, **32**
relative trace, 35
residual subsystem, **247**
  automorphisms in subsystem, 252
resistant, **216**
  abelian groups are, 25
  $\mathrm{PSU}_3(2^n)$, 221
  Suzuki 2-groups are, 220
  Suzuki simple groups, 220
restriction of a biset, **254**
Rosenberg's lemma, 30

satisfies Frobenius reciprocity, 257
saturated, 22, **100**
  for finite groups, 100
  other definitions, 106, 107
saturated $\mathcal{F}$-conjugacy class, **103**
Schur multiplier, **170**
second isomorphism theorem, 143
simple fusion system, **160**
  and normal subsystems, 279
  on 2-groups of 2-rank 2, 180
simple groups with strongly closed
    abelian subgroups, 185
simplex, **58**
  degenerate, 58
  non-degenerate, 58
simplicial groups are fibrant, 75
simplicial map, **59**
simplicial object, **75**
simplicial R-module
  $R \otimes X$, 78
  RX, 78
simplicial set, **58**
  category, **59**
  path connected, 89
  from poset, 60
  from simplicial complex, 58
singular simplex, **66**
singular simplices functor, **66**
  and geometric realization, 66
Solomon fusion system, 171
  and blocks of finite groups, 343
soluble fusion system, **181**
  is constrained, 183
  and p-soluble groups, 185
sparse fusion system, **232**
stabilizer of a biset, **253**
stable
  biset, 256

under a group action, **34**
strongly closed, **127**
  abelian subgroup, 153
  for finite groups, 127
  and intersections, 128
  and products, 129, 147
  under quotients, 145
strongly p-embedded, 14, **119**
  classification for $p = 2$, 120
subgroup
  essential, 119
  fully automized, 97
  fully centralized, 101
  fully normalized, 101
  normal in a fusion system, 148
  radical, 118
  receptive, 99
  strongly closed, 127
  strongly p-embedded, 119
  weakly closed, 127
subnormal, **286**
subsystem, **95**
  be defined on, 95
  generated by an invariant map, 202
  normal, 272
  $\mathrm{N}_P(Q)\,\mathrm{C}_{\mathcal{F}}(Q)$, **115**
  $Q\,\mathrm{C}_{\mathcal{F}}(Q)$, **115**
surjective morphism of fusion systems,
  136
surjectivity property, **190**, 202
Suzuki 2-group, 175, **220**

tame intersection, **13**
third isomorphism theorem, 144
Thompson subgroup, **10**
Thompson's $A \times B$ lemma, 310
Thompson's normal p-complement
    theorem
  for finite groups, **11**
  for fusion systems, **235**
three subgroup lemma, 302, 310
TI subgroup, **24**
total space, **77**
transfer
  control of, 264
  control of constrained, 264
  for fusion systems, **261**
transitive biset, **253**
transitive factor of a biset, **253**
transporter, **84**
transporter system, **84**

trivial fusion system, **232**
trivial intersection, **24**

universal fusion system, **95**
universally weakly closed, **357**

weak conjugation family, **14**
weak equivalence
  for fibrant spaces, 74
  simplicial sets, **70**
weak homotopy equivalence, **56**
weak intersection, **281**
  of subnormal subsystems, **287**
weakly characteristic subsystem, **159**
weakly closed, **127**
  for finite groups, 127
  and products, 128
  under quotients, 145
weakly normal map, **205**
weakly normal subsystem, **151**
  equivalent conditions, 156
  generated by a weakly normal map,
    207
  intersection of, 281
  and normal subgroups of fusion
    systems, 158
  and normal subsystems, 279
  not normal, 271
  transitivity, 157
weakly subnormal, **286**
well placed, **224**
wreathed 2-group, **176**